Reinforced and P...
Design to EC2

Concrete is an integral part of twenty-first century structural engineering and an understanding of how to analyse and design concrete structures is a vital element in the subject. With Eurocodes having replaced European national standards it's important to get to know the new codes which, for most countries, are more complex than before. Newly revised to Eurocode 2, this second edition retains the original's emphasis on qualitative understanding of the overall behaviour of concrete structures.

A unique feature of the new edition is a whole chapter on case studies. This provides a unique insight into the way a structure is put together and the alternative structural schemes that are considered in the early stages of design. The concept has been used very successfully in a Problem Based Learning environment at University College Dublin. Groups of students are assigned case study problems, and after a few days they present their solutions to expert judges who give feedback on the practicalities of their chosen solutions. This highlight of the undergraduate experience teaches the students communication as well as design skills.

This book provides civil and structural engineering students and graduates with complete coverage of the analysis and design of reinforced and prestressed concrete structures. Great emphasis is placed on developing a qualitative understanding of the overall behaviour of structures and on bringing together all the strands in the design process – load paths, developing a structural scheme, preliminary sizing, analysis and detailed design.

Eugene OBrien is Professor of Civil Engineering at University College Dublin.

Andrew Dixon is Director of Downes Associates Consulting Civil & Structural Engineers.

Emma Sheils is employed by Roughan & O'Donovan, a Civil and Structural Engineering Consultancy and is a former Postdoctoral Research Fellow of University College Dublin.

Reinforced and Prestressed Concrete Design to EC2

The complete process

Second edition

Eugene OBrien, Andrew Dixon and Emma Sheils

 Spon Press
an imprint of Taylor & Francis
LONDON AND NEW YORK

First published 1995 by Longman Group.
Reprinted 1999.

This edition published 2012
by Spon Press
2 Park Square, Milton Park, Abingdon, Oxon OX14 4RN

Simultaneously published in the USA and Canada
by Spon Press
711 Third Avenue, New York, NY 10017

Spon Press is an imprint of the Taylor & Francis Group, an informa business

British Library Cataloguing in Publication Data
A catalogue record for this book is available from the British Library

Library of Congress Cataloging in Publication Data
A catalog record has been requested for this book

ISBN: 978-0-415-57194-4 (hbk)
ISBN: 978-0-415-57195-1 (pbk)
ISBN: 978-0-203-85664-2 (ebk)

Typeset in Sabon
by Integra Software Services Pvt. Ltd, Pondicherry, India

MIX
Paper from
responsible sources
FSC
www.fsc.org   FSC® C004839

Printed and bound in Great Britain by the MPG Books Group

Contents

Preface

In his early years as a designer, the first author was asked to design a reinforced concrete floor slab for a plant room in the attic of a hotel. The plant room had no door, being accessed through a square opening in the slab from the room below, and the only location available for this opening was in an area where the slab moments were very high. He grappled with this problem for a great deal of time, looking for structural solutions, until he realized that he needed an overview of the problem – why was this opening needed anyway? A number of telephone calls later, the alternative emerged. Access to the plant room could be achieved through a doorway from elsewhere in the attic instead of through a hatch from below and the troublesome opening could in fact be omitted.

All of this taught him the philosophy behind this book, namely that every member of the design team must understand the complete design process – the thinking behind all the decisions that relate in any way to his/her contribution. Thus we have, in one volume, covered all aspects of the design process from initial conception of the structural alternatives through the process of analysis and on to the detailed design traditionally taught in concrete design courses. The second edition brings the text in line with the relevant Eurocodes and National Annex documents. What is most significant about this second edition, however, is that we have now included the final piece of the jigsaw – the case studies. The first edition included traditional material plus an explanation of how loads are carried by structures and how members are sized. This edition extends this to include, in Chapter 7, six detailed examples of how to go from the initial brief through to a full preliminary design. Alternative schemes are explored and the relative advantages compared, just as happens in the design office.

We have sought to strike a balance between the very practical knowledge of how to apply code clauses and a theoretical understanding of structural behaviour. The Eurocode can be tedious to apply, with seemingly endless notation and definition. We have tried to focus on the important principles, sparing those unfamiliar with the code from some of the more obscure detail. However, much of the detail was necessary as we wanted to ensure that readers have enough information to design most concrete structures.

We have compiled what might, at first sight, appear to be a rather strange collection of material – loading, some qualitative design (load paths, etc.), quite a lot of analysis, rules of thumb and methods of sizing up members, as well as some conventional reinforced and prestressed concrete design. We did this because we see the traditional separation of analysis and design as artificial and have found that many of our students graduating are confused about the distinction between an applied load and a capacity to resist it. We feel that they need to have all the material in one book in order to understand the interrelationships between conceptual design, analysis and detailed design of concrete structures.

There is much in this book that is unique or unusual in a concrete text – how to calculate the distribution of wind loads between cores and shear walls is not something found in most of our competitor books. In this we have been encouraged by feedback from young graduates who have found it most useful. We also feel that many textbooks are lacking on some very essential practical points. For example, an explanation of the calculation of wind load is somewhat unsatisfactory, being largely based on empirical evidence. Nevertheless, it is a very necessary evil in the design office and we feel that all students should have some exposure to such basic essentials before graduating. This will give them some familiarity with the concepts before they are faced with real structures and will reduce the risk of a misinterpretation of the code.

Since 2010, EN 1992 (and its associated National Annexes) has been adopted in Europe as the legal standard for concrete design. They are not easy documents, especially for beginners! We hope that our book will help to ease your way as gently as possible into these new codes of practice.

Acknowledgements

The following publishers are acknowledged for permission to reproduce tables and/or figures: E&FN Spon for material from *Bridge Deck Behaviour* by E. C. Hambly and from *Reinforced Concrete Designer's Handbook* by C. E. Reynolds and J. C. Steedman; Prentice Hall for material from *Reinforced Concrete: Mechanics and Design* by J. G. McGregor and J. K. Wight; The Concrete Society for material from *Concrete Buildings Scheme Design Manual*, *Residential Cellular Concrete Buildings*, *Economic Concrete Frame Elements to Eurocode 2*, *Concise EC2 and High Performance Buildings – Using Tunnel Form Concrete Construction*; Penguin for an adapted figure from *Structures: or Why Things Don't Fall Down* by J. E. Gordon. Extracts from the *Manual for the Design of Reinforced Concrete Building Structures to Eurocode 2* are reproduced with the kind permission of the Institution of Structural Engineers, from whom complete copies can also be obtained. Permission to reproduce extracts from BS EN 1990, BS EN 1991-1-1, BS EN 1991-1-3, BS EN 1991-1-4, BS EN 1992-1-1, BS EN 1992-1-2 is granted by BSI, from whom complete copies can be obtained.

Special thanks go to Peter Flynn, Arup Consulting Engineers; John Considine, Barrett Mahony Consulting Engineers and Michael Quilligan, University of Limerick (formerly of PUNCH Consulting Engineers) for their guidance on the case studies in Chapter 7. Photographs and images are reproduced with the kind permission of Corus Panels and Profiles, Ancon, Keegan Quarries Ltd., CCL and Spiroll, the Board of Trinity College Dublin and Hy-Ten Ltd. The cover photograph is courtesy of the Irish Management Institute and Arthur Gibney & Partners, Architects. Many others have helped in various ways: Aidan Duffy of Dublin Institute of Technology with the carbon footprint of different materials, Ciaran McNally of University College Dublin with durability of concrete and FRC, and Tony Dempsey and Aonghus O'Keeffe of Roughan & O'Donovan with prestressed concrete.

Special thanks go to the authors of the STRAP analysis package which has been used extensively, particularly in Chapters 4 and 5.

This book is dedicated to Sheena, Milo, Bevin and Kevin.

Part 1

Structural loading and qualitative design

Introduction

The first part of this book deals with what is arguably the most important aspect of structural design – qualitative design. Structural failures are rarely a result of a calculation error in which, say, a stress is thought to be 10 per cent less than its actual value. More commonly, failure results from the omission of reinforcement altogether from a critical connection or confusion over the role that each structural member plays in resisting the applied loads.

Chapter 1 presents an overview of the complete design process from conception to finished drawing. In addition, a description is given of the function of various types of structural member and structural system and the ways in which they resist load. The chapter also presents the factors that affect the choice of reinforced or prestressed concrete as an appropriate structural material. In Chapter 2, the ways of combining structural members to form complete structures are described. Also, a qualitative explanation is given of how loads are carried through the various structural members to the ground. Finally, in Chapter 3, the principal sources of loading are described. Poor decisions in the provision for load can result in a structure that is either unsafe at the one extreme or is uneconomical at the other. An explanation is given of the nature of loads and the means by which they are quantified.

It is hoped that, from Part I of the book, the reader will develop a qualitative understanding of how a structure resists loads.

Chapter 1

Fundamentals of qualitative design

1.1   The design process

Design in any field is a logical, creative process that requires a wide amalgamation of skills. As a complete process, structural engineering design can be divided into three main stages:

1. Conceptual design
2. Preliminary analysis and design
3. Detailed analysis and design.

The three stages are dealt with here in Parts I, II and III, respectively. The first stage, described in detail below, consists of the drawing up of one or a number of structural schemes that are safe, buildable, economical and robust. The second stage consists of performing preliminary calculations to determine if the proposed structural schemes are feasible. Rules of thumb are used to determine preliminary sizes for the various members and approximate methods are used to check these sizes and to estimate the quantities of reinforcement required. In the third stage of the design process, the adequacy of the preliminary member sizes is verified and the quantities of reinforcement calculated accurately. The whole process begins with **analysis** of the structure. That is to say, the distributions of bending moment, shear force, etc. due to all possible combinations of applied loading, are found. The various structural members are then **designed**. This is the process by which the capacity of each member to **resist** moment, shear, etc. is compared with the values due to applied loading. If the capacity is inadequate, the quantity of reinforcement and/or the member size is increased. Following completion of these three stages, drawings and specifications are prepared.

Conceptual design, which forms the first stage of the design process, involves the identification of design constraints and the putting together of structural schemes that comply with these constraints. The fundamental constraints are things like the allowable budget, site and size restrictions, provision for safe access, final appearance and utility. These are normally specified by the client's brief or by the body, such as the local planning council, that has given permission for the project. More detailed constraints emerge during the multidisciplinary design process. For example, the architect involved in the development of a large hotel may request a large open space in the central foyer to enable free movement of people in the area. If columns are allowed, as illustrated in Fig. 1.1(a), the span lengths and hence the beam depths will be modest (the required beam depth tends to be proportional to the span length). If, however, no columns are to be used, the span length and hence, the beam depth will be large, as shown in Fig. 1.1(b). This results in additional cost and increases the height of the overall structure. The spherical dome scheme illustrated in Fig. 1.1(c) would tend to be the most expensive of the three schemes as it would involve curved shuttering to form the shape of the concrete during construction. Nevertheless, it may be chosen over the alternatives for its superior aesthetic qualities.

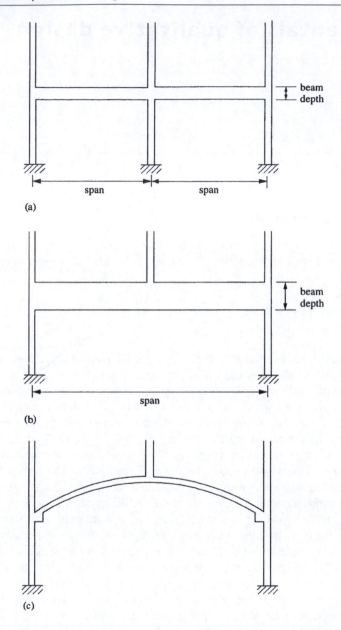

Figure 1.1 Alternative structural schemes: (a) two short spans; (b) one long span; (c) spherical dome

The form of a structure often emerges as a result of discussion between various parties, and good solutions are the result of clear lines of communication and the skill of the individuals involved in the project.

1.2   Structural materials

At an early stage in the design of a modern structure, the structural engineer chooses a suitable material or group of materials that will form the main structural elements. The selection of materials for construction depends on many factors. Due to improvements in technology, communication and transport worldwide, there are a wide variety of materials available for construction. However, the sustainability of a design is also important and factors such as the use of locally sourced materials or renewable resources and the recyclability of construction materials must also be considered. Due to these issues and the range of construction materials available, it is often necessary to carry out approximate preliminary designs for a number of material options to determine the most appropriate for the project. More often, however, the choice of material is founded on a knowledge of the properties of alternative materials and on experience gained from previous design projects.

The principal structural materials

The principal **raw materials** of structural design are steel, concrete (including concrete block units), timber and clay fired bricks. Of these, steel and concrete are the most widely used in practice. The main advantage of steel over other construction materials is its great strength, both in tension and compression. The strength of concrete is dependent on the type, quality and relative proportions of its constituents. To grade the strength of concrete, the compressive strength of simple cylinder (according to Eurocode) or cube samples at 28 days, is generally used. Values of the compressive strength of concrete at 28 days can vary from 5 N/mm^2 to 90 N/mm^2 but typically range between 30 N/mm^2 and 55 N/mm^2. An important characteristic of a hardened concrete is that its tensile strength is much less than its strength in compression, generally being between 1 N/mm^2 and 3 N/mm^2. For simplicity, designers will often assume the tensile strength to be equal to 0 N/mm^2.

The raw materials of construction are often combined to form what are loosely referred to as **structural materials**. In this way, the distinctive properties of the different raw materials can be used to the greatest advantage. The principal structural materials are described in the following sections.

Ordinary reinforced concrete

Concrete reinforced with steel is perhaps the most widespread structural material presently in use around the world. Concrete has many advantages such as its low cost, versatility and high compressive strength but it has the great disadvantage of being weak in tension. Steel has considerably higher tensile strength but tends to be more expensive per unit weight. In ordinary reinforced concrete (reinforced concrete for short), the advantages of both raw materials are utilized when the concrete resists compressive stresses while the steel resists tensile stresses. A typical reinforced concrete beam is illustrated in Fig. 1.2. Reinforced concrete members can be fabricated *in situ*, that is, directly at the site of construction. Reinforced concrete members that are prepared and fabricated offsite and then assembled on site are known as **precast** concrete members. The choice between precast and *in situ* concrete depends on a number of factors that are discussed later in this section.

Fibre reinforced concrete

Fibre reinforcement in concrete has been used in the past to limit plastic shrinkage and settlement cracking on the surface of a concrete member. Fibres are now being used in flat slabs in place of longitudinal and transverse reinforcing steel, and can span up to about 6 m. They are also used in the concrete section as part of a composite design (see below). The fibres are scattered randomly into the concrete during mixing, as illustrated in Fig. 1.3. They are typically about 50 mm in length with a 1 mm diameter and can be made of steel or polymer. The polymer fibres have the added advantage of being non-corroding. However, steel fibres are generally used to reduce/avoid having a steel mesh, whereas polymer fibres are weaker and are usually only used for crack control.

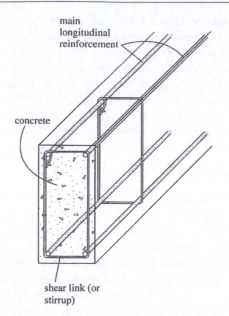

main
longitudinal
reinforcement

concrete

shear link (or
stirrup)

Figure 1.2 Reinforced concrete

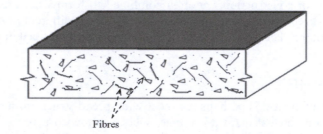

Fibres

Figure 1.3 Fibre reinforced concrete

Prestressed concrete

Like ordinary reinforced concrete, prestressed concrete consists of concrete resisting compression and reinforcing steel resisting tension. However, unlike reinforced concrete, the concrete is compressed during construction and is held in this state throughout its design life by the reinforcing steel. Having the concrete in a compressed state avoids tensile cracking which prevents contaminants from getting into the steel, thereby increasing the resistance of the steel to corrosion. In addition, prestressing of the concrete increases the overall stiffness of the member and reduces deflections. A typical prestressed concrete member is illustrated in Fig. 1.4. In this member, the prestressed tendon is at the centroid of the cross-section. This causes compression throughout the concrete and results in a deflected shape as shown in Fig. 1.4. Like ordinary reinforced concrete, prestressed concrete members can be fabricated *in situ* or as precast units.

Structural steel

Unlike concrete, steel can be used by itself as a structural material for most types of member. Structural steel is available in many shapes, some illustrated in Fig. 1.5, which have evolved over

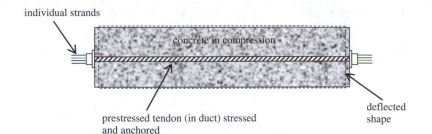

Figure 1.4 Prestressed concrete

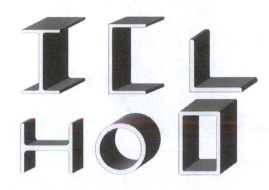

Figure 1.5 Structural steel sections

the years to produce sections that are efficient in resisting bending and buckling. Structural steel has approximately the same stress–strain relationship in tension and compression and so steel sections that carry their loads in bending will generally be symmetrical about the neutral axis. However, local buckling due to large shear forces often places further restrictions on the allowable compressive stresses in such members. The yield strength of structural steel depends on the steel grade but is typically in the range of 200 N/mm^2 to 400 N/mm^2.

Composite construction

The advantages of reinforced concrete and structural steel can be combined in what is known as composite construction. Fig. 1.6 illustrates a typical example of this increasingly popular structural 'material'. The cheaper reinforced concrete slab is used to span locally to create floor space while a combination of the structural steel beam and the concrete is used to support the slab and the loads applied to it.

Timber

Timber from mature trees is one of the earliest construction materials used by man. The strength of timber is directly related to the variety of tree. In addition, its strength will be affected by its density, moisture content, grain structure and a number of inherent defects such as cracks, knots and insect infestations. Typical permissible stresses for softwoods loaded parallel to the grain orientation are less than 6 N/mm^2 for members in compression, tension and bending. For members loaded normal to the grain, the permissible stress is even less. However, with the use of laminating techniques, in which

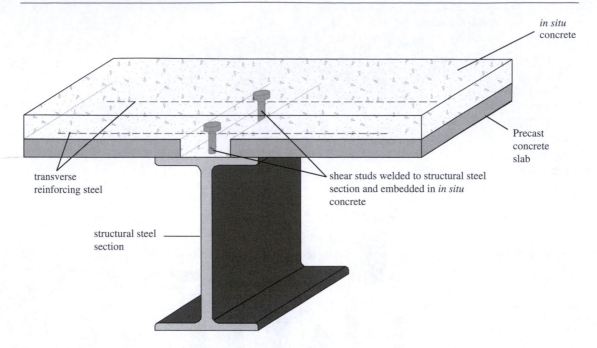

Figure 1.6 Composite construction

thin strips of timber are glued together to form hefty sections, permissible stresses of up to 20 N/mm^2 can be achieved.

Masonry

Masonry is made up of a combination of clay bricks or concrete blocks fixed together with mortar (cement or lime mixture). Masonry structures are characterized by being strong in compression but weak in tension. In compression, among the factors that affect strength are the strength and shape of the units (bricks or blocks), the composition and thickness of the mortar joint and the bond between the mortar and the unit. With a strong clay brick and a mortar with a high cement content, the compressive strength of masonry can reach 15 N/mm^2 or more. However, high variability in the quality of both manufacture and construction results in safe design strengths much less than this. In flexure, the strength of masonry is limited by the low tensile strength. However, this can be overcome by the provision of steel reinforcement in the bed joints and/or by prestressing.

Factors affecting choice of structural material

The principal criteria that may influence the choice of structural material for a given project are:

(a) strength
(b) durability
(c) aesthetics
(d) versatility
(e) safety
(f) speed of erection

(g) maintenance
(h) cost
(i) transport/craneage.

The properties of reinforced and prestressed concrete are compared below with the properties of structural steel, timber and masonry under each of these nine headings. It should be noted that the number of structural materials tends to be limited in any given construction project. This is to minimize the diversity of skills required from the workforce.

Strength

The relative strengths of the main structural materials have already been discussed above. However, it should also be noted that the ability of a material to sustain external loads is dependent on the mechanisms by which the loads are carried in a member. For example, members which are in pure compression or tension will carry their loads more efficiently than members in bending since the stress is evenly distributed across the section (this will be seen in the following section). For this reason, the available strength of a structural material depends as much on the method of load transfer as its characteristic strength. Nevertheless, it can in general be stated that reinforced and prestressed concrete and structural steel are strong materials. Relative to these, timber and masonry are generally rather weak and are more suitable for short spans and/or light loads.

Durability

The durability of a material can be defined as its ability to resist deterioration under the action of the environment during its design life. Of the four raw materials used in construction, steel has by far the least resistance to such corrosion (or 'rusting' as it is more commonly known), particularly in aggressive humid environments. Hence, the durability of a structural material that is wholly or partly made from steel will largely be governed by how well the steel is protected.

A significant advantage of reinforced and prestressed concrete over other structural materials is their superior durability. The durability of the concrete itself is related to the proportions of its constituents, the methods of curing and the level of workmanship in the mixing and placing of the wet concrete. The composition of a concrete mix can be adjusted so that its durability specifically suits the particular environment. The protection of the steel in reinforced and prestressed concrete against the external environment is also dependent on the concrete properties, especially the permeability. The durability of concrete can also be enhanced with the addition of secondary cementitious materials, such as pulverized fuel ash (PFA) or ground granulated blastfurnace slag (GGBS). This results in a lower permeability, thus slowing down the rate of ingress of aggressive agents into the concrete. However, its resistance to corrosion is also related to the amount of surrounding concrete, known as the cover, and the crack widths under day-to-day service loads.

Structural steel, like concrete, is considered to be very durable against the agents of wear and physical weathering (such as abrasion). However, one of its greatest drawbacks is its lack of resistance to corrosion. Severe rusting of steel members will result in a loss in strength and, eventually, to collapse. The detrimental effect of rusting is found to be negligible when the relative humidity of the atmosphere is less than ~70 per cent and therefore protection is only required in unheated temperate environments. Where corrosion is likely to be a problem, it can often be prevented by protective paints or other coatings. Although protective coatings are very effective in preventing corrosion, they do add significantly to the capital and maintenance costs (unlike concrete, for which maintenance costs are minimal).

For timber to be sufficiently durable in most environments, it must be able to resist the natural elements, insect infestation, fungal attack (wet and dry rot) and extremes in temperature. Some timbers, such as cedar and oak, possess natural resistance against deterioration due to their density and the presence of natural oils and resins. However, for the types of timber most commonly used in construction, namely softwoods, some form of preservative is required to increase their durability. When suitably treated, timber has excellent properties of durability.

Masonry, like concrete, can also be adapted to suit specific environments by selecting more resistant types of blocks/bricks for harsh environments. Unreinforced masonry is particularly durable and can last well beyond the typical 50-year design life.

Aesthetics

The aesthetic quality of a completed structure is strongly influenced by the finish on the external faces. For concrete, this final appearance depends on the standards of placement and compaction and the quality of the formwork. Badly finished concrete faces, with little or no variation in colour or texture over large areas, can form the most unsightly views. Concrete is a versatile material, however, and when properly placed, it is possible to produce structures with a wide variety of visually appealing finishes. In the case of precast concrete, an excellent finished appearance can usually be assured since manufacture is carried out in a controlled environment. To improve the appearance of concrete, certain blended cements or admixtures can be added when mixing. For example, the addition of GGBS results in a lighter coloured concrete that is more aesthetically pleasing as well as improving the durability.

Exposed structural steel in buildings can be displeasing to the eye in many settings and so is often covered in cladding in order to provide an acceptable finish. In other applications, the use of brightly painted, closed, hollow, circular or rectangular sections can enhance the appearance of a building.

Timber and brick faced structures will generally have an excellent finished appearance, provided a high quality of workmanship is achieved. Masonry also offers a sense of scale and is available in a wide variety of colours, textures and shapes. However, in many instances, concrete block units need to be finished with a sand/cement render for appearance.

In addition to their aesthetic properties, concrete and masonry structures also have the advantage of possessing good sound and thermal insulation properties.

Versatility

The versatility of a material is defined as its ability (a) to be fabricated in diverse forms and shapes and (b) to undergo substantial last-minute alterations on site without detriment to the overall design. Steel can be worked into many efficient shapes on fabrication but is only readily available from suppliers in standard sections. Concrete is far more versatile in this respect as it can be formed by moulds into very complex shapes. Timber is the most limited as it is only available from suppliers in a limited number of standard sizes. Laminated timber, on the other hand, can be profiled and bent into complex shapes. Masonry can be quite versatile since the dimensions of walls and columns can readily be changed at any time up to construction. The disadvantage of steel, timber and precast concrete construction is their lack of versatility on site compared with *in situ* reinforced concrete and masonry to which substantial last-minute changes can be made. *In situ* prestressed concrete is not very versatile as changes can require substantial rechecking of stresses.

Safety

The raw material of concrete is very brittle and failure at its ultimate strength can often occur with little or no warning. Steel, being a very ductile material, will undergo large plastic deformations before collapse, thus giving adequate warning of failure. The safety of reinforced concrete structures is increased by providing 'under-reinforced' concrete members (the concepts of 'under-reinforced' and over-reinforced concrete are discussed in Chapter 8). In such members, the ductile steel reinforcement fails in tension before the concrete fails in compression, and there is considerable deformation of the member before complete failure. Although timber is a purely elastic material, it has a very low stiffness ($\sim1/20^{th}$ that of steel) and hence, like steel, it will generally undergo considerable deflection before collapse.

An equally important aspect of safety is the resistance of structures to fire. Steel loses its strength rapidly as its temperature increases and so steel members must often be protected from fire to prevent collapse before the occupants of the structure have time to escape. For structural steel, protection in the form of intumescent paints, spray-applied cement-binded fibres or encasing systems, is expensive and can often be unsightly. Concrete and masonry possess fire-resisting properties far superior to most other materials. In reinforced and prestressed concrete members, the concrete acts as a protective barrier to the reinforcement, provided there is sufficient cover. Hence, concrete members can retain their strength in a fire for sufficient time to allow the occupants to escape safely from a building. Timber, although combustible, does not ignite spontaneously below a temperature of about 500°C. At lower temperatures, timber is only charred by direct contact with flames. The charcoal layer which builds up on the surface of timber during a fire protects the underlying wood from further deterioration and the structural properties of this 'residual' timber remain unchanged.

Health and safety considerations during the construction process can also influence the choice of structural material. A risk assessment for the construction and operation of a structure must be carried out for each project by the design team. Factors such as transport/delivery, sequence of placement and erection, maintenance operations, etc. must be considered at the design stage. If a material is selected for a project based on its relative merits outlined above, the designer must then be satisfied that the construction and in-service maintenance can be carried out in a safe manner. If it is found that an alternative material has less risk associated with its use, consideration must then be given to its addition.

Speed of erection

In many projects, the speed at which the structure can be erected is of paramount importance due to restrictions on access to the site or completion deadlines. In such circumstances, the preparation and fabrication of units offsite will significantly reduce the erection time. Thus, where precast concrete (reinforced and/or prestressed) and structural steel are used, the construction tends to be very fast. Complex timber units, such as laminated members and roof trusses, can also be fabricated offsite and quickly erected. The construction of *in situ* concrete structures requires the fixing of reinforcement, the erection of shuttering, and the casting, compaction and curing of the concrete. The shutters can only be removed or 'struck' when the concrete has achieved sufficient strength to sustain its self-weight. During the period before the shutters can be struck, which can be several days, very little other construction work can take place (on that part of the structure) and hence, the overall erection of the complete structure tends to be slow. Masonry, although labour intensive, can be erected very rapidly and the structure can often be built on after as little as a day.

Maintenance

Less durable materials such as structural steel and timber require treatment to prevent deterioration. The fact that the treatment must be repeated at intervals during the life of the structure means that there is a maintenance requirement associated with these materials. In fact, for some of the very large exposed steel structures, protective paints are applied on a continuous basis. Most concrete and masonry structures require virtually no maintenance. An exception to this is structures in particularly harsh environments, such as coastal regions and areas where de-icing salts are used (bridges supporting roads). In such cases, regular inspections of reinforced and prestressed concrete members are now a standard part of many maintenance programmes.

Cost

The cost of a structural material is of primary interest when choosing a suitable material for construction. The relative cost per unit volume of the main construction materials will vary between countries. However, the overall cost of a construction project is not solely a function of the unit cost of the material. For example, although concrete is cheaper per unit volume than structural steel, reinforced concrete members generally require a greater volume than their equivalent structural steel members because of the lower strength of concrete. As a consequence, reinforced concrete can become the more expensive structural material. If reinforced concrete members are cast *in situ*, construction costs tend to be greater than for the steel structure because of the longer erection time and the labour requirements. However, the high cost of structural steel and its protection from corrosion and fire counteract any initial saving with the result that either material can be more cost-effective. In general, it is only by comparing the complete cost of a project that the most favourable material can be determined. As a general guide, however, it can be said that reinforced concrete and structural steel will incur approximately the same costs, masonry will often prove cheaper than both where it is feasible while the cost of timber is very variable.

There is also an environmental cost associated with materials. The carbon footprint of cement is considerable. Cement production is thought to account for up to 5 per cent of global anthropogenic carbon dioxide. A great deal of heat/energy is required to pulverize, sinter and grind materials at various stages of the manufacturing process, thus resulting in the release of greenhouse gases (GHGs). In addition, the release of carbon from the rock during the formation of clinker accounts for approximately 60 per cent of total GHG emissions. Although the situation is improving as alternative technologies emerge, there is a limit to the emission reductions given the dominance of chemical reactions (which require heat) in the manufacturing process. In particular, the substitution of a certain percentage of ordinary Portland cement with by-products from other industries such as GGBS (which is a by-product from the production of steel) reduces the overall carbon footprint.

The rolling of structural steel sections is also a highly energy intensive process so this material too has a high carbon footprint. Timber is a renewable material and, when sourced from sustainably managed forests, its use in construction has the effect of sequestering carbon long term in the structure. In relation to masonry, blocks are made from concrete which has a high carbon footprint. However, the cement content of concrete blocks is lower than typical structural reinforced or prestressed concrete, with associated lower levels of CO_2 emissions. Although clay is a renewable resource, brick manufacture is energy intensive (requiring temperatures >1,000°C), and therefore, has a high environmental impact.

Transport/Craneage

In certain circumstances, the choice of structural material and construction method may be determined by the accessibility of the site and the availability of craneage. For example, in a small project, it may be possible to avoid the need for cranes by the use of load-bearing masonry walls and timber floors. Depending on their weight and size, structural steel and precast concrete units may require substantial craneage and it is often the limit on available craneage that dictates the size of such units.

Table 1.1 Comparison of the structural properties of concrete (reinforced and prestressed), structural steel, timber and masonry

	Reinforced and prestressed concrete	Structural steel	Timber	Masonry
Strength	Very good	Excellent	Fair	Good except in tension
Durability	Excellent	Poor against corrosion[*]	Poor[*]	Excellent
Appearance	Fair	Fair	Excellent	Excellent/good (bricks/ concrete blocks)
Safety	Excellent	Poor fire resistance[*]	Good	Excellent for fire, but not ductile
Speed of erection	Slow for *in situ*	Very fast	Very fast	Very fast but labour intensive
On site versatility	Excellent for *in situ* reinforced concrete, poor otherwise	Poor	Fair	Very good

[*] Unless protected

In general, *in situ* concrete requires little craneage although cranes, when available, can be used for moving large shutters. The time for which a crane is needed is also important. For example, precast concrete floors require craneage, but assembly on site can often be carried out using a mobile crane which is hired for only a few hours.

Table 1.1 serves as a summary of the relative advantages and disadvantages of the main structural materials under the categories discussed above. At this stage, it should be appreciated that the choice of any structural material is heavily dependent on the particular structure and the conditions under which it is constructed. The following examples briefly illustrate the process of selecting an appropriate structural material.

Example 1.1 Multi-storey warehouse

Problem: Your client requires a multi-storey warehouse in an industrial estate. In order to have it operational for the Christmas rush, construction time must be kept to a minimum.

Solution: From the location and function of the proposed building, appearance is assumed to be non-critical. To ensure a minimum erection time, structural steel is used for the main structural members. For fire resistance, the structural steel is sprayed with cement-binded fibre.

Example 1.2 Grandstand

Problem: Your client requires a grandstand to be constructed between football seasons. This is to be a prestigious structure so its appearance is of primary concern. Adequate safety, especially fire resistance, is also a major issue.

Solution: Precast concrete is selected for the main structural members of the grandstand because it is fast to erect and efficient in carrying the loads. In addition, it has a high natural fire resistance and good appearance. Any members that are too large for precasting are constructed *in situ*. Structural steel is chosen for the roof which is to cover the stands because of its low self-weight and high strength.

Example 1.3 Apartment building

Problem: Your client requires a four-storey apartment building to be constructed in the centre of a town. Appearance and a minimum running maintenance are the governing design constraints.

Solution: Brickwork-faced masonry is chosen as the main structural material since it requires the minimum of maintenance and (for external façades) has an excellent appearance. In addition, blockwork (for internal walls) is inexpensive compared with other materials and requires little or no craneage (this may be a factor on such a constrained site). Precast prestressed concrete floors are installed using a mobile crane. This combination of masonry and prestressed concrete is designed to provide adequate fire resistance and sound insulation.

1.3   Basic structural members

A complete structure is essentially a combination of members that can be categorized by their main function. Some structures can be broken down into sub-systems in which groups of these members act together to perform a specific function. However, before such complex systems are considered, it is necessary first to review the five basic mechanisms of load transfer that arise in members.

Tension

When a member is being stretched by forces parallel to its axis, as illustrated in Fig. 1.7(a), the stress produced is known as tension. Members used primarily to resist tension (such as ropes) need not have a capacity to resist transverse forces or bending moments. Homogeneous members in tension only have a uniform stress distribution over their cross-section as illustrated in Fig. 1.7(b) and hence

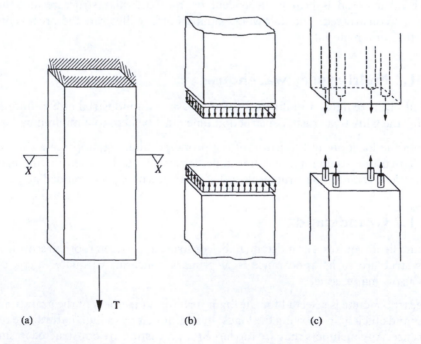

(a)   (b)   (c)

Figure 1.7 (a) Tension member; (b) stress distribution at section X–X; (c) stress distribution in reinforced concrete after cracking

can utilize the available material strength most efficiently. In reinforced concrete, the concrete cracks under the smallest tensile force. Once this occurs, the tensile force is carried solely by the reinforcement crossing the cracks, as illustrated in Fig. 1.7(c).

Compression

If the loads on a tension member were to be reversed so as to squeeze rather than to stretch, then the member would be subjected to compression. Unlike tension members, those in compression must have some flexural rigidity to prevent failure through buckling. In addition to its material properties, the buckling strength of a compression member is dependent on its length, its cross-sectional geometry and the type of supports in which it is held (pinned, fixed or otherwise). The compressive stress distribution in a homogeneous member is illustrated in Fig. 1.8(b). In reinforced concrete, however, the steel carries a greater portion of the stress, as shown in Fig. 1.8(c), because it is stiffer (i.e. has a higher modulus of elasticity) than the surrounding concrete.

Shear

The application of transverse forces perpendicular to the axis of a member results in the development of shear stresses. Consider the horizontal member in Fig. 1.9(a) which is supported on a column and has a homogeneous cross-section. The internal shear force, V, at any section X–X along its length is equal to the applied vertical force, P, by equilibrium. The distribution of the shear stress in the member (Fig. 1.9(b)) is not uniform over its cross-sectional area but, for linear elastic materials,

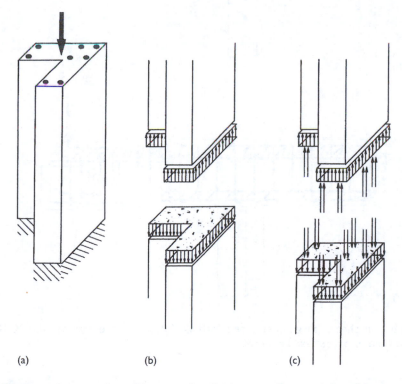

(a)   (b)   (c)

Figure 1.8 (a) Compression member; (b) stress distribution – uniform section; (c) stress distribution in reinforced concrete member

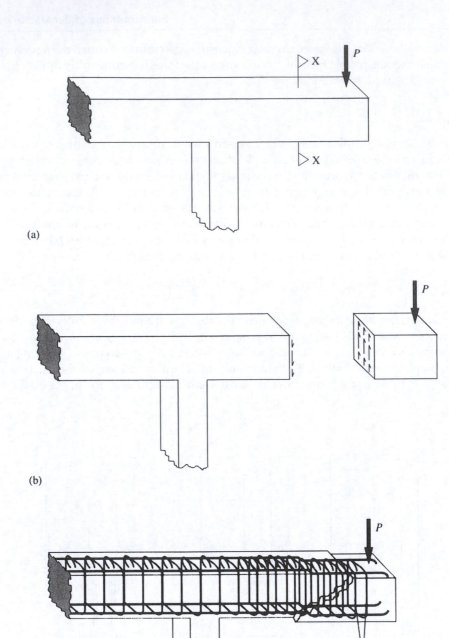

(a)

(b)

(c)

links

Figure 1.9 Member in shear: (a) geometry and loading; (b) shear stress at section X–X; (c) shear in reinforced or prestressed concrete

varies parabolically from zero at the top and bottom surfaces to a maximum at the centre. However, reinforced and prestressed concrete are not homogeneous and when failing in shear, are neither linear nor elastic. The shear failure of a typical reinforced concrete member is illustrated in Fig. 1.9(c). The inclined tensile cracks that are formed in the concrete are held closed by vertical shear reinforcement known as **links** (or stirrups).

Torsion

Torsion occurs in members when a transverse external force acts outside the plane containing the axis of the member (i.e. an eccentric force), such as in Fig. 1.10(a). The effect of torsion, illustrated in Fig. 1.10(b), is to cause a twisting action in the loaded member. The magnitude of this twisting, θ, is dependent on the applied torque, Pe, the length of the member, l, the cross-sectional geometry and the elastic shear modulus of the material. For linear elastic homogeneous members, the distribution of torsional stress is illustrated in Fig. 1.10(c). Unlike the distribution of shear stress, the torsional stress at a section increases from zero at the centre to a maximum at the edges. For this reason, torsional failure in concrete is initiated by tensile cracking at the surface of the member. Torsional cracking in reinforced concrete is resisted by closed links, as illustrated in Fig. 1.10(d). The lapping of the links in this way ensures a continuity of reinforcement all around the section.

Flexure

It has been shown above that all members that transmit transverse loads laterally to one or more supports do so, at least partially, by shear force mechanisms. If the loads are applied outside the plane of the member's axis then the loads are also transferred by torsion. In addition to these two mechanisms, transversely loaded members transmit their loads by bending action (flexure). The central point load in Fig. 1.11(a) exerts a bending moment and causes the member to sag. In a linear elastic homogeneous member, the longitudinal fibres at the top become shorter due to the bending and are, therefore, stressed in compression, while the fibres at the bottom face become longer and are

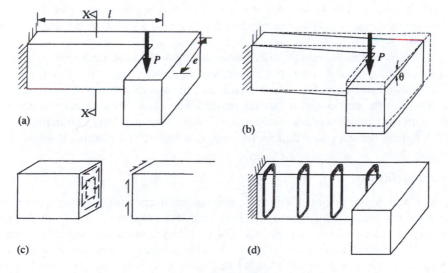

Figure 1.10 Member in torsion: (a) geometry and loading; (b) deformed shape; (c) stress at section X–X; (d) torsion in reinforced or prestressed concrete

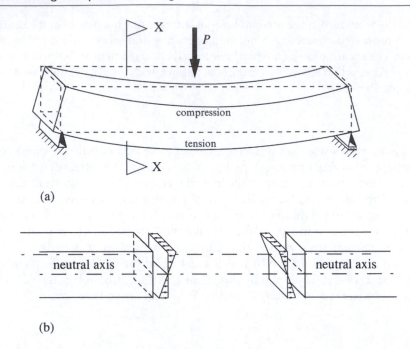

Figure 1.11 Homogeneous linear elastic member in flexure: (a) geometry, loading and deflected shape; (b) stress distribution at section X–X

stressed in tension (apart from timber, most materials are not fibrous and the concept of fibres in a bending member is only used as a helpful analogy for the behaviour of the material). From Fig. 1.11(b), it can be seen that the outer fibres in both tension and compression will extend or shorten more than the internal fibres, and for an elastic material the stress distribution will be triangular, as shown. The location within the member where the bending stress is zero, between the tensile and compressive zones, is known as the neutral axis of the member. Of course, in three-dimensional structures this is a surface, as illustrated in the figure, not an axis. For homogeneous members (remember that concrete with reinforcement is not homogeneous) the neutral axis passes through the centroid of the cross-section.

Due to its minimal tensile strength, concrete is assumed to crack under the smallest of tensile stresses. Thus, for a reinforced concrete member in bending, cracks extend through the tension zone to the neutral axis, as illustrated in Fig. 1.12(a). Failure of the member is prevented by the longitudinal steel which traverses the cracks and resists the tensile forces. Reinforced concrete only behaves as a linear elastic material under everyday service loads. Under the much larger ultimate loads for which sections are designed, the stress distribution becomes non-linear, as illustrated in Fig. 1.12(b).

Member nomenclature

It has been seen above that flexure, shear and torsion often occur simultaneously in many transversely loaded members. In one-dimensional members, cantilevers transfer load to only one support and beams transfer load to two or more supports. The two-dimensional equivalents of cantilevers and beams, where the applied loads are carried by bending, shear, etc. in two directions, are **slabs**. Members such as beams that carry their loads by bending and shear may also be subjected to either tensile or compressive forces. An example of such a member is a column with moment-resisting connections (see Fig. 1.13). The two-dimensional equivalent of a column is known as a **wall**. The

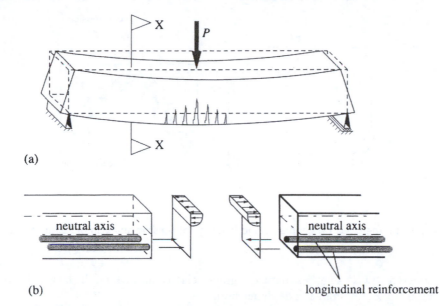

(a)

(b)   longitudinal reinforcement

Figure 1.12 Reinforced concrete member in flexure: (a) geometry, loading and deflected shape; (b) stress distribution at section X–X under ultimate loading

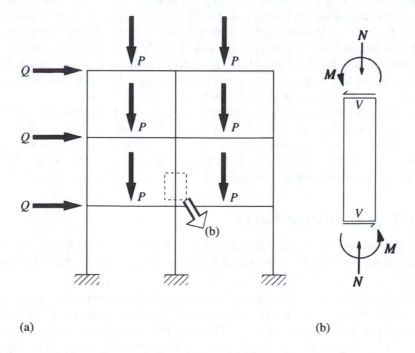

(a)   (b)

Figure 1.13 Column with shear, flexure and axial force: (a) frame; (b) column from frame

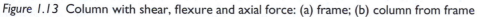

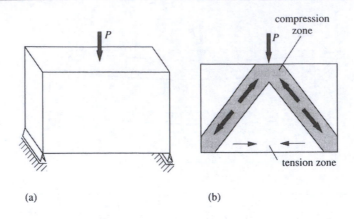

Figure 1.14 Deep beam member: (a) geometry and loading; (b) tension and compression zones

Eurocode for concrete, 'EC2', suggests that a column be classed as a wall if the larger side dimension is greater than four times the smaller side dimension.

By increasing the depth of a beam, while keeping the span length constant, as illustrated in Fig. 1.14(a), the member becomes very stiff in the plane of bending. If the depth is increased substantially, the member becomes so stiff that the applied load is effectively carried through tension and compression zones, as shown in Fig. 1.14(b), rather than by bending and shear. This can be referred to as **membrane action**, although historically such members are aptly named **deep beams**. EC2, the Eurocode for concrete design, defines a deep beam as one in which the span is less than three times the overall section depth. A cantilever with a particularly deep section might also resist load by membrane action and be termed a **deep cantilever**. A deep cantilever is one in which the span is less than 1.5 times the depth. Deep cantilevers taking horizontal loads in buildings, such as that illustrated in Fig. 1.15, are more commonly known as **shear walls**.

For the purposes of design, it is often convenient to label individual members by the mechanism of load transfer that governs the design of the member, rather than by their historical names. This mechanism is termed the 'primary' mechanism of load transfer and the other, less critical mechanisms are termed 'secondary' mechanisms. For instance, beams and slabs where bending, is the primary mechanism are sometimes termed flexural members.

Example 1.4 Flexural members I

Problem: Fig. 1.16(a) illustrates a reinforced concrete beam. For the applied load illustrated in Fig. 1.16(a), determine the primary methods of load transfer and suggest how the member should be reinforced.

Solution: As the span lengths are much greater than the depth of the member, the primary methods of load transfer for this example are bending and shear. Fig. 1.16(b–d) illustrate the deflected shape and the shear force and bending moment diagrams. The shear force is constant in each span and is greater in the shorter right-hand span. The bending moment varies across each span and reaches a maximum over the internal support. Note from the deflected shape that the member is hogging (as opposed to sagging) throughout its length. This results in tension in the top of the member. Thus, to prevent flexural cracking, longitudinal steel reinforcement must be provided along the top of the member in both spans, as illustrated in Fig. 1.16(e). More steel is provided over the internal support since the bending moment is greatest here. For practical purposes in order to fix the links, some longitudinal

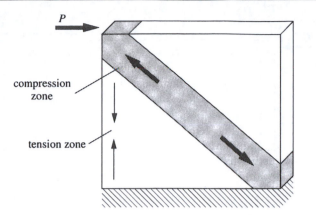

Figure 1.15 Shear wall

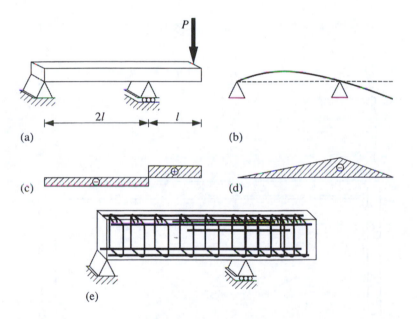

Figure 1.16 Reinforced concrete beam: (a) geometry; (b) deflected shape; (c) shear force diagram; (d) bending moment diagram; (e) beam reinforcement

reinforcement would also be placed in the bottom of the section. Links are provided in each span to prevent shear cracking. The spacing of the links in the right-hand span is less than that in the left-hand span as the shear force is higher.

Example 1.5 Flexural members II

Problem: Fig. 1.17(a) illustrates a reinforced concrete frame. For the applied horizontal load, determine the primary methods of load transfer and suggest how the member should be reinforced.

Solution: For the member of Fig. 1.17(a), a small amount of the load is carried by the overturning effect, i.e. tension in AB and compression in CD, but primarily the load is carried by bending in all members. Fig. 1.17(b) illustrates the deflected shape and bending moment diagram. Steel

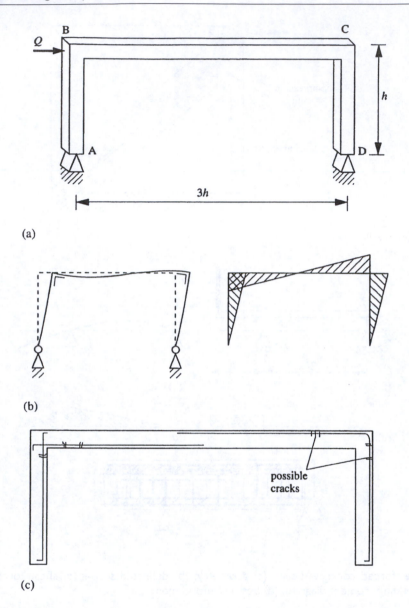

Figure 1.17 Reinforced concrete frame: (a) geometry; (b) deflected shape and bending moment diagram; (c) longitudinal reinforcement

reinforcement is provided where tension cracks are likely to occur, as shown in Fig. 1.17(c). In addition, some links should be provided to resist the shear forces in AB and CD.

1.4   Structural systems

Structural systems are defined as groups of structural members (as defined in Section 1.3) that act together to perform a specific function. A complete structure can incorporate any number of independent systems all of which act together to transfer the applied loads to the foundations and provide overall stability to the structure. Two of the more basic systems considered here are trusses and frames.

Trusses

The truss is a simple structural concept comprised solely of tension and compression members connected together by hinges or pins. The role of the hinges is to prevent the transmission of any bending moment through the members of the truss. In their simplest form, trusses are only loaded at their joints to preclude bending of individual members. In practice, hinge connections are rarely used and even steel truss members are in fact welded or bolted together. However, idealized hinge connections give fairly accurate results, and for the purposes of simplified analyses, such pin-joints are assumed in the design of trusses. Truss systems are commonly used as an alternative to flexural members where large distances need to be spanned. Thus, truss systems transmit their load by bending overall, but are comprised of members acting together which only transfer the load axially. For stability, trusses are built up in a triangular rather than rectangular configuration, with each triangle comprising three members. Two of the most common forms of truss, the 'N' or Pratt truss and the Warren truss, are illustrated in Fig. 1.18. These perform the function of cantilevers and

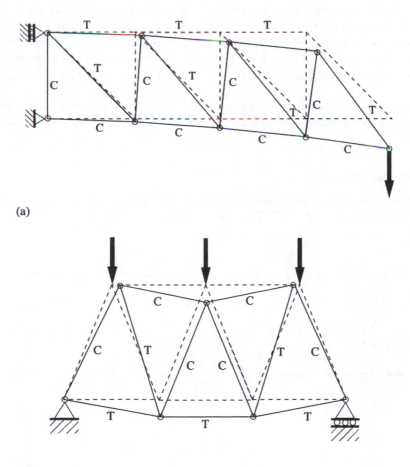

(a)

(b)

Figure 1.18 Truss systems: (a) cantilevered 'N' truss; (b) Warren truss; (T = tension, C = compression)

beams, respectively, for particularly long spans. The tension and compression members in each configuration are indicated in the figure. More complex three-dimensional trusses, known as space trusses, can be built up so that the applied loads are carried to supports in two directions (like a two-way spanning slab for particularly large spans). While trusses are far more commonly constructed using structural steel and timber, concrete trusses are feasible for very large spans.

Frames

Frames are similar to trusses in that they are comprised of straight members joined together at their ends. In fact, the structure of Fig. 1.17(a) is a simple two-dimensional (plane) frame. Unlike trusses, however, the members in a frame are rigidly connected together. This can be seen in Fig. 1.17(b). The strength of the overall frame is thus derived from the bending and shear resistance of its members. Axial forces are also present in frames but their influence is usually small compared with the moments and shear forces and so frames are generally considered as systems of flexural members.

The frame of Fig. 1.19 consists of ten flexural members. The vertical forces at J and K are transmitted by bending and shear in members CF and FI, respectively, to the tops of the vertical members. The loads are carried from there to the supports by compression in the vertical members and bending/shear in all members. The horizontal loads at B and C are transmitted to the supports by bending action in all of the members. Although the vertical point loads exert compressive forces on the vertical members and the horizontal forces cause compression in the horizontal members, the section requirements of all members in a frame are generally governed by the bending and shear capacity.

Frame action also occurs in Vierendeel girders, Fig. 1.20, where the connections allow the bending moments in individual members due to applied loads to be transferred to adjoining members.

Arches and Catenaries

Arches are one of the oldest forms of construction and were used in many historic buildings such as the Notre-Dame Cathedral. The predominant stress in an arch is axial compression,

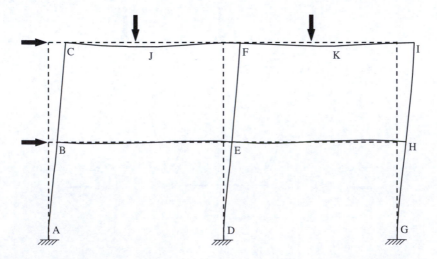

Figure 1.19 Frame

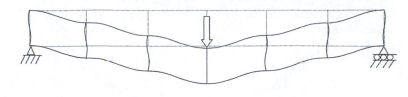

Figure 1.20 Vierendeel girder

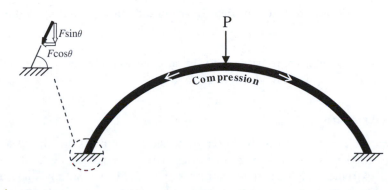

Figure 1.21 Arch structure

which makes it a very efficient structural system. Therefore, arches can be used to span large distances. However, due to the horizontal thrust at the base of the arch (see Fig. 1.21), the design of the foundations is a vital part of the process, especially when the span is large in relation to the height. In a masonry or stone arch, the bricks or stones are usually tapered (wedge shaped) and are fixed closely together with mortar so that forces are transmitted from one segment to the other across the entire contact area. When downward pressures are applied to the arch, the stones or bricks are pushed together, resulting in axial compression throughout (Fig. 1.22).

Catenary structures received their name from the shape that a hanging cable (or chain) forms under its own weight when it is pinned at both ends, as illustrated in Fig. 1.23. A catenary structure is

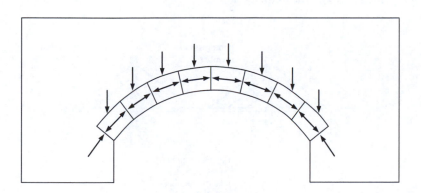

Figure 1.22 Compression in masonry arch structure (adapted from Gordon J.E. (1991) *Structures: or Why Things Don't Fall Down*, Penguin)

Figure 1.23 Catenary curve formed by chain which is pinned and hanging under its own weight

essentially the opposite of an arch, resisting loads in pure tension. However, due to their flexibility, catenary structures often require some form of stiffening.

Cable-supported structure

A cable-supported structure is a combination of tension, compression and flexural members. Such systems are used to span very large distances where the use of flexural systems alone is infeasible due to extremely high bending moments. Fig. 1.24 is a simple diagrammatic representation of a cable-stayed structure that is frequently adopted for the design of long bridges. The main cables form the tension members, compression occurs in the two towers and small bending moments are developed in the deck members. Great spans can be achieved with such a system (200 m to >400 m) because of the exceptional efficiency of the tension and compression members and because the bending moments in the span are minimized by the support provided by the tension members. Similar systems have also been used successfully in the construction of cantilevered stadium roofs and large aircraft hangars.

Another form of cable supported structure is the suspension bridge, Fig. 1.25. Since the main cable is in pure tension, this is a very efficient structural system and is capable of spanning large distances. A suspension bridge consists of four essential components. These are:

1.   The main cable and vertical hangers, which support the deck and stiffening girder
2.   The deck and stiffening girder
3.   The anchor blocks, which support the main cable both vertically and horizontally
4.   The pylons, which support the main cables

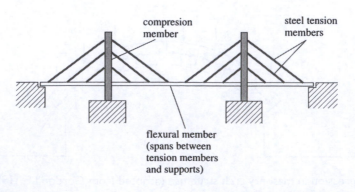

Figure 1.24 Cable-stayed system

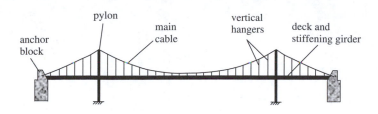

Figure 1.25 The essential components of a suspension bridge

The suspended structure, roadway, and traffic all hang from the main cables. The vertical hangers from the main cable are also in pure tension. The stiffening girder carries the bridge deck between the hangers and prevents excessive vibration. The loading generally consists of self-weight, traffic and wind. The pylon is basically a tower structure, where the main load is the axial compression from the cables attached to it. The main task of the pylon is to transmit the reactions from the cable system to the substructure, although the stress in the pylon due to wind loading must also be considered.

Problems

Section 1.2

1.1 Discuss the relative merits of the main structural materials in the context of the construction of domestic houses.

1.2 A prestigious seven-storey building is to be constructed in the business district of a major city. The client wishes to occupy only three storeys and to secure tenants for the remaining four. Specify the preferred structural material, giving your reasons. The client has already made some late changes to the plans and it is possible that further changes may be made during construction.

1.3 A jetty is to be constructed in a marine environment for industrial use. Discuss the merits of alternative structural materials.

1.4 A bridge is to be constructed on level ground to carry a minor road over a proposed new motorway. Recommend a suitable structural material.

1.5 A pedestrian bridge is to be constructed over an existing motorway. Suggest a suitable construction material.

1.6 A large single-storey supermarket is to be constructed in which there are to be very long clear spans with no internal obstructions. Suggest a suitable structural material.

Section 1.3

1.7 For the member illustrated in Fig. 1.26, how are the loads transferred to the support? If this member is made from reinforced concrete, where would you expect cracking to be most severe:

(a) if $H \gg P$;
(b) if $P \gg H$

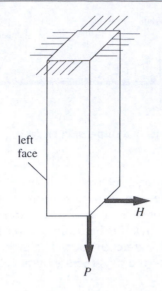

Figure 1.26 Member of Problem 1.7

1.8 How is the load transferred to the supports in the reinforced concrete member illustrated in Fig. 1.27?

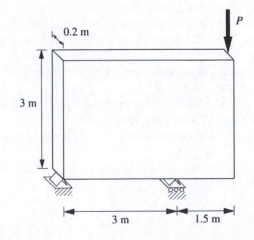

Figure 1.27 Member of Problem 1.8

1.9 What are the mechanisms of load transfer in each of the structures illustrated in Fig. 1.28?

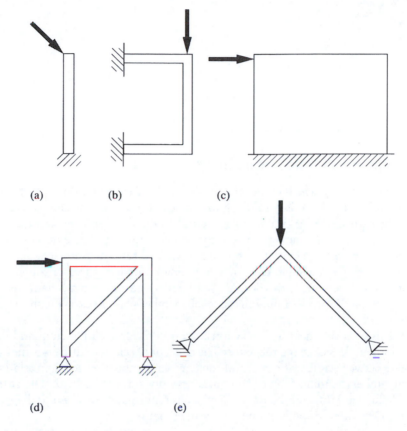

(a)   (b)   (c)

(d)   (e)

Figure 1.28 Structures for Problem 1.9

Chapter 2

Basic layout of concrete structures

2.1 Identification of load paths in structures

Load paths are the routes by which external actions (loads) are carried through the members of a structure to its foundations. An ability to recognize these paths and the mechanisms by which loads are transmitted through individual members is central to the development of a qualitative under-standing of the behaviour of a complete three-dimensional structure. The external loads applied to a structure can usually be resolved into horizontal and vertical force components that are resisted by structural members acting in bending, tension, compression, torsion, shear or some combination of these mechanisms. Identification of the primary load paths and transfer mechanisms in a structure provides information on the precise function that each member plays in carrying the external loads to the foundations.

Load paths are most easily identified by a consideration of force equilibrium and the concept of relative stiffness of members carrying the loads. The concept of relative stiffness is that applied loads tend to be shared between adjacent structural members in proportion to their relative stiffnesses, with the stiffer members tending to carry the larger proportion of the load. The stiffness, k, of a member can be defined as either (a) the force, P, which is required to cause a unit displacement, δ, or (b) the moment, M, required to cause a unit rotation, θ; that is:

$$k = \frac{P}{\delta} \tag{2.1}$$

or:

$$k = \frac{M}{\theta} \tag{2.2}$$

For a particular member, stiffness clearly depends on where the force/moment is applied among other things. Stiffnesses for some of the more common arrangements of load on members are given in Appendix A.

The examples below illustrate the techniques, based on the concepts of force equilibrium and relative stiffness, that are used to identify the load paths and, hence, the primary functions of members in a variety of structures.

Example 2.1 Concept of relative stiffness

Problem: For the structure of Fig. 2.1, determine the expressions for the portions of the total load, P, carried by the two members, beam ABC and column BD. Assume a constant Young's modulus, E, throughout.

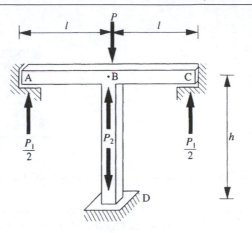

Figure 2.1 Geometry and loading

Solution: In the structure of Fig. 2.1, part of the load, *P*, is carried by beam ABC to supports at A and C and the rest is carried by column BD to the support at D. From Appendix A, No. 6, the beam stiffness is:

$$k_1 = \frac{48EI}{(2l)^3} = \frac{6EI}{l^3}$$

where *I* is the second moment of area of the beam. From Appendix A, No. 7, the column stiffness is:

$$k_2 = \frac{AE}{h}$$

where *A* is the cross-sectional area of the column. For typical values of *A, E, I, h* and *L, k_2* is much greater in magnitude than k_1. Thus, the column will generally carry the greater portion of the load (in fact the column will carry over 90 per cent of *P* in most practical cases). By relative stiffness, the precise portion of the load carried by compression in the column is P_2 where:

$$P_2 = \left(\frac{k_2}{k_1 + k_2} \right) P$$

Similarly, the portion of the load carried by bending and shear in the beam is:

$$P_1 = \left(\frac{k_1}{k_1 + k_2} \right) P$$

Example 2.2 Portal frame

Problem: In the frame of Fig. 2.2, by what primary mechanism is the horizontal force carried to the supports?

Solution: The solution is best determined by a consideration of the relative stiffnesses of the members of the structure. In this case, the horizontal thrust on the structure is transferred to the foundation

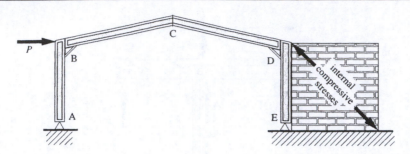

Figure 2.2 Rigidly constructed portal frame with masonry shear wall buttress

primarily by deep beam action in the shear wall rather than by bending in the frame, ABCDE. This is because of the greater stiffness of the shear wall against horizontal force.

Example 2.3 Sports stadium roof loads

Problem: In the reinforced concrete structure of Fig. 2.3, by what mechanisms are the forces, F_1 and F_2, transferred to the ground?

Solution: Fig. 2.3 illustrates a section through a sports stadium. Members AD, DF, FH, BE, EG, GI, DE and FG are rigidly connected at their ends. All other members are effectively pinned at their ends and hence do not transfer bending moment.

This problem is approached initially using the basic principles of equilibrium and force resolution. In Fig. 2.4(a), the force, F_1, is resolved into components parallel and perpendicular to the centroid of member ABC (the curve through the arrow representing F_1 indicates that this is to be replaced with

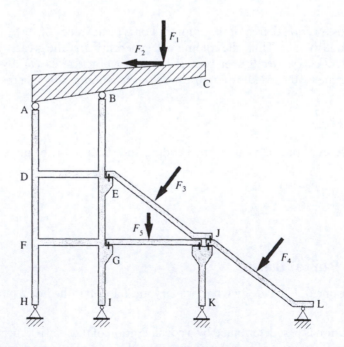

Figure 2.3 Section through a sports stadium showing roof and stand loads

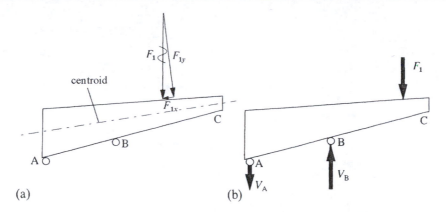

Figure 2.4 Effects of force F_1: (a) resolution of F_1 into components; (b) reactions due to F_1

the two other arrows). It is clear from the figure that the perpendicular component, F_{1y}, is greater than F_{1x}. This component is carried to points A and B by flexure and shear mechanisms. The smaller parallel component, F_{1x}, is transferred to the supports by compression of member ABC. The fact that the perpendicular component, F_{1y}, is larger, combined with the fact that reinforced concrete members are much weaker in flexure than in compression, means that the primary mechanism of transfer of F_1 to A and B is bending and shear in member ABC.

The reactions at the pinned connections, A and B, due to F_1 are illustrated in Fig. 2.4(b). These reactions can readily be determined by static equilibrium of the external forces. Alternatively, resolution of the internal forces from their local coordinates (i.e. their components parallel and perpendicular to the centroid of the member) to global horizontal and vertical components at points A and B will yield the same results. It can be seen that the total force, F_1, tends to cause a clockwise rotation of member ABC about point B. This generates compression in member BE and tension in member AD. In fact, because the connections are pinned, F_1 is transferred purely by tension and compression from points A and B down through the vertical members to the supports at H and I.

The applied horizontal force, F_2, on member ABC can similarly be resolved into components parallel and perpendicular to the neutral axis as illustrated in Fig. 2.5. The component of force perpendicular to the centroid is clearly smaller than that parallel to it. However, the capacity of a reinforced concrete member to resist bending is also smaller. Hence, in this case, either axial compression or bending/shear could be the primary mechanism of load transfer.

The horizontal force, F_2, also tends to cause rotation of member ABC about point B but in an anti-clockwise direction and this results in a small tension in member BE and compression in member AD. However, the primary effect of the load, F_2, is to cause bending in members AD and BE as illustrated in Fig. 2.6. The ratio in which the load is carried by the two members is determined by their relative stiffnesses where the stiffness of a cantilevered member in bending by definition is

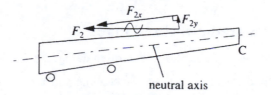

Figure 2.5 Resolution of force F_2 into components

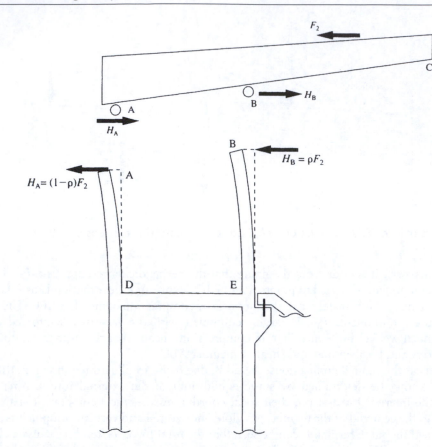

Figure 2.6 Bending in members AD and BE due to load F_2

inversely proportional to the cube of the length of the member (see Appendix A, No. 4). In this case, member AD is shorter than member BE which makes it significantly stiffer. Hence, member AD carries the larger proportion of the load, F_2, by bending and shear. The internal forces due to the horizontal load, F_2, may be transferred from points D and E to the foundations by the 'frame' action illustrated in Fig. 2.7 which primarily involves bending of the members. However, concrete members of normal proportions are much stiffer against axial force than against flexure. Thus, the greater proportion of the horizontal load is transmitted from points D and E to the foundations by the 'truss' action shown in Fig. 2.8. Under this action, the primary mechanism by which the applied load, F_2, is carried from points D and E is tension in members EJ and JL and compression in members EG, GI and JK. In addition, there will be some bending in members DE and EG as these will take the moment reaction to the cantilever, BE. Similarly, some bending will be induced in members DE and DF by the moment reaction to member AD.

Example 2.4 Sports stadium stand loads

Problem: How are the forces, F_3, F_4 and F_5, illustrated in Fig. 2.3, carried to the foundation?

Solution: The external loads, F_3, F_4 and F_5, cause bending and shear force in the members to which they are directly applied and in this way are transferred to the ends of the members. The reactions at the ends can be resolved into horizontal and vertical components as illustrated in Fig. 2.9 for the

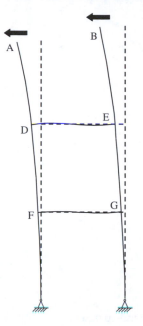

Figure 2.7 Frame action due to load F_2

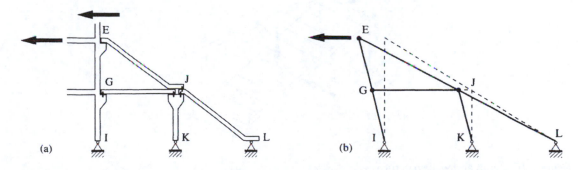

Figure 2.8 Truss action due to load F_2: (a) geometry and loading; (b) deflected shape

load, F_3. The horizontal components of these reactions are carried to the foundation by the truss action illustrated in Fig. 2.8. Some bending will occur at E due to the eccentricity, e, of the vertical reaction due to F_3 (see Fig. 2.9). The horizontal reaction at point J due to F_4 is also carried to ground primarily by truss action. The vertical reactions at point J from F_3, F_4 and F_5 are carried to the ground primarily by compression in member JK.

Example 2.5 Ramp and wall

Problem: How are the forces, F_1, F_2 and F_3, in the *in situ* reinforced concrete structure of Fig. 2.10 carried to ground?

Solution: The ramp and wall system in Fig. 2.10 is a simple, but interesting, three-dimensional concrete structure. The system is comprised of three members which, because they are cast *in situ*, are fully connected to form a continuous structure. It is rigidly supported at all points along its base. The

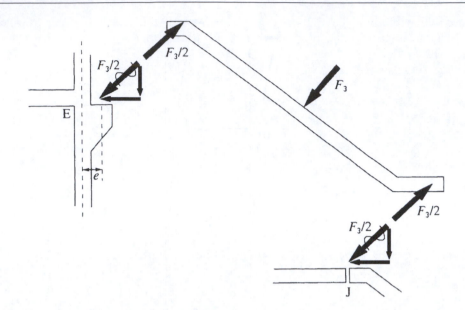

Figure 2.9 Transfer of F_3 into grandstand structure

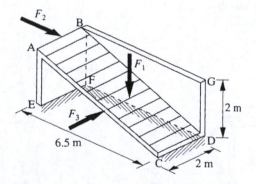

Figure 2.10 Ramp and wall structure

mechanisms of load transfer in three-dimensional structures are not much more complex than in two-dimensional structures and the load paths in this system can be identified using the same techniques that were used for the previous examples.

Consider first the external point load, F_1, which is applied at the mid-point of member ABCD. This load is transferred primarily by bending and shear force mechanisms to the three supported sides of the slab member. Bending in two directions is taking place simultaneously in this member as the load spans between AB and CD but is also supported by cantilever action from BD. The vertical force reaction at AB is transferred to the foundations by compression in ABEF. There is also bending in ABEF as the rigid connection at AB will result in a moment reaction. Similarly, the force and moment reactions along BD result in compression and bending in BFD. The rigid connection between members ABEF and BGFD reduces the bending action in each member near this support. Thus, in ABEF, the moment near A is considerably greater than at B. This tends to cause twisting between A and B. Further details of plate theory are given in Chapter 5 and in specialist texts such as that of Reddy (2006).

Resistance to the applied force, F_2, is provided by member BGFD acting as a deep cantilever and member ABCD acting in compression combined with ABEF acting in tension. As it is rigid at EF, member ABEF can also act as a cantilever. However, the other mechanisms are far stiffer than member ABEF in bending and so the force, F_2, is primarily carried to the ground by membrane action (i.e. axial force).

Member ABCD also acts in bending and shear to transfer the applied load, F_3, to points A, B, C and D. However, the member does not act as a slab in this case and is in fact acting as a simple beam (note that it is not a deep beam since its span/depth ratio is greater than three). At end AB the reaction due to F_3 is carried to the ground by member ABEF acting as a deep cantilever (shear wall) and so the tension and compression zones generated in the member are the primary load-carrying mechanisms. Member ABEF will also tend to warp due to the moment reaction from ABCD, causing torsion in the slab.

Example 2.6 Three-dimensional skeletal frame

Problem: How does the *in situ* concrete frame illustrated in Fig. 2.11 deform under the applied horizontal load, F_1?

Solution: Fig. 2.11 represents a two-storey skeletal concrete structure that has rigidly connected members. Under the external horizontal load, F_1, which is applied at joint I, the space frame can be thought of as three interconnected single-bay plane frames like frame ABCDEF. The horizontal force is primarily transmitted to the foundations of the structure by a 'frame' action in each of the three plane frames. The largest portion of the force, F_1, is transferred directly through the central frame, GHIJKL, rather than being dispersed by transverse bending (bending in the horizontal plane) through members such as CI and IO to the two outer frames. This is because a frame at the point of application of a load provides more stiffness than frames connected to that point by flexible

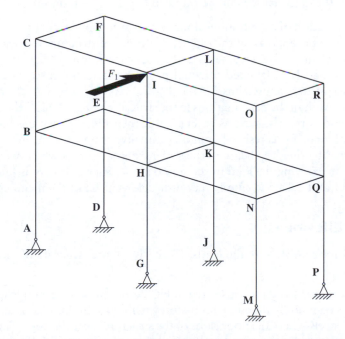

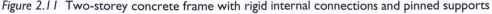

Figure 2.11 Two-storey concrete frame with rigid internal connections and pinned supports

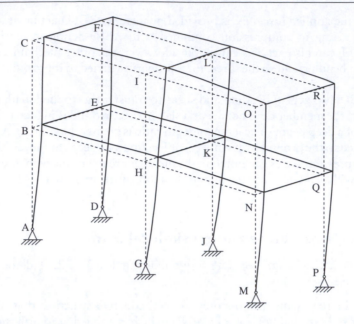

Figure 2.12 Deflected shape of frame of Fig. 2.11

members. The central frame, therefore, deforms more than the outer frames as can be seen in Fig. 2.12.

Example 2.7 Frame stiffening

Problem: How can the deformations in the frame in Fig. 2.11 be reduced?

Solution: The introduction of a continuous floor system to the top level of the frame, as illustrated in Fig. 2.13, has the effect of increasing the rigidity of the entire structure against the horizontal load, F_1. The slab effectively acts as a deep beam and forces each frame to deform by the same amount. Hence, the applied load is distributed in equal proportions to the three plane frames. Thus, the deformation of the entire structure, illustrated in Fig. 2.14, is uniform at any horizontal cross-section and is significantly less than the deformations in the structure of Fig. 2.11. If the external frames were replaced by continuous panels, such as in Fig. 2.15, then the primary mechanisms by which the external horizontal load, F_1, is resisted would not involve frame action. Instead, most of the load would be transferred by the deep beam to the stiffer external panels which would then act as shear walls (cantilevers) in carrying the load to the foundations. Such a system is far more rigid than the frames in Figs. 2.11 and 2.13 and entails very little deformation of the members.

Example 2.8 Lift core

Problem: By what mechanisms are the loads, F_1 and F_2, in the structure of Fig. 2.16 carried to the foundations?

Solution: The continuous structure illustrated in Fig. 2.16 represents a simple model of a reinforced concrete lift core as commonly found in multi-storey buildings. In this example, horizontal loads are applied at the two levels where the floor slabs of the structure meet the core. The two applied forces, F_1 and F_2, are transferred by two-way bending and shear in member FBHD to members ABCD and

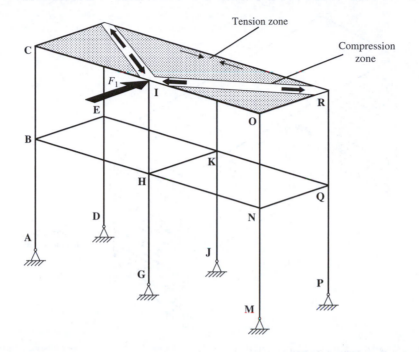

Figure 2.13 Two-storey concrete frame with second floor slab connected to the frame

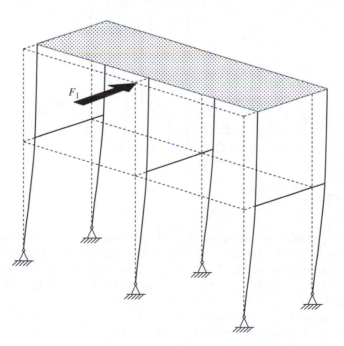

Figure 2.14 Deflected shape of frame of Fig. 2.13

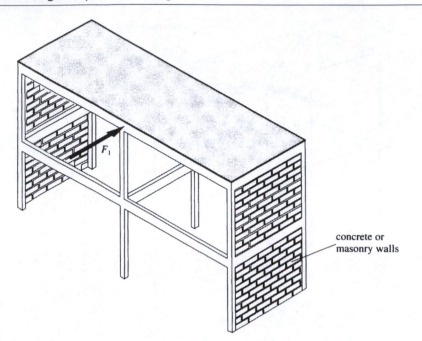

Figure 2.15 Frame with rigid external panels

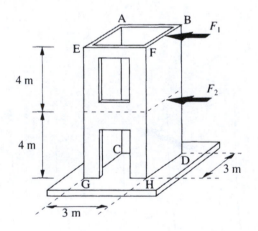

Figure 2.16 Lift core

EFGH. The greater portion of each force is carried to the stiffer solid member with no door openings, ABCD, and this could cause torsion of the core. However, if the building as a whole is symmetrical and is joined together with horizontal floor slabs, both members will deflect by the same amount and, consequently, any twisting of the core due to the unequal stiffness of these vertical members will be prevented. For the dimensions of the core given in Fig. 2.16, member ABCD acts as a cantilever beam for the reactions due to the load, F_1, and carries the force by bending and shear. At the lower level, however, the member, being short but deep, acts as a deep cantilever in transmitting the reaction due to F_2 to the foundation. The mechanism by which member EFGH transmits the applied forces to the foundation pad is dependent on the size of the lift access openings in the member. If the dimensions of

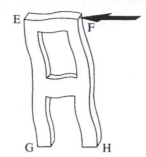

Figure 2.17 Frame action of member EFGH when the openings are large

the openings are small relative to the size of the member, they have little effect on the behaviour except in that they generate concentrations of stress at edges for which extra reinforcement needs to be provided. If, on the other hand, the openings are relatively large, the loads are transmitted by the frame action illustrated in Fig. 2.17. Thus, the rigidity of member EFGH and the stresses generated within it are dependent upon the size of the openings at each level of the core.

Finally, it is important to bear in mind another mechanism by which lift cores can resist horizontal load, particularly for taller buildings. This is by the complete core acting as a cantilever of hollow box section. In this example, tension would develop in FBHD and compression in EAGC with the walls ABCD and EFGH acting as webs of the cantilever that join the top and bottom flanges together.

2.2   Vertical load resisting systems

The gravity loads that act on a building are normally applied directly to the floors and roof. For a multi-storey structure, the loads are transferred from the roof and floors to the foundations by a system of compression and bending/shear members. The precise vertical load resisting system used is dependent on the specific function of the structure, its layout and the magnitude of the vertical loads.

Many multi-storey structures will have reinforced or prestressed concrete floor slabs because of the excellent fire resistance and sound insulation properties of concrete. Concrete floor slabs can be divided into the following four general categories:

1.   One-way spanning slabs
2.   Two-way spanning slabs
3.   Flat (beamless) slabs
4.   Composite slabs

One-way spanning slabs

Rectangular slabs that are only supported at two opposite edges by beams or walls are classed as one-way spanning slabs since the loads are carried by a combination of bending and shear in one direction only (Fig. 2.18). One-way slabs can be either simply supported over one span or continuous over a number of spans and the slab cross-section can be of uniform solid, voided or ribbed construction. Fig. 2.19 illustrates some typical cross-sections that might be used at section X–X in Fig. 2.18. It is important to note that for voided and ribbed slabs the voids/ribs run from one supported end to the other.

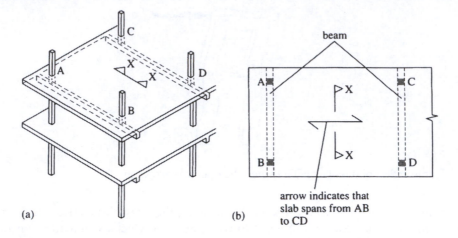

Figure 2.18 One-way spanning slabs: (a) geometry of structure; (b) plan view of slab ABCD

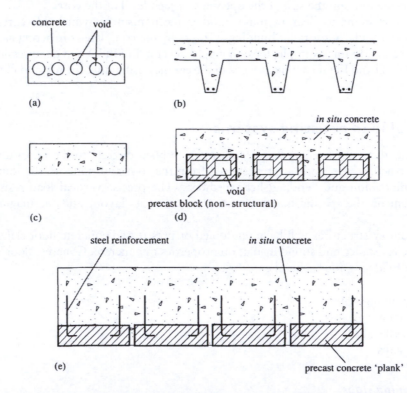

Figure 2.19 Alternative sections through slab (section X–X in Fig. 2.18): (a) voided (usually precast and prestressed); (b) ribbed; (c) solid; (d) ribbed with block infill; (e) precast planks with *in situ* infill

Two-way spanning slabs

Slabs that are supported along all four edges by beams or walls are known as two-way spanning slabs since the applied loads are effectively transferred in two directions to the supported edges (Fig. 2.20). Two-way spanning slabs are normally either solid uniform slabs or, for longer spans, 'waffle' slabs

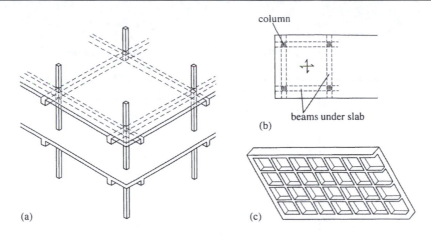

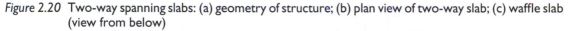

Figure 2.20 Two-way spanning slabs: (a) geometry of structure; (b) plan view of two-way slab; (c) waffle slab (view from below)

(Fig. 2.20(c)) of a shape not unlike edible waffles. In waffle slab construction, the slab is often made solid near the supports to increase the shear and bending moment capacity.

Square two-way spanning slabs tend to be the most economical shape since each supporting beam or wall carries the same proportion of the total load from the slab and this results in the minimum required slab and beam depth. It is a basic principle relating to slabs that load tends to be transferred to the nearest support (more details in Section 5.6). Thus, in a rectangular slab, the greater proportion of the load spans across the shortest distance. If the length of the slab is greater than twice its width, the slab effectively spans one way.

Flat slabs

Slabs that are not supported by beams or walls along the edges but are supported directly by columns are known as flat slabs (Fig. 2.21). The required slab depth for such a system is generally less than for

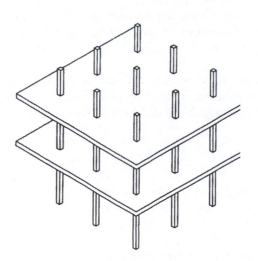

Figure 2.21 Flat slab construction

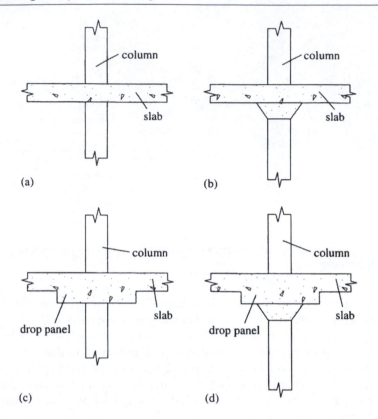

Figure 2.22 Flat slab construction details: (a) no column head; (b) flared column head; (c) slab with drop panel; (d) flared column head and drop panel

a one-way spanning slab system but greater than for a two-way spanning slab system. However, the use of flat slabs results in a reduced depth of structure overall as beams, when they are present, govern the structural depth. Another great advantage of this system is that it only requires simple shuttering – there are no beams for which formwork must be prepared. One disadvantage of flat slab construction is that the arrangement of reinforcement can be complex, particularly adjacent to columns where punching shear reinforcement is often required if the slab depth is kept to a minimum (see Chapter 11).

Flat slabs are sometimes provided with 'drops' or enlarged column heads, as illustrated in Fig. 2.22, to increase the shear strength of the slab around the column supports. However, this substantially increases the complexity of the shuttering, countering one of the principal advantages of the system.

Composite slabs

Slabs that consist of profiled steel decking and *in situ* concrete are known as composite slabs, Fig. 2.23. The steel decking constitutes reinforcement for the slab as well as permanent formwork for the concrete. The profiled steel deck is fixed in place before the wet concrete is poured. The steel deck must be designed to carry the weight of the wet concrete or propped during construction. Shear studs are welded to the steel beam before the concrete is poured. The function of the shear studs is to transmit shear stresses between the hardened concrete and the steel beam, facilitating composite

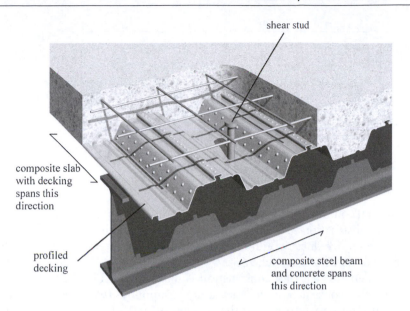

Figure 2.23 Composite floor slab (ComFlor 60®; figure courtesy of Corus Panels and Profiles)

action for the beam. A reinforcing mesh can also be contained within the concrete section to limit cracking (e.g. due to shrinkage). Alternatively, fibre reinforced concrete can be used. In a composite slab, the steel decking provides resistance against tension and the concrete resists compression forces.

Example 2.9 Floor slabs

Problem: Fig. 2.24 represents the plan of a typical floor in a five-storey office building. The office space is to have non-permanent partition walls and a minimum number of internal load-bearing walls. Determine an appropriate floor system for the building if:

(a)   The main service ducts are to run through the corridors between a suspended ceiling and the floor above.
(b)   The floor-to-floor height is to be kept to a minimum.

Solution: For the plan dimensions given in Fig. 2.24, a two-way spanning slab system would need to be supported in two bays, that is, on a practical minimum of three rows of columns in the N–S

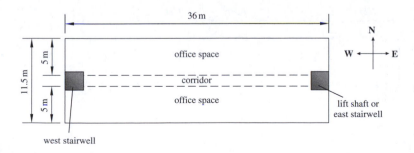

Figure 2.24 Layout of proposed office block

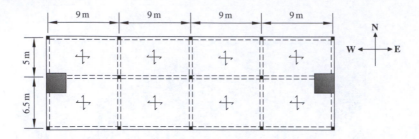

Figure 2.25 Two-way spanning slabs

direction. The internal columns are located along one wall of the corridor, as illustrated in Fig. 2.25, to minimize internal obstructions. However, the scheme illustrated in Fig. 2.25 is not a very effective solution because the service ducts would have to run under the lateral beams, which increases the total floor height over that for other schemes.

An alternative scheme of flat slab construction is illustrated in Fig. 2.26. The maximum span/depth ratios for flat slab construction are less than for continuous two-way spanning slabs. For this reason, it is typically necessary to reduce the spacing of the columns of the building as shown. However, the overall storey depth of the flat slab scheme is less and this proves advantageous where the services are to be fixed below the floor.

A second alternative scheme, this time with one-way spanning slabs, is illustrated in Fig. 2.27. For this building layout, a scheme of one-way spanning slabs would usually be supported in two bays.

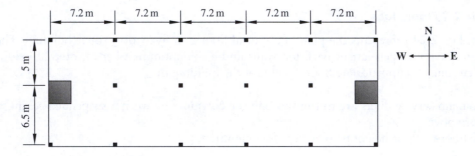

Figure 2.26 Flat slab scheme

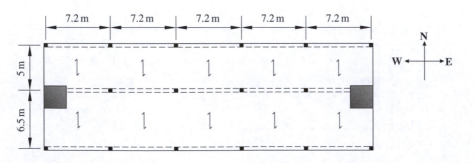

Figure 2.27 One-way spanning slab scheme

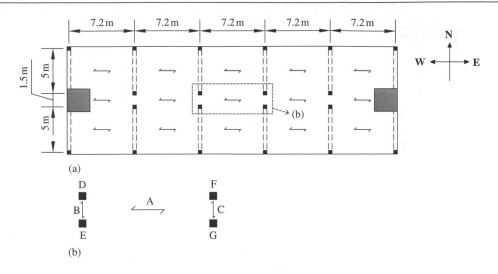

Figure 2.28 Combined flat slab and one-way spanning slab scheme

The slabs are to span in a N–S direction, as illustrated in Fig. 2.27, to minimize obstruction of the main service ducts by the beams. However, the beams still obstruct the secondary service ducts leading from the main ducts in the corridors into the office space which results in a small increase in storey height.

A third alternative scheme is to use a combination of flat slab and one-way spanning slabs, as illustrated in Fig. 2.28. This scheme minimizes the obstruction to both the main and secondary service ducts. For this scheme, the load on regions of the slab, such as region A illustrated in Fig. 2.28(b), spans to other strips of slab, B and C, which, in turn, span between columns (D, E and F, G, respectively). Thus the strips of slab over the corridor between the columns act as beams and are heavily reinforced.

2.3 Horizontal load resisting systems

Horizontal forces (e.g. wind loads, horizontal loading due to geometric imperfections) are generally smaller in magnitude than vertically applied gravity loads. However, the resistance of structural systems to horizontal load is often considerably less than to gravity loads. A qualitative appreciation of the alternative methods of resisting horizontal load allows the engineer to select the most suitable structural layout.

One adverse effect of horizontal forces is their tendency to cause overturning of the entire structure, which can occur with tall, slender structures. For example, overturning occurs in the structure of Fig. 2.29(a) if the overturning moment, Fh, due to the horizontal force is greater than the stabilizing or restoring moment, Wb, where W is the self-weight of the structure and its foundations. Overturning is prevented in practice either by tying the structure down to heavy foundations (Fig. 2.29(b)) or by providing the structure with a more expansive foundation pad (Fig. 2.29(c)). The restoring moment in the former method is increased by the added weight of the heavy foundations, and in the latter method the lever arm of the restoring moment is increased.

More critical for most concrete structures is the prevention of collapse by the 'racking' effects of horizontal forces, illustrated for a precast concrete frame in Fig. 2.30(a). A relatively small applied horizontal force causes rotation at the dowel connections and can lead to collapse of the structure (Fig. 2.30(b)). The racking effect of loads in a practical structure is opposed by effectively increasing its rigidity against horizontal forces. The actual method by which a structure is stiffened is normally

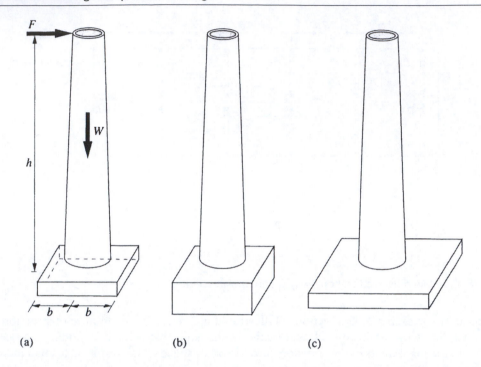

Figure 2.29 Overturning of a structure due to a horizontal load: (a) geometry and loading; (b) heavy foundation; (c) wide foundation

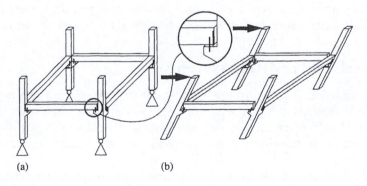

Figure 2.30 Collapse of a dowel-jointed precast concrete frame

dependent on its height (and hence the magnitude of the horizontal forces) and on the material from which it is constructed.

One common method of stiffening a reinforced concrete frame against horizontal forces is to provide rigid connections between the members of the frame to prevent any relative rotation of the connected members. Theoretically, only a small number of rigid joints are required in a frame to provide horizontal stability. However, in practice most, if not all, members are rigidly connected so that resistance to horizontal loads is shared by all members of the frame. A typical rigid joint construction detail for a reinforced concrete frame is illustrated in Fig. 2.31. Rigidly connected frames are quite effective in supporting vertical loads. However, frames of typical proportions are a

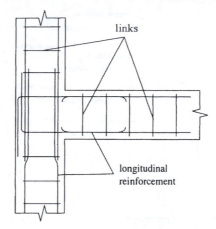

Figure 2.31 Typical detail for a rigid reinforced concrete connection

relatively inefficient means of providing lateral stability. Thus, it is unusual to find such frames providing stability against horizontal forces in buildings of more than a few storeys.

Framing can be used to provide stability to significantly taller buildings if particularly deep members are utilized. Member stiffnesses in frames are proportional to their second moments of area and are inversely proportional to the cube of their length, as can be seen from Appendix A, Nos 4 and 5. Gross second moments of area are, in turn, proportional to the cube of the member depth. Thus, members with low span/depth ratios offer substantial resistance to rotation. When the span/depth ratios of the beams are of similar magnitude to those of the columns, the frame of Fig. 2.32(a) deflects in the manner illustrated in Fig. 2.32(b). Now, if the beams in such a frame are greatly deepened, the beam members become very stiff and restrict the rotation at the joints (A, B, etc.). This substantially reduces the horizontal deflection as it restricts the rotation at the ends of the columns as

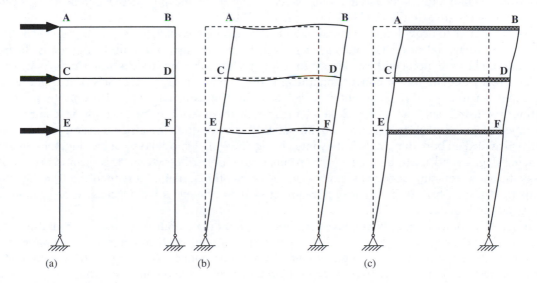

Figure 2.32 Frame action: (a) simple frame; (b) deflected shape for beams of typical stiffness; (c) deflected shape when beams are stiff

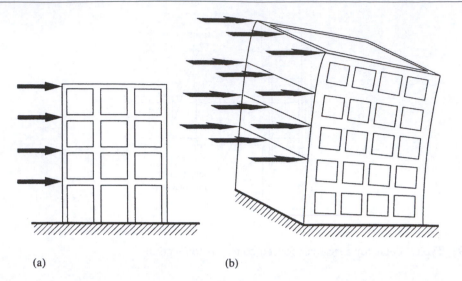

(a)   (b)

Figure 2.33 Hollow tube structure: (a) short and deep beams and columns; (b) very short and deep beams and columns

illustrated in Fig. 2.32(c). Often, the depth of internal beams is restricted and hence such deep members can only readily be provided on the perimeters. In these structures, load is transferred transversely across the frame through the floor slabs to the perimeter from where it is taken by frame action to the foundations. When both beams and columns on the perimeter of a building have very low span/depth ratios and the openings between them become small (see Fig. 2.33), the overall behaviour becomes more like that of a solid cantilever member than that of a frame. In three-dimensional structures of this type, the perimeter of the building can act like a hollow tube to resist horizontal load as illustrated in Fig. 2.33(b). This principle is used to provide lateral stability in particularly tall buildings such as the John Hancock Tower in Chicago. Further details on the design of tall buildings are given in the book by Taranath (2009).

An alternative to providing horizontal stability through rigid connections and/or deep beams is to incorporate diagonal or 'cross' bracing in skeletal frames to carry the bulk of the loads. In fact, bracing can be used along with rigid joints to reduce greatly the magnitude of frame deformations in taller structures. The use of conventional diagonal bracing is limited to panels of a frame where openings, such as windows, are not required. Steel bracing is usually provided only by members acting in tension, such as in Fig. 2.34(a), because of the possibility of buckling in compression members (Fig. 2.34(b)). In order to provide stability in both directions, steel tension members are often provided on both diagonals as illustrated in Fig. 2.34(c). A special type of bracing, known as K-bracing (Fig. 2.34(d)), can be used where openings are required. However, this form is structurally less efficient at resisting horizontal loads than the conventional diagonal type. If the frame is fitted with rigid floors, bracing is often only required in the perimeter of the structure or around the stairwells and lift cores.

Bracing is not a commonly employed method of providing stability against horizontal loads in concrete buildings because of the unsuitability of ordinary reinforced concrete for tension members and the high cost of prestressed concrete. While it is not very common to mix structural materials on site, it is perfectly conceivable to provide steel bracing members in a concrete frame.

Shear walls, in reinforced concrete or masonry, are the most popular method of providing lateral stability in concrete structures. The precise effect of introducing such rigid panels into a skeletal

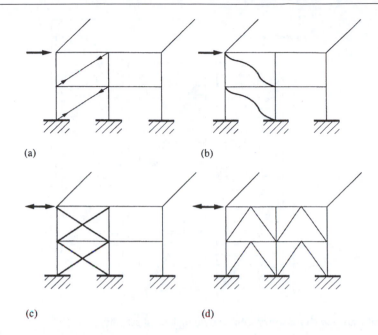

Figure 2.34 Horizontal stability through bracing: (a) tension members; (b) compression members buckle; (c) tension members for load in either direction; (d) K-bracing

frame has been discussed in Example 2.7. Unlike most braced frames, small openings can be provided in concrete shear walls, often with little detriment to their strength. Concrete shear walls can be used effectively to provide horizontal stability for structures of up to 20 storeys. Masonry shear walls are also good but cannot be used to the same height because low tensile strength leads to diagonal cracking in taller masonry structures as illustrated in Fig. 2.35. However, masonry used as an infill for a skeletal frame, as illustrated in Fig. 2.36, can be used successfully in taller structures since the frame confines the masonry and reduces the stresses which would cause diagonal cracking. Due to the tendency of clay masonry to slowly expand with time as moisture from the atmosphere is absorbed, expansion joints must be provided at regular intervals to avoid cracking. Concrete masonry units, however, experience shrinkage over time as a result of moisture loss, resulting in the formation of gaps in the panels which may need to be filled. Other disadvantages of infill masonry panels are an increase in construction time and a sensitivity of the panels to openings (such as for doors and windows). Shear walls can be provided either internally, typically as walls to the stairway,

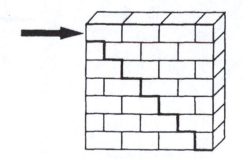

Figure 2.35 Diagonal cracking of masonry shear wall

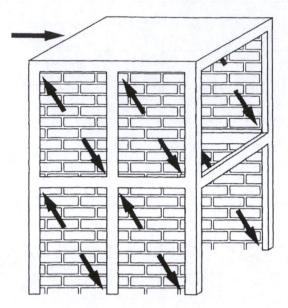

Figure 2.36 Skeletal frame with masonry infill providing lateral stability

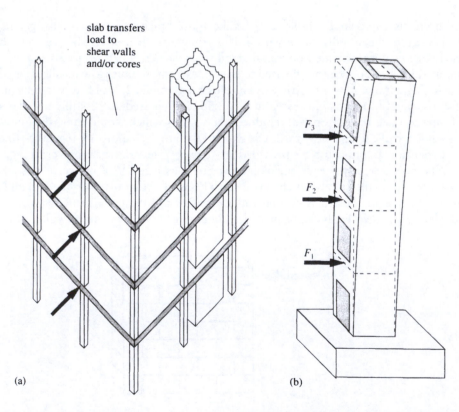

slab transfers
load to
shear walls
and/or cores

F_3

F_2

F_1

(a)

(b)

Figure 2.37 Stability using concrete core: (a) portion of structure; (b) deflected shape of core

or on the perimeter, where they form the outer panels of the structure. However, external panels are not always acceptable aesthetically.

Another system commonly used to provide lateral stability in concrete structures is the reinforced concrete core, an example of which is illustrated in Fig. 2.37. Core systems are conveniently provided in multi-storey structures as enclosures for lift shafts and stairwells. In such a capacity, the core not only provides lateral stability to the structure but also serves as a fire-resisting shell in the case of staircases and as a rigid supporting tube in the case of lifts. The core illustrated in Fig. 2.37(a) does not simply act as four shear walls providing stability in two perpendicular directions – it acts as a rigid hollow box in bending, cantilevered from the foundations as shown in Fig. 2.37(b). As for shear walls, the horizontal loads are carried to the core by slabs acting as deep beams.

Example 2.10 Lateral stability I

Problem: Fig. 2.38 shows the floor plan and typical cross-section for a two-storey office building. There is to be minimum structural obstruction internally and on the perimeter except on the west face where an existing building adjoins. Planning restrictions require that the building height be kept to a minimum. Devise a structural scheme to resist horizontal wind loading.

Solution: Lateral stability of this structure can be achieved by any one of the methods described above. Alternatively, a combination of different methods may be incorporated into the structure to resist horizontal loads from different directions and to satisfy the design specifications. The scheme illustrated in Fig. 2.39 is one such solution. To keep structural depth to a minimum, flat slab construction will be used. To resist E–W load with the minimum of internal obstruction, rigidly connected frames are provided in which slabs form the horizontal members. As the slabs are not very stiff, columns are provided at 7 m intervals along the north and south perimeters. The 7 m × 9 m grid is also acceptable for a flat slab of reasonable depth. To resist N–S wind, the cheaper alternative of infill shear walls is selected. Wall 1 is located on the west perimeter as no windows are possible on this face. Walls 2 and 3 are placed on either side of the stairs. For wind from the south on bay A, say, the

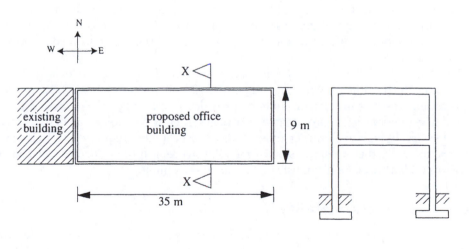

(a)   (b)

Figure 2.38 Two-storey office building: (a) floor plan; (b) typical section X–X

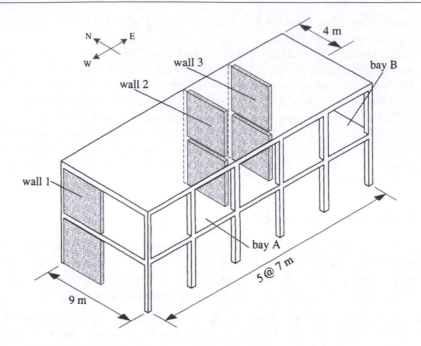

Figure 2.39 Structural scheme for the provision of lateral stability

slabs at each level act as beams to transfer the load to walls 1, 2 and 3. For wind on bay B, the slabs act as cantilevers to again transfer the load to walls 1, 2 and 3.

Movement joints

Movement joints are used to divide a large structure into a number of independent sections. They should pass through the entire structure above foundation level in one plane. In reinforced concrete frame structures, movement joints should be provided at approximately 50 m centres in both directions. In addition, movement joints should be provided where there is any significant change in the type of foundation or of the height of the structure and should be located, if possible, at a change in plan geometry.

Note: The *Manual for the Design of Reinforced Concrete Building Structures to Eurocode 2*, published by the Institution of Structural Engineers (2006) (henceforth referred to as the ISE manual), recommends that movement joints be provided in buildings where there are significant changes in plan geometry or at approximately 50 m centres. However, joints are best avoided if at all possible as they are difficult to construct and complicate the cladding requirements in the immediate vicinity of the joint. Further, the provision of the joint makes it more difficult to provide stability against horizontal loads. This is demonstrated in the next example.

Example 2.11 Lateral stability II

Problem: Devise a scheme to resist the horizontal wind loads on the five-storey concrete structure of Fig. 2.40(a).

Solution: Due to the significant change in geometry, a movement joint is positioned between the north and south wings, and the lateral stability of each wing must be considered separately. Lateral

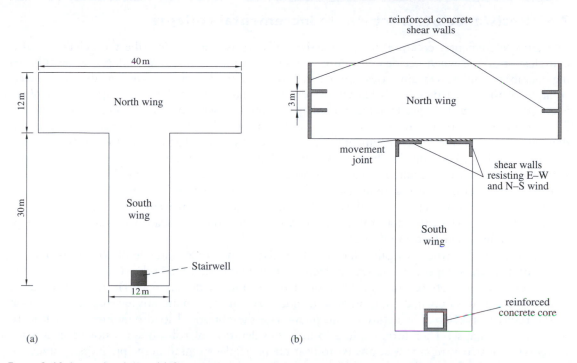

Figure 2.40 Lateral stability: (a) five-storey concrete structure; (b) provision of lateral stability in North and South wings

stability of the north wing against wind from all directions is best achieved by providing some combination of shear walls and/or cores constructed in either masonry or concrete. In the scheme of Fig. 2.40(b), wind loads are transferred by the floor slabs to the shear walls and/or cores which then carry the load to the foundations in proportion to their relative stiffnesses. If the shear walls/cores are of unequal stiffness, some twisting of the structure will occur under the action of a northerly wind. When such twisting effects are substantial, they must be considered and allowed for in the design (see Section 5.7).

To resist horizontal loads in the south wing, a reinforced concrete core is provided around the south stairwell. However, to prevent twisting of the structure under the action of wind from the east or west, some form of lateral support is also required at the northern end of the south wing. The provision of masonry or concrete shear walls at the position illustrated in Fig. 2.40 (b) satisfies this requirement but does cause significant internal obstruction. Fame action could be utilized to give the necessary E–W stability at the north end of the south wing. However, the difference in stiffness between the frame and the core would tend to result in twisting of the building and is best avoided.

Connections

In the provision of lateral stability it is of vital importance to ensure that the connections between the adjoining members in a structure are adequate to transfer the applied horizontal loads. For rigidly connected frames, the moment capacity of the connections, such as that illustrated in Fig. 2.31, must be sufficient to resist the bending moment induced by frame action. In shear wall and beam-slab details, each member, where appropriate, should be adequately anchored or tied to its adjoining member to provide robustness.

2.4   Resistance of structures to incremental collapse

The term 'incremental collapse' or 'progressive collapse' is used to describe the behaviour of a complete structure when accidental failure of a single member causes the collapse of neighbouring members and, in certain cases, the entire structure. In the case of such a 'domino' effect, collapse of the initial member is known as primary damage and collapse of adjoining members is called secondary damage. Incremental collapse is often described as disproportionate collapse because the extent of the overall damage is out of proportion to the magnitude of the initial damage. Probably the best documented case of incremental collapse was the disaster at the Ronan Point flats, England, in 1968, in which the failure of one vertical concrete panel due to a localized gas explosion resulted in extensive secondary damage to the whole high-rise structure. The vertical panel had been supporting the floor slab above at one corner and its failure led to collapse of that floor. The external wall slabs of the floors above collapsed in turn and the weight of their impact caused a progressive collapse of the floor and wall panels in one corner of the block right down to ground level.

The integrity of a structure against incremental collapse due to accidental loads, otherwise known as its robustness, is dependent on several factors, including design methods, structural layout, the type of connections between members and the nature of the accidental load. The current trends of reducing factors of safety in design and making more efficient use of materials have led to a reduction in the reserve capacity of structures to accommodate the abnormal load conditions that lead to incremental collapse. Therefore, while a formal consideration of robustness is not necessary for many types of structure, it is wise always to bear the principles in mind in the preliminary stages of design.

The best approach to take when making a qualitative assessment of the robustness of a structure is to determine the effect on the stability of the structure of removing each member in turn. Statically determinate structures, such as that illustrated in Fig. 2.41, are inherently non-robust because the primary damage of any one member by accidental loads will generally result in disproportionate collapse. Statically indeterminate or 'redundant' structures, however, are robust as they contain more support restraints and/or internal members than are required for stability. This allows a redistribution of force through secondary load paths in members adjacent to the damaged member and the structure is often capable of absorbing the accidental load without incurring significant secondary damage. For instance, if member JL is removed from the structure in Fig. 2.3, the horizontal reactions from force F_3 are taken by frame action in DEFGHI. However, the failure of the inclined member in the determinate frame of Fig. 2.41 would lead to collapse of the entire structure. Further information on incremental collapse and guidelines for designing robust structures are given in the ISE publication *Stability of Buildings* (1988).

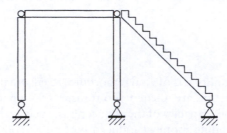

Figure 2.41 Removal of any member of the pin-jointed structure will result in disproportionate collapse

Example 2.12 Multi-storey plane frame

As an example of qualitative assessment of the robustness of a redundant frame, consider the two-dimensional concrete structure in Fig. 2.42(a). The elimination of the external column, DH, due to a

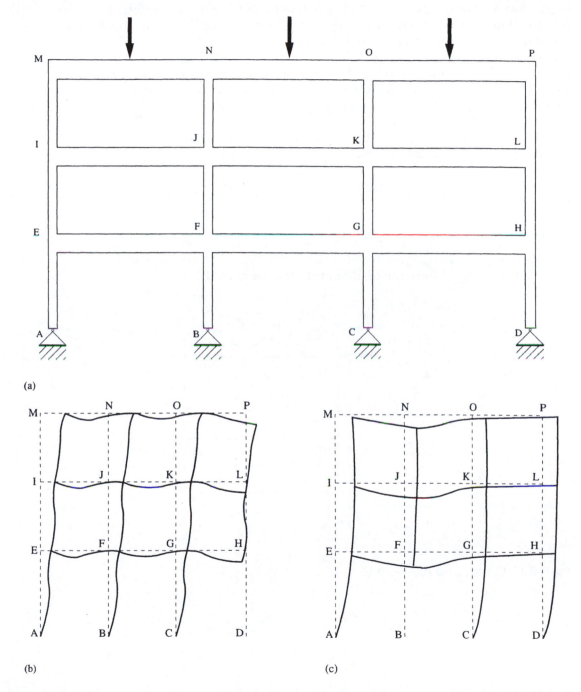

(a)

(b)   (c)

Figure 2.42 Robustness of rigid frame: (a) original frame; (b) effect of removing member DH; (c) effect of removing member BF

vehicle impact for example, results in the deformation of the frame illustrated in Fig. 2.42(b). Specifically, the removal of this member causes a substantial change in the distribution of bending moments and axial forces within the frames, particularly in members GH, KL, OP, PL and LH which now transfer the vertical load into the rest of the structure by frame action, and a significant increase in the axial load in member CG, which may lead to buckling of the member. Thus, incremental collapse of the structure is prevented only if these members and their connections are sufficiently strong to resist the new distribution of force/moment.

Similarly, the removal of member BF, as illustrated in Fig. 2.42(c), results in complete structural failure if members AE and CG are unable to carry the extra compressive force and members EFG, IJK and MNO are not strong enough to carry the increased bending moment. However, in this case when deformations are very large, members EFG, IJK and MNO could still act as tension members even if they have yielded in bending.

By the hypothetical removal of other individual members, the requirements of each member in the structure to prevent total collapse due to the failure of a single member can be determined.

Problems

Section 2.1

2.1 (a) How is the force, F, illustrated in Fig. 2.43, transferred to the supports?
     (b) Find the axial force and bending moment in member BD.

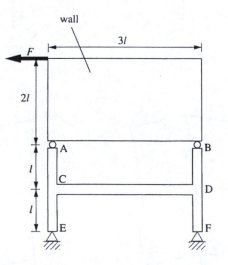

Figure 2.43 Structure for Problem 2.1

2.2 By what mechanisms of load transfer is the horizontal force illustrated in Fig. 2.44 transferred to the supports? Where is the bending moment due to this force at maximum?

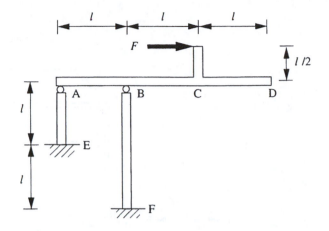

Figure 2.44 Structure for Problem 2.2

2.3 What percentage of the total vertical force, *F*, in the structure in Fig. 2.45, is carried by the member AC?

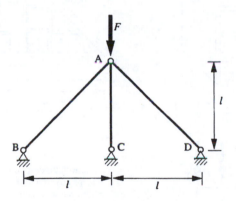

Figure 2.45 Structure for Problem 2.3

Sections 2.2 and 2.3

2.4 A single-storey factory is to be designed in which the structure may be exposed to corrosive gases. Propose a preliminary structural solution given that the dimensions in plan are 15 m × 40 m and that there are to be no internal obstructions.

2.5 State, giving reasons, whether the proposals in Fig. 2.46 are structurally stable and robust.

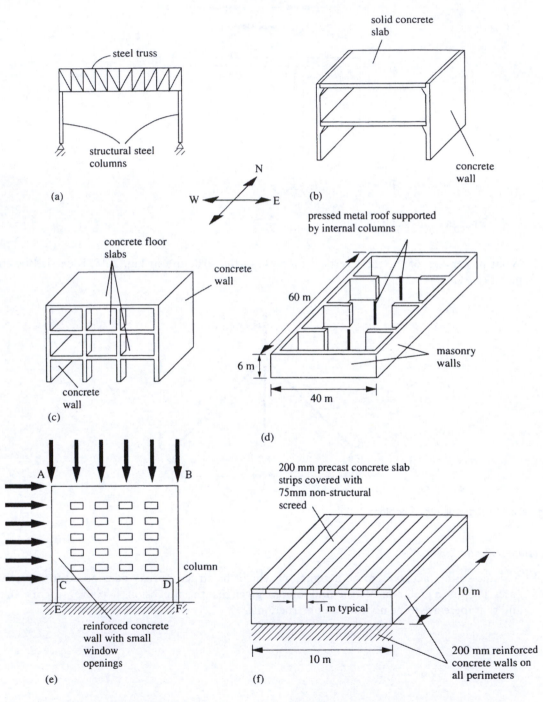

Figure 2.46 Structures for Problem 2.5

Chapter 3

Loads and load effects

3.1 Loads and load effects

The aim of structural design is to ensure that, with an acceptably high probability, a structure will remain fully functional during its intended life. The expected lifetime of a structure is formally known as its **design life** and for permanent buildings is usually 50 years. During its design life, a structure must be capable of safely carrying all applied loads and other stress-inducing actions that might reasonably be expected to occur. Thus, it is necessary to identify and, more importantly, quantify the various types of load that act on its members. The different classes of loads that commonly act on structures and must be considered in design are the subject of this chapter. Due to the variable nature of most loads and materials, structural design philosophy relies heavily on the use of statistical principles. The application of these principles to ensure structural safety is discussed in Section 3.7.

It is worth pointing out at this stage that in contrast to construction procedures, in which the structure is erected from the ground upwards, building structures are generally *designed* from the top downwards. This is because the magnitudes of the internal gravity forces in members at any level depend on the forces being transferred from the levels above as well as the loads applied directly at that level.

3.2 Classification of loads

The term **action** is used in the Eurocodes collectively to describe forces and environmental effects on structures. An action can be defined as anything that gives rise to internal stresses in a structure and can be direct or indirect in nature. **Direct actions** are forces applied to the structure by external agents such as wind or traffic whereas **indirect actions** are imposed deformations in the structure that do not result from external forces. Temperature variation, settlement of supports and shrinkage of concrete members are examples of indirect actions that give rise to internal stresses in structures.

Actions can be classified further by their variation with time. Those that have little or no variation in magnitude over the design life of the structure are known as **permanent actions** (sometimes called dead loads). On the other hand, actions that are unlikely to maintain a constant magnitude over the design life are termed **variable actions**. Two important types of variable action are those due to wind and the gravity loading due to the occupants of the structure. The weight of occupants is known as occupancy load, or **imposed load** (also called live load). In a typical building, the self-weight of the floor slab is an example of a permanent action while the weight of people and furniture is occupancy load, a variable action. **Accidental actions**, such as the impact of a vehicle against a column, are also time variant. However, in contrast to variable actions, accidental actions are those which are unlikely to occur very often in significant magnitude and duration. It is worth noting that indirect actions are also classified by their variation (or lack of variation) with time. Shrinkage of a beam can induce stress in the columns to which it is attached. As this is long term in nature, shrinkage is classified as an (indirect) permanent action.

Table 3.1 Classification of loads

Class	Action	Examples	
		Direct	Indirect
Time variation	Permanent	Soil pressure, self-weight of structure and fixed equipment	Settlement, shrinkage, creep (results from direct permanent actions)
	Variable	People, wind, furniture, snow, traffic, construction loads	Temperature effects
	Accidental	Explosion, vehicular impact	Temperature rise during fire
Spatial variation	Fixed	Self-weight (generally), trains (fixed in direction normal to rails)	–
	Free	Persons, office furniture, vehicles	–
Static/dynamic	Static	All gravity loads	–
	Dynamic	Engines, turbines, wind on slender structures	

Actions can also be classified by their spatial variation, that is, variation in the area of application. **Fixed actions** are actions that have no freedom of movement within or on the structure. Clearly, structural self-weight falls into this category. **Free actions**, on the other hand, are actions that can occur at arbitrary locations on or within the structure. Furniture, for example, is a free action.

Two other terms that are applied to the classification of the nature of a load or structural response are **static** and **dynamic**. Direct actions are known as 'dynamic' actions if they cause significant acceleration or vibration of the structure to which they are applied, and as 'static' actions if they cause no significant acceleration. In sufficiently slender structures many different types of action can be dynamic. For example, the surge due to acceleration in a gantry crane can be dynamic. In tall, slender towers, wind can also be dynamic, but for most concrete structures it is essentially a static action. Wind actions on non-slender structures can be known as quasi-static actions, where a dynamic action is represented by an equivalent static action.

Table 3.1 summarizes the different load classifications, giving some common examples of each type of action. It is important to realize that most actions can be categorized under each class. For example, self-weight can be classified as a direct action that is permanent, fixed and static.

The next section deals with loaded areas in buildings. In the three sections following the next, the more familiar types of direct action, namely permanent gravity loads, variable gravity loads and the static effect of wind loads, are treated in greater detail. More specialized actions, such as seismic loads and vehicle loads, are not considered further.

3.3   Tributary areas

The tributary area of a beam is defined as the area which contributes to the loading of that beam. When a uniformly loaded floor slab is simply supported by beams on two opposite sides, half of the load is carried by each beam. The slabs of Fig. 3.1(a–b) are simply supported on beams AD, BE and CF. Thus, all of the load on the shaded area in Fig. 3.1(a), which is the tributary area of beam BE, is supported by that beam. This follows from the fact that the reaction at each end of a uniformly loaded simply supported beam is half the total load (see Appendix C, No. 1). If the slab of Fig. 3.1(a) were continuous over beam BE, as illustrated in Fig. 3.1(c), the reaction on BE due to a uniformly distributed load throughout the floor would be greater with this beam carrying a greater portion of the total load. It can be seen from Appendix C, No. 4, that, in this situation, beam BE would take $\frac{5}{4}$ of the total load on one span ($\frac{5}{8}$ from each span) and the tributary area would be $\frac{5}{4}$ (6 m) $\times$ 10 m.

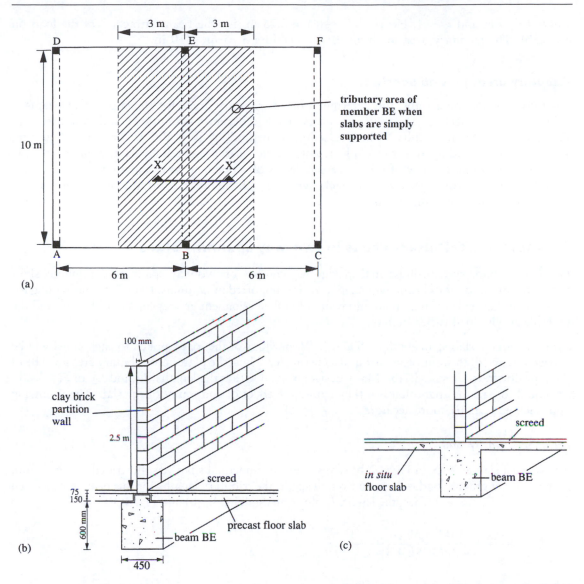

Figure 3.1 Calculation of loads: (a) floor plan; (b) section X–X; (c) alternative section X–X if slab were continuous over BE

When loading is not uniformly distributed on all spans, as is the case for variable gravity load, the reaction and hence the tributary length is affected. For example, if the slab of Fig. 3.1(a) is continuous over beam BE, as illustrated in Fig. 3.1(c), and is loaded uniformly on the panel, ABDE only, then the tributary length is calculated from Appendix C, No. 6. The tributary length for support BE is then:

$$l_b = \frac{l(4k+1)}{8k}$$

where $l = 6$ m and $k = 1$. Hence, $l_b = \frac{5}{8}(6) = 3.75$ m (i.e. the beam attracts $\frac{5}{8}$ of the load on the slab). The tributary area for beam BE is therefore 3.75 m × 10 m.

Tributary areas for slab panels

For two-way spanning slabs, the determination of tributary areas for the supporting beams is a little more subjective. Different engineers make different assumptions and can arrive at significantly different results. When selecting a method, it is useful to bear in mind that actual variable loading due to people furniture, etc. is represented by equivalent uniformly distributed loading. In view of this, excessive levels of refinement in calculating tributary areas would seem inappropriate. One approach, which is sufficiently accurate for most purposes, is illustrated in the following example.

Example 3.1 Tributary areas in two-way spanning slabs

Problem: A floor system, illustrated in Fig. 3.2, consists of two-way spanning continuous slabs supported by continuous beams which, in turn, are supported by columns. Determine the loading for beam abcd that will give maximum moment at b if the permanent gravity load is 10 kN/m² and the variable gravity load varies from 0 to 7 kN/m².

Solution: As the slab spans both N–S and E–W, portions of the load applied to each panel will be carried by each of the four beams around its perimeter. Thus, the pattern of tributary areas will be of the type illustrated in Fig. 3.2(b). Many designers assume that the lengths, L_1 and L_2, of Fig. 3.2(b) are equal to half the span, that is – they ignore the effect of continuity of the slab. The following approach is, perhaps, more accurate.

Permanent gravity load

The strip of slab at X–X in Fig. 3.2(b) spans over three unequal spans, as illustrated in Fig. 3.3(a). As permanent gravity load is uniform over all spans, the tributary lengths for this beam are as given in Appendix C, No. 8. Hence, the length L_1 illustrated in Fig. 3.2(b) and Fig. 3.3(a) is:

$$L_1 = \frac{L(6k^4 + 15k^3 + 6k^2 - 2k - 1)}{4k(4k^2 + 8k + 3)}$$

where $k = \frac{6}{8} = 0.75$ and $L = 8$ m. Thus:

$$L_1 = 2.157 \text{ m}$$

The length, L_3, also illustrated in Fig. 3.2(b) and 3.3(a) is half the span length, that is 4 m. Similarly, considering section Y–Y of Fig. 3.2(b), illustrated in Fig. 3.1(b), we find from Appendix C, No. 5:

$$L_5 = L\left(\frac{3k^2 + k - 1}{8k}\right)$$

where, again (by coincidence), $k = \frac{6}{8} = 0.75$ and $L = 8$ m. Thus:

$$L_5 = 1.917 \text{ m}$$

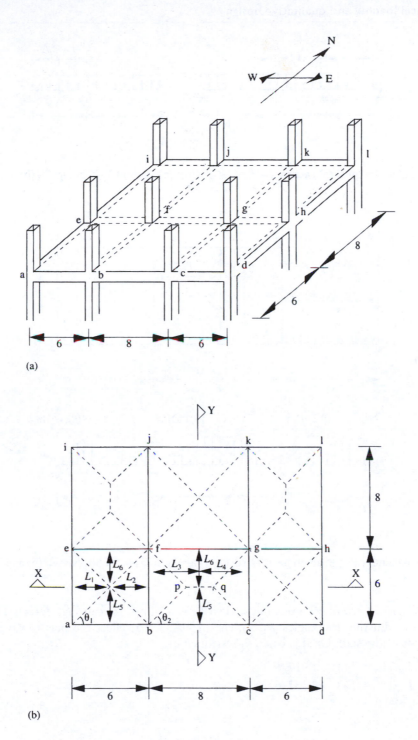

(a)

(b)

Figure 3.2 Two-way spanning slab on continuous beams: (a) geometry; (b) pattern of tributary areas (plan)

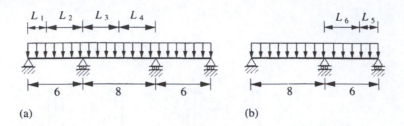

Figure 3.3 Tributary lengths: (a) section X–X of Fig. 3.2(b); (b) section Y–Y of Fig. 3.2(b)

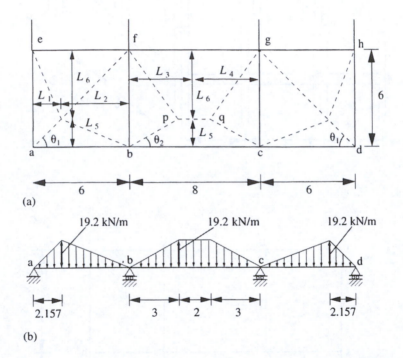

Figure 3.4 Permanent gravity load on beam abcd: (a) tributary areas (part plan view); (b) applied loading on beam

A sensible pattern of tributary areas might then be that illustrated in Fig. 3.4(a). This has been constructed by adjusting the angles, θ_1, θ_2, etc., from 45° to reflect the bias towards the simply supported ends of the slab. On this basis, θ_1 is defined by:

$$\tan \theta_1 = \frac{L_5/(L_5 + L_6)}{L_1/(L_1 + L_2)} = \frac{1.917/6}{2.157/6}$$

$$= 0.889$$

$$\Rightarrow \theta_1 = 42°$$

Similarly:

$$\theta_2 = \tan^{-1}\left(\frac{L_5/(L_5 + L_6)}{L_3/(L_3 + L_4)}\right) = \tan^{-1}\left(\frac{1.917/6}{4/8}\right)$$

$$= 33°$$

In practice, the tributary area for beam abcd can often be constructed without calculation of such angles. This is done by first constructing the tributary areas on the assumption that all slabs are simply supported and all angles are 45° (as illustrated in Fig. 3.2(b)) and then editing this pattern. For example, points p and q (Fig. 3.2(b) and Fig. 3.4(a)) are adjusted in the ratio $L_3 / (L_3 + L_4)$:0.5 for the E–W direction and $L_5 / (L_5 + L_6)$:0.5 for the N–S direction. In the E–W direction, p is 3 m east of b in Fig. 3.2(b) and there is no adjustment as $L_3 / (L_3 + L_4)$ is equal to 0.5. The result of this is the pattern illustrated in Fig. 3.4(a).

Having determined the pattern of tributary areas, the loading on each beam can be found. Thus the permanent gravity loading on beam abcd varies linearly in span ab from zero at a to a maximum intensity of $(10L_5 =)19.2$ kN/m and back to zero again at b as illustrated in Fig. 3.4(b). Similarly, there is a trapezoidal distribution of load on span bc with a maximum intensity of 19.2 kN/m.

Variable gravity load

The variable gravity load on this floor can have any intensity of loading from 0 to 7 kN/m². Load in spans ab and bc will cause a hogging moment at b while load in span cd will actually reduce that moment. Thus, the load case which maximizes the moment at b will include the full loading on panels abef and bcfg and zero loading on panel cdgh. Load on other panels will not greatly influence the moment at b. For such panels, it is reasonable (and allowed by the Eurocode) to assume the minimum level of variable gravity loading (i.e. zero). Thus, variable gravity loading for maximum moment at b is taken to be that illustrated in Fig. 3.5(a). The strip of slab at section Y–Y is now loaded on one span only and the tributary lengths are calculated from the formulae in Appendix C, No. 6. Evaluation of the formulae in the appendix with $k = 8/6$ gives reactions of 45 and 59 per cent of the applied load at supports A and B, respectively (refer to figure in Appendix C). The fact that the sum of reactions equals 104 per cent is explained by a downward reaction at the third support equal to 4 per cent of the applied load. The tributary lengths, l_A and l_B, are 45 and 55 per cent of the span length. While a complete assigning of the loading in these proportions would underestimate the reaction at B, it is reasonably accurate for the reaction at A, which is all that is of interest for this load case. Hence:

$$L_5' = (0.45)(6) = 2.678 \text{ m}$$
$$L_6' = (0.55)(6) = 3.322 \text{ m}$$

Similarly, from Appendix C, No. 9:

$$L_1' = 2.057 \text{ m} \qquad L_2' = 3.943 \text{ m}$$
$$L_3' = 4.338 \text{ m} \qquad L_4' = 3.662 \text{ m}$$

Thus, the variable gravity loading on beam abcd is, from Fig. 3.5(a), as illustrated in Fig. 3.5(b).

The total loading on a beam such as that considered in Example 3.1 is the sum of the loadings due to permanent and variable gravity load. As an alternative to adding the loading intensities, it is generally more convenient to analyse separately to find the moment due to permanent and variable loading and to add the results.

3.4 Permanent gravity loads on structures

Permanent gravity loads consist of the self-weight of the structural members plus the weight of objects permanently fixed to the structure over its design life. These materials include permanent partitions and walls, windows, ceilings, cladding, heating ducts, plumbing and electrical installations and any other essentially unmovable equipment.

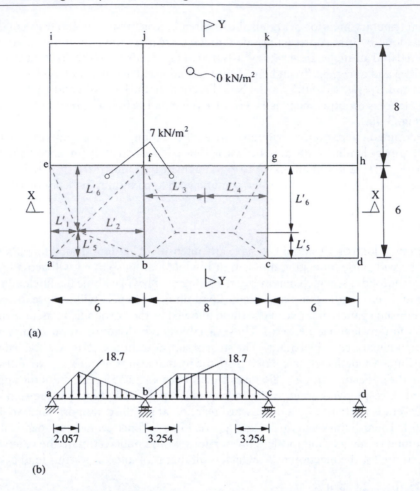

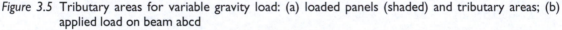

Figure 3.5 Tributary areas for variable gravity load: (a) loaded panels (shaded) and tributary areas; (b) applied load on beam abcd

The self-weight of individual structural members is normally determined from their dimensions and the weight density of the materials used. The weight of non-structural objects is estimated either from their approximate dimensions and the density of their constituent materials (given in EN 1991-1-1, Annex A) or from manufacturers' specifications. The characteristic weight densities for various materials that can be used to determine the magnitude of permanent gravity loads in a structure, are listed in Table 3.2.

Example 3.2 Permanent gravity loads

Problem: For the structure of Fig. 3.1(a–b), determine the permanent gravity load acting on the support at point B.

Solution: The permanent gravity load, W, acting on the member, BE, is the combined weight of the permanent partition wall, a portion of the slab and screed and the self-weight of the member itself. The precast floor slab units are supported solely by the three members, AD, BE and CF, in two simply supported spans. The weight on each span is distributed equally between these beams, with each member carrying the weight of 3 m by 10 m of the slab. The total tributary area supported by the

Table 3.2 Recommended values of weight density for common construction materials (from BS EN 1991-1-1, Annex A)

Construction material	Weight density (kN/m^3)
Concrete	
reinforced or prestressed	25
unreinforced (natural aggregate)	24
cement mortar	19–23
Masonry	
concrete blocks[*]	14–20
brickwork[*]	18–24
stone masonry	
granite	27–30
limestone	20–29
Metals	
aluminium	27
cast iron	71–72.5
copper	87–89
lead	112–114
steel	77–78.5
zinc	71–72
Timber	
softwood	3.5–5
hardwood	6.4–10.8
chipboard	7–8
plywood	4.5–7
glued laminated timber	3.5–4.4

[*] Typical values obtained from Brooker (2006) in the *Concrete Buildings Scheme Design Manual*, published by *The Concrete Centre*.

member BE is shaded in Fig. 3.1(a). The total permanent gravity load acting on the member BE is (assuming the weight density of the partition wall and screed is 22 kN/m^3 and 23 kN/m^3, respectively):

$W =$ weight of partition wall

$\qquad +$ weight of slabs and screed

$\qquad +$ self-weight of beam BE

$$W = 22 \times 0.1 \times 2.5 \times 10 + 2(25 \times 0.15 \times 3 \times 10)$$
$$+ 2(23 \times 0.075 \times 3 \times 10)$$
$$+ 25 \times 10 \times (0.6 \times 045) \text{ kN}$$

(*Note* that the top 0.15 m of the beam has been allowed for in the calculation of the slab weight)

$\Rightarrow W = 451$ kN

Since member BE is itself simply supported and the load is uniform, half the total load is supported on each column. Therefore, the total permanent gravity load at point B is:

$$\frac{W}{2} = 225.5 \text{ kN}$$

Permanent gravity loads for initial sizing of members

The magnitudes of the permanent gravity loads on a structure are dependent on the size of the structural members which are, in turn, governed by the applied loads. Thus, in order to determine the permanent gravity loads acting on a structure, it is necessary to obtain initial estimates for the member sizes. Methods for determining initial estimates for the sizes of structural concrete members are given in Chapter 6. In checking the adequacy of such initial designs, the ISE manual provides the following guidelines for complete components:

(a) For floors where the finish is not specified, add 1.8 kN/m^2 to the weight of the slab to allow for a screed finish of about 75 mm depth (density of screed taken as 23 kN/m^3).
(b) The weight of ceilings and services for normal use can be taken initially as 0.5 kN/m^2, acting on the entire plan area of the floor.

3.5 Variable gravity loads on structures

Variable gravity loads that act on the floors and roof of a structure are those which vary in magnitude during the design life. They include the weight of persons, furniture, movable partitions, stored material, snow, vehicles and any other objects that are removable.

Imposed loads on floors

Imposed loads on floors can be classified as those which relate to the occupancy of a structure. They may be caused by people, furniture, movable partitions, stored material or vehicles. When light-weight non-permanent partitions are to be present but their location is not known, they can be represented by an imposed load intensity of approximately 1.0 kN/m^2 acting on the entire plan area of the floor (from the ISE manual). More specific values for moveable partitions, which depend on the self-weight of the partitions, can be obtained from Eurocode 1 (EC1).

The magnitude of the variable gravity loads due to occupancy is determined by the specific functions for which different areas of the building are used. When determining imposed loads on floors, a distinction can be made between the following types of areas in a building:

- Assembly areas
- Commercial and administration areas
- Escape routes
- Parking areas
- Maintenance areas
- Residential areas
- Storage and production areas.

The spatial and time variation of occupancy floor loads is generally different for each type of use. For instance, the imposed load in a hospital waiting room is greater than that in a hospital ward because the former is classed as an assembly area which is susceptible to overcrowding.

In recent years, probabilistic models have been developed for the derivation of suitable intensities of imposed loads that take account of their time-varying and space-varying nature. For the variation in time, the imposed load is separated into two components: a sustained component and an extraordinary (short-term) component (Fig. 3.6). The sustained component represents the mean loading for normal day-to-day use (persons, miscellaneous stored material, furniture, etc.). This changes, for example, with the tenant using the floor area – some tenants will use more or heavier

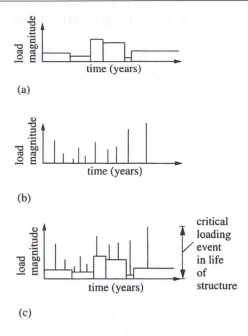

Figure 3.6 Variable gravity loads: (a) sustained load; (b) extraordinary load; (c) total load (sum of (a) and (b))

furniture than others. The extraordinary component represents unusual extremes in loading such as concentrations of persons (emergency situations, parties, ceremonies) and short-term stacking of furniture as might occur during redecoration work.

It is impractical to represent all possible spatial arrangements of occupancy loads at all times. For design purposes, an equivalent uniformly distributed load (EUDL) is used. Using probabilistic load models, values of load obtained from load surveys and engineering judgement, intensities of EUDL have been derived for a wide range of building types and uses. It is intended that the bending moment, shear force, etc. due to the EUDLs will equal the values that would actually occur in buildings under the same use with an acceptably remote probability of occurrence. Values of EUDLs for different occupancies are given in EN 1991-1-1 for actions on structures, some of which are given in Table 3.3. The recommended values are underlined where a range of values are provided. More detailed descriptions and recommended values are also given in the *United Kingdom National Annex* (UK NA).

The figures in Table 3.3 represent nominal values which are adequate for most structures. However, where it is likely that these values are exceeded, such as in storage areas, precise figures should be obtained from the weight densities of the stored material. When a storage area is used for books or other documentation (i.e. in category E) the loading may be determined based on the loaded area and the height of the stored material. An extensive list of recommended values for the weight densities of stored material is given in EN 1991-1-1, Annex A. For determining local effects due to actions, the Eurocode also provides equivalent values of concentrated loads.

Example 3.3 Imposed loads on floors

Problem: The two-storey building illustrated in Fig. 3.7 is intended for use as a library. The ground floor and second-floor levels are to be used as reading rooms only while the first floor is to be used for

Table 3.3 Imposed loads on buildings (from BS EN 1991-1-1, refer to the National Annex for recommended values for each country)

Category	Description	Examples	Intensity of EUDL (kN/m²)
A	Areas for domestic and residential activities	Rooms in houses, rooms and wards in hospitals, bedrooms in hotels, kitchens and toilets	
	Floors		1.5 – 2.0
	Stairs		2.0 – 4.0
	Balconies		2.5 – 4.0
B	Office areas		2.0 – 3.0
C	Areas where people may congregate		
	C1 Areas with tables	In schools, restaurants, reading rooms	2.0 – 3.0
	C2 Areas with fixed seats	In churches, theatres, lecture halls, waiting rooms	3.0 – 4.0
	C3 Areas for moving people	In museums. Access areas in hotels and in public buildings	3.0 – 5.0
	C4 Areas for physical activities	Dance halls, stages, gymnasia	4.5 – 5.0
	C5 Areas susceptible to overcrowding	Concert halls, grandstands	5.0 – 7.5
D	Shopping areas		
	D1 Retails shops		4.0 – 5.0
	D2 Department stores	Areas in warehouses	4.0 – 5.0
E	Areas susceptible to accumulation of goods	Areas of storage use, libraries	7.5

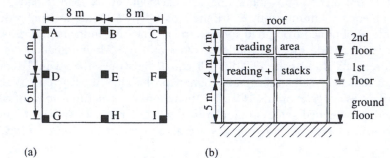

(a)

(b)

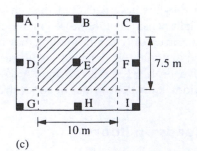

(c)

Figure 3.7 Calculation of imposed gravity loads: (a) floor plan; (b) end elevation; (c) tributary areas for columns

reading and book storage. Lightweight partitions are to be present at unspecified locations. The total variable gravity load on the roof (snow load) is 1.2 kN/m^2. Determine the total variable gravity load on column E due to the two floors and the roof. The floors are of flat slab (beamless) construction.

Solution: From Table 3.3, the characteristic variable gravity load on each floor is:

> First floor (reading room + stacks) = 7.5 kN/m^2
> Second floor (reading room only)   = 3 kN/m^2
> Roof   = 1.2 kN/m^2

The partition loading consists of a further 1 kN/m^2 on the first and second floors. The tributary areas are found by assuming two-span continuous beam behaviour in both directions. Hence, from Appendix C, No. 4, column E supports an area of $\frac{5}{4}$ (8) $\times$ $\frac{5}{4}$ (6) = 75 m^2. The tributary areas of all columns are illustrated in Fig. 3.7(c). The total variable gravity load in column E is the sum of the loads acting on the roof and the first and second floors (the load from the ground floor is assumed to be transferred directly to the foundations), that is:

$$P_E = 75[(7.5 + 1) + (3 + 1) + (1.2)]$$
$$P_E = 1027.5 \text{ kN}$$

Under certain circumstances, it is possible to reduce the recommended value of the variable gravity loads for buildings over two storeys in height and/or buildings with a large floor area. This is allowed because, in such conditions, it is unlikely that an extraordinary loading event will occur simultaneously throughout the structure. EC1 allows a reduction in variable gravity load as a function of the number of storeys and the loaded area.

Snow and other variable gravity loads on roofs

Roofs can be classified as being either accessible or non-accessible except for maintenance and repair. In the case of an accessible roof, the intensity of the variable gravity load due to occupancy is taken from the appropriate class in Table 3.3. When access is restricted, EC1 provides a range of recommended values for the load intensity. In the UK NA, the load intensity on a roof with a slope of less than 30 ° with restricted access is taken as 0.6 kN/m^2. In addition to the variable gravity loads associated with occupancy, roofs are also exposed to gravity snow loads which must be considered in design. However, according to EC1, imposed load should not be applied together with snow or wind load.

The deposition and accumulation of snow on roofs is dependent on a number of factors, including roof geometry and surface properties, the proximity of adjacent buildings, local climate and local surface terrain. In EN 1991-1-3, design values for the intensity of snow loads on roofs for persistent/transient design situations are determined using the formula:

$$s = \mu_i C_e C_t s_k \ [\text{kN/m}^2] \tag{3.1}$$

where C_e and C_t are exposure and thermal coefficients, respectively, and usually have values of unity. The parameter s_k is the characteristic value of snow load on the ground, and its intensity depends on geographic location and the altitude of the site. Methods of calculating s_k vary between countries. In the UK, for example, it is calculated as follows:

$$s_k = [0.15 + (0.1Z + 0.05)] + \left(\frac{A - 100}{525}\right) \tag{3.2}$$

where A is the altitude of the site in metres above mean sea level and Z is the zone number given in Fig. 3.8.

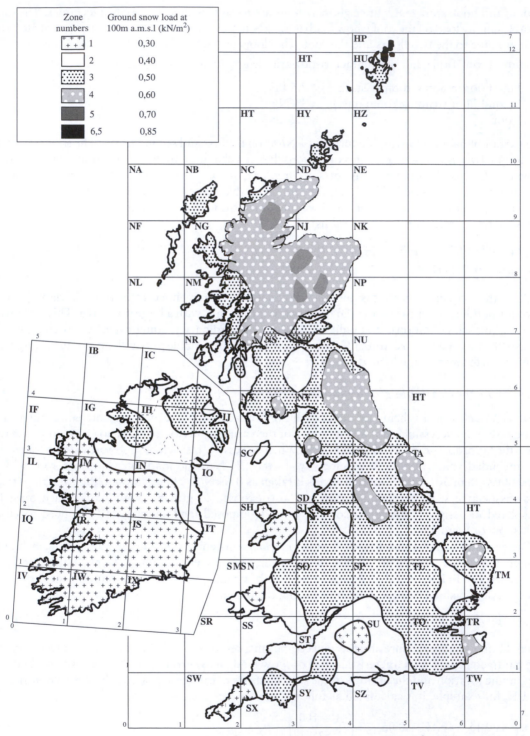

Jersey - Zone 3, 0,50 kN/m² at 100m a.m.s.l, Guernsey - Zone 2, 0,40 kN/M² at 100m a.m.s.l

Figure 3.8 Zone number, Z, for UK and Ireland (from UK NA to BS EN 1991-1-3)

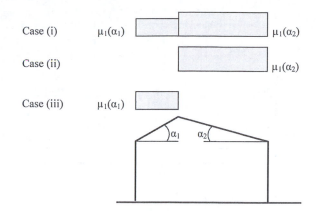

Figure 3.9 Snow load arrangements for single bay pitched roofs (from BS EN 1991-1-3 and UK NA to BS EN 1991-1-3)

Table 3.4 Shape coefficients, μ_1, for pitched roofs (from BS EN 1991-1-3 and UK NA to BS EN 1991-1-3)

Roof pitch, α_i	μ_1			
	$0° \leq \alpha_i \leq 15°$	$15° < \alpha_i\ 30°$	$30° < \alpha_i < 60°$	$\alpha_i \geq 60°$
undrifted load arrangements (Case (i))	0.8	0.8	0.8 $(60 - \alpha_i) / 30$	0.0
drifted load arrangements (Case (ii) and Case (iii))	0.8	0.8 + 0.4 $(\alpha_i-15) / 15$	1.2 $(60 - \alpha_i) / 30$	0.0

The parameter μ_i is a snow load shape coefficient that is dependent on the geometry of the structure. It is obtained from tables and formulae. For example, for a duopitch roof, EC1 recommends that both drifted and undrifted load arrangements be considered. One load case is specified for the undrifted load arrangement, Case (i), and two load cases are specified for a drifted load arrangement, Cases (ii) and (iii). However, the Eurocode also allows alternative load arrangements for the drifted case to be specified in the National Annex, and the drifted load arrangements and shape coefficients specified in the UK NA are different from those in the Eurocode. For a duopitch roof (see Fig. 3.9), the shape coefficients to be used in the UK for undrifted and drifted load arrangements are given in Table 3.4. A summary of the procedure for calculating snow load intensity is illustrated in Fig. 3.10. Alternative procedures are provided in the Eurocode for other design situations (e.g. when snow is prevented from sliding off the roof, for exceptional snow falls, for multi-span roofs, for circular roofs).

Example 3.4 Snow load on roofs

Problem: Determine the snow load for the duopitch roof of a single bay building (pitch $= 20°$), which is located at sea level in the city of Norwich. Exposure and thermal coefficients of unity may be assumed.

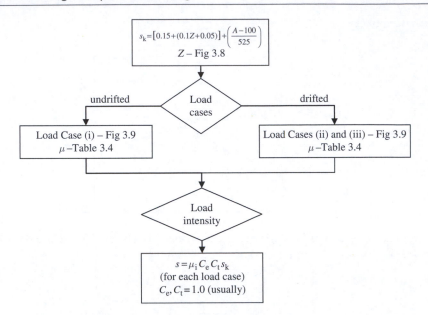

Figure 3.10 Flow chart for the calculation of snow load intensity

Solution: From Fig. 3.8, Norwich is in Zone 4, which results in a snow load of $(A = 0)$:

$$s_k = [0.15 + (0.1(4) + 0.05)] + \left(\frac{0 - 100}{525}\right)$$

$$s_k = 0.41 \text{ kN/m}^2$$

The shape coefficients are taken from Table 3.4. For the undrifted load arrangement, Case (i):

$$\mu_1(20°) = 0.8$$

From Equation (3.1), the load intensity for Case (i) on both pitches is:

$$s = \mu_i C_e C_t s_k$$
$$s = 0.8 \times 1 \times 1 \times 0.41$$
$$s = 0.33 \text{ kN/m}^2$$

Since the roof is symmetrical, the shape coefficients for the drifted load arrangement for Cases (ii) and (iii) are the same, and are given by:

$$\mu_1(20°) = 0.8 + 0.4(\alpha_i - 15)/15 = 0.93$$

For Cases (ii) and (iii), there is no load on one pitch while the load intensity on the other pitch is:

$$s = \mu_i C_e C_t s_k$$
$$s = 0.93 \times 1 \times 1 \times 0.41$$
$$s = 0.38 \text{ kN/m}^2$$

Note that for this building, the snow load intensity is less than the minimum 0.6 kN/m^2 imposed load for a roof with restricted access, and therefore is a less onerous load case for the roof design.

3.6   Wind load on structures

Wind forces are variable loads that act directly on the internal and external surfaces of structures. The intensity of wind load on a structure is related to the square of the wind velocity and the dimensions of the surface that is resisting the wind (frontal area). Wind velocity is dependent on geographical location, the height of the structure, the topography of the area and the roughness of the surrounding terrain.

The response of a structure to the variable action of wind can be separated into two components, a background component and a resonant component. The background component involves static deflection of the structure under the wind pressure. The resonant component, on the other hand, involves dynamic vibration of the structure in response to changes in wind pressure. In most structures, the resonant component is relatively small and structural response to wind forces is treated using static methods of analysis alone. However, for tall or otherwise flexible structures, the resonant component of wind should be calculated using dynamic methods of analysis. Such structures are not considered further here.

Static effects of wind load on buildings

Basic wind velocity

The fundamental value of the basic wind velocity, $v_{b,0}$, for a locality is defined as the mean wind velocity at 10 m above flat open country terrain averaged over a period of 10 min with a return period of 50 years, irrespective of wind direction but accounting for altitude. EC1 (EN 1991-1-4) allows the value for $v_{b,0}$ to be specified in the National Annex. In the UK NA, for example, it is calculated using:

$$v_{b,0} = v_{b,map} c_{alt} \tag{3.3}$$

where the values for $v_{b,map}$ for the UK are given in Fig. 3.11 and c_{alt} is the altitude factor which allows for the altitude of the site on which the structure is located. Wind speeds tend to be greater in sites located at high altitudes. c_{alt} is calculated using:

$$c_{alt} = 1 + 0.001A \qquad z \leq 10\text{m} \tag{3.4}$$

$$c_{alt} = 1 + 0.001A(10/z)^{0.2} \quad z > 10\text{m} \tag{3.5}$$

where A is the altitude of the site in metres above mean sea level and z is the height above ground. Conservatively, Equation (3.4) can be used for buildings of any height. The height above ground can be taken as z_s which is the reference height of the structure. For buildings, the value of z_s is calculated using:

$$z_s = 0.6h \geq z_{min} \tag{3.6}$$

where h is the height of the building and z_{min} is a minimum height above ground (given in Table 3.5 for different terrain categories). The basic wind velocity, v_b, is found using fundamental value of the basic wind velocity, $v_{b,0}$:

$$v_b = c_{dir} c_{season} c_{prob} v_{b,0} \tag{3.7}$$

where c_{dir}, c_{season} and c_{prob} are directional, season and probability factors. The recommended values for c_{dir} and c_{season} are 1.0 in EC1. However, tables are provided in the UK NA for determining more

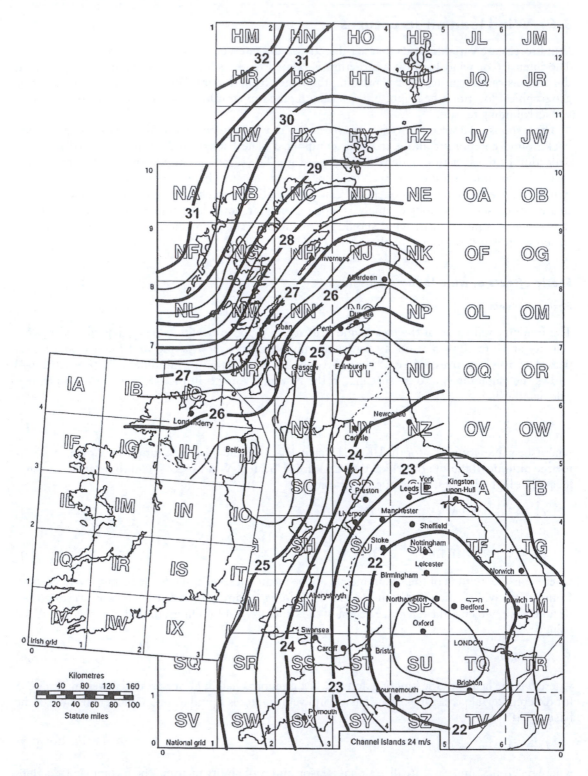

Figure 3.11 $V_{b, map}$ for the UK (from UK NA to BS EN 1991-1-4)

Table 3.5 Terrain categories and parameter values (from BS EN 1991-1-4)

UK NA Category	ECI Category	Terrain description	z_0 (m)	z_{min} (m)
Sea	0	Sea or coastal area exposed to the open sea	0.003	1
Country	I	Lakes or flat and horizontal area with negligible vegetation and without obstacles	0.01	1
	II	Area with low vegetation such as grass and isolated obstacles (trees, buildings) with separations of at least 20 obstacle heights	0.05	2
Town	III	Area with regular cover of vegetation or buildings or with isolated obstacles with separations of maximum 20 obstacle heights (such as villages, suburban terrain, permanent forest)	0.3	5
	IV	Area in which at least 15 per cent of the surface is covered with buildings and their average height exceeds 15 m	1.0	10

accurate values of these factors for the UK. The direction factor, c_{dir}, allows for the orientation of the structure in relation to the direction of the prevailing wind. Conservatively, a value of 1.0 may be used for c_{dir}. The seasonal variation factor, c_{season}, may be applied to structures of a temporary nature that are exposed to wind for only part of a given year. It reflects the fact that storm winds are less likely in the summer months in most European countries. (Temporary structures are subjected to a reduced risk of exposure to strong winds simply by virtue of their reduced design life. This phenomenon can be allowed for by means of a separate adjustment to the wind reference velocity.) The probability factor allows the annual probability of being exceeded to be varied and has a value of 1.0 for a 50 year return period (i.e. annual probability of exceedence of 0.02).

Using the basic wind velocity, v_b, the basic wind pressure, q_b, is calculated from:

$$q_b = 0.5\rho v_b^2$$

where ρ is the density of air (recommended value in UK is 1.226 kg/m^3).

Terrain roughness

To take account of the nature of the terrain upwind of the structure, five terrain categories are defined in EC1, and are given in Table 3.5. The recommended values for the minimum height above ground, z_{min}, and the roughness length, z_0, are also provided for each of the terrain categories. In the UK NA, these categories are simplified and combined into just three terrain categories, '*sea*' terrain (category 0 in EC1), '*country*' terrain (categories I and II in EC1) and '*town*' terrain (categories III and IV in EC1).

Peak velocity pressure

Expressions for the calculation of the peak velocity pressure are provided in EC1. However, alternative procedures may be provided in the National Annex. The methods presented here are in accordance with the UK NA, which differ from the EC1 methods. The specific methods which are used depend on whether the orography (cliffs, hills etc.) is significant or not. EC1 states that orography should be taken into account if the average slope of the upwind terrain is >3°. In the UK NA, further illustrations are provided for the definition of significant orography.

In cases where the orography is not significant the peak velocity pressure is calculated from:

$$q_p(z) = c_e(z)q_b \quad \text{(for country terrain)} \tag{3.8}$$

$$q_p(z) = c_e(z)c_{e,T}(z)q_b \quad \text{(for town terrain)} \tag{3.9}$$

where $c_e(z)$ is the exposure factor, $c_{e,T}(z)$ is an additional exposure correction factor for town terrain and q_b is the basic wind pressure. The exposure factor takes account of the variation from the basic wind pressure due to the roughness of the terrain, the height at which q_p is sought (z) and, the distance upwind from the shoreline. For town terrain there is an additional exposure correction factor, $c_{e,T}(z)$, that depends on the height of the structure and the distance to the town perimeter. $c_e(z)$ and $c_{e,T}(z)$ can be determined from Fig. 3.12 and Fig. 3.13 respectively. It is also recognized that in terrain category IV closely spaced buildings cause the wind to behave as though the ground level has been raised by a particular height, h_{dis}. This is known as the displacement height and the procedure for the determination of this parameter is given in the EN 1991-1-4 Annex A.5.

In cases where orography is significant, a simplified procedure is provided for structures where the height is less than 50 m. The peak velocity pressure is calculated using the following:

$$q_p(z) = c_e(z)q_b \left[\frac{(c_o(z) + 0.6)}{1.6}\right]^2 \quad \text{(for country terrain, } z \leq 50\,\text{m)} \tag{3.10}$$

$$q_p(z) = c_e(z)c_{e,T}(z)q_b \left[\frac{(c_o(z) + 0.6)}{1.6}\right]^2 \quad \text{(for town terrain, } z \leq 50\,\text{m)} \tag{3.11}$$

where $c_o(z)$ is the orography factor. This factor accounts for the increase of mean wind speed over isolated hills and cliffs and depends on the upwind slope, the length and height of the feature and the orographic location factor. The procedure for calculating the orography factor is detailed in EN 1991-1-4 Annex A.3.

For buildings that are higher than 50 m, the peak velocity pressure is calculated using:

$$q_p = [1 + (3I_v(z)_{flat}/c_o(z))]^2 0.5\rho v_m^2 \quad \text{(for country terrain, } z > 50\,\text{m)} \tag{3.12}$$

$$q_p = [1 + (3I_v(z)_{flat}k_{I,T}/c_o(z))]^2 0.5\rho v_m^2 \quad \text{(for town terrain, } z > 50\,\text{m)} \tag{3.13}$$

where $I_v(z)_{flat}$ is the turbulence intensity at height z for sites in country terrain, $k_{I,T}$ is the turbulence correction factor for town terrain and v_m is the mean wind velocity. Both $I_v(z)_{flat}$ and $k_{I,T}$ can be determined from figures in the UK NA. The mean wind velocity is calculated using:

$$v_m = c_r(z)c_o(z)v_b \quad \text{(for country terrain)} \tag{3.14}$$

$$v_m = c_r(z)c_{r,T}(z)c_o(z)v_b \quad \text{(for town terrain)} \tag{3.15}$$

where $c_r(z)$ is the roughness factor which accounts for the variability of mean wind velocity due to the height of the structure above ground level and the roughness of the terrain and $c_{r,T}(z)$ is a roughness

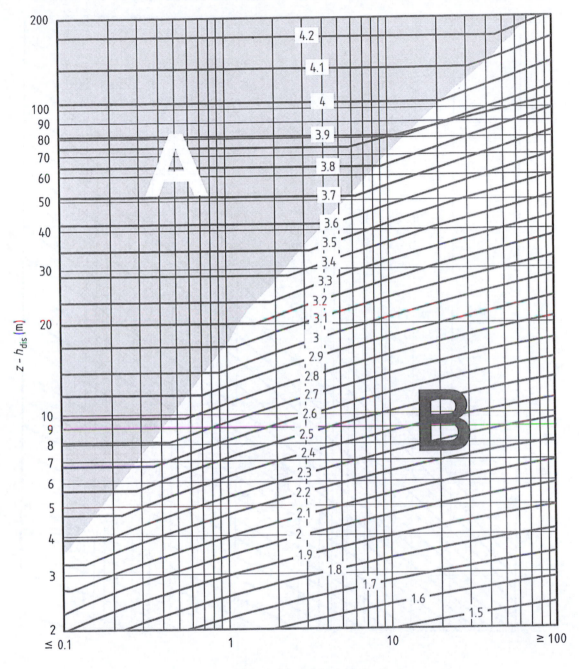

Figure 3.12 Exposure factor, $c_e(z)$ (from UK NA to BS EN 1991-1-4)

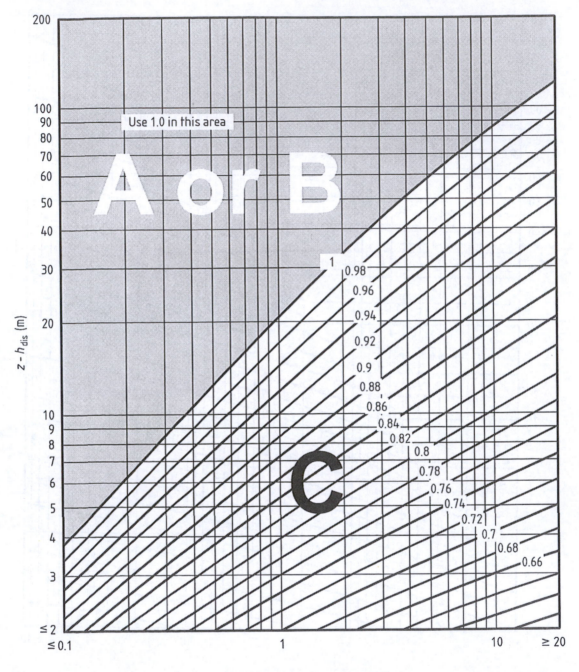

Use 1.0 in this area

A or B

C

1 0.98
0.96
0.94
0.92
0.9
0.88
0.86
0.84
0.82
0.8
0.78
0.76
0.74
0.72
0.7
0.68
0.66

$z - h_{dis}$ (m)

Distance inside town terrain (km)

Figure 3.13 Exposure correction factor for town terrain, $c_{e,T}(z)$ (from UK NA to BS EN 1991-1-4)

correction factor for sites in town terrain. Both $c_r(z)$ and $c_{r,T}(z)$ can be determined from figures in the UK NA.

External wind pressure

The wind pressure acting on the external surface of a structure is a function of the peak velocity pressure. In order to determine the contact pressure on the outside of a structure or part of a structure, the peak velocity pressure, $q_p(z)$, of the wind must be multiplied by an external pressure coefficient, c_{pe}. Thus, the external pressure is:

$$w_e = q_p(z_e)c_{pe} \tag{3.16}$$

where $q_p(z_e)$ is the peak velocity pressure evaluated at a reference height, z_e. Reference heights for the calculation of external pressure coefficients for the windward wall depend on the breadth to height ratio of the structure. For rectangular buildings whose breadth, b, is greater than or equal to their height, h, as illustrated in Fig. 3.14(a), the reference height equals the actual height. When h exceeds b but is less than or equal to $2b$, the building is considered in the two parts illustrated in Fig. 3.14(b). When h exceeds $2b$, the building is considered in multiple parts. A lower part extends upwards from the ground a distance b. An upper part extends downwards from the top a distance b. The rest of the building can be divided into any number of parts, with the reference height in each case calculated as the distance from the ground to the top of the part. The recommended reference height for leeward and side walls is the height of the building (i.e. uniform pressure distribution over the whole height). For buildings that are in Terrain Category IV, the reference height z_e can be replaced with $z_e - h_{dis}$ due

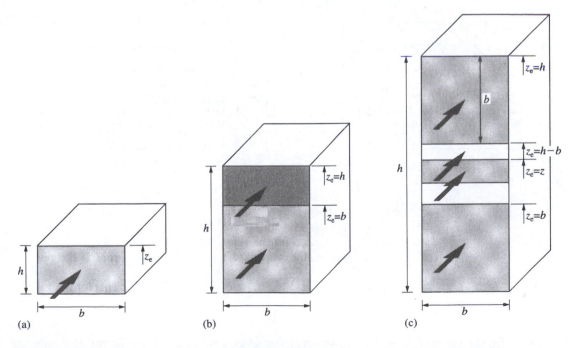

Figure 3.14 Reference height, z_e, for rectangular buildings: (a) $h \le b$; (b) $b < h \le 2b$; (c) $h > 2b$

to the effect of closely spaced buildings. Further information can be obtained from EN 1991-1-4 Annex A.5. Conservatively, the value of h_{dis} can be taken as 0.

The external pressure coefficient, c_{pe}, accounts for the variation in dynamic pressure on different zones of the structure due to its geometry, area and proximity to other structures. For instance, the wind acting on the structure in Fig. 3.15 is slowed down by the windward face and generates a pressure on that face. The wind is then forced around the sides and over the top of the structure, causing suction on the sides and on all leeward faces. Suction can also be generated on the windward slope of a pitched roof if the pitch is sufficiently small.

With reference to Fig. 3.16, the external pressure coefficients for the various façade zones of a rectangular building are given in Table 3.6. Similar tables are given in EC1 for other building shapes. The values in Table 3.6 are valid for surface areas in excess of 10 m^2 only. Values for lesser surface

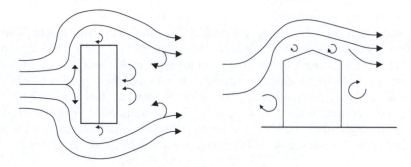

Figure 3.15 Wind flow past a rectangular building: (a) plan; (b) end elevation

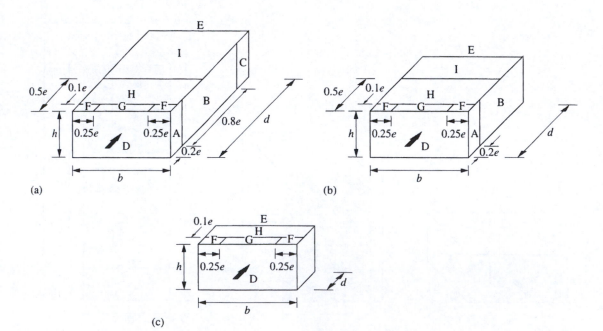

Figure 3.16 Façade and roof zones for the determination of external pressure coefficient (e = lesser of b and $2h$): (a) $d > e$; (b) $d < e$; (c) $e \geq 5d$ (use linear interpolation for intermediate values)

Table 3.6 External pressure coefficients for the walls of a rectangular building (from BS EN 1991-1-4)

Façade zone (Fig. 3.16)	$h/d \leq 0.25$	$h/d = 1$	$h/d = 5$
A	−1.2	−1.2	−1.2
B	−0.8	−0.8	−0.8
C	−0.5	−0.5	−0.5
D	+0.7	+0.8	+0.8
E	−0.3	−0.5	−0.7

areas are given in the Eurocode when local effects are to be considered. When determining the overall wind loads on a building, the UK NA provides a table of net pressure coefficients to be used as an alternative to the sum of pressure coefficients for façade zones D and E (windward and leeward). Further guidance is also given in the UK NA for buildings that are affected by funnelling (i.e. where the flow is accelerated due to a small gap between adjacent buildings).

External pressure coefficients for the roof zones in a flat-roofed building are given in Table 3.7. Other values are specified for areas less than 10 m², when parapets are present or when the eaves are curved. Although national choice is not permitted in the Eurocode for external pressure coefficients, the UK NA has included an advisory note recommending the use of alternative tables for the values of external pressure coefficients for roofs. The UK NA to BS EN 1991-1-4 should be consulted for more information. Pressure coefficients are considered positive when the pressure is acting on to the surface of the structure and negative when the pressure is acting away from that surface. Thus, the external pressure coefficient is positive when acting inwards.

Internal wind pressure

Internal pressure arises due to openings, such as windows, doors and vents, in the cladding. In general, if the windward panel has a greater proportion of openings than the leeward panel, then the interior of the structure is subjected to positive (outward) pressure as illustrated in Fig. 3.17(a). Conversely, if the leeward face has more openings, then the interior is subjected to a negative (inward) pressure as illustrated in Fig. 3.17(b). Like external pressure, internal pressure is considered positive when acting on to the surface of the structure. Thus, internal pressure is positive when the pressure acts outwards.

Internal pressure on a building or panel is given by:

$$w_i = q_p(z_i)c_{pi} \tag{3.17}$$

where z_i is the reference height for internal pressure and c_{pi} is the internal pressure coefficient. The reference height for internal pressure is equal to the reference height for external pressure (z_e) on the faces where openings in the walls result in the creation of internal pressure. If there are several

Table 3.7 External pressure coefficients for a flat roof with sharp eaves (from BS EN 1991-1-4)

Roof zone (Fig. 3.16)	Coefficient
F	−1.8
G	−1.2
H	−0.7
I	±0.2

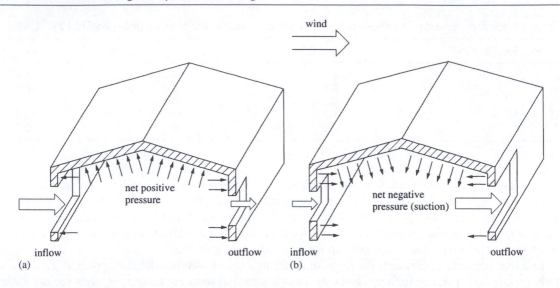

Figure 3.17 Internal pressures in structures

openings, the largest value of z_e on that face should be used in the determination of z_i. The magnitude of c_{pi} depends on the distribution of openings around the building. A building is considered to have a dominant face if the area of openings on one face is at least twice the area of openings on the remaining faces. Where the building has a dominant face, the internal pressure coefficient is taken as a fraction of the external pressure coefficient at the dominant opening. If the area of openings on one face is twice the others $c_{pi} = 0.75\, c_{pe}$, whereas, if the area of openings on one face is at least three times the others $c_{pi} = 0.9\, c_{pe}$. For intermediate values, linear interpolation may be used. For buildings without a dominant face, the values of c_{pi} are obtained from Fig. 3.18, depending on the height to depth ratio of the building, h/d, and the opening ratio, μ:

$$\mu = \frac{\sum \text{area of openings where } c_{pe} \leq 0}{\sum \text{area of all openings}} \tag{3.18}$$

For a flat roof building, the area of openings where $c_{pe} \leq 0$ will generally be the area of all openings other than those on the windward wall. In cases where μ cannot be determined, c_{pi} is taken as the more onerous of +0.2 and −0.3.

Wind forces on structures

The total wind force (F_w) acting on a structure or structural element can be calculated by summing the external, internal and friction forces. The external force is given by:

$$F_{w,e} = c_s c_d \sum_{\text{surfaces}} w_e A_{ref} \tag{3.19}$$

where c_s is the size factor which accounts for the non-simultaneous occurrence of peak pressures over the surface, c_d is the dynamic factor that accounts for the dynamic response, w_e is the external wind pressure and A_{ref} is the reference area of the individual surface. In the Eurocode, an expression is given for the calculation of the product $c_s c_d$, also called the structural factor. However, in the UK NA, these values are determined separately. The values of c_s can be obtained from Table 3.8. The columns

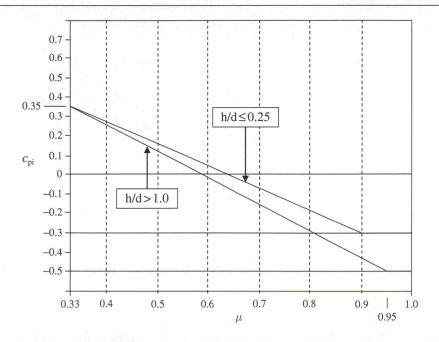

Figure 3.18 Internal pressure coefficients, c_{pi}, in buildings with no dominant face (from BS EN 1991-1-4)

Table 3.8 Size factor c_s (from UK NA to BS EN 1991-1-4)

$b + h$	$z - h_{dis} = 6$ m			$z - h_{dis} = 10$ m			$z - h_{dis} = 30$ m			$z - h_{dis} = 50$ m			$z - h_{dis} = 200$ m		
(m)	A	B	C	A	B	C	A	B	C	A	B	C	A	B	C
1	0.99	0.98	0.97	0.99	0.99	0.97	0.99	0.99	0.98	0.99	0.99	0.99	0.99	0.99	0.99
5	0.96	0.96	0.92	0.97	0.96	0.93	0.98	0.97	0.95	0.98	0.98	0.96	0.98	0.98	0.98
10	0.95	0.94	0.88	0.95	0.95	0.90	0.96	0.96	0.93	0.97	0.96	0.94	0.98	0.97	0.97
20	0.93	0.91	0.84	0.93	0.92	0.87	0.95	0.94	0.90	0.95	0.95	0.92	0.96	0.96	0.95
30	0.91	0.89	0.81	0.92	0.91	0.84	0.94	0.93	0.88	0.94	0.93	0.90	0.96	0.95	0.93
40	0.90	0.88	0.79	0.91	0.89	0.82	0.93	0.91	0.86	0.93	0.92	0.88	0.95	0.94	0.92
50	0.89	0.86	0.77	0.90	0.88	0.80	0.92	0.90	0.85	0.92	0.91	0.87	0.94	0.94	0.91
70	0.87	0.84	0.74	0.88	0.86	0.77	0.90	0.89	0.83	0.91	0.90	0.85	0.93	0.92	0.90
100	0.85	0.82	0.71	0.86	0.84	0.74	0.89	0.87	0.80	0.90	0.88	0.82	0.92	0.91	0.88
150	0.83	0.80	0.67	0.84	0.82	0.71	0.87	0.85	0.77	0.88	0.86	0.79	0.90	0.89	0.85
200	0.81	0.78	0.65	0.83	0.80	0.69	0.85	0.83	0.74	0.86	0.84	0.77	0.89	0.88	0.83
300	0.79	0.75	0.62	0.80	0.77	0.65	0.83	0.80	0.71	0.84	0.82	0.73	0.87	0.85	0.80

A, B and C in this table correspond to the exposure zones in Fig. 3.12 (for country terrain) and Fig. 3.13 (for town terrain) that are identified when determining the values of $c_e(z)$ and $c_{e,T}(z)$, respectively. In the UK NA, the value of c_d depends on the logarithmic decrement of structural damping, δ_s, of the structure. For reinforced concrete buildings ($\delta_s = \sim 0.1$) and Fig. 3.19 can be used to determine c_d.

The internal force is given by:

$$F_{w,i} = \sum_{surfaces} w_i A_{ref} \tag{3.20}$$

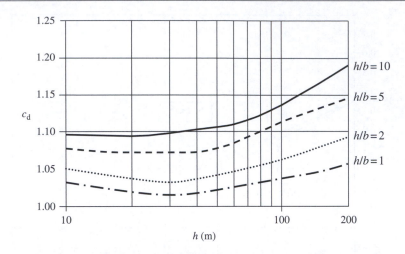

Figure 3.19 Dynamic factor c_d for reinforced concrete buildings (from UK NA to BS EN 1991-1-4)

where w_i is the internal wind pressure and A_{ref} is the reference area of the individual surface. For individual wall panels, the net wind pressure on the wall is the difference between the external and internal wind pressures. However, internal pressure within a structure is usually self-equilibrating, where the internal forces on opposite faces of a building cancel out. Thus, it results in no net force on the overall structure.

The frictional force, F_{fr}, is caused by wind friction on external surfaces which are parallel to the wind (i.e. side walls). This force acts in the direction of the wind, parallel to the external surface. The friction force needs to be considered only if the sum of the total area of surfaces parallel to the wind is greater than or equal to four times the sum of the total area of surfaces perpendicular to the wind (i.e. the windward and leeward faces). Details of the calculation of F_{fr} are given in the Eurocode.

According to the Eurocode, when summing the wind forces acting on a structure, the lack of correlation between the windward and the leeward faces of a building may be taken into account. This is due to the non-simultaneous occurrence of maximum pressures on the windward and leeward faces. When calculating the overall force on a building, a reduction factor may be applied to the force, depending on the geometry of the building. For $h/d \leq 1$ a factor of 0.85 may be applied and for $h/d \geq 5$ a factor of 1.0 may be applied. For intermediate values of h/d, linear interpolation may be used.

In the above method, the total wind force acting on a structure or structural element is calculated from the summation of individual wind forces acting on a structure. Fig. 3.20 summarizes the steps involved in the calculation of wind pressures and forces acting on a structure or structural element where orography is not significant. As an alternative to calculating individual surface pressures and forces, the total wind force on a structure can be calculated directly from Equation (3.21) using an additional force coefficient, c_f.

$$F_w = c_s c_d c_f q_p(z_e) A_{ref} \tag{3.21}$$

The Eurocode provides figures for the determination of this force coefficient for common forms of structures and sections used in structural frames.

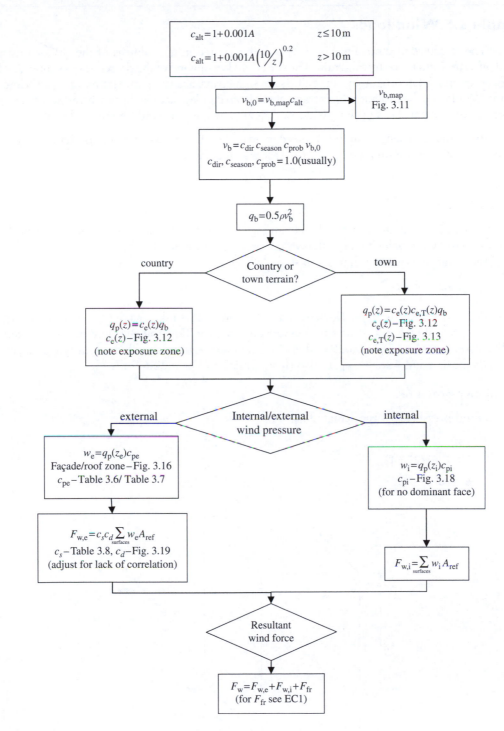

$$c_{alt} = 1 + 0.001A \qquad z \leq 10\,\text{m}$$
$$c_{alt} = 1 + 0.001A \left(\frac{10}{z}\right)^{0.2} \qquad z > 10\,\text{m}$$

$$v_{b,0} = v_{b,map} c_{alt}$$

$v_{b,map}$
Fig. 3.11

$$v_b = c_{dir} c_{season} c_{prob} v_{b,0}$$
$$c_{dir}, c_{season}, c_{prob} = 1.0\,(\text{usually})$$

$$q_b = 0.5 \rho v_b^2$$

country

Country or
town terrain?

town

$$q_p(z) = c_e(z) q_b$$
$c_e(z)$ – Fig. 3.12
(note exposure zone)

$$q_p(z) = c_e(z) c_{e,T}(z) q_b$$
$c_e(z)$ – Fig. 3.12
$c_{e,T}(z)$ – Fig. 3.13
(note exposure zone)

external

Internal/external
wind pressure

internal

$$w_e = q_p(z_e) c_{pe}$$
Façade/roof zone – Fig. 3.16
c_{pe} – Table 3.6 / Table 3.7

$$w_i = q_p(z_i) c_{pi}$$
c_{pi} – Fig. 3.18
(for no dominant face)

$$F_{w,e} = c_s c_d \sum_{\text{surfaces}} w_e A_{ref}$$
c_s – Table 3.8, c_d – Fig. 3.19
(adjust for lack of correlation)

$$F_{w,i} = \sum_{\text{surfaces}} w_i A_{ref}$$

Resultant
wind force

$$F_w = F_{w,e} + F_{w,i} + F_{fr}$$
(for F_{fr} see EC1)

Figure 3.20 Flow chart for the calculation of wind forces (where orography is not significant)

Example 3.5 Wind loads

Problem: The structure illustrated in Fig. 3.21 is to be located in the suburbs of the city of Bristol in England on a site 2 km inside town terrain. The site is 240 km upwind of the shoreline at an altitude of 30 m. It can be assumed that orography at the site is not significant. It is an apartment building with internal partitions. Wind from the east and west is transmitted from the clad faces to the north and south masonry walls. Each external panel has opening windows equal in area to one tenth of the total wall area.

(a) Determine the total bending moment due to wind at the base of the north and south masonry walls.
(b) Calculate the maximum pressure on the east masonry wall.

Solution:

Basic wind velocity

The value of $v_{b,map}$ for Bristol can be taken from Fig. 3.11 and is 22 m/s. To obtain the fundamental value of the basic wind velocity, $v_{b,0}$, this value must be multiplied by the altitude factor, c_{alt}. The height of the building is greater than 10 m so c_{alt} is given by:

$$c_{alt} = 1 + 0.001A(10/z)^{0.2}$$

The altitude, A, of the site is 30 m. The height above ground, z, is taken as the reference height of the structure (i.e., $z_s = 0.6h \geq z_{min}$). The site is in terrain category III (Table 3.5) and has a z_{min} value of 5 m. Therefore, $z = 12$ m, $c_{alt} = 1.029$ and $v_{b,0} = 22.64$ m/s. Assuming values of unity for c_{dir}, c_{season} and c_{prob} the basic wind velocity, v_b, is also (from Equation (3.7)) 22.64 m/s.

Peak velocity pressure

The basic wind pressure, q_b, is calculated as:

$$\begin{aligned} q_b &= 0.5\rho v_b^2 \\ &= 0.5(1.226)(22.64)^2 \\ &= 314\,\text{N/m}^2 \end{aligned}$$

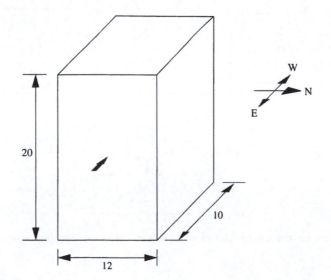

Figure 3.21 Building of Example 3.5

Since the site is in town terrain, the peak velocity pressure is, from Equation (3.9):

$$q_p(z) = c_e(z)c_{e,T}(z)q_b$$

As the height exceeds the breadth but is less than twice its value, the building is considered in two parts, as illustrated in Fig. 3.14(b). The reference heights for external pressure are thus:

$$z_e = h = 20\,\text{m}$$

and

$$z_e = b = 12\,\text{m}$$

Since the site is in terrain category III, the displacement height due to closely spaced buildings is not considered (i.e., $h_{dis} = 0$). With a distance upwind to the shoreline of 240 km (i.e. > 100 km), the values of $c_e(z)$ from Fig. 3.12 are:

$$c_e(20) = 2.77$$

and

$$c_e(12) = 2.45$$

For a site 2 km inside town terrain, the values of $c_{e,T}(z)$ from Fig. 3.13 are:

$$c_{e,T}(20) = 0.95 \;(\text{Exposure zone C})$$

and

$$c_{e,T}(12) = 0.895 \;(\text{Exposure zone C})$$

This gives peak velocity pressures of:

$$q_p(20) = (2.77)(0.95)(314) = 826\,\text{N/m}^2$$

and

$$q_p(12) = (2.45)(0.895)(314) = 689\,\text{N/m}^2$$

External wind pressure

It can be seen from Fig. 3.16 that only façade zones D (front face) and E (rear face) are of interest in this example. The ratio h/d is 20/10 = 2. Hence, from Table 3.6 (using linear interpolation):

$$c_{pe}(\text{Façade zone D}) = +0.8$$

$$c_{pe}(\text{Façade zone E}) = -0.5 + (-0.2)\left(\frac{2-1}{5-1}\right) = -0.55$$

At the reference height of 20 m, the external pressure on façade zone D is:

$$\begin{aligned}
w_e &= q_p(20)c_{pe} \\
&= (826)(0.8) \\
&= 661\,\text{N/m}^2
\end{aligned}$$

While at the reference height of 12 m, the external pressure on façade zone D is:

$$w_e = q_p(12)c_{pe}$$
$$= (689)(0.8)$$
$$= 551 \text{ N/m}^2$$

Note: Keep in mind that the forces *towards* the face of the building are always positive.

For the leeward face (façade zone E) the reference height to be used is the height of the building. Therefore, at the reference height of 20 m, the external pressure on façade zone E is:

$$w_e = q_p(20)c_{pe}$$
$$= (826)(-0.55)$$
$$= -454 \text{ N/m}^2$$

(this pressure is negative as it acts away from the face of the building)

The wind forces due to the external pressures are given by:

$$F_{w,e} = c_s c_d \sum_{\text{surfaces}} w_e A_{\text{ref}}$$

Using linear interpolation, the size factor for each reference height is obtained from Table 3.8 for $b + h = 32$ m (using exposure zone C from Fig. 3.13). Since this involves a double interpolation, the size factor is first determined for heights of 10 m and 30 m using $b + h = 32$ m:

$$c_s(10) = 0.836; c_s(30) = 0.876$$

The size factor for the required reference heights are then determined using another linear interpolation:

$$c_s(20) = 0.856; c_s(12) = 0.84$$

The dynamic factor, $c_d = 1.03$, is obtained from Fig. 3.19 for a ratio of $h/b = 1.66$ and a height of 20 m. An additional factor is applied to the wind forces to take account of the lack of correlation between the windward and leeward faces of the building. This factor is 1.0 for $h/d \geq 5$ and 0.85 for $h/d \leq 1$. Using linear interpolation, for this case the factor is 0.89 for a ratio of $h/d = 2$.

From this analysis, the resulting wind forces are calculated using Equation (3.19) and are illustrated in Fig. 3.22. Two forces are applied to the windward face (façade zone D) as it is split into two areas:

$$F_{w,e}(20) = (0.856)(1.03)(661)(8 \times 12) \times (0.89)$$
$$= 50\text{kN}$$

$$F_{w,e}(12) = (0.84)(1.03)(551)(12 \times 12) \times (0.89)$$
$$= 61 \text{ kN}$$

Each force is applied at the mid-height of the corresponding area. One force is applied to the leeward face (façade zone E).

$$F_{w,e}(20) = (0.856)(1.03)(-454)(20 \times 12) \times (0.89)$$
$$= 86 \text{ kN}$$

(a) Internal pressure within this structure is self equilibrating. Thus, while it can cause significant pressures on individual wall panels, it results in no net force on the structure overall.

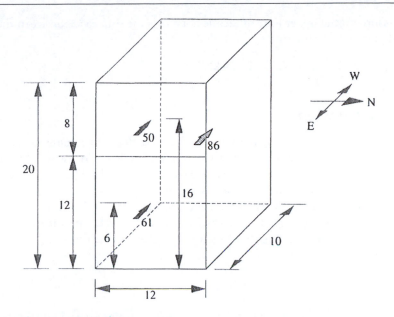

Figure 3.22 Forces due to east wind

Accordingly, the overturning moment at the base of the north and south walls due to wind is unaffected by internal pressure and is given by:

Moment $= (50)16 + (61)6 + 86(10) = 2026\,\text{kNm}$

Of this, half will apply at the base of each of the two walls.

(b) To determine the total pressure on the east wall, it is necessary to calculate the internal as well as the external pressure. The maximum pressure will occur in the upper part of façade zone D. In this part of the building, the reference height for internal pressure, z_i, is equal to the reference height for external pressure (i.e. $z_e = 20$ m). Hence: $z_i = 20$ m

The peak velocity pressure at this height, $q_p(z_i)$, is calculated as before and is 826 N/m². There is no dominant face in this building since each face has openings equal in area to one tenth of the total wall area. The opening ratio, μ, is calculated based on the area of openings on walls where $c_{pe} \le 0$ (i.e. on walls where there is external suction), which is on the leeward and side walls:

$$
\mu = \frac{\sum \text{area of openings where } c_{pe} \le 0}{\sum \text{area of openings}}
$$
$$
= \frac{0.1[(12 \times 20) + (2 \times 10 \times 20)]}{0.1[(2 \times 12 \times 20) + (2 \times 10 \times 20)]}
$$
$$
= 0.73
$$

A value of c_{pi} equal to -0.21 is obtained from Fig. 3.18 (where $h/d > 1$). Thus:

$$
w_i = q_p(z_i)c_{pi}
$$
$$
= (826)(-0.21)
$$
$$
= -173\,\text{N/m}^2
$$

The net pressure on the upper part of façade zone D is the difference between the external and the internal pressures, that is:

$$w_e - w_i = 0.661 - (-0.173) = 0.834 \, \text{kN/m}^2$$

3.7   Limit-state design

There is considerable random variation in the strength and stiffness of concrete structures and in the actions applied to them. Thus, a statistical analysis of the complete process would seem logical and indeed considerable progress has been made on research in this field. However, it is not at present practical to carry out a complete statistical analysis for the purposes of everyday design. What is current practice in most parts of the world is **load and resistance factor design** which is a deterministic design approach using factors to reflect the statistical variability in the parameters. In simple terms, load and resistance factor design consists of factoring up applied actions and factoring down the material resistances.

Limit-state theory

Limit-state theory is a method of design under which structures are designed to meet a number of functions or conditions. A limit state is defined as a situation where the structure ceases to fulfil one of the specific functions or conditions for which it was originally designed.

For concrete structures, two main groups of limit state exist:

1.   **The ultimate limit state (ULS)**
     This is reached when the structure, or part of the structure, collapses. The collapse may be due either to a loss of equilibrium or stability, or to failure by rupture of structural members.
2.   **The serviceability limit state (SLS)**
     This is reached when the structure, while remaining safe, becomes unfit for everyday use due to phenomena such as excessive deformation, cracking or vibration.

It is clear that the two groups of limit state have different degrees of importance. For example, the cracking of a concrete member affects its appearance and durability but is far less serious than collapse of the structure due to the rupture of a critical member. Thus, in limit-state design, the factors applied for safety (ULS) are significantly greater than the factors applied for serviceability (SLS), reflecting the greater importance of preventing collapse of the structure.

Safety factors for loads and materials

A structural member will support its applied loads only if its ability to resist load is greater than the load applied. All loads and material strengths are variable and estimates of each must be made in design. Limit-state design uses the concept of 'characteristic' loads and resistance. For loads, these are the values that have a minimal probability (usually 2 per cent) of being exceeded even once in a given year (Fig. 3.23(a)). For material resistance, these are the values below which a maximum number (usually 5 per cent) of specimens are expected to fall (Fig. 3.23(b)).

To allow for the possibility that the material resistance is less than its characteristic value and the load is greater than its characteristic value, partial safety factors are introduced to the characteristic resistance and load before being applied as design values. Thus, the design load, F_d, is defined as:

$$F_d = \gamma_f F_k$$

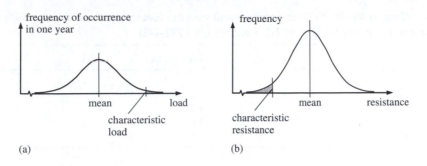

Figure 3.23 Characteristic values of: (a) loads; (b) material resistance

where γ_f is the partial safety factor for actions and F_k is the characteristic value of an action. Throughout the Eurocodes, G is used to denote permanent actions, Q is used to denote variable actions and A is used to denote accidental actions. Hence for permanent loads, for example, the characteristic load, F_k, becomes G_k and the design load, F_d, becomes G_d. In the case of variable loads, F_k becomes Q_k and is also known as the principal representative value. For accidental actions the design value, denoted A_d, is normally specified directly.

The factor, γ_f, accounts for possible deviations of the load beyond its characteristic value, inaccurate modelling of the load, uncertainty in the assessment of the load effects and uncertainty in the assessment of the limit state being considered.

Similarly, the design resistance, X_d, is given by:

$$X_d = X_k/\gamma_m$$

where X_k is the characteristic resistance and γ_m is the partial safety factor for material resistance. The factor, γ_m, takes account of uncertainties in the strength and dimensions of the material along with any inaccuracy due to the methods of modelling member behaviour.

For the ultimate limit state (ULS), values for γ_f and γ_m are recommended by the Eurocode for two principal design situations, fundamental and accidental. Fundamental design situations correspond to normal conditions of use of the structure (known as persistent situations) and to transient situations, such as during construction or repair. Accidental design situations involve the consideration of accidental loads (vehicular impact, fire, explosion, etc.) that are generally greater than the fundamental permanent and variable loads but shorter in duration. The recommended values for γ_f for both design situations are given in Table 3.9. In Table 3.9, γ_f becomes γ_g or γ_q for permanent and variable actions, respectively. For the serviceability limit state, the partial safety factor is 1.0 for all

Table 3.9 Partial safety factors (γ_f) for permanent loads and variable loads at the ultimate limit state (from BS EN 1990)

Design situation	Permanent action (γ_g*)	Variable action (γ_q)
Fundamental		
Favourable effect	1.0	0
Unfavourable effect	1.35	1.5
Accidental	1.0	1.0

* The value of γ_g should be the same throughout a structural member

Table 3.10 Partial safety factors (γ_m) for concrete and steel reinforcement at the serviceability and ultimate limit states (from BS EN 1992-1-1 and BS EN 1992-1-2)

Design situation	Concrete (γ_c)	Steel (γ_s)
ULS		
Persistent and transient	1.5	1.15
Accidental (non-fire)	1.2	1.0
Accidental (fire)	1.0	1.0
SLS	1.0	1.0

actions and design situations. The values of and γ_m for ULS and SLS are given in Table 3.10. In Table 3.10, γ_m becomes γ_c or γ_s for concrete and steel, respectively.

In the case of safety factors for loads, the Eurocode also makes a distinction between the favourable and unfavourable effects of load for fundamental design situations. A load is favourable if its effect is to reduce the effect of other loads. For instance, if permanent and imposed gravity loads act against the overturning effect of wind load on a structure, their effects are considered favourable. However, only permanent loads can be applied favourably in design since the presence of variable loads cannot be depended upon. Hence, the safety factor for favourable variable loads is zero.

Representative values of variable loads

In the case of variable loads, portions of the characteristic values, known as **representative values**, are considered for various design situations. The representative values take account of the differing probabilities of a variable load reaching its full characteristic value in different circumstances. The characteristic value, Q_k is known as the **principal representative value**. Other representative values for combinations of loads are found by applying factors ψ_i (where $\psi_i < 1$) to the characteristic value, Q_k. The relevant design values used in EN 1990 are given by:

combination values: $Q_{rep} = \psi_0 Q_k$
frequent values: $Q_{rep} = \psi_1 Q_k$
quasi-permanent values: $Q_{rep} = \psi_2 Q_k$

where the design load $Q_d = \gamma_q Q_{rep}$. The terms 'combination', 'frequent' and 'quasi-permanent' have specific meanings as outlined below.

In a fundamental design situation, it is unlikely that all the variable loads (e.g. imposed gravity, wind and snow) will reach their characteristic values simultaneously. Therefore, in such situations, only the principal load (i.e. that which causes the greatest load effect) is applied at its main representative value, Q_k. The remaining variable loads are applied at their combination values, $\psi_0 Q_k$.

In design situations where an accidental load occurs in combination with other variable loads, it is unlikely that the other variable loads will be at their full combination values (accidental loading is very unlikely). Thus, in an accidental design situation, the accidental load is applied at its design (accidental load) value, A_d. Of the other variable loads, the principal one is applied at its frequent value, $\psi_1 Q_k$, and the remaining variable loads are applied at their quasi-permanent values, $\psi_2 Q_k$.

The Eurocode provides recommended values of the factors, ψ_0, ψ_1 and ψ_2 for variable gravity and wind loads to be applied at both the serviceability and the ultimate limit states, which are given in Table 3.11. Alternative values can also be specified in the National annexes.

Taking the design values for γ_g and γ_q from Table 3.9 and the representative load factors from Table 3.11, the values for actions for use in combination with other actions at the ultimate limit state have been summarized in Table 3.12. Alternative combinations that may be less conservative are also given in the Eurocode. For more information on these combinations, refer to EN 1990.

For design at the serviceability state, three combinations are specified for consideration by EC2, namely 'characteristic', 'frequent' and 'quasi-permanent'. The design values of loads for each combination are given in Table 3.13. For example, the characteristic combination would be used in the calculation of maximum short-term deflections while the quasi-permanent combination would be used to determine the long-term deflections.

Table 3.11 Representative load factors, ψ_0, ψ_1 and ψ_2 (from BS EN 1990 and UK NA to BS EN 1990) (see Table 3.3 for category description)

Action	ψ_0	ψ_1	ψ_2
Imposed loads			
category A, B	0.7	0.5	0.3
category C, D	0.7	0.7	0.6
Wind	0.6 (0.5*)	0.2	0
Snow			
site where altitude > 1,000 m	0.7	0.5	0.2
site where altitude ≤ 1,000 m	0.5	0.2	0

* Value in UK NA which differs from recommended value in Eurocode

Table 3.12 Design values for actions for use in combination with other actions at ultimate limit states

Design situation	Permanent actions	Accidental actions	Variable actions	
			Principal action	All other actions
Fundamental				
favourable	$1.0G_k$	–	0	0
unfavourable	$1.35G_k$	–	$1.5Q_k$	$1.5\psi_0 Q_k$
Accidental	$1.0G_k$	A_d	$1.0\psi_1 Q_k$	$1.0\psi_2 Q_k$

Table 3.13 Design values for actions for use in the three combinations at serviceability limit states

Combination	Permanent actions	Variable actions	
		Principal action	All other actions
Characteristic	$1.0G_k$	$1.0Q_k$	$1.0\psi_0 Q_k$
Frequent	$1.0G_k$	$1.0\psi_1 Q_k$	$1.0\psi_2 Q_k$
Quasi-permanent	$1.0G_k$	$1.0\psi_2 Q_k$	$1.0\psi_2 Q_k$

Example 3.6 Limit-state design

Problem: Determine the design ULS value of the internal bending moment for a beam in an apartment building in the UK given the following results from a computer analysis:

	Internal moment (kNm)
Permanent gravity	200[*]
Imposed	150[*]
Wind	50[*]
Design accidental	100
(*characteristic value)	

Solution: The design moment due to permanent gravity loading is $200\,\gamma_g$. For the fundamental design situation, imposed is the principal variable loading. Hence, the design moment due to variable loading is $\gamma_q(150 + 50\psi_0)$ and the total design moment is:

$$200\gamma_g + \gamma_q(150 + 50\psi_0)$$
$$= 200(1.35) + 1.5(150 + 50 \times 0.5)$$
$$= 533 \text{ kNm}$$

For the accidental design situation, the design moment due to variable loading is:

$$100 + \gamma_q(150\psi_1 + 50\psi_2)$$

An apartment building is residential so it is in Category A of Table 3.3. Hence, $\psi_1 = 0.5$ for imposed load and the total design moment is:

$$200\gamma_g + 100 + \gamma_q(150\psi_1 + 50\psi_2)$$
$$= 200(1.0) + 100 + 1.0(150 \times 0.5 + 50 \times 0)$$
$$= 375 \text{ kNm}$$

Design procedure

For the design of reinforced concrete structures, it is normal practice to design first for the ultimate limit state as this tends to govern the design. Then, conditions of serviceability, which are usually less critical, are checked. For prestressed concrete structures, either the serviceability limit state or the ultimate limit state can govern.

The critical (maximum and minimum) effects of the applied loads on each structural member are established by a consideration of all possible realistic combinations of individual factored design loads. For loading arrangements, the critical design situations for all members need to be established and the Eurocode recommends simplified loading arrangements that can be used for buildings. To determine the maximum sag moment at mid span, the design permanent and variable loads are applied to alternate spans and only the design permanent load is applied to the other spans. To determine the maximum hogging moment over the supports, the design permanent and variable loads are applied to two adjacent spans and only the design permanent load is applied to the other spans. Additional simplified loading arrangements are recommended in the UK NA to EC2.

Example 3.7 Continuous beam

Problem: The reinforced concrete beam illustrated in Fig. 3.24 is subject to its self-weight, a permanent gravity load of characteristic value 25 kN/m and an imposed load with characteristic value 20 kN/m. Use a computer analysis program to determine the maximum and minimum ULS bending moments.

Solution: It is necessary to consider alternative configurations of applied load to determine the critical bending moments in each span and at the supports.

Critical moments between supports

To find the maximum bending moment in span AB, say, we put the maximum load (i.e. design permanent plus variable load) in spans AB and CD and the minimum load (design permanent load only) in span BC, as shown in Fig. 3.25(a). Although the load in span BC acts as a favourable permanent load in this case, the Eurocode states that the same value of γ_g should be used throughout the member. Therefore, a partial safety factor of 1.35 is applied. It can be seen from the corresponding deflected shape (Fig. 3.25(b)) that the load in the centre span tends to reduce the curvature in span AB. As the moment is proportional to the curvature, the load in the centre span has the effect of reducing the bending moment in span AB. This is why minimum load is applied in the centre span. This configuration of load also gives the maximum moment in span CD and the minimum moment in span BC. In a similar manner it can be seen that the load configuration illustrated in Fig. 3.25(c) gives the maximum interior moment in the centre span and the minimum interior moments in the two outer spans.

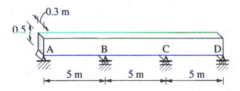

Figure 3.24 Continuous beam

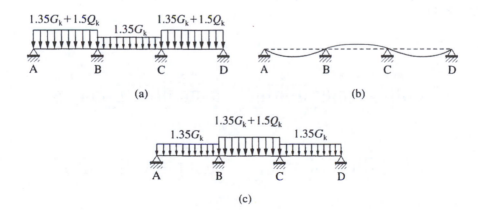

Figure 3.25 Design loads for critical sag moments: (a) maximum moment in AB; (b) deflected shape for loading of (a); (c) maximum moment in BC

Critical support moments

To get the maximum hog moment at support B (i.e., the minimum value of moment if hog is taken as negative), we put the maximum load in spans AB and BC and the minimum load in span CD, as illustrated in Fig. 3.26(a). A similar load configuration gives the maximum hog moment at support C.

Thus, the total of four load cases must be analysed for this example. The self-weight is given by the product of concrete density and cross-sectional area. Thus, the total characteristic permanent load is given by:

$$G_k = 25 \times 0.3 \times 0.5 + 25 = 28.75 \, \text{kN/m}$$

Design loads for each load case are given in Fig. 3.27. The beam is next analysed for these load cases to determine the bending moments. Alternative methods of analysis are discussed in Chapter 4 but typically, this beam might be analysed using a program based on the stiffness method. The results of such an analysis are given in Fig. 3.28. Any of these four distributions of bending moment could occur in the beam depending on the distribution of load and the beam must be capable of resisting the extremes of all four. The four diagrams of Fig. 3.28 can be superimposed to yield the 'bending moment envelope' of Fig. 3.29. This envelope gives the maximum and minimum applied bending moments at each section of the beam. In span BC, for instance, the maximum (sag) mid-span bending moment is 80 kNm for which substantial bottom reinforcement is required. The minimum mid-span bending moment (i.e. greatest hog moment) is –13 kNm. Thus, in addition to the bottom reinforcement required to resist the sag moment, top reinforcement is required at this point to resist hog moment.

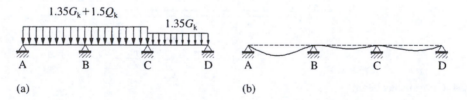

Figure 3.26 (a) Design load for maximum hogging moment at support B; (b) deflected shape for loading of (a)

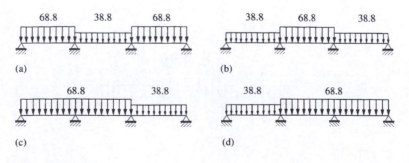

Figure 3.27 Design loads in kN/m

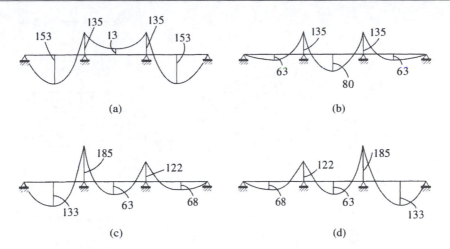

(a)

(b)

(c)

(d)

Figure 3.28 Bending moment diagrams for load configurations of Fig. 3.27 in kNm

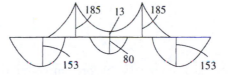

Figure 3.29 Bending moment envelope (kNm)

Example 3.8 Columns

Problem: The structural frame in Fig. 3.30 is to form the skeleton of a residential building in the UK (altitude <1000 m), where access to the roof is restricted except for maintenance and repair. Determine the design load combinations that yield the worst combinations of maximum ultimate axial force and maximum bending moment in column HJ, given the characteristic permanent (G_k), imposed (Q_{ik}), snow (Q_{Snk}) and wind (Q_{wk}) loading shown. (Note: No loading is applied to N–S beams.) For this building, the snow load intensity is greater than the minimum 0.6 kN/m² imposed load for a roof with restricted access. According to EC1, imposed load should not be applied together with wind load or snow load. Therefore, the snow load is applied to the roof as this is the more onerous load case.

In reinforced concrete, applied axial force and bending moment combine to cause the failure of columns. Any increase in the bending moment is always detrimental. However, the axial force component can sometimes be beneficial. For this reason, we must determine the combinations of applied load that yield the following conditions for column HJ:

(a)   maximum axial force plus coexistent bending moment
(b)   minimum axial force plus coexistent bending moment
(c)   maximum bending moment and coexistent axial force

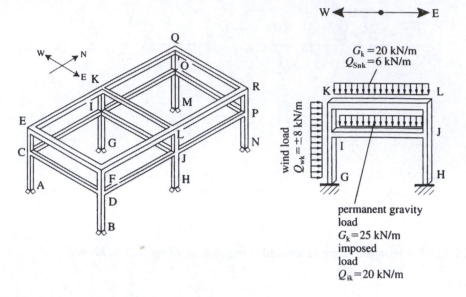

Figure 3.30 Skeletal frame: (a) geometry; (b) central frame

Maximum axial force

To determine the maximum axial (compressive) force in column HJ and the coexistent bending moment, the permanent and variable gravity loads on members IJ and KL are applied at their maximum unfavourable design values. The wind load on members GI and IK tends to cause the building to overturn, which results in tension in the windward columns and compression in the leeward columns. Thus, westerly wind load (i.e. wind from the west) has the effect of increasing the axial compression in columns HJ and JL and is applied at its unfavourable design value. The design load combination for maximum axial compression in column HJ is, therefore, unfavourable permanent and variable gravity load plus unfavourable westerly wind load. If the imposed load is considered to have the greatest effect on the magnitude of the compressive force then, from Tables 3.10 and 3.11, the factored combination is:

$$1.35G_k + 1.5Q_{ik} + 1.5\psi_0 Q_{wk} + 1.5\psi_0 Q_{Snk}$$
$$= 1.35G_k + 1.5Q_{ik} + 1.5(0.5)Q_{wk} + 1.5(0.5)Q_{Snk}$$

If, on the other hand, the wind load is considered to be the principal variable action, the factored combination is:

$$1.35G_k + 1.5\psi_0 Q_{ik} + 1.5Q_{wk} + 1.5\psi_{Sn} Q_{Snk}$$
$$= 1.35G_k + 1.5(0.7)Q_{ik} + 1.5Q_{wk} + 1.5(0.5)Q_{Snk}$$

The corresponding design loads are shown in Fig. 3.31(a–b). The design maximum compressive force and coexistent moment in column HJ are taken from the results of the more onerous load case. A third load case could also be considered, where snow load is taken as the principal variable action. However, it is not considered here as it will be less onerous than the first load case.

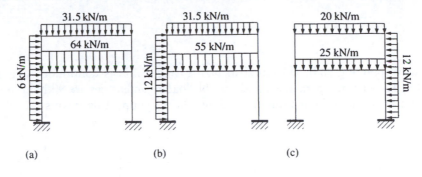

Figure 3.31 Design loads

Minimum axial force

To determine the minimum axial force in column HJ and the coexistent bending moment, the permanent and variable gravity loads on members IJ and KL are applied at their favourable design values. Since easterly wind load will cause tension in column HJ it is applied at its unfavourable value. From Table 3.11, the combination of load that gives rise to the minimum compressive force in column HJ is $1.0\,G_k + 1.5\,Q_{wk}$ (design values for favourable variable actions are zero). Fig. 3.31(c) gives the corresponding design loads.

Maximum bending moment

Fig. 3.32(a) illustrates the general shape of the bending moment diagram for the frame resulting from the gravity loads. Similarly, Fig. 3.32(b) illustrates the bending moment diagram for the frame resulting from the westerly wind load only. It can be seen from the figure that, in column HJ, the bending moments due to the gravity loads and the westerly wind load are additive. Therefore, the design load combination that yields the maximum bending moment in column HJ is unfavourable permanent and variable gravity load plus unfavourable wind load, that is, the same combinations that are used to find the maximum axial force in column HJ (Fig. 3.31(a–b)).

Therefore, the structure is analysed for the three loading cases of Fig. 3.31 and the capacity of the column is checked for all three moment/force conditions.

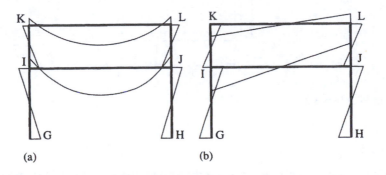

Figure 3.32 Bending moment diagrams: (a) for gravity loads; (b) for wind loads

Problems

Section 3.3

3.1 The one-way spanning slab, illustrated in plan in Fig. 3.33, is continuous over the central support, EB. The factored permanent and variable loading intensities are 9 kN/m^2 and 5 kN/m^2, respectively. Determine the maximum loadings on the supporting beams.

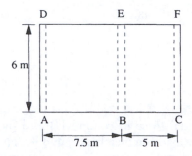

Figure 3.33 Slab of Problem 3.1

3.2 The two-way spanning continuous slab illustrated in Fig. 3.34 is subjected to factored permanent and variable loads of 10 kN/m^2 and 6 kN/m^2 respectively. Determine the maximum loadings on beam BE.

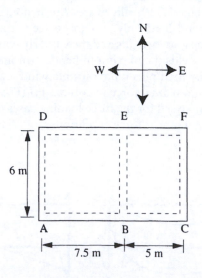

Figure 3.34 Slab of Problem 3.2

Sections 3.4 and 3.5

3.3 For preliminary design purposes, estimate the gravity loadings for a 200 mm thick reinforced concrete floor in an office with non-permanent partitions.

3.4 A hotel is to be constructed at a location near Birmingham, 150 m above sea level. The roof is to be used as a fire escape route. It is made of a reinforced concrete slab of thickness 175 mm which is covered in a lightweight sealant. The duopitch roof is sloped at 1:10 to facilitate the runoff of water. Calculate the permanent and variable gravity loading intensities. Vertical loading due to wind may be assumed not to govern the roof design.

3.5 For the hotel of Problem 3.4, it becomes apparent (during the detailed design stages) that designing the complete roof as an escape route results in an unduly large variable gravity loading. Accordingly, the escape route is limited to one portion of the roof, with access to the remaining area restricted by a masonry wall. Determine the gravity loadings for the latter part of the roof.

Section 3.6

3.6 The building illustrated in Fig. 3.35 is located in a suburban area outside London. The altitude is 25 m, the distance upwind from the shoreline is 60 km and the distance inside town terrain is 8 km. It can be assumed that orography at the site is not significant. In order to check the capacity of the structure to resist applied horizontal forces, check the wind forces in the N–S direction.

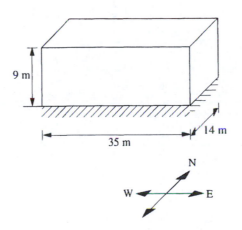

Figure 3.35 Building of Problem 3.6

3.7 For the building of Problem 3.6, the interior is open-plan and windows and doors are located approximately uniformly around the perimeter. In order to check the capacity of the masonry wall panels, determine the maximum wind pressure on the south wall.

Section 3.7

3.8 For the floor of Problem 3.3, determine the maximum design ULS loading intensity for the fundamental design situation and the maximum design SLS loading intensity for the characteristic design combination.

3.9 For a roof with a permanent gravity loading of 4.5 kN/m^2, a wind loading of 0.65 kN/m^2 and a snow loading of 0.5 kN/m^2 determine the minimum and maximum ULS loading intensities for the fundamental design situation. The altitude of the site is less than 1,000 m.

Part II

Preliminary analysis and design

Introduction

The conceptual design stage discussed in Part I may end with only one feasible solution that satisfies the design constraints being selected by the design team. More often, however, a handful of alternative schemes will be proposed that must be considered in greater detail before a final choice is made on the most appropriate solution. This second stage of the design process is known as preliminary analysis and design and it is carried out by the structural engineer in order to find more reliable estimates of structural actions and section requirements. The techniques involved at this stage are based on simple calculations that can be done on a hand calculator. The results provide the design team with more reliable information on the projected cost and structural efficiency of each proposed scheme. The results may also bring to light further design constraints that have to be checked before the scheme is provisionally accepted or finally rejected. Some of this part of the book deals with this stage in the design process. However, there are also two chapters on the analysis of concrete structures. In order to appreciate the assumptions commonly made in an approximate preliminary analysis, it is necessary to review the fundamentals of structural analysis. In addition, these chapters provide a sound basis for an understanding of a number of concepts used in detailed design (such as plastic moment redistribution).

Thus, Chapter 4 deals with the analysis of determinate and indeterminate structures to find bending moment, shear and axial forces due to applied loads. In addition to linear elastic analysis, concepts of elastic-plastic and fully plastic analysis are dealt with as they are of particular relevance to the detailed design of ductile concrete structures such as beams and slabs. In Chapter 5, the applications of analysis fundamentals to concrete structures are considered and assumptions commonly made in both preliminary and detailed analysis are outlined. Of relevance to detailed design is the analysis of slabs which is also dealt with in this chapter. Shear wall systems are also a special application of an analysis technique to concrete buildings.

In Chapter 6, the rules of thumb used to determine preliminary sizes for concrete members are described and formulae are presented for the estimation of quantities of reinforcement. Finally, Chapter 7 brings together all of the topics that were covered in previous chapters with examples of various scheme designs.

Chapter 4

Fundamentals of structural analysis

4.1 Introduction

Structural analysis, in the context of this book, is the process of calculating the effects of loads on a structure. The principal load effects of interest are the internal shear forces, axial forces, bending moments and torsion. In some cases, structures are also analysed to determine deflections. Other load effects are directly related to the principal effects. For example, crack widths in concrete members are related to internal bending moment and axial force.

Estimates of the load effects are used in the preliminary design stage to find approximate values for the quantities of reinforcement required to resist them. For the same reason, accurate values for load effects are required at the detailed design stage to allow accurate calculation of reinforcement requirements. To perform an approximate analysis, the properties of the members, such as the second moment of area, must be known. These are calculated using preliminary member sizes which, in turn, are found from simple formulae and rules of thumb as described in Chapter 6. The preliminary sizes are also necessary for the calculation of structural self-weight. This chapter describes some of the fundamental concepts of analysis that are necessary for an understanding of concrete design. The principles underlying the more important methods of analysis are also outlined.

For the methods of analysis presented, a distinction is made between statically **determinate** and **indeterminate** structures. Statically determinate structures can be analysed solely by consideration of equilibrium. Statically indeterminate structures, on the other hand, require further information for their analysis and the solution process is more complex than for determinate structures. A distinction is also made between **linear elastic** and **non-linear** methods. Linear elastic methods of analysis are based on the assumption that deformation is proportional to the applied load (linear) and that deformation will go if the load is removed (elastic). Linear elastic methods are very important for studying the performance of structures under relatively small loads – in practice, serviceability limit state (SLS) loads. Non-linear methods of analysis, on the other hand, consider the performance of structures in which some of the members have yielded. In theory, therefore, we would expect people to use non-linear methods to analyse structures at the ultimate limit state (ULS). However, in practice, linear elastic methods are sometimes used for ULS as well as SLS conditions.

Sign conventions and free-body diagrams

Fig. 4.1 illustrates the positive sign conventions used throughout this book for internal bending moments, shear forces and axial forces, which will be discussed in detail below.

For internal bending moments in horizontal members, sag moment is taken as positive, as illustrated in Fig. 4.1. The internal moment at any point in a member can be represented by a pair of equal and opposite arrows. **These arrows should always emanate from the tension face.** The

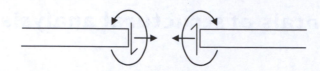

Figure 4.1 Positive sign conventions for internal bending moments, shear forces and axial forces

results of analysis, such as the distribution of internal bending moment in a structure, are usually represented by diagrams superimposed on the structure. The convention adopted for the illustration of internal moments throughout this book is that **moment is shown on the tension side** of members, that is, the side where cracks would form (see Fig. 4.2). **Therefore, arrows representing internal moment should always come from the side on which the bending moment diagram is drawn.** This is illustrated in Fig. 4.2(c) which has areas with positive and negative bending moments.

As for internal moments, a shear force can be represented by a pair of equal and opposite arrows at any point. The positive sign convention for shear in a horizontal member is shown in Fig. 4.1. Fig. 4.3 illustrates this sign convention for the frame of Fig. 4.2(a), which has positive and negative internal shear forces. Negative shear in horizontal members is represented by a pair of arrows which, if joined, would form the shape of the letter 'N', as illustrated in Fig. 4.3(b) (see right end of CD). Arrows representing negative shear in vertical members can be joined to form the letter 'Z' (see member AC). **Positive shear is represented by arrows which form a backward 'N' (horizontal member) or a backward 'Z' (vertical member).**

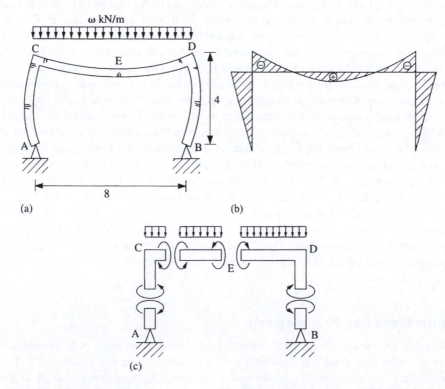

Figure 4.2 Sign convention for bending moment: (a) loading and deflected shape; (b) bending moment diagram; (c) internal moments

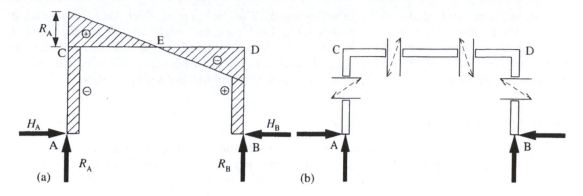

Figure 4.3 Sign convention for shear forces: (a) shear force diagram; (b) reactions and internal shear forces

Shear force diagrams are drawn by following the directions of the arrows that represent the reactions and applied loads. The vertical reactions at A and B in this example are clearly both acting upwards. They are transmitted through compression in AC and BD to give vertically upward forces at each end of CD. Starting at C (one should always start at the left-hand end or bottom) and drawing a line upwards a distance proportional to the magnitude of this vertical force gives the start of the diagram (Fig. 4.3(a)). From this elevated point, the diagram descends linearly in the direction of the applied uniformly distributed load by a total amount equal to the total load applied to this member, that is, by 8ω. At point D the upward reaction completes the shear force diagram on CD. From the deflected shape (Fig. 4.2(a)), it can be seen that the horizontal reactions at A and B are inward. Thus, the shear force in AC is drawn by moving to the right at A by a distance proportional to the horizontal reaction. The arrows representing the internal shear force form a 'Z' shape, indicating a negative internal shear force. As no load is applied to AC, the shear remains constant throughout the member length. Similarly, the shear force diagram for BD is drawn by moving to the left (positive shear force).

The sign convention for axial force is **positive tension** and **negative compression**. For the example of Fig. 4.2(a), all members are in compression. Arrows representing moment, shear and axial force are illustrated together in Fig. 4.4(a). The **pairs of arrows used to represent positive (tensile) axial force at a point should point towards each other.** The opposite rule applies to arrows representing negative (compressive) forces.

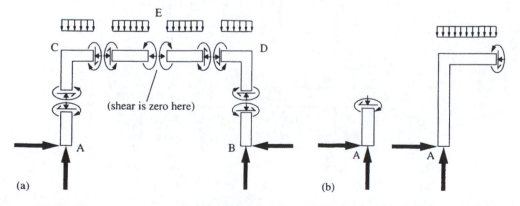

Figure 4.4 Break-up of frame into free-body diagrams: (a) complete frame showing internal forces and moments at break points; (b) typical free-body diagrams which must satisfy equilibrium

Moment, shear and axial force together represent all of the ways in which load is transmitted through two-dimensional frames such as that of Fig. 4.4. Hence, each of the segments of the frame of Fig. 4.4 constitutes a free-body diagram and must independently satisfy equilibrium. Regardless of where structures are 'broken' to form free-body diagrams, this principle stands. Thus, for each of the segments illustrated in Fig. 4.4(b), equilibrium of moments and forces must be satisfied.

Example 4.1 Frame example

Problem: The bending moment diagram for a three-bay frame is illustrated in Fig. 4.5(a). From the bending moment diagram, calculate (a) the moment M_1 in the 6 m high column BF at B, and (b) the distribution of shear force in that column given that no horizontal load is applied to it.

Solution:
(a) Calculation of M_1
The moments adjacent to joint B are represented by pairs of arrows in Fig. 4.5(b), remembering that the arrows must emanate from the side on which the moment is drawn on the bending moment diagram. Moment equilibrium at the joint gives:

$$M_1 = 200{-}165 = 35\,\text{kNm}$$

(b) Shear force distribution in BF
The free-body diagram for column BF is illustrated in Fig. 4.5(c). Taking moments about point B, we have:

$$R_1 \times 6 = 35$$

$$\Rightarrow R_1 = 5.83\,\text{kN}$$

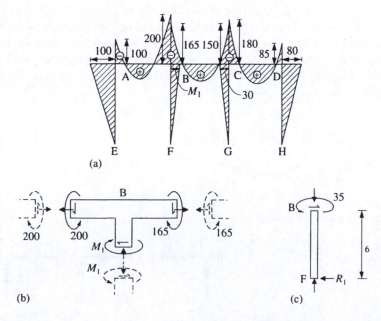

(a)

(b)

(c)

Figure 4.5 Frame example: (a) bending moment diagram; (b) free-body diagram; (c) moment equilibrium at B

Hence, the distribution of shear force in column BF is uniform over its length and has a magnitude of 5.83 kN. As the arrow representing shear in Fig. 4.5(c) forms the lower part of a backward 'Z', shear in the member is positive.

4.2 Finding moment and shear in determinate linear elastic structures

In the majority of *in situ* reinforced concrete structures, the structural members, such as beams, columns and slabs, are cast as one to give a continuous structure. Thus, rigid connections between members are a principal characteristic of *in situ* reinforced concrete structures. To provide robustness in precast concrete structures, it is common practice to connect separate precast units together using rigid joints. One method of providing a rigid connection is the insertion of an *in situ* concrete strip between the ends of each unit, as illustrated in Fig. 4.6.

As a consequence of the methods of construction, there are few examples of concrete structures that are statically determinate. Possibly the only common examples of determinate concrete structures are the simply supported and cantilevered beam and slab. An example of a simply supported beam is illustrated in Fig. 4.7(a). The idealized model of this beam that is used in its analysis is illustrated in Fig. 4.7(b). It is common in such an idealized model to show only one roller support since, if both were on rollers, the structure would be unstable.

Example 4.2 Simply supported beam

Problem: If the beam of Fig. 4.7 has an ultimate factored uniformly distributed load, ω, of 20 kN/m (included self-weight) and a span length, L, of 10 m, determine the distribution of internal bending moment and shear force in the member.

Solution: By equilibrium, the sum of the vertical reaction R_1 at support A and the vertical reaction R_2 at support B must equal the total applied load, that is:

$$R_1 + R_2 = \omega L \tag{4.1}$$

By symmetry:

$$R_1 = R_2 = \frac{\omega L}{2} \tag{4.2}$$

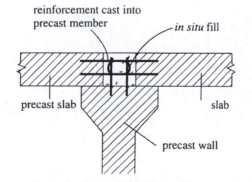

reinforcement cast into precast member

in situ fill

precast slab

slab

precast wall

Figure 4.6 In situ connection of precast members

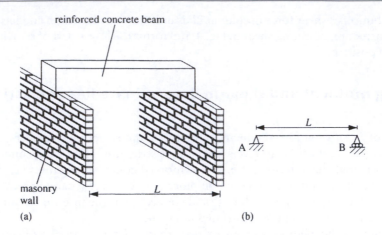

Figure 4.7 Simply supported beam: (a) actual beam; (b) idealized model

Therefore:

$$R_1 = R_2 = \frac{20 \times 10}{2} = 100\text{kN} \tag{4.3}$$

The free-body diagram for a length x of the beam is illustrated in Fig. 4.8(a). The internal bending moment, M_x $(0 < x \le L)$, at a distance x from the left support is found by summing external bending moments about that point. Thus:

$$R_1(x) - \omega(x)\left(\frac{x}{2}\right) - M_x = 0$$

$$\Rightarrow M_x = R_1 x - \frac{\omega x^2}{2} \tag{4.4}$$

$$\Rightarrow M_x = 100x - 10x^2 \tag{4.5}$$

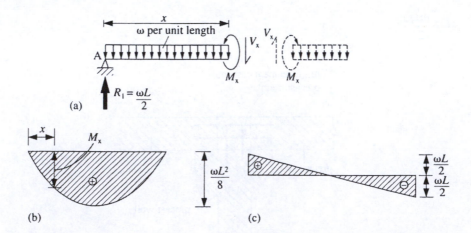

Figure 4.8 Uniformly loaded simply supported beam: (a) free-body diagram; (b) bending moment diagram; (c) shear force diagram

Using Equation (4.5), the bending moment diagram for the beam can be drawn and is illustrated in Fig. 4.8(b). The maximum bending moment occurs at the centre ($x = 5$ m) and is equal to 250 kNm. It can be shown by substitution of Equation (4.2) into Equation (4.4) with $x = L/2$ that the maximum bending moment is given by:

$$M_{max} = \frac{\omega L^2}{8} \tag{4.6}$$

The shear force diagram for this beam is drawn by moving upwards from A by an amount proportional to R_1 ($= \omega L/2$), and then moving linearly downwards across the beam by a total vertical distance proportional to the total load, that is ωL. Finally, the diagram is completed at B by moving back up by a distance proportional to R_2 ($= \omega L/2$). The full shear force diagram is illustrated in Fig. 4.8(c).

Example 4.3 Balcony slab

Problem: Determine the distributions of bending moment and shear force in the one-way spanning slab of Fig. 4.9(a) for an applied uniformly distributed loading of ω kN/m^2.

Solution: If we take a 1 m wide strip through the slab, it will act like a continuous beam with an overhang. The idealized model for this strip is illustrated in Fig. 4.9(b) where the supports at A, B and C represent the support provided by the beams to the slab. Although rotation at supports A, B, and C is restrained by torsion in the beams, the torsional effects are small and are commonly ignored for the idealized model.

There are four unknown reactions (H_1, R_1, R_2 and R_3) from which it follows that the slab is indeterminate, with the degree of indeterminacy equal to one (three equations of equilibrium and four reactions). Thus, we are unable to determine the distribution of bending moment throughout the member by simple statics. However, the overhang portion of the slab acts as a cantilever and so this part is determinate. The free-body diagram for this portion of the member is illustrated in Fig. 4.10(a).

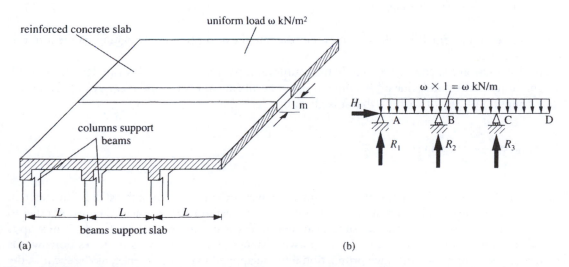

Figure 4.9 Balcony slab: (a) actual slab; (b) idealized model (1 m wide strip)

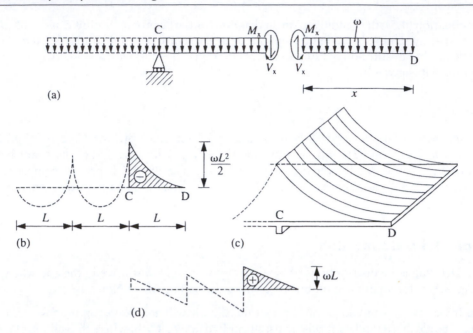

Figure 4.10 Bending moment and shear force diagrams: (a) free-body diagram for cantilever; (b) bending moment diagram for 1 m strip; (c) bending moment diagram for whole cantilever; (d) shear force diagram for 1 m strip

In the free-body diagram it is assumed that there is a hogging moment in span CD. By summing moments about a section a distance x from the free end of the cantilever, we have:

$$M_x - \omega x \left(\frac{x}{2} \right) = 0$$

$$\Rightarrow M_x = + \frac{\omega x^2}{2} \text{ (i.e. hogging)} \tag{4.7}$$

The positive sign in Equation (4.7) implies that the initial assumption of a hogging moment in this section is correct.

The distribution of bending moment in the cantilever is parabolic and is illustrated in Fig. 4.10(b). The maximum (absolute value of) bending moment occurs at support C and it can readily be shown from Equation (4.7) (with $x = L$) that it is given by:

$$M_{\text{max}} = \frac{\omega L^2}{2} \text{ (hogging)} \tag{4.8}$$

The distribution of bending moment in the cantilever over the entire width of the slab is illustrated in Fig. 4.10(c). From this diagram, it can be seen that steel tension reinforcement, aligned parallel to CD, is required near the top surface of the cantilever. The shear force diagram for the 1 m strip is started by moving upwards at C by an unknown distance. Then the diagram moves downwards linearly to the right a total distance proportional to the applied load, ωL. Hence, as the shear at the free end is zero, the shear just right of C must be ωL (Fig. 4.10(d)).

4.3   Finding internal bending moment in indeterminate linear elastic structures

There are many methods of linear elastic analysis of indeterminate structures that can be used to determine the internal bending moments. Two of these methods, the flexibility/force method and the stiffness/displacement method, reduce the problem to one of matrix algebra that can readily be solved by computer. Of these, the flexibility/force method consists of introducing 'releases' in the structure that reduce it to a familiar one. However, it is not always obvious (especially to a computer) where the releases should be introduced. For this reason, programmers tend to favour the stiffness/displacement method that can be readily applied to a greater range of structural geometries. Because of the widespread use of the stiffness method in computer programs, it is described briefly below.

Of the many hand methods for finding the internal bending moment in structures, moment distribution is by far the most popular. An understanding of moment distribution also serves to better one's understanding of structural behaviour in general. For these reasons it too is described below.

Both the stiffness method and moment distribution are applicable to skeletal structures, that is structures in two or three dimensions made up of one-dimensional members such as beams or columns (Fig. 4.11(a)). Other methods are more suitable for the analysis of structures that incorporate two- or three-dimensional continuous members, such as slabs or walls (Fig. 4.11(b)). Of these, the computer-based finite element method (FEM) is the most popular and is quite widely used in design offices. The finite element method may be viewed as an extension of the stiffness/displacement method to two and three dimensions. It can be used, among other things, to find the internal bending moments in slabs and walls.

Stiffness method

The stiffness or displacement method of analysis is an extremely powerful method for the analysis of indeterminate frames to find internal bending moments. It is based on the use of a number of simple formulae and the application of the principle of superposition. The procedure is briefly explained here through the example of Fig. 4.12. Interested readers are referred to the book by Bhatt and Nelson (1990) for a more detailed explanation of the method.

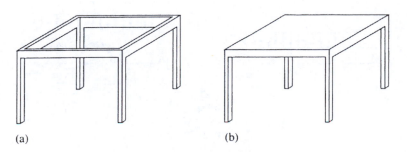

(a)   (b)

Figure 4.11   Skeletal and non-skeletal structures: (a) three-dimensional skeletal structure; (b) three-dimensional structure with one-dimensional members and a two-dimensional continuous member (slab)

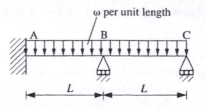

Figure 4.12 Two-span beam

Example 4.4 Continuous beam

Problem: For the beam of Fig. 4.12, use the stiffness method of analysis to find the internal bending moment at B and plot the bending moment diagram for the span BC.

Solution: Fundamental to the stiffness method is a knowledge of the reactions and the distributions of the internal bending moment in simple members such as those of Appendix B. In addition, a knowledge of member stiffnesses (force/unit deflection or moment/unit rotation) is required (see Appendix A). Based on this knowledge, individual structural members are isolated by imposing **external** restraining moments to give a series of independent members. The structure is then equivalent to a series of isolated structural members subjected to the applied loading (Fig. 4.13(b)) plus the effect of 'unfixing' the restraint(s) (Fig. 4.13(c)). For this equivalent structure, the moment at each fixing point must equal zero; that is, the sum of:

(a)   the moment reaction at the fixing point due to applied load (M_1 in Fig. 4.13(b)), and
(b)   the external moment required to distort the structure into its final shape (M_2 in Fig. 4.13(c)) must equal zero. This principle is applied at each fixing point to establish a set of simultaneous equations in the unknown rotations. For this example, there is only one such equation and so the simultaneous solution of equations becomes trivial. The reaction at the fixing point, B, due to the applied load is, from Appendix B (Nos 6 and 8):

$$M_1 = \frac{\omega L^2}{12} - \frac{\omega L^2}{8} = -\frac{\omega L^2}{24}$$

It is important to remember here that the moment is an **external** reaction. Internal moments are represented by pairs of equal and opposite arrows and are positive if sagging. An external

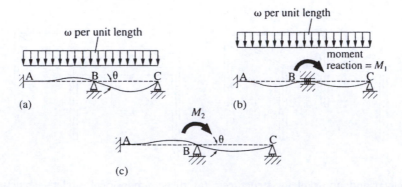

Figure 4.13 (a) Original structure; (b) fixed structure with applied loading; (c) imposed rotation to unfix structure. Basis of stiffness method is that (a) = (b) + (c)

moment, on the other hand, is represented by a single arrow that here is taken as **positive if it is clockwise**.

The external moment required to distort the structure into its final shape is, from Appendix A (Nos 1 and 2):

$$M_2 = \frac{4EI\theta}{L} + \frac{3EI\theta}{L} = \frac{7EI\theta}{L}$$

where E is the modulus of elasticity and I is the second moment of area. The combined term, EI, is commonly known as the flexural rigidity of the member. As there is no external moment reaction at B:

$$M_1 + M_2 = 0$$

$$\Rightarrow -\frac{\omega L^2}{24} + \frac{7EI\theta}{L} = 0 \tag{4.9}$$

$$\Rightarrow \theta = \frac{\omega L^3}{168EI} \tag{4.10}$$

Having found the rotation(s) at the fixing point(s), the moment at B (or at any other point) is found by superimposing the effects of the rotation(s), shown in Fig. 4.14(b), on the moment in the fixed beam shown in Fig. 4.14(a). In the fixed beam BC:

$$M_B(\text{fixed}) = \frac{-\omega L^2}{8} \text{ (hogging moment)} \tag{4.11}$$

(This is negative as it is a hogging internal moment.) The effect of rotating point B through θ is:

$$M_B(\text{due to } \theta) = \frac{3EI}{L}\theta \text{ (sagging moment)}$$

$$= \frac{3EI}{L}\frac{\omega L^3}{168EI} = \frac{\omega L^2}{56} \tag{4.12}$$

Summing Equations (4.11) and (4.12) gives the internal moment at B due to all effects:

$$M_B(\text{total}) = \frac{-3\omega L^2}{28} \text{ (i.e. net internal hogging moment)} \tag{4.13}$$

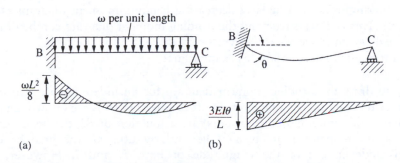

Figure 4.14 Superposition of moments: (a) fixed beam and associated bending moment diagram (b) rotation at B and associated bending moment diagram

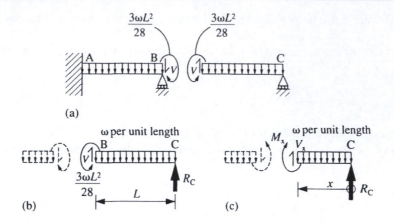

Figure 4.15 Free-body diagrams to determine distribution of internal moment in span BC: (a) internal moments just to the right of B; (b) free-body diagram for span BC; (c) free-body diagram for portion of span BC

The distribution of bending moment in span BC can now be found using equilibrium of the free-body diagrams illustrated in Fig. 4.15(b–c). Taking moments about B in the free-body diagram of Fig. 4.15(b), the reaction at C is found from:

$$R_C L - \omega L (L/2) + M_B = 0$$
$$\Rightarrow R_C = \frac{\omega L}{2} - \frac{M_B}{L} = \frac{\omega L}{2} - \frac{3\omega L}{28} \qquad (4.14)$$

Note: Since M_B is already drawn as a hogging moment (negative) in the free-body diagram of Fig. 4.15(b) which is then used to derive Equation (4.14), the absolute value of the moment is substituted into the equation for M_B.

Taking moments at a point a distance x to the left of C in the free-body diagram of Fig. 4.14(c) gives:

$$M_x = R_C x - \omega x^2 / 2$$
$$= \frac{\omega L x}{2} - \frac{3\omega L x}{28} - \frac{\omega x^2}{2} \qquad (4.15)$$

Hence, the bending moment diagram can be drawn and is illustrated in Fig. 4.16(a). It can be seen that this diagram is identical to the bending moment diagram in a simply supported beam (Fig. 4.16(b)) superimposed on a linearly decreasing bending moment diagram associated with M_B (Fig. 4.16(c)), This is always true and the bending moment diagram can be drawn by superposition of these diagrams once M_B is known.

Two important points to note about this example are:

(a) the ability to draw the bending moment diagram for an indeterminate structure by superposition once the internal bending moments at the member ends are known; and
(b) the fact that, while the imposed rotation(s) are a function of EI, the bending moments are not; that is, for a member of constant flexural rigidity, EI, the final bending moment diagram is independent of the magnitude of both E and I. However, the bending distribution is dependent on the relative magnitudes of E and I, as can be seen in the following example.

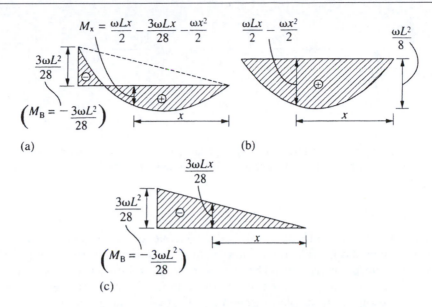

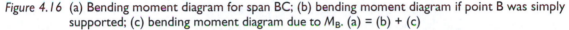

Figure 4.16 (a) Bending moment diagram for span BC; (b) bending moment diagram if point B was simply supported; (c) bending moment diagram due to M_B. (a) = (b) + (c)

Example 4.5 Continuous beam with varying stiffness

Problem: For the beam of Fig. 4.12, determine the internal bending moment at B if the flexural rigidity of span AB is $2EI$ and that of span BC is EI.

Solution: In this case, the moment required to distort the structure into its final shape becomes:

$$M_2 = \frac{4(2EI)\theta}{L} + \frac{3EI\theta}{L} = \frac{11EI\theta}{L}$$

Adding this to M_1 and equating to zero gives:

$$\theta = \frac{\omega L^3}{264EI} \tag{4.16}$$

In a manner similar to Example 4.4, the combined internal moment is found by superimposing the effects of the rotation on the moment in the fixed beam. Summing the components as before gives the internal moment at B due to all effects:

$$M_B(\text{total}) = \frac{-5\omega L^2}{44} \tag{4.17}$$

Thus, while the internal moment is not proportional to the absolute magnitude of E and I, it is significantly affected by their relative magnitudes.

The stiffness method is extremely versatile. In addition to being fixed against rotation, structures can be fixed against translation. This allows the method to be applied to more complex structures, such as the frame of Fig. 4.17. The method can also be applied to three-dimensional structures using identical techniques.

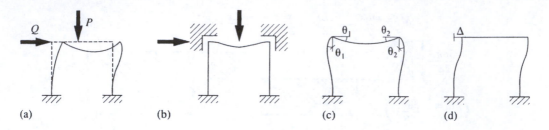

Figure 4.17 Stiffness method applied to frame: (a) original structure; (b) fixed structure with applied loads; (c) imposed rotation to unfix structure; (d) imposed translation to unfix structure. (a) = (b) + (c) + (d)

Moment distribution

The moment distribution method is based on similar principles to the stiffness method. However, the moment distribution method, unlike the stiffness method, is particularly easy to apply by hand even if the number of fixing points is large. The method is often an iterative process and, as such, the solution can be derived to any required degree of accuracy. Like the stiffness method, moment distribution is based on the principle of isolating individual members by fixing the joints of the structure. Using moment distribution, these joints are released and the internal moments at the joints are 'distributed' and balanced until the joints have rotated to their correct orientations. The distribution of the internal moments is carried out in proportion to the relative rotational stiffnesses of adjacent members.

Moment distribution can be carried out using two methods, the diagrammatic method or the tabular method. Using the diagrammatic method each joint is released in succession, whereas, using the tabular method several fixities can be released simultaneously, making it a quicker process when a lot of iterations are required to balance the moments at the supports. Although this method is often quicker, the examples in this book are carried out primarily using the diagrammatic method as it clearly describes the process involved in moment distribution analysis. Moment distribution tables are also provided for completeness.

Rotational stiffness and carry-over factors

Rotational stiffness, in the context of moment distribution, is defined as the moment required to induce a unit rotation at one joint of a beam. The ratio of the moment induced at the other joint of the beam to this moment is defined as the carry-over factor. For the general cases of fixed- and pin-ended beams having uniform flexural rigidity, EI, the stiffness formulae are given in Appendix A (Nos 1 and 2). For the fixed-ended beam, the stiffness is $4EI/L$ and the carry-over factor from A to B is $(-2EI/L)/(4EI/L) = -0.5$. In the pin-ended beam, the stiffness is $3EI/L$ and the carry-over factor is zero.

Distribution factors

It was stated above that the distribution of the internal moments is carried out in proportion to the relative rotational stiffness of adjacent members meeting at a joint. More specifically, the proportion of moment carried by each member meeting at a joint is the total moment at the joint, M, multiplied by the distribution factor for each member. The distribution factor (DF) for member j is defined as the ratio of the rotational stiffness K_j of member j to the sum of the rotational stiffnesses of all members, that is:

Distribution factor, member

$$j = \frac{K_j}{\sum\limits_i K_i} \qquad (4.18)$$

where $i = 1, 2, 3, ..., m$ and m is the total number of members meeting at this point. Thus, the proportion of the total moment M taken by member j is given by:

$$M_j = \frac{K_j}{\sum\limits_i K_i} M \qquad (4.19)$$

Note that the sum of the distribution factors for all members meeting at a joint is always equal to unity. The techniques involved in moment distribution are explained here for the example illustrated in Fig. 4.12.

Example 4.6 Continuous beam

Problem: Using the moment distribution method, determine the (internal) bending moment diagram for the continuous beam of Fig. 4.12 (*EI* constant).

Solution: As with the stiffness method, it is sufficient to fix this structure at B, as a fixity at that point isolates the beams AB and BC from one another. Thus, the fixed structure and the associated bending moment diagram are, from Appendix B, Nos 6 and 8, as illustrated in Fig. 4.18. The external fixity at B corresponds to the discontinuity in the bending moment diagram at B. It can be seen from Fig. 4.18(c) that the moments in the free-body diagram would generate an external moment reaction of $\omega L^2/24$ at B. To remove the external reaction at this point, the discontinuity in the bending moment diagram must be removed. This is done by applying a correction at B. Essentially, the hog moment to the left

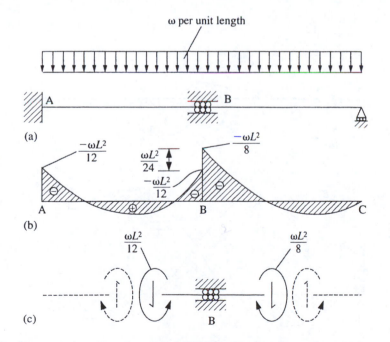

Figure 4.18 (a) Fixed structure; (b) bending moment diagram for fixed structure; (c) free-body diagram at B

of B is increased (add negative number) while the hog moment to the right of B is reduced (add positive number).

For end B of member AB, the rotational stiffness is $4EI/L$, while for end B of member BC the rotational stiffness is $3EI/L$ (see Appendix A, Nos 1 and 2). From Equation (4.18), the distribution factor to the left of B is given by:

$$\frac{4EI/L}{(4EI/L) + (3EI/L)} = \frac{4}{7}$$

Similarly, the distribution factor to the right of B can be shown to be $\frac{3}{7}$. Thus, a correction of $-\frac{4}{7}(\omega L^2/24) = -(4\omega L^2/168)$ is made to the left of B and a correction of $+\frac{3}{7}(\omega L^2/24) = 3\omega L^2/168$ is made to the right of B. In addition, the carry-over moments to A and C due to these corrections must be added to the bending moment diagram. The carry-over factor for member AB (fixed end) is -0.5 and so a correction of $-0.5(-4\omega L^2/168)$ is carried over to A. The carry-over factor for member BC (pinned end) is zero and so there is no correction at C. The total correction to the bending moment diagram of Fig. 4.18(b) is given in Fig. 4.19(a) and the bending moment diagram found by adding these two diagrams is as illustrated in Fig. 4.19(b). After this correction no discontinuity exists in the bending moment diagram. Thus, no iteration is required and the bending moment diagram of Fig. 4.19(b) is the final solution.

The moment distribution table for this example is given in Table 4.1. Using the tabular method it is assumed that **all moments are positive if the arrows at the member ends are clockwise**. As above, the bending moments at the member ends are determined assuming external fixity at B. For this example, Fig. 4.20 illustrates the loading and fixed end moments for each part of the beam separately. This figure is used to enter the fixed end moments into Table 4.1. The convention of *positive clockwise* is continued throughout the table when correcting moment discontinuities. When the iterations are complete the balanced bending moment at each support is found by summing the moments in that column.

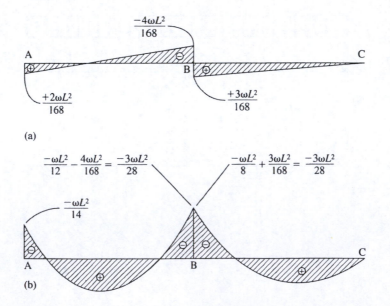

Figure 4.19 Moment distribution: (a) correction of moment discontinuity at B; (b) corrected bending moment diagram (equal to sum of (a) and Fig. 4.18(b))

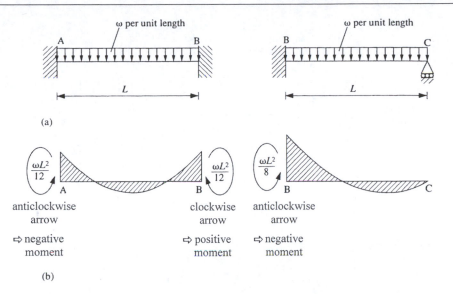

Figure 4.20 Fixed end moments for moment distribution table: (a) loading on each part of the beam (with assumed external fixity at B); (b) fixed end moments

Table 4.1 Tabular method of moment distribution for Example 4.6

	A		B	C
Distribution factors		$^4/_7$	$^3/_7$	
Carry-over factors	0.5 (B→A)			0 (B→C)
Fixed end moments	$-\omega l^2/12$	$+\omega l^2/12$	$-\omega l^2/8$	
Balance B (clockwise +ve)		$+4\omega l^2/168$	$+3\omega l^2/168$	
Carry-over (clockwise +ve)	$+2\omega l^2/168$			0
Balanced moments	$-\omega l^2/14$	$+3\omega l^2/28$	$-3\omega l^2/28$	0

Unlike Example 4.6, most problems require iteration of the moment distribution process before the discontinuities in the bending moment diagram are removed. One such problem is considered in the following example.

Example 4.7 Braced frame (no sidesway)

Problem: Using moment distribution, determine the bending moment diagram for the structure and loading of Fig. 4.21 if the second moment of area of the columns, I_c, is half that of the beams, I_b.

Solution: The structure of Fig. 4.21 is a 'braced' frame, that is a frame in which a structure other than the frame itself is provided to resist horizontal deflection. In this case, a masonry wall is present. Of course, the frame will carry some moment due to the horizontal load at the eaves but, as the wall is far

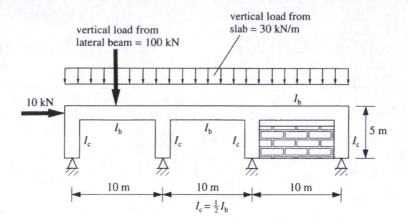

Figure 4.21 Braced frame

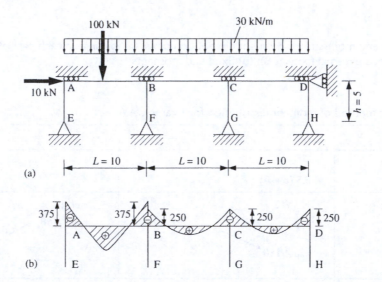

Figure 4.22 Fixed structure and associated bending moment diagram

stiffer than the frame, this moment is small and, for a structure such as this, it is reasonable to ignore it.

The fixed frame and the associated bending moment diagram are illustrated in Fig. 4.22. It can be seen that there are discontinuities in the bending moment diagram at A (375 kNm), B (125 kNm) and D (250 kNm).

Using the diagrammatic method, the moment distribution process is started by releasing the fixity at A. The stiffness of member AE is $3EI_c/L$ and the stiffness of member AB is $4EI_b/L$ where I_c and I_b are the second moments of area of the columns and the beams respectively. Thus, the distribution factors at A are given by:

$$DF_{AE} = \frac{3EI_c/h}{(4EI_b/h) + (3EI_c/h)} = \frac{3EI_c/5}{(4E(2I_c)/10) + (3EI_c/5)} = \frac{3}{7}$$

and:

$$DF_{AB} = 1 - DF_{AE} = \frac{4}{7}$$

Therefore, to remove the discontinuity of 375kNm at A, a correction of $-\frac{3}{7}(375)$ kNm is made to the left of (i.e. below) A and a correction of $+\frac{4}{7}(375)$ kNm is made to the right of A. In addition, the carry-over moments to E and B due to these corrections must be added to the bending moment diagram. The carry-over factor for member AB (Appendix A, No. l) is –0.5 and so a 'correction' of $-0.5\frac{4}{7}(375)$ kNm is carried over to B. The carry-over factor for member AE (pinned end) is zero and so there is no correction at E. Thus, the first correction to the bending moment diagram of Fig. 4.22(b) is given in Fig. 4.23.

Next, joint B is released while keeping all the other fixities, including A, in place. This means that some moment will be 'carried over' to A and a discontinuity reintroduced there (this is how the process becomes iterative). From the bending moment diagram of Fig. 4.23(b), the correction which is now required at B to remove the discontinuity there is (482–250=) 232 kNm. The distribution factors at B are found in a similar manner to those for joint A. They are:

$$DF_{BA} = \frac{4}{11}$$
$$DF_{BF} = \frac{3}{11}$$
$$DF_{BC} = \frac{4}{11}$$

The carry-over factor for the two fixed-ended members, AB and BC, is –0.5 and the carry-over factor for the pin-ended member, BF, is zero. Thus, in order to remove the discontinuity at B, a sag moment must be introduced just left of B and a hog moment just right of it. This corresponds to a clockwise rotation of the joint as illustrated in Fig. 4.24. The total correction to the bending moment

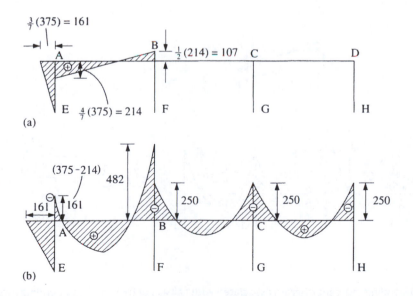

Figure 4.23 (a) Bending moment diagram associated with release of fixity at A; (b) bending moment diagram after release of fixity at A (equal to the sum of (a) and Fig. 4.22(b))

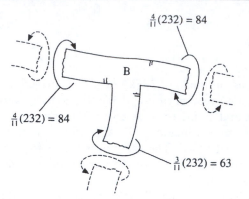

Figure 4.24 Correcting moments at B

diagram associated with the release of B is shown in Fig. 4.25(a). This yields the new bending moment diagram illustrated in Fig. 4.25(b). Moment equilibrium is now satisfied at B (see Fig. 4.25(c)) even though there is a discontinuity in beam moment there. This discontinuity is due to the moment taken by the column. Now, however, joints A, C and D are out of equilibrium.

The completion of the first iteration requires the release of the fixities at the remaining joints, namely C and D. The entire process, that is, relaxing the fixities at each of A, B, C and D, must then be repeated a number of times until moment equilibrium is established simultaneously at all joints in the frame.

The final solution for this example is given in Fig. 4.26. The moment distribution table for this example is also given in Table 4.2. Using this method several fixities can be released simultaneously, making it a quicker process when a lot of iterations are required to balance the moments at the supports. However, if moment distribution is being used for an approximate analysis or to check the output from a computer analysis, it is often unnecessary to achieve complete convergence of the solution, and a result of sufficient accuracy can be found quickly using the diagrammatic method.

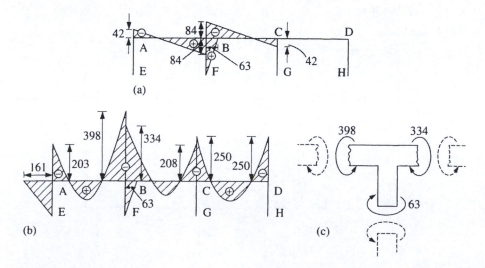

Figure 4.25 (a) Bending moment diagram associated with release of fixity at B; (b) bending moment diagram after release of fixity at B (equal to sum of (a) and Fig. 4.23(b)); (c) free-body diagram at B after release there

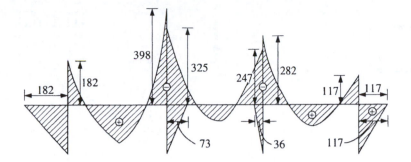

Figure 4.26 Final bending moment diagram after convergence

Table 4.2 Tabular method of moment distribution for Example 4.7

	A		B				C		D	
			A	F	C	B	G	D		
Distribution factors	3/7	4/7	4/11	3/11	4/11	4/11	3/11	4/11	4/7	3/7
Carry-over factors		0.5	0.5		0.5	0.5		0.5	0.5	
		(B→A)	(A→B)		(C→B)	(B→C)		(D→C)	(C→D)	
Fixed end moments	0.0	−375.0	375.0	0.0	−250.0	250.0	0.0	−250.0	250.0	0.0
Balance A, B, D	160.7	214.3	−45.5	−34.1	−45.5	0.0	0.0	0.0	−142.9	−107.1
Carry-over		−22.7	107.1		0.0	−22.7		−71.4		
Balance A, B, C	9.7	13.0	−39.0	−29.2	−39.0	34.2	25.7	34.2		
Carry-over		−19.5	6.5		17.1	−19.5		0.0	17.1	
Balance A, B, C, D	8.3	11.1	−8.6	−6.4	−8.6	7.1	5.3	7.1	−9.8	−7.3
Carry-over		−4.3	5.6		3.5	−4.3		−4.9	3.5	
Balance A, B, C, D	1.8	2.5	−3.3	−2.5	−3.3	3.3	2.5	3.3	−2.0	−1.5
Carry-over		−1.7	1.2		1.7	−1.7		−1.0	1.7	
Balance A, B, C, D	0.7	0.9	−1.1	−0.8	−1.1	1.0	0.7	1.0	−1.0	−0.7
Carry-over		−0.5	0.5		0.5	−0.5		−0.5	0.5	
Balance A, B, C, D	0.2	0.3	−0.3	−0.3	−0.3	0.4	0.3	0.4	−0.3	−0.2
Carry-over		−0.2	0.2		0.2	−0.2		−0.1	0.2	
Balance A, B, C, D	0.1	0.1	−0.1	−0.1	−0.1	0.1	0.1	0.1	−0.1	−0.1
Carry-over		−0.1	0.0		0.1	−0.1		−0.1	0.1	
Balanced moments	181.7	−181.7	398.3	−73.4	−324.8	247.2	34.6	−281.9	117.1	−117.0

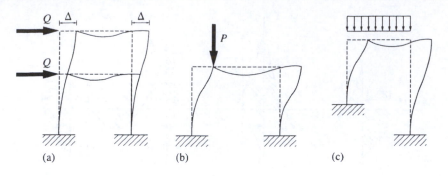

Figure 4.27 Unbraced frames

Moment distribution can also be used for the analysis of 'unbraced' or 'sway' frames where the frame itself must resist horizontal displacement. The horizontal displacement can be produced by the application of horizontal loads (Fig. 4.27(a)), by the application of non-symmetrical vertical loads (Fig. 4.27(b)) or by lack of symmetry in the structure (Fig. 4.27(c)). The moment distribution procedure for such structures is illustrated in Example 4.8.

Example 4.8 Unbraced frame

Problem: Using moment distribution, determine the internal bending moments in the frame of Fig. 4.28(a) for the applied ultimate loads specified. The bending moment diagram for the corresponding braced frame is given in Fig. 4.28(b).

Solution: The forces applied to the structure of Fig. 4.28(a) produce a sidesway in the frame. The joints B and C will be displaced by an amount, Δ, as illustrated in Fig. 4.29(a). This sway frame is equivalent to a frame that is restrained against displacement at C (Fig. 4.29(b)) with a horizontal reaction of R_d, plus a frame subject to an external force, F_d, equal and opposite to R_d, at C (Fig. 4.29(c)). The approach then is to apply moment distribution to the restrained frame of Fig. 4.29(b) to find the internal moments and the external reaction, R_d. The internal moments in the frame of Fig. 4.29(c), for the horizontal force F_d, are then found again using moment distribution. Subsequently, the principle of superposition is used to combine the moments in each frame (Fig. 4.29(b–c)) to yield the actual internal moments in the sway frame.

The internal bending moments for the restrained frame of Fig. 4.29(b) are given in Fig. 4.28(b). Thus, the external reaction, R_d, can be determined from the bending moment diagram of Fig. 4.28(b).

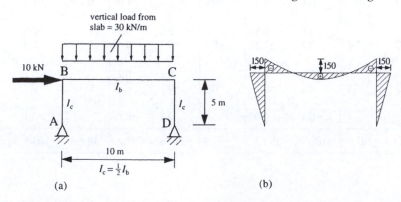

Figure 4.28 (a) Unbraced frame; (b) bending moment diagram for braced frame

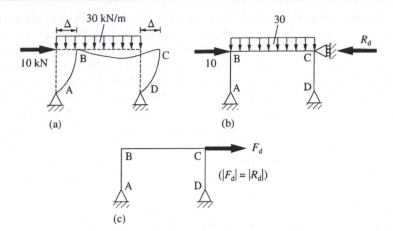

Figure 4.29 Sway frame: (a) deflected shape of original frame; (b) frame restricted against sway; (c) frame subjected to sway force only. (a) = (b) + (c)

In this case, equilibrium of horizontal forces dictates that the reaction, R_d, is equal to the sum of the applied horizontal load plus the horizontal reactions at the bases of the columns. As the applied vertical loading is symmetrical, the horizontal reactions at the bases of the columns are equal in magnitude and opposite in direction. Therefore, the external reaction, R_d, is equal to the applied horizontal load of 10 kN.

The internal bending moments for the frame of Fig. 4.29(c) must now be determined for $F_d =$ 10 kN. There is no simple direct method by which moment distribution can be used to find the internal moments due to a sway force such as F_d. However, for any given sway, Δ', the internal moments can be found and from these the force required to induce that sway, F'_d, can be calculated. The internal moments due to F_d are these moments scaled by the ratio F_d / F'_d.

The first step in this process is to fix the joints B and C against rotation and to apply the force, F'_d, and thus the sway, Δ', as illustrated in Fig. 4.30(a). Each column is now like a cantilever with the fixed

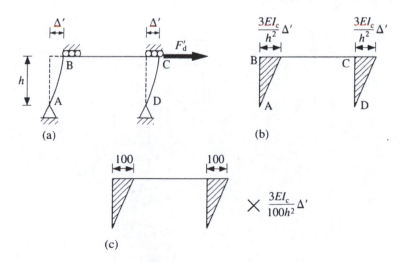

Figure 4.30 Sway frame fixed against joint rotation: (a) deflected shape; (b) bending moment diagram; (c) bending moment diagram where moments displayed are to be multiplied by $3EI_c\Delta'/(100h^2)$

end at the top, so the bending moment diagram is, from Appendix A, No. 4, as illustrated in Fig. 4.30 (b). An equivalent, but more convenient, form of displaying the bending moment diagram is illustrated in Fig. 4.30(c). The distribution factors are the same as in Example 4.7. Hence, to release the moment fixity at B, a correction of $-\frac{3}{7}(100)$ is made to the left of (below) B and a correction of $+\frac{4}{7}(100)$ is made to the right of B. In addition, a correction of $-0.5\frac{4}{7}(100)$ is carried over to C. The carry-over factor for member BA (pinned end) is zero and so there is no correction at A. Thus, the first correction to the bending moment diagram of Fig. 4.30(c) is as given in Fig. 4.31.

Next, joint C is released while keeping all the other fixities, including B, in place. From the bending moment diagram of Fig. 4.31(b), the free-body diagram for joint C is given in Fig. 4.32. By moment equilibrium at this joint, it can be seen that the correction that is now required at C is 71 kNm. Thus, the correction to the bending moment diagram of Fig. 4.31(b) is as illustrated in Fig. 4.33(a). This yields the new bending moment diagram of Fig. 4.33(b), in which equilibrium is satisfied at C.

As in Example 4.7, the problem is completed by iterating the entire process until complete convergence is achieved. The final internal moments due to F'_d are illustrated in Fig. 4.34(a). The moment distribution table is also given in Table 4.3.

Thus, the force required to induce a sway of Δ', F'_d, can be determined from the bending moment diagram of Fig. 4.34(a). In this case, equilibrium of horizontal forces dictates that F'_d is equal to the sum of the horizontal reactions at the bases of the columns. The column reactions are found by applying the equations of equilibrium to the free-body diagrams of each column, illustrated in

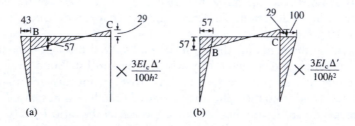

Figure 4.31 (a) Release of fixity at B; (b) bending moment diagram after release of fixity at A (equal to the sum of (a) and Fig. 4.30(c))

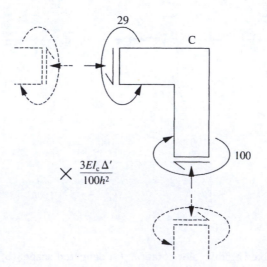

Figure 4.32 Moment equilibrium at C

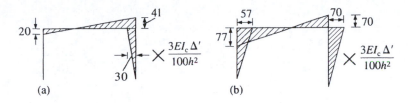

(a)   (b)

Figure 4.33 (a) Release of fixity at C; (b) bending moment diagram after release of fixity at B (equal to sum of (a) and Fig. 4.31(b))

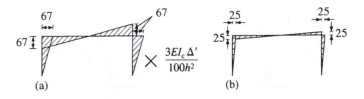

(a)   (b)

Figure 4.34 Bending moment diagrams due to sway forces: (a) due to F'_d; (b) due to F_d

Table 4.3 Tabular method of moment distribution for Example 4.8

	A	B		C	D
Distribution factors		$3/7$	$4/7$	$4/7$	$3/7$	
Carry-over factors	0	0.5		0.5	0
	(B→A)	(C→B)		(B→C)	(C→D)
Fixed end moments	0.00	−100.0	0.0	0.0	−100.0	0.00
Balance B, C		42.9	57.1	57.1	42.9	
Carry-over	0.00		28.6	28.6	0.0	0.00
Balance B, C		−12.2	−16.3	−16.3	−12.2	
Carry-over	0.00		−8.2	−8.2	0.0	0.00
Balance B, C		3.5	4.7	4.7	3.5	
Carry-over	0.00		2.3	2.3	0.0	0.00
Balance B, C		−1.0	−1.3	−1.3	−1.0	
Carry-over	0.00		−0.7	−0.7	0.0	0.00
Balance B, C		0.3	0.4	0.4	0.3	
Carry-over	0.00		0.2	0.2	0.0	0.00
Balance B, C		−0.1	−0.1	−0.1	−0.1	
Carry-over	0.00		−0.1	−0.1	0.0	0.00
Balanced moments	0.00	−66.7	66.6	66.6	−66.7	0.00

Note: All moments in the table are to be multiplied by $3EI_c\Delta'/(100h^2)$.

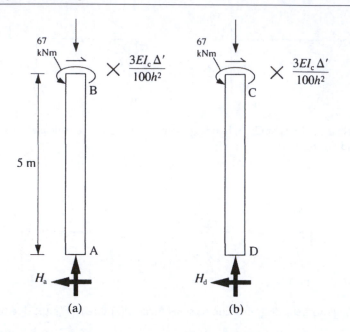

Figure 4.35 Free-body diagrams to determine reactions: (a) column AB; (b) column DC

Fig. 4.35(a–b). For instance, by taking moments about point B of Fig. 4.35(a), the reaction at A is given by:

$$5H_a = 67(3EI_c\Delta'/100h^2)$$

$$\Rightarrow H_a = 13.4(3EI_c\Delta'/100h^2)\text{kN}$$

Similarly, the column reaction at D is found to be:

$$H_d = 13.4(3EI_c\Delta'/100h^2)\text{kN}$$

By horizontal equilibrium of the entire structure:

$$H_a + H_d - F'_d = 0$$
$$\Rightarrow\ F'_d = H_a + H_d$$
$$\Rightarrow\ F'_d = 26.8\,(3EI_c\Delta'/100h^2)\ \text{kN}$$

An internal force of this magnitude causes a deflection of Δ' and induces the internal moments given in Fig. 4.34(a). Since the deflection is linear elastic, the force F_d will develop moments in the frame that are proportional to those developed by F'_d. Therefore, the internal moments induced by the force F_d are equal to those in Fig. 4.34(a) factored by an amount:

$$\frac{F_d}{F'_d} = \frac{10}{26.8(3EI_c\Delta'/100h^2)}$$

Thus, the bending moment diagram due to F_d is found by removing the 'multiplier' of $3EI_c\Delta'/100h^2$ from Fig. 4.34(a) and scaling the numbers in the diagram by 10/26.8. The result is illustrated in Fig. 4.34(b).

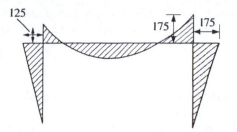

Figure 4.36 Bending moment diagram for sway frame of Fig. 4.28(a) (equal to sum of Fig. 4.28(b) and Fig. 4.34(b))

The final bending moment for the frame of Fig. 4.28(a) is found by direct superposition of Fig. 4.34(b) and Fig. 4.28(b), and is shown in Fig. 4.36.

4.4 Non-linear analysis of indeterminate structures

The linear elastic analysis of structures is based on the assumption that there is a linear relationship between the stress and strain in a member, that is to say that stress is given by:

$$\sigma = E\varepsilon \tag{4.20}$$

where ε is strain and E is the elastic modulus (Young's modulus) for the material in question. Also, for a section in bending (Fig. 4.37), the linear elastic moment-curvature relationship is given by:

$$\frac{M}{I} = \frac{E}{R} \tag{4.21}$$

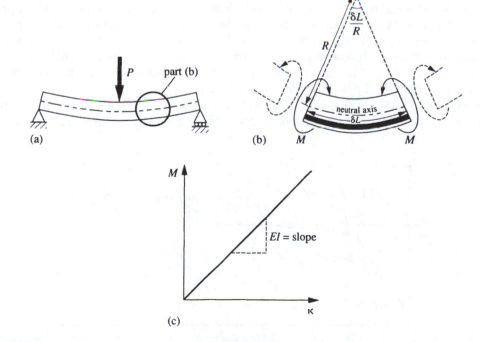

Figure 4.37 Linear elastic bending: (a) beam in bending; (b) segment of beam in bending; (c) moment/curvature relationship for linear elastic and homogeneous beam

where I is the second moment of area of the cross-section and R is the radius of curvature (see Fig. 4.37(b)). Alternatively, the linear elastic relationship can be expressed as:

$$M = EI\kappa \tag{4.22}$$

where $\kappa = 1/R$ is known as the curvature. Thus, for a linear elastic section, the moment-curvature relationship is linear as illustrated in Fig. 4.37(c).

Equation (4.22) and the second moment of area in this equation are applicable to beams of homogeneous cross-section. Reinforced concrete is not homogeneous as it consists of two materials (concrete and steel) that have considerably different values for the elastic modulus. However, it will be shown in Chapter 8 that it is possible to transform a reinforced concrete section into an equivalent homogeneous concrete section and to calculate an equivalent second moment of area. When the internal moment, M, is very small, the concrete is uncracked and the equivalent second moment of area is denoted I_u. Denoting the elastic modulus for concrete as E_c, Equation (4.22) becomes:

$$M = E_c I_u \kappa \tag{4.23}$$

However, at quite low moment, the section cracks, the equivalent second moment of area drops to a much lower value, and Equation (4.23) becomes:

$$M = E_c I_c \kappa \tag{4.24}$$

where I_c is the equivalent second moment of area of the cracked section. This relationship is represented diagrammatically in Fig. 4.38.

Equation (4.24) remains valid until the material behaviour becomes non-linear. For a properly designed reinforced concrete section, the steel yields before the concrete crushes. This happens at an applied moment of M_y, as illustrated in Fig. 4.38. As steel is a ductile material, the section too is ductile, and beyond the yield point the curvature increases greatly for a relatively small increase in the applied moment. Complete failure of the section occurs when the concrete at the extreme fibre in compression crushes. The curvature at this stage is denoted κ_{ult}.

The moment-curvature relationship of Fig. 4.38 can be idealized by the simplified relationship illustrated in Fig. 4.39. Linear elastic analysis is based on the assumption that $\kappa \leq \kappa_y$ at all sections of

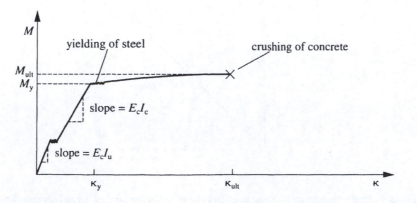

Figure 4.38 Moment/curvature relationship for reinforced concrete

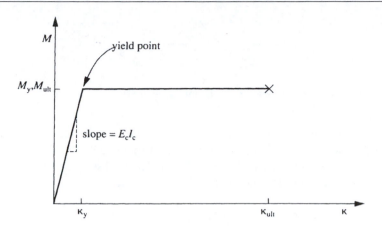

Figure 4.39 Idealized moment/curvature relationship for reinforced concrete

all members. Thus, linear elastic analysis techniques, such as the stiffness method and moment distribution, are based on this assumption. In this section, non-linear methods of analysis are considered for which there is no such restriction on κ.

Elastic-plastic analysis

In reinforced concrete structures close to collapse, the curvatures at certain sections will exceed κ_y. In such situations, methods of elastic-plastic analysis can be used to accurately predict the behaviour of the structure. The general procedure is illustrated with Example 4.9.

Example 4.9 Continuous beam

Problem: Using elastic-plastic analysis, determine the critical value for the load factor, λ, to cause failure of the beam of Fig. 4.40. The beam has sufficient reinforcement to resist a maximum bending moment, both in sag and hog, of M_y.

Solution: We assume that the loading is proportional, that is, that the factor λ increases linearly with time. Thus, the beam starts with zero load and both loads increase simultaneously until the beam fails. Initially, for a load of $\lambda_1 P$, say, where λ_1 is small, all sections of the beam are in the elastic zone. For this situation, linear elastic analysis is perfectly accurate. The bending moment diagram (illustrated in Fig. 4.41) can be found using a method of analysis such as moment distribution. Alternatively, since this structure is symmetrical, there is no rotation at node C (i.e., it behaves as

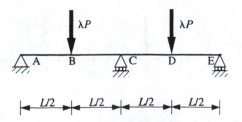

Figure 4.40 Beam of Example 4.9

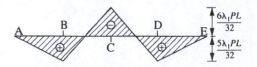

Figure 4.41 Elastic bending moment diagram for beam of Fig. 4.40

if fixed at C) and the beam is effectively two propped cantilevers back to back. In this case, the bending moment diagram can be found by referring to Appendix B, No. 7. It can be seen from the figure that a maximum moment of $6\lambda_1 PL/32$ occurs at the internal support. If we define λ_y as the load factor at which the first yield occurs, then point C becomes plastic (i.e. yields) when:

$$\frac{6\lambda_1 PL}{32} = M_y$$

from which:

$$\lambda_y = \frac{32M_y}{6PL}$$

For point C, we are then at the yield point illustrated in Fig. 4.39. For any further increase in load, point C can continue to resist the moment M_y but can provide no further resistance to rotation. Thus, for any additional increase, λ_a, in the load factor, point C is effectively hinged (see Fig. 4.42) and is termed a 'plastic' hinge. The structure of Fig. 4.42(c) is statically determinate and can readily be analysed to give the bending moment diagram of Fig. 4.43(c). Thus, for a load factor of $\lambda = \lambda_y + \lambda_a$, the total bending moment diagram is found by adding the bending moments associated with the loads of Fig. 4.42(b–c). This is illustrated in Fig. 4.43.

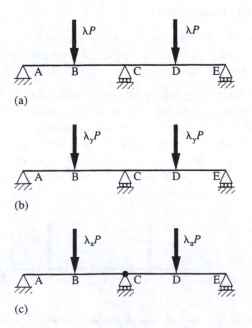

Figure 4.42 Beam after plastic hinge has formed: (a) original load where $\lambda > \lambda_y$; (b) portion of load for which beam is elastic; (c) additional load after yield at C ($\lambda_a = \lambda - \lambda_y$). (a) = (b) + (c)

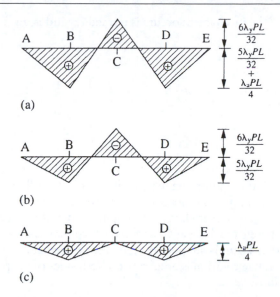

(a)

(b)

(c)

Figure 4.43 Components of bending moment diagram for $\lambda > \lambda_y$ corresponding to loads of Fig. 4.42: (a) total bending moment diagram equal to sum of (b) and (c) ($\lambda = \lambda_y + \lambda_a$); (b) elastic bending moment diagram (λ_y); (c) bending moment diagram due to additional load beyond the elastic (λ_a)

If the beam is ductile at C, that is, $\kappa_{ult} \gg \kappa_y$, the load factor can be increased beyond λ_y until the moment at points B and D reaches the yield moment. At this stage, each span forms a mechanism and collapses as illustrated in Fig. 4.44. From Fig. 4.43(a), the moments at B and D are given by:

$$M_B = M_D = \frac{5\lambda_y PL}{32} + \frac{\lambda_a PL}{4}$$

At collapse, $M_B = M_D = M_y$, from which;

$$\lambda_a = \frac{4}{PL}\left(M_y - \frac{5\lambda_y PL}{32}\right)$$

But:

$$\lambda_y = \frac{32M_y}{6PL}$$

$$\Rightarrow \lambda_a = \frac{4}{PL}\left[M_y - \frac{5PL}{32}\left(\frac{32M_y}{6PL}\right)\right]$$

$$\Rightarrow \lambda_a = \frac{2M_y}{3PL}$$

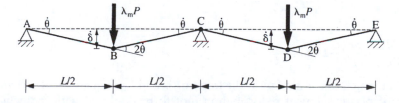

Figure 4.44 Collapse mechanism, $\lambda = \lambda_m$

The load factor at which the mechanism occurs is termed λ_m and is given by:

$$\lambda_m = \lambda_y + \lambda_a$$

$$\Rightarrow \lambda_m = \frac{32M_y}{6PL} + \frac{2M_y}{3PL}$$

$$\Rightarrow \lambda_m = \frac{6M_y}{PL} \tag{4.25}$$

All of the above is based on the assumption that, at all times up to collapse, the curvature at point C is less than κ_{ult}. In fact, this assumption is not necessarily true, as for concrete beams (especially deeper ones) the difference, $\kappa_{ult} - \kappa_y$, is not very large. The implication of this is that it is possible for the beam to collapse due to the concrete crushing at C for some λ such that $\lambda_y < \lambda < \lambda_m$. From Appendix A, No. 6, the rotation at the end of a simply supported beam due to a central point load, F, is $FL^2/16E_cI_c$. At C in Fig. 4.42(c), plastic hinge rotation is occurring on both sides of the point with the result that the rotation after initial yield is:

$$\theta = 2\left(\frac{\lambda_a PL^2}{16E_cI_c}\right) = \frac{\lambda_a PL^2}{8E_cI_c}$$

If the plastic hinge forms over a length, L_h, of beam near C, the rotation is related to the radius of curvature by:

$$\theta = L_h/R$$
$$\Rightarrow \theta = L_h\kappa$$

Hence, the capacity to rotate after formation of the plastic hinge is:

$$\theta_{ult} = L_h(\kappa_{ult} - \kappa_y)$$

Thus, rotation at C reaches its ultimate capacity when:

$$\frac{\lambda_a PL^2}{8E_cI_c} = L_h(\kappa_{ult} - \kappa_y)$$

$$\Rightarrow \lambda_a = \frac{8E_cI_c}{PL^2}L_h(\kappa_{ult} - \kappa_y)$$

The load factor at which this beam fails through inadequate rotational capacity is:

$$\lambda_\theta = \lambda_y + \lambda_a$$

$$\Rightarrow \lambda_\theta = \frac{32M_y}{6PL} + \frac{8E_cI_c}{PL^2}L_h(\kappa_{ult} - \kappa_y) \tag{4.26}$$

The value for λ at which this beam fails is the lesser of λ_θ and λ_m (Equations (4.25) and (4.26)).

The concept of plastic moment redistribution (not to be confused with moment distribution), which is described in Chapter 5, is an approximation to elastic-plastic analysis.

Plastic analysis

Plastic analysis is totally different in its approach from linear elastic or elastic-plastic analysis. The value of λ_m is determined from a consideration of the conditions at collapse with no regard to the deformations that occur prior to this. For the structure of Example 4.9, the conditions **during** collapse are as illustrated in Fig. 4.44. Thus, during collapse, the external load at B is moving at a rate of $\dot{\delta}$ (deflection per unit time) while the pin at A is rotating at a rate of $\dot{\theta}$ where $\dot{\delta} = (L/2)\dot{\theta}$. By simple geometry, the rate of rotation of the hinge at C is $2\dot{\theta}$ ($\dot{\theta}$ in each span) and hence the rates of rotation at B and D are also $2\dot{\theta}$.

Note that elastic deformations are also present between the plastic hinges but these have no influence on the calculations. It has been proven that, during collapse of the frame by this mechanism, the internal rate of work done at the plastic hinges equals the external rate of work done by the loads, $\lambda_m P$. The internal rate of work done at a plastic hinge during collapse is equal to the rate of rotation of the hinge multiplied by the ultimate moment capacity of the member, M_{ult} (taken to be equal to the yield moment). Thus, the total internal rate of work done, $\dot{W}_i$, by the three hinges during collapse is:

$$\dot{W}_i = 3M_{ult}(2\dot{\theta}) = 6M_{ult}\dot{\theta}$$

The external rate of work done by each load during collapse is its magnitude multiplied by its rate of deflection. Thus, the total external rate of work done, $\dot{W}_e$, during collapse is:

$$\dot{W}_e = 2\lambda_m P\dot{\delta} = 2\lambda_m P(L/2)\dot{\theta}$$

Equating the internal and external rate of work done gives:

$$6M_{ult}\dot{\theta} = 2\lambda_m P(L/2)\dot{\theta}$$

$$\Rightarrow \lambda_m = \frac{6M_{ult}}{PL}$$

As can be seen from the simplified moment-curvature diagram of Fig. 4.39, the yield moment, M_y, is assumed to be the same as the ultimate moment capacity, M_{ult}. Thus, the result is the same as the result of elastic-plastic analysis (Equation (4.25)).

It should be noted that there is often more than a single possible collapse mechanism for a given structure. When this is the case, each mechanism must be considered in turn and the lesser of the load factors is taken as λ_m.

Plastic analysis is very easy to perform but the results give no information on how much plastic hinge rotation occurs at the plastic hinges before the structure collapses. This renders it unsuitable for concrete beams and, in particular, concrete beams for which ductility is relatively limited. However, concrete slabs, not being deep, are far more ductile than concrete beams and a method of plastic analysis known as yield line theory (see Chapter 5) is used in their analysis.

4.5   Finding shear force, axial force and deflection

A complete analysis of a structure involves not only the determination of internal moments but also other quantities such as shear force, axial force and deflection. However, as will be seen from the following examples, it is a relatively simple matter to determine these quantities once the distribution of bending moment has been determined.

Shear force

Shear is an important phenomenon in reinforced and prestressed concrete. For determinate structures, shear can be found using the principles of equilibrium. For indeterminate structures, the stiffness method can be used to determine the shear distribution directly. Alternatively, the distribution of moment can be found using either the stiffness method or moment distribution, and the shear distribution found from this. The first step in finding the shear force at any section in any member of a structure is to determine the reactions. This can be done by considering the equilibrium of free-body diagrams. Once the reactions are known, the shear force diagram can readily be drawn.

Example 4.10 Shear distribution in continuous beam

Problem: The linear elastic bending moment diagram for the beam of Fig. 4.45(a) has been found using moment distribution and is given in Fig. 4.45(b). Determine the shear force diagram for this beam.

Solution: The free-body diagrams for member AB, joint B and member BC are illustrated in Fig. 4.46 (a–c), respectively. The reactions R_1, R_2 and R_3 can be found by taking moments about the point B in both Fig. 4.46(a) and (c). By equilibrium, the clockwise moments must equal the anticlockwise moments.

Hence, in Fig. 4.46(a):

$$6R_1 + 97.5 = 25 \times 6 \times 3$$

$$\Rightarrow R_1 = 58.8\,\text{kN}$$

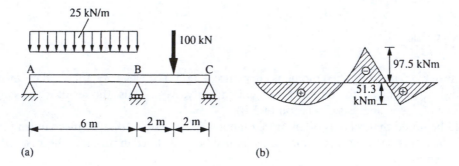

(a)   (b)

Figure 4.45 Continuous beam: (a) geometry and loading; (b) bending moment diagram

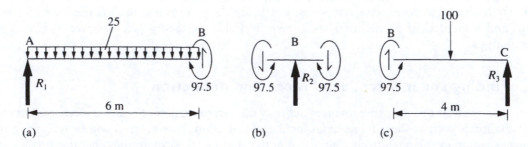

(a)   (b)   (c)

Figure 4.46 Free-body diagrams for beam of Fig. 4.45: (a) member AB; (b) joint B; (c) member BC

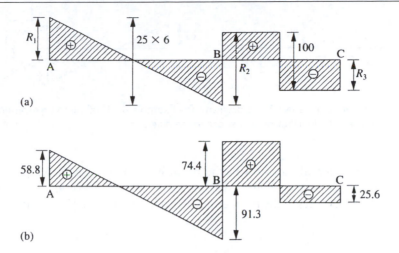

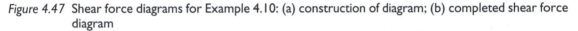

Figure 4.47 Shear force diagrams for Example 4.10: (a) construction of diagram; (b) completed shear force diagram

and in Fig. 4.46(c):

$$4R_3 + 97.5 = 100 \times 2$$

$$\Rightarrow R_3 = 25.6\,\text{kN}$$

By equilibrium of vertical forces in the entire structure:

$$R_1 + R_2 + R_3 = 25 \times 6 + 100$$

$$\Rightarrow R_2 = 165.6\,\text{kN}$$

Once the reactions are known, the shear force diagram can be plotted in the same way as for determinate structures (see Sections 4.1 and 4.2), that is by starting at the left-hand side of the beam and 'following the arrows'. Thus, start at point A, by going up by R_1, as shown in Fig. 4.47(a). The uniformly distributed loading of 25 kN/m acts downwards and so the diagram next goes down linearly from R_1 to a minimum of $R_1 -(25 \times 6)$ just to the left of point B. At point B, the diagram again goes up, by an amount R_2. Shear then remains constant in the section of span BC where there is no applied loading. At mid-span, however, the diagram goes down by 100 kN because of the point load. The diagram remains level again from mid-span to point C. The rise in the diagram at point C is equal to the reaction R_3. The complete shear force diagram thus constructed is illustrated in Fig. 4.47(b).

Example 4.11 Shear distribution and non-linear analysis

Problem: For the continuous beam of Fig. 4.45(a), the internal bending moments have been found using plastic moment redistribution (an approximation to elastic-plastic analysis) and are given in Fig. 4.48(a) (note that the bending moments from this non-linear analysis are different from those from the linear elastic analysis illustrated in Fig. 4.45(b)). Using this new bending moment diagram, determine the distribution of shear force for the beam.

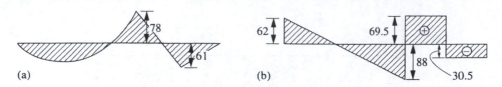

Figure 4.48 Bending moment and shear force diagrams for Example 4.11: (a) bending moment diagram after plastic moment redistribution; (b) shear force diagram

Solution: Although the internal moments in the beam have changed somewhat, equilibrium still holds (equilibrium always holds). Hence, the reactions can be determined from the free-body diagrams of individual members in the manner described in Example 4.10. In this case, the support reactions are:

$$R_1 = 62 \text{ kN}$$
$$R_2 = 157.5 \text{ kN}$$
$$R_3 = 30.5 \text{ kN}$$

The corresponding shear force diagram is illustrated in Fig. 4.48(b). By comparison of Fig. 4.47(b) and Fig. 4.48 (b), it can be seen that the change in the bending moment diagram has the effect of changing (redistributing) the shear forces in the beam.

Axial force

As for shear force, axial force can be found directly using the stiffness method or can be determined by applying the equations of equilibrium to free-body diagrams once the distribution of internal bending moment is known. Axial force can govern the design of columns and walls. On the other hand, the axial forces due to horizontal wind loads that act in beams and slabs tend to be small and can usually be ignored in design. The following example illustrates the procedure by which the axial force in members can be found once the internal moments are known.

Example 4.12 Axial forces in a braced frame

Problem: For the braced frame of Fig. 4.49(a), linear elastic analysis followed by plastic moment redistribution was used to determine the internal bending moments illustrated in Fig. 4.49(b). Use the redistributed bending moment diagram to find the axial force in each of the columns. All horizontal forces are transferred by a floor slab, which is attached to ABC, to shear walls elsewhere in the building.

Solution: The free-body diagrams for member AB, member AD and the joint between the two members are illustrated in Fig. 4.50. From the free-body diagram for the joint, it can be seen that the shear force, V_2, at the left end of member AB is transferred to member AD as an axial (compressive) force, A_1, and that $V_2 = A_1$. The magnitude of this force cannot be determined directly from a consideration of the equilibrium of member AD. However, V_2 can be found by taking moments about the point B in the free-body diagram for member AB. By equilibrium:

$$10V_2 + 214 = 25 \times 10 \times 5 + 89.5$$
$$\Rightarrow V_2 = 112.5 \text{ kN}$$

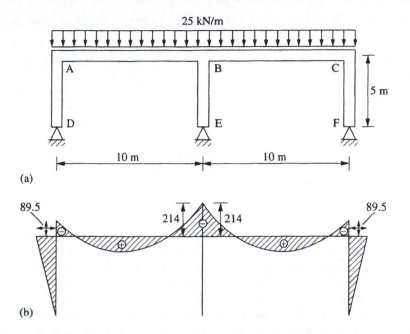

(a)

(b)

Figure 4.49 Frame example: (a) geometry and loading; (b) redistributed bending moment diagram

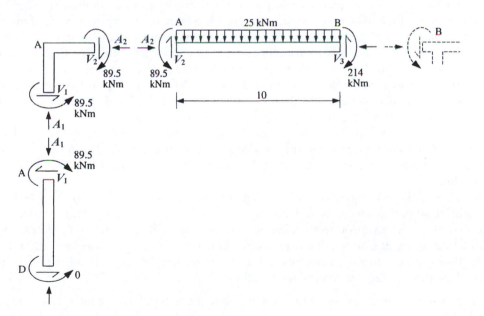

Figure 4.50 Free-body diagram for AD, A and AB

Thus, the axial force, A_1, in member AD is 112.5 kN. By symmetry, the axial force, A_5, in member CF is also 112.5 kN.

By equilibrium of the entire frame, the reaction at E and hence the axial force in the centre column, A_3, is given by:

$$A_3 = 2 \times 25 \times 10 - A_1 - A_5$$

$$\Rightarrow A_3 = 275 \text{kN}$$

By taking moments about D in the free-body diagram for member AD, it can be shown that the shear force, V_1, is 17.9 kN, Thus, by equilibrium of joint A the axial force in member AB is 17.9 kN.

Deflection

It is necessary to limit serviceability limit state deflections in many types of concrete structures as excessive deflection can cause adverse effects such as the cracking of partitions. Therefore, much time is devoted in structural analysis courses to the calculation of deflection in linear elastic structures. However, all of these methods are limited for a number of reasons. Not least of these are the problems of selecting the appropriate second moment of area for reinforced concrete and making allowance for time-dependent deflection in both reinforced and prestressed concrete. Prestressed concrete beams experience little or no cracking and it is usually sufficiently accurate to calculate the second moment of area ignoring the presence of reinforcement. Reinforced concrete beams generally do crack and so it is necessary to calculate the equivalent second moment of area of the cracked section taking account of the reinforcement present. However, calculation of the cracked second moment of area requires a knowledge of the quantities of steel reinforcement present in the beam and this is not known at the early stages of design. It is also necessary to know which parts of the beam are in hog and which are in sag as the cracked second moments of area will, in general, be different for each case. Finally, the beam will not usually be cracked everywhere, the degree of cracking being dependent on the internal bending moment, so it would be excessively conservative to use a cracked second moment of area everywhere for the calculation of deflection. The time-dependent deflection due to creep is normally accounted for by reducing the modulus of elasticity. Such calculations are often hugely inaccurate as creep is extremely difficult to predict.

Fortunately, the Eurocode for concrete, EC2, specifies that it is not necessary to calculate the deflections in beams and slabs if certain span/depth ratios (listed in Table 6.5) are not exceeded. Where these span/depth ratios are exceeded, deflections need to be calculated and checked against allowable limits.

Once values of the modulus of elasticity (E) and the second moment of area (I) have been selected, the calculation of deflection can be done in many ways. The stiffness method, for instance, can be used to determine deflections directly. However, some computer programs only determine deflections and rotations at the nodes, that is the points at which two or more members meet. From this data, the moment-area theorems can be used to determine the deflection at any other point in the structure. The moment-area theorems are as follows:

Theorem 1: The difference in slope between two points in a linear elastic member is equal to the area under the M/EI diagram between these two points. Thus, for two points, A and B, the difference in slope is given by (see Fig. 4.51):

$$\theta_A - \theta_B = \int_A^B \frac{M}{EI} dx \qquad (4.27)$$

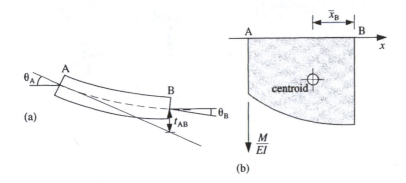

Figure 4.51 Moment area theorems: (a) segment of beam in bending; (b) bending moment diagram divided by EI

Theorem 2: The vertical deviation, t_{AB}, of the tangent to a point A on a linear elastic member from a point B (see Fig. 4.51) is equal to the 'moment' of the M/EI diagram between the two points about point B, that is:

$$t_{AB} = \bar{x}_B \int_A^B \frac{M}{EI} dx \tag{4.28}$$

where $\bar{x}_B$ is the distance from point B to the centroid of the portion of the M/EI diagram between the two points.

The application of these two theorems is illustrated in Example 4.13.

Example 4.13 Deflections in a propped cantilever

Problem: The bending moment diagram for the propped cantilever in Fig. 4.52(a) is illustrated in Fig. 4.52(b). The deformation of the member under the applied load is shown in Fig. 4.52(c). If the second moment of area is constant throughout its length and is 1.3×10^{-3} m^4, determine the maximum vertical deflection in the member. Assume that the modulus of elasticity, E, equals 27×10^6 kN/m^2.

Solution: The point of contraflexure of the beam (i.e. where the bending moment is zero) is found to be 2.73 m from point A. From the deflected shape of the beam, it can be seen that the maximum vertical deflection occurs in the right half of member AC at a point D, say. Assume that the location of point D is a distance y from B. From Fig. 4.53(a), the area under the bending moment diagram, A_M, between the points A and D is found to be:

$$A_M = -\frac{1}{2}(93.75)(2.73) + \frac{1}{2}(78.13)(5-2.73) + \frac{1}{2}(y)(78.13 + M_y)$$

but, by linear interpolation of the bending moment diagram:

$$M_y = 78.13(5-y)/5$$

$$= 78.13 - 15.63y$$

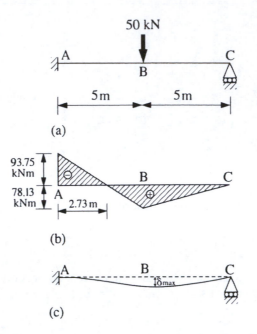

(a)

(b)

(c)

Figure 4.52 Beam of Example 4.13: (a) geometry and loading; (b) bending moment diagram; (c) deflected shape

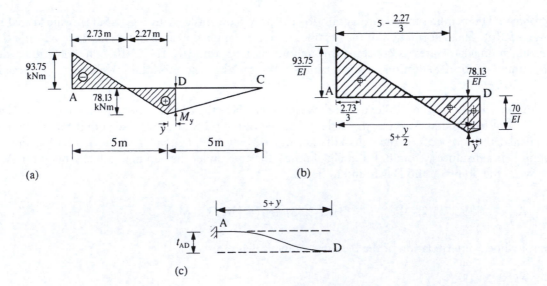

(a)

(b)

(c)

Figure 4.53 (a) bending moment diagram for AC; (b) segment of M/EI diagram for AD; (c) deflected shape of AD

Therefore:

$$A_M = -\frac{1}{2}(93.75)(2.73) + \frac{1}{2}(78.13)(5-2.73) + \frac{1}{2}(y)(78.13 + 78.13 - 15.63y)$$

$$A_M = -7.81y^2 + 78.1y - 39.29$$

By Theorem 1, the difference in slope between the two points A and D is given by Equation (4.27), that is:

$$\theta_A - \theta_D = \int_A^D \frac{M}{EI}dx = \frac{A_M}{EI} \qquad (4.29)$$

The slope at the fixed support A is zero (i.e. $\theta_A = 0$) and the slope, θ_D, at the point of maximum deflection is also zero. Thus Equation (4.29) becomes:

$$0 = \frac{A_M}{EI}$$
$$\Rightarrow -7.81y^2 + 78.1y - 39.29 = 0$$
$$\Rightarrow y^2 - 10y + 5 = 0$$

This quadratic equation is solved to give the distance from point B at which maximum deflection occurs, that is:

$$y = 0.53\,\text{m}$$

The corresponding bending moment, M_y at y, is:

$$M_y = 78.13 - 15.63(0.53) = 70\,\text{kNm(sag)}$$

By Theorem 2, the vertical deviation, t_{AD} (Fig. 4.53(c)), of the tangent at point D with respect to the tangent at point A is given by:

$$t_{AD} = \bar{x}_A \int_A^D \frac{M}{EI}dx$$

Since the tangent to D is horizontal, the total deflection at D is equal to the vertical deviation, t_{AD}. Thus:

$$\delta_{max} = \bar{x}_A \int_A^D \frac{M}{EI}dx$$

It is convenient to consider each portion of area under the M/EI diagram separately and then calculate the distance between the centroid of this portion and A. Hence, referring to Fig. 4.53(b):

$$\delta_{max} = -\frac{1}{2}\left(\frac{93.75}{EI}\right)(2.73)\left(\frac{2.73}{3}\right) + \frac{1}{2}\left(\frac{78.13}{EI}\right)(2.27)\left(5-\frac{2.27}{3}\right)$$
$$+ \left(\frac{70}{EI}\right)(y)\left(5+\frac{y}{2}\right) + \frac{1}{2}\left(\frac{78.13-70}{EI}\right)(y)\left(5+\frac{y}{3}\right)$$

Substitution for $I = 1.3 \times 10^{-3}$ m^4, $E = 27 \times 10^6$ kN/m^2 and $y = 0.53$ m gives:

$\delta_{max} = 0.0133$ m

$\quad\quad\; = 13.3$ mm

Problems

Sections 4.1 and 4.2

4.1 For the beam illustrated in Appendix B, No. 6, verify that equilibrium is satisfied and draw the shear force diagram for the beam.

4.2 Verify the reactions and the bending moment diagram and construct the shear force diagram for the beam of Appendix B, No. 3.

Section 4.3

4.3 Use the stiffness method to find the bending moment diagram for the structure illustrated in Fig. 4.54.

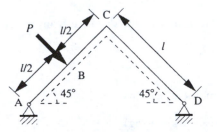

Figure 4.54 Structure for Problem 4.3

4.4 Use moment distribution to find the bending moment diagram for the structure of Problem 4.3.

4.5 Find the bending moment diagram for the beam whose geometry and loading are as illustrated in Fig. 4.55.

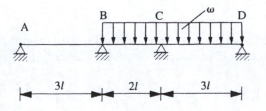

Figure 4.55 Beam of Problem 4.5

4.6   The structure illustrated in Fig. 4.56 resists horizontal forces by frame action. Find the bending moment diagram due to the horizontal loading given that the second moment of area of the beam, I_b, is twice that of the columns, I_c.

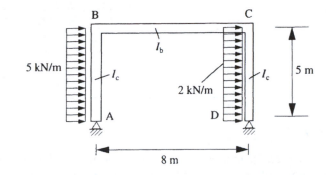

Figure 4.56 Structure for Problem 4.6

Section 4.4

4.7   For the structure of Problem 4.4, the yield moment for both sag and hog is $Pl/5$. Use elastic-plastic analysis to determine the yield load factor and the mechanism load factor.

4.8   For the structure of Problem 4.7, use plastic analysis to determine the mechanism load factor.

4.9   For the beam of Problem 4.5, the yield moment in sag is $0.8\ ql^2$, while that in hog is $0.6\ ql^2$. Determine the mechanism load factor.

Section 4.5

4.10   Find the shear force and axial force diagrams for the structure of Problem 4.7:
   (a)   for linear elastic bending;
   (b)   for the mechanism collapse condition.

Chapter 5

Applications of structural analysis to concrete structures

5.1   Introduction

The majority of *in situ* concrete structures are cast in large pours which results in continuity between structural members (beams, columns, etc.). It follows that, *in situ* concrete structures are generally indeterminate. Precise methods of analysis of such three-dimensional structures can be time consuming. In practice, concrete structures are often assumed to be made up of smaller, more manageable, two-dimensional sub-structures. These sub-structures are derived in such a way that they are easier to analyse (by hand or by computer) even though they may still be indeterminate. At the same time, these sub-structures model the behaviour of the actual structure within acceptable margins of accuracy. This chapter presents some of the more popular approaches to the modelling and analysis of concrete structures.

The more popular methods of analysis of two-dimensional indeterminate structures, such as the stiffness method or the finite element method, are most conveniently executed on computer. It is widely accepted that these methods are, in most cases, too laborious to be carried out by hand. However, even when using computer programs, it is important to be able to check the results. For this reason, less exact or 'approximate' methods of analysis are required that are simple enough to be carried out by hand. These methods of analysis also give the designer insight into the behaviour of a structure. Finally, approximate methods of analysis can be used where an exact solution is not required or where time does not permit a more thorough analysis. For these reasons, this chapter concentrates on hand methods, rather than computer methods, suitable for the approximate analysis of concrete structures.

Apart from the yield line method for slabs, the methods of analysis presented in this chapter assume linear elastic behaviour of the structural material. To allow for the actual inelastic behaviour of reinforced concrete under ultimate loads, the results of the elastic analyses are adjusted using a method known as plastic moment redistribution (not to be confused with the moment distribution method of analysis).

5.2   Continuous beam analysis

Continuous beam analysis is perhaps the most common analysis problem in concrete buildings. It can be applied to the approximate analysis of strips through one-way spanning slabs in addition to beams. Where a number of load cases have to be considered for continuous beams, computer programs (using the stiffness method) are probably the most convenient method of analysis. However, to check the results from a computer, or to carry out a quick analysis, formulae such as those given in Appendix D can be applied to the member. Alternatively, a solution can be found using moment distribution, as illustrated in the following example.

Example 5.1 Continuous beam analysis

Problem: For the floor system of Fig. 5.1, find upper limits for (a) the maximum moment in a 1 m strip of slab at section X–X or Y–Y and (b) the vertical load in column H due to a factored uniformly distributed loading of 10 kN/m² acting throughout the slab.

Solution: (a) Maximum moment in slab

The slab at section X–X of Fig. 5.1 is spanning one way between beams AB, DE and GH (the slab will also span N–S to some degree but it is common practice in such situations to assume it to span between beams). Each of the beams acts as a support to the slab and, in turn, transfers the reactions from the slab to the columns. While the torsional stiffness of the supporting beams provides some resistance to rotation at points J, K and L, this is not generally very significant and it is common to assume that the strip of slab is simply supported by the beams as illustrated in Fig. 5.2. The moment distribution method can now be used to find the bending moment diagram due to the applied loading on a 1 m wide strip.

The fixed structure and the associated bending moment diagram are as illustrated in Fig. 5.3(a–b). The bending moment at the fixed end of a pinned-fixed beam is $\omega l^2/8$ (see Appendix B, No. 6). The process of moment distribution is started by releasing point K. The distribution factors at this point are equal to:

$$DF_{KJ} = \frac{3EI/L_{JK}}{(3EI/L_{JK}) + (3EI/L_{KL})} = \frac{1/5}{1/5 + 1/7} = \frac{7}{12}$$

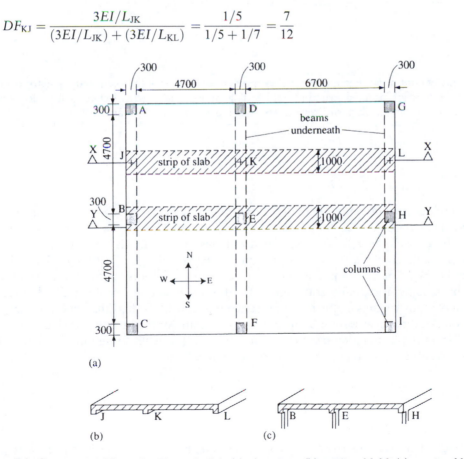

(a)

(b)   (c)

Figure 5.1 Geometry of floor for Example 5.1: (a) plan view; (b) section X–X; (c) section Y–Y

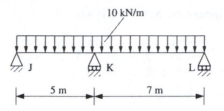

Figure 5.2 Strip of slab at section X–X

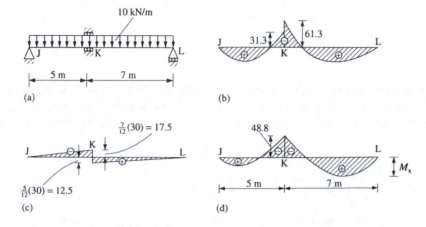

Figure 5.3 Analysis of continuous beam of Fig. 5.2 by moment distribution: (a) fixed structure; (b) bending moment diagram for fixed structure; (c) bending moment diagram associated with release of K; (d) bending moment diagram after release of K (equal to sum of (b) and (c))

$$DF_{KL} = 1 - DF_{KJ} = 1 - \frac{7}{12} = \frac{5}{12}$$

Therefore, to release the discontinuity of 30 kNm at point K, a correction of $-\frac{7}{12}(30)$ is made to the left of K and a correction of $+\frac{5}{12}(30)$ is made to the right of K. The associated carry-over factors are zero and the bending moment diagram associated with this correction is as illustrated in Fig. 5.3(c). The bending moment diagram after the release of point K is the sum of those illustrated in Fig. 5.3(b–c) and is illustrated in Fig. 5.3(d). As equilibrium is now satisfied at all joints, no further distribution is required and the bending moment diagram of Fig. 5.3(d) is the final one. From Fig. 5.3(d) it can be seen that the maximum sagging moment occurs between K and L. To find the maximum sagging moment, the point of zero shear is found. The free-body diagram for this strip of the slab between K and L, derived from the bending moment diagram of Fig. 5.3(d), is illustrated in Fig. 5.4(a). The reaction at L is determined by taking moments about point K.

$$7R_L + 48.8 = 10 \times 7 \times \frac{7}{2}$$
$$\Rightarrow R_L = 28.0\,\text{kN}$$

Similarly the reaction at J is found to be 15.3 kN and by equilibrium of the vertical forces acting on the beam the reaction at K is 76.7 kN. The shear force diagram for the strip of slab at section X–X is

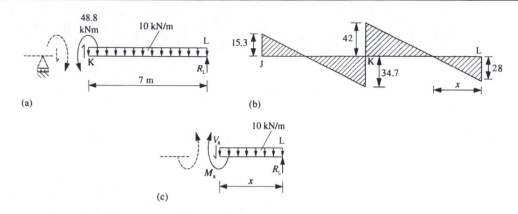

Figure 5.4 (a) free-body diagram for part of section X–X; (b) shear force diagram for strip of section X–X; (c) free-body diagram to calculate M_x

illustrated in Fig. 5.4(b). The shear force is zero between K and L at a distance x from support L. The distance x is found using the free-body diagram in Fig. 5.4(c). By equilibrium of the vertical forces:

$$V_x = R_L - \omega x = 0$$
$$\Rightarrow x = \frac{R_L}{\omega} = \frac{28}{10}$$
$$= 2.8 \text{ m}$$

Thus, the maximum sagging moment in the beam, M_x, is given by:

$$M_x = R_L x - \omega x^2 / 2$$
$$= (28)(2.8) - (10)(2.8)^2 / 2$$
$$= 39.2 \text{ kNm}$$

Therefore, with reference to Fig. 5.3(d), it can be seen that the maximum hogging moment in the strip of slab X–X is the hogging moment of 48.8 kNm at point K. Since the ends of the slab are not really simply supported, there will in fact be a small hogging moment along the edge and, as a result, a nominal amount of top reinforcement should be provided along the edges of the slab.

For the strip of slab at section X–X (see Fig. 5.1), mid-way between the supporting columns, the model of Fig. 5.2 is sufficiently accurate. However, for strips nearer the supporting columns, such as the strip at section Y–Y, the resistance to rotation at points B, E and H is more significant. Therefore, a more appropriate model for the strip of section Y–Y would be one which would allow for the rotational stiffness of the columns at these points. If the columns were very stiff, a suitable model would be that illustrated in Fig. 5.5(a) and the bending moment diagram would be as illustrated in Fig. 5.5(b) (see Appendix B, No. 8). The actual bending moment diagram at section Y–Y will, in fact, be something between those illustrated in Fig. 5.3(d) and Fig. 5.5(b). For preliminary purposes, the envelope of moments from these figures could be used for design. For a more exact result, a frame analysis is necessary, as described in Section 5.5.

It can be seen that the central support and in-span moments in Fig. 5.5(b) are approximately the same as or smaller than those for the model of Fig. 5.2 and so the previous results for the simply supported model can be used for the design of all 1 m strips through the slab. An exception to this is

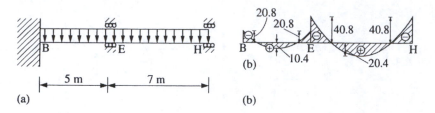

Figure 5.5 (a) beam with infinitely stiff columns; (b) associated bending moment diagram

at the outer supports where, as can be seen at B and H in Fig. 5.5(b), there exists a substantial hogging moment. For this reason, it is necessary to design the slab for a hog moment (up to 20.8 or 40.8 kNm/m, as appropriate) along its edges in the vicinity of all external columns. The precise distance away from each column in the North–South direction within which top reinforcement is provided is a matter for engineering judgement.

EC2 states that where there is partial fixity along the support of a slab and this is not considered in the analysis, top reinforcement should be provided to resist a hogging moment at least 25 per cent of the maximum moment in the adjacent span. This may be reduced to 15 per cent at an end support.

(b) Vertical load in column H

To find the maximum vertical load in column H, it is first necessary to find the reactions from each 1 m wide strip of slab in the East–West direction. From above, the reaction at L for a strip at section X–X is 28 kN. The reaction for a strip at section Y–Y can be found by simple equilibrium. If the columns at section Y–Y are assumed to be infinitely stiff, as in the model of Fig. 5.5(a), half the reaction is taken at each support and:

$$R = \frac{10 \times 7}{2} = 35 \text{ kN} \qquad (\text{reaction at H from 1 m strip at section Y–Y})$$

Thus, the reaction transferred to beam GHI from any strip parallel to section X–X is something between 28 kN and 35 kN. For the calculation of the maximum force in column H, it is conservative (perhaps too much so) to assume the greater of these values for all strips. The resulting loading on beam GHI is a uniformly distributed load of 35 kN/m. The beam can now be analysed to determine the reaction at H.

If the columns were assumed to have infinite rigidity, each span (GH and HI) would be fixed at either end and the reaction at H would equal half the load from each span, that is:

$$R_H = 2\left(\frac{35 \times 5}{2}\right) = 175 \text{ kN}$$

If, on the other hand, the column stiffnesses were ignored, the continuous beam would be on simple supports at G, H and I and the reaction at H (from Appendix C, No. 4) would be:

$$R_H = \frac{5\omega L}{4} = \frac{5 \times 35 \times 5}{4} = 219 \text{ kN}$$

Hence, a conservative estimate for the maximum axial compression in column H is 219 kN. The ISE manual recommends that where beams and slabs are assumed to be simply supported, the axial load in the columns obtained from this analysis should be increased by 10 per cent. Loads may need to be increased further if adjacent spans and/or loads vary greatly.

5.3   Plastic moment redistribution

Reinforced and prestressed concrete members with bending (i.e. beams and slabs) are designed to have a certain ductility under ultimate loads. This ductility ensures that the member is capable of undergoing a certain amount of rotation after yielding of the tension steel reinforcement and before crushing of the concrete in compression. The idealized moment-curvature relationship assumed for a member in bending can be seen in the graph of Fig. 4.39. The portion of this graph between κ_y and κ_{ult} represents the ductility. It will be shown in Chapter 8 that a limitation on the neutral axis depth is the mechanism by which ductility is guaranteed by EC2. As was seen in Section 4.4, when members in bending have this ductility, they have the potential to continue to resist load beyond the time of initial yield. This feature was demonstrated in Example 4.9. If a member in bending was to be designed using the linear elastic assumption only, the maximum load factor would be the value at which the applied moment first equals M_{ult}. For the member of Example 4.9, this occurs at:

$$\lambda_y = \frac{32M_{ult}}{6PL} = \frac{5.33M_{ult}}{PL}$$

while the mechanism load factor for this beam has been shown to be:

$$\lambda_y = \frac{6M_{ult}}{PL}$$

Clearly, when a member has sufficient ductility to achieve the mechanism load factor, considerable savings can be achieved by recognizing this extra load-carrying capacity, Elastic-plastic analysis is impractical for everyday design because it requires a knowledge of the moment capacity of all members in advance of the analysis. A much simpler approach which avoids this requirement is plastic moment redistribution. This is an approximate method by which the elastic bending moment diagram is adjusted to account for the ductile behaviour of reinforced and prestressed members in bending. EC2 allows the original elastic moment in continuous members to be reduced by up to 30 per cent. The precise amount of redistribution allowed is dependent on the grade of the concrete and on the ductility characteristics of the reinforcement as well as the neutral axis depth. Specifically, the following limits are imposed on the ratio of the redistributed moment to the moment before redistribution, δ. Concrete is graded by the strength of test cylinders or cubes. The UK NA to EC2 specifies that for concrete grades with cylinder compressive test strength up to 50 N/mm^2:

$$\delta \geq 0.4 + 1.0x_u/d \tag{5.1}$$

where x_u is the distance from the extreme compressive fibre in the section to the neutral axis after redistribution, and d is the effective depth to the tension reinforcement. Reinforcing steel is available in three classes of ductility, A, B and C, where Class A has the lowest ductility and class C has the highest. Where Class B or Class C reinforcement is used, δ must be greater than or equal to 0.7 (i.e. maximum of 30 per cent redistribution). Where Class A is used, δ must be greater than or equal to 0.80.

The saving that results from reducing the elastic moment at a given section in a member using plastic moment redistribution is invariably offset by the fact that equilibrium requires the bending moment at other points in the beam to be increased. For this reason, the optimum amount of moment redistribution that should be carried out depends greatly on the geometry and loading of the member. The application of the technique is illustrated in Example 5.2.

Example 5.2 Plastic moment redistribution

Problem: The bending moment diagram for the two-span beam of Fig. 5.6(a), derived using an elastic method of analysis, is illustrated in Fig. 5.6(b). Two arrangements of reinforcement are available for design: two 20 mm diameter bars giving a moment capacity of 254 kNm, or two 25 mm diameter bars giving a moment capacity of 306 kNm. As the beam cross-section is rectangular, either pair of bars can be placed near the top of the section to resist hog or near the bottom to resist sag. Using plastic moment redistribution, determine the most suitable arrangement of reinforcement if the factored ULS load, P, is 150 kN and L is 10 m. It may be assumed that up to 30 per cent redistribution is allowed.

Solution: For a factored ULS load of 150 kN, the original elastic bending moment over the central support is equal to 281 kNm (hog) and the maximum elastic in-span moment is equal to 235 kNm (sag). To resist these design elastic moments, two 20 mm diameter bars (2H20) of combined moment capacity 254 kNm are required in the interior of each span and two 25 mm diameter bars (2H25) of combined moment capacity 306 kNm are required over the central support (see Fig. 5.6(c)).

In Chapter 4, it was shown that a bending moment diagram can be constructed by superimposing the bending moment diagrams for simply supported spans on the bending moment diagram associated with the support moments. Thus, the elastic bending moment diagram of Fig. 5.6(b) is the sum of the bending moment diagrams illustrated in Fig. 5.7(a–b).

When a plastic hinge forms at a support it prevents any further increase in moment at that point. Now, if the moment capacity at B were to be limited to 254 kNm (two 20 mm diameter bars), a hinge would form there for an applied load less than 150 kN and, for any further increase in load, spans AB and BC would behave as if simply supported. The new bending moment diagram is constructed simply by superimposing the diagram associated with $M_B = -254$ kNm (Fig. 5.7(c)) on the diagram

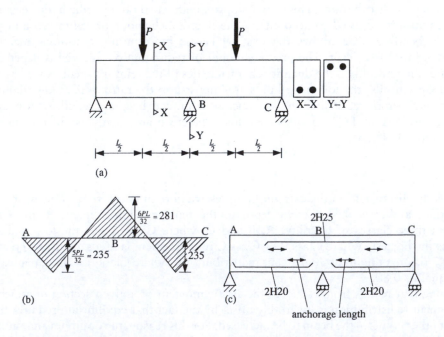

Figure 5.6 Two-span beam: (a) geometry and loading; (b) elastic bending moment diagram; (c) reinforcement to resist elastic moments

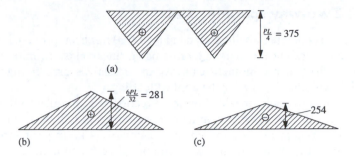

$$\frac{PL}{4} = 375$$

(a)

$$\frac{6PL}{32} = 281$$

254

(b)   (c)

Figure 5.7 Components of bending moment diagram: (a) elastic bending moment diagram if spans are simply supported; (b) elastic bending moment diagram associated with support moment; (c) bending moment diagram associated with reduced support moment

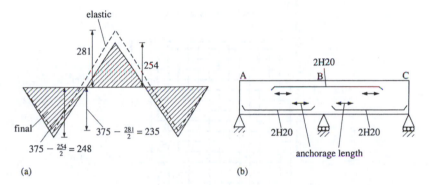

elastic

281

254

final

$$375 - \frac{281}{2} = 235$$

$$375 - \frac{254}{2} = 248$$

2H20

A   B   C

2H20   2H20

anchorage length

(a)   (b)

Figure 5.8 (a) bending moment diagram before and after plastic redistribution; (b) reinforcement to resist moments after plastic redistribution

of Fig. 5.7(a). The result of this 10 per cent redistribution at B is illustrated in Fig. 5.8 (the elastic bending moment is shown dotted for comparison). It can be seen that, even though the in-span moment has increased as a result of the redistribution, it can still be resisted using two 20 mm bars (capacity = 254 kNm). Hence, for a redistribution of 10 per cent, a significant saving in reinforcement has been made. It is clear that any further increase in the amount of moment redistribution would result in greater steel requirements at the interior of each span (i.e. redistribution moment would exceed 254 kNm).

5.4   The implications of lower-bound methods

Lower-bound methods are methods of design where a distribution of bending moment is assumed that satisfies the requirements of equilibrium but not necessarily compatibility. Thus, it may not be the actual distribution of internal moment in the structure that would be found by linear elastic analysis. It has been proven by a theorem known as the static theorem (Heyman 1971) that a structure of **infinite ductility** is safe under its applied loads if a distribution of internal bending moment can be found that satisfies the equilibrium requirements and nowhere exceeds the ultimate moment capacity. The following examples illustrate the application of lower-bound methods.

Example 5.3 Two-way spanning slab

The square reinforced concrete slab illustrated in Fig. 5.9(a) has been provided with reinforcement in both directions. The bending moment diagrams using linear elastic (finite element) analysis are illustrated in Fig. 5.9(b). If the **moment capacity** of the slab is greater than 18.4 kNm in both directions, the slab will not yield under the applied loading.

Now, if reinforcement is only provided in one direction (e.g. parallel to the Y-axis), the bending moment diagram for a 1 m wide strip spanning in one direction will be as illustrated in Fig. 5.9(c). Thus, if bands of reinforcement with a moment capacity of 54 kNm per metre width are provided parallel to the Y-axis, then the structure will remain safe even though there is no reinforcement parallel to the X-axis. Therefore, since a distribution of internal bending moment can be found that satisfies the equilibrium requirements and nowhere exceeds the ultimate moment capacity, this is an example of a lower-bound approach.

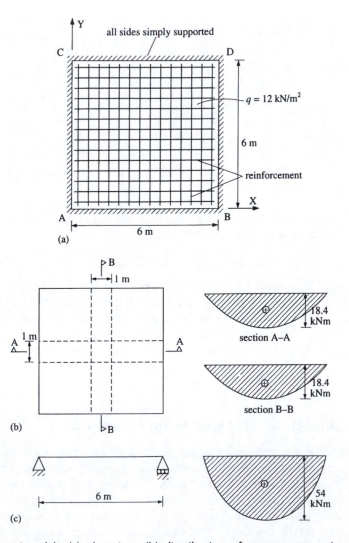

Figure 5.9 Two-way spanning slab: (a) plan view; (b) distribution of moment per unit width from linear elastic analysis; (c) distribution of moment in one-way spanning beam

This kind of lower-bound approach is only reasonable if the structure has **sufficient ductility** to allow the assumed bending moment distribution to occur. In this case where the slab is only reinforced in one direction, there would be considerable cracking parallel to the reinforcement (i.e. yielding in unreinforced direction) before the load is taken by the bands of reinforcement.

A certain minimum quantity of reinforcement is provided in all concrete structures (more details in Chapter 8). It is generally felt that if the assumed distribution of bending moment is reasonably close to the linear elastic distribution, then slabs will have sufficient ductility to allow design by lower-bound methods.

Example 5.4 Portal frame

For the portal frame of Fig. 5.10(a), preliminary sizing rules for the members result in the following preliminary dimensions (rules for preliminary sizing are given in Chapter 6):

beam depth, $h = 450$ mm
beam breadth, $b = 250$ mm
column depth and breadth $= 250$ mm

It is common practice to calculate the second moments of area of members neglecting the reinforcement and cracking of the members. Thus:

$$I_{beam} = bh^3/12 = 1.9 \times 10^{-3}\,\text{m}^4$$
$$I_{col} = h^4/12 = 0.325 \times 10^{-3}\,\text{m}^4$$

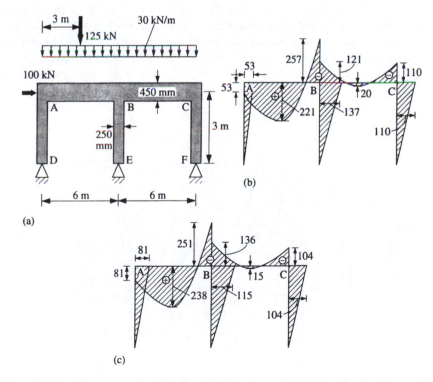

Figure 5.10 Portal frame: (a) geometry and loading; (b) bending moment diagram for $I_{beam} = 1.9 \times 10^{-3}$ m⁴; (c) bending moment diagram for $I_{beam} = 5.72 \times 10^{-3}$ m⁴

A computer analysis (stiffness method) of the frame under the applied loads, using these second moments of area, gives the bending moment diagram of Fig. 5.10(b). In checking the moment capacity of each member against these applied moments, it is found that the preliminary size of the beam is inadequate for the applied moment of 257 kNm at B. Therefore, the beam section throughout ABC is increased to 650 mm, which increases the second moment of area to $I_{beam} = 5.72 \times 10^{-3}$ m^4. It has been shown in Example 4.5 that the relative values for the second moments of area in members affect the distribution of moment. However, the distribution of Fig. 5.10(b) does satisfy equilibrium even if it is now incorrect. Therefore, by the static theorem, it is not necessary to reanalyse the structure if it has sufficient ductility in the members to facilitate redistribution of the moments. For this example, the increase in beam stiffness is so large that there may not be adequate ductility.

Note: Linear elastic analysis for the same geometry and loads but using the new $I_{beam} = 5.72 \times 10^{-3}$ m^4 results in the bending moment diagram of Fig. 5.10(c). It can be seen that, at some points (point A, for example), the moment has increased significantly.

5.5   Analysis of frames

Apart from flat slab (beamless) construction, most concrete buildings contain a structure of beams and columns which, when rigidly connected (as is usually the case with concrete), make up a continuous frame. Fig. 5.11 illustrates the structural framework of one such multi-storey concrete building. The framework of this building would usually be concealed behind cladding panels that protect the occupants of the building from the external environment. As can be seen from the figure, lateral stability of this frame in one direction (E–W) is provided by shear walls. In the other direction, no shear walls are present and the horizontal stability of the frame is achieved through frame action.

The analysis of a complete three-dimensional frame, such as that in Fig. 5.11, can be carried out by hand or by computer using any appropriate method such as, for example, the stiffness method. However, the mathematical complexity of the solution process generally makes it infeasible to analyse a complete three-dimensional structure by hand. Even when analysing by computer, the

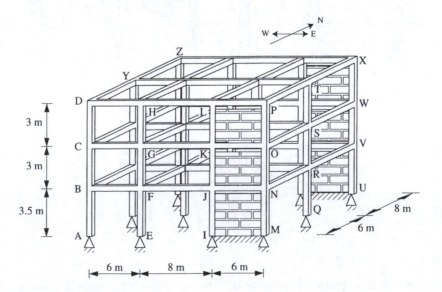

Figure 5.11 Three-dimensional skeletal frame

solution may become unduly complex. One particular aspect of analysis that complicates the design of a complete three-dimensional structure is the need to consider all possible arrangements of load. In theory, every possible combination of permanent, variable and wind loading must be considered to determine the critical load effects in each member. The greater the number of members in the frame, the greater the number of possible combinations of applied load. For this reason, certain assumptions and simplifications are commonly made before the structure is analysed.

To overcome the complexity of considering the full multi-storey skeletal structure and to facilitate a solution by hand, a common simplification is to represent the three-dimensional frame with smaller, two-dimensional sub-frames. This substantially reduces the total number of load cases that must be considered for each sub-frame and simplifies the process of describing the structural model to the computer. The precise method of simplification depends on whether or not the original frame is braced against horizontal loads. A frame that is braced against horizontal loads using substantial bracing members such as cores and/or shear walls is termed a non-sway frame. Due to the presence of such stiff bracing members, there is little or no lateral deflection in a non-sway frame. For this reason, such a frame is designed to resist only the applied vertical loads (i.e. the bracing members are designed to carry the horizontal loads). A frame that does not have shear walls, cores or other bracing members transmits the horizontal loads to its foundations by frame action. This type of frame undergoes significant horizontal deflection under applied horizontal loads and hence is known as a sway frame. Sway frames must be designed to resist both vertical and horizontal loads.

Analysis of non-sway frames

Consider the frame of Fig. 5.11, which is braced against lateral loads in the E–W direction by masonry shear walls and contains *in situ* concrete floor slabs at each floor level. The first simplification that can be made is to assume, in the E–W direction, the frame can be represented by three two-dimensional non-sway frames, as illustrated in Fig. 5.12. Note that horizontal loads applied to the central frame are transmitted by membrane action in the floor slabs to the shear walls. Note also that the vertical loadings for the two outer plane frames are the same and hence only one needs to be analysed. The central plane frame carries a greater vertical load since it supports a greater floor area. The masonry shear walls are designed to resist horizontal loads and are detailed so as not to resist any vertical loads.

The plane frame of Fig. 5.12 can readily be analysed by computer for each possible arrangement of load. However, two alternative methods are available for further simplifying the plane frame to facilitate a hand solution. The first of these methods is to divide the plane frame into a set of

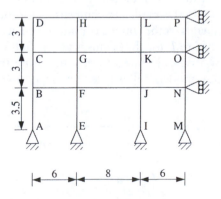

Figure 5.12 Two-dimensional frame

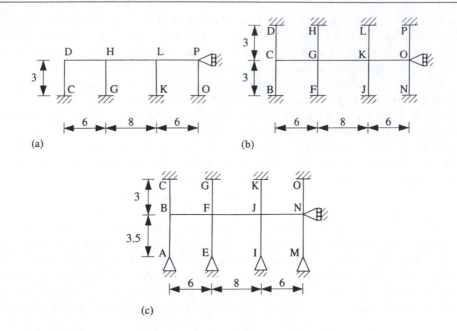

Figure 5.13 Sub-frames for the frame of Fig. 5.12: (a) top; (b) middle; (c) bottom

sub-frames, each of which is analysed separately. Each sub-frame is made up of the beams at one level together with the columns connected to these beams. For example, the plane frame of Fig. 5.12 can be divided into the three sub-frames of Fig. 5.13. The columns meeting the beams are assumed to be fixed at their ends as illustrated (except when the assumption of a pin-ended column is more suitable as would often be the case at foundation level). These sub-frames can readily be analysed by hand using, say, the moment distribution method to give the moments, shears, etc., in both the beams and the columns.

A less exact, and not necessarily more conservative, simplification than the sub-frame is to assume that the beams are continuous over the supporting columns and that the columns provide no restraint to the rotation of the beams. Therefore, the beams in each plane frame can be analysed as continuous beams on simple supports. Fig. 5.14 shows one such continuous beam for the plane frame of Fig. 5.12. The moment distribution method may then be used to determine the moments, shears and reactions in each beam. The moments in the columns can be determined by analysing models such as that of Fig. 5.15. In this model, to compensate for the increased stiffness of the beams due to the assumption of fixed ends (rather than a continuous structure) and to account for cracking effects in the beam, the ratio of I to L for the beams is at one-half of their actual values.

The analysis procedure for a typical braced frame, indicating the arrangements of loading that must be considered for a typical model, is illustrated by the following example.

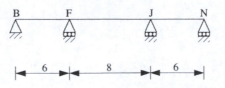

Figure 5.14 Continuous beam model

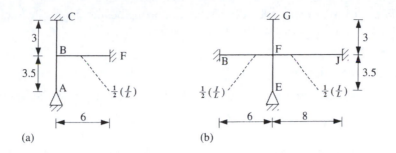

Figure 5.15 Sub-frames for column analysis

Example 5.5 Analysis of non-sway frame

Problem: For the structure of Fig. 5.11, determine the maximum sagging moment in member FJ. The characteristic floor loads are:

permanent gravity (incl. self-weight) = 6.0 kN/m²
imposed = 4.0 kN/m²
wind (any direction) = 1.0 kN/m²

The flexural rigidity, *EI*, of the columns is half that of the beams. The floor consists of precast prestressed concrete slab units spanning N–S (only). The distribution of horizontal wind load through the slab to the shear walls is achieved through a 75 mm thick structural screed on top of the precast units.

Solution: The maximum bending moment in member FJ is found by analysing the plane frame of Fig. 5.12. Due to the presence of the shear wall in the end bays, the frame is effectively braced against wind from an easterly or westerly direction, as illustrated in the figure. Thus, to facilitate a hand solution, the frame can be simplified into the three sub-frames of Fig. 5.13 of which only that of Fig. 5.13(c) is of interest for member FJ.

The specified permanent load acting at each level of the frame of Fig. 5.12 includes the weight of the portion of slab carried by the frame and the self-weight of the beams. Fig. 5.16 illustrates two sections, E–W and N–S, through a typical slab in the frame of Fig. 5.11. The weight of the slab

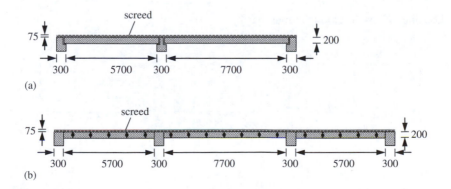

Figure 5.16 Sections through slab: (a) N–S (view from east); (b) E–W (view from north)

spanning N–S (Fig. 5.16(a)) is carried equally by the supporting beams. Thus, the total characteristic permanent load acting on the beams in the outer frame is:

$$g_k = (0.5)(6m)(6kN/m^2)$$
$$= 18.0 \, kN \text{ per metre length}$$

In a similar manner, the characteristic variable gravity floor load carried by the frame is given by:

$$q_{ik} = (0.5)(6m)(4kN/m^2) = 12.0 \, kN/m$$

Load can have either a favourable or an unfavourable effect on a member and it is necessary to determine which, before the appropriate load factors can be selected. For the sub-frame of Fig. 5.13(c), the maximum distributed load for design at ULS is:

$$1.35(18.0) + 1.5(12.0) = 42.3 \, kN/m$$

The Eurocode specifies that the safety factor for permanent loads should be the same throughout the beam. Therefore, the minimum distributed load for design at ULS is:

$$1.35(18.0) = 24.3 \, kN/m$$

The loading arrangement which gives the maximum sagging moment in member FJ is maximum distributed load on member FJ and minimum distributed load on adjacent members, as illustrated in Fig. 5.17. Using the moment distribution method, the sub-frame of Fig. 5.17 can readily be analysed by hand. The resulting bending moment diagram is illustrated in Fig. 5.18. From this figure, it can be seen that the maximum sagging moment in member FJ is approximately equal to 131 kNm.

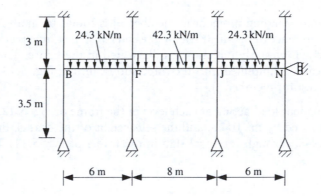

Figure 5.17 Loading for maximum sag in member FJ

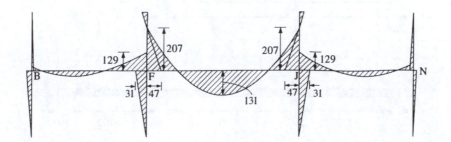

Figure 5.18 Bending moment diagram for loading of Fig. 5.17

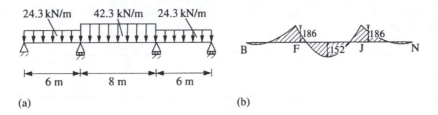

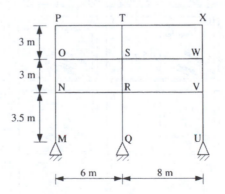

Figure 5.19 Continuous beam model: (a) geometry and loading; (b) bending moment diagram

Alternatively, the maximum sagging moment in member FJ can be estimated using the continuous beam approximation of Fig. 5.19(a). The bending moment diagram for this model is illustrated in Fig. 5.19(b). Comparing the diagrams of Fig. 5.18 and Fig. 5.19(b), it can be seen that, for this example, there is a significant variation in results between the two models with the simpler model of Fig. 5.19(a) giving more conservative results for the maximum sagging moment in member FJ but less conservative results for the maximum hogging moment at F and J.

Analysis of sway frames

Consider the frame of Fig. 5.11 which is not braced against lateral loads in the N–S direction. As for non-sway frames, the first simplification which can be made is to assume that, in the N–S direction, the frame can be represented by four two-dimensional sway frames, one of which is illustrated in Fig. 5.20. Each of the four plane frames is then analysed for all realistic combinations of load. However, as the slabs in this example span N–S, the only significant loading is that due to wind. The values of the moments, shears and axial forces to be used in the design of individual members are the critical values that result from applied wind load.

Where vertical loads are present in a sway frame, analysis to determine the effect of vertical load only may be carried out in the same manner as described above for non-sway frames. Thus sub-frame models, such as those illustrated in Fig. 5.13, can be used. The effects of all applied loads, horizontal and vertical, acting together are then found by superposing the effects of the vertical loads with those of the horizontal loads.

If the effects of the horizontal loads are to be calculated by hand, the plane frames may be simplified by assuming that points of contraflexure occur at the centre of all members, as illustrated in Fig. 5.21. As the moment at a point of contraflexure is zero, a hinge can be assumed at the centre of each member, as illustrated in Fig. 5.21(b). This simplification then allows the frame to be analysed

Figure 5.20 Sway frame

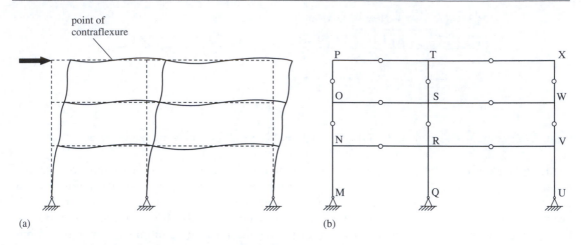

(a)   (b)

Figure 5.21 (a) deflected shape; (b) points of contraflexure

using either the portal method or the cantilever method. The following example illustrates the analysis of a sway frame using the portal method.

Example 5.6 Analysis of sway frame

Problem: With reference to the data from Example 5.5, estimate the maximum hogging moments in member SW, assuming the southerly wind results in a characteristic load intensity of 1.0 kN/m^2.

Solution: Using the same approach as in Example 5.5, the maximum bending moment in member SW is found by analysing the plane frame of Fig. 5.20. However, since the structure is not braced against horizontal loads, the plane frame of Fig. 5.20 must be analysed as a sway frame.

Fig. 5.22 illustrates the assumed load paths for the horizontal wind loads acting on the external south-facing cladding panels. The characteristic force acting on each panel is given by:

$$Q_{wk} = q_{wk}A_p$$

where q_{wk} is the intensity of the load, given as 1.0 kN/m^2, and A_p is the total area of the panel.

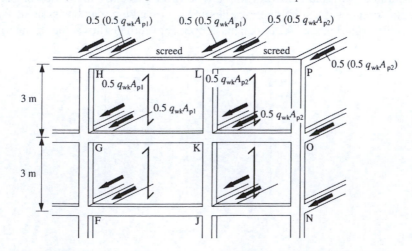

Figure 5.22 Assumed load paths (A_{p1} = area of panel HLKG; A_{p2} = area of panel LPOK)

Each panel is assumed to act as a simply supported, one-way spanning vertical slab, transferring the load to the beams at the top and bottom edges of the panel. In the external panels, half of the load is then transferred by the structural screed to the outer frames, as illustrated in Fig. 5.22. Thus, the characteristic force acting at the top of the external frames is:

$$0.5\left(0.5q_{wk}A_{p2}\right) = 0.25q_{wk}A_{p2}$$

where A_{p2} is the area of the external panel. That is, the tributary area is one-quarter the area of the complete external panel. For point P, the applied force is thus:

$$0.25q_{wk}A_{p2} = 0.25(1)(3 \times 6) = 4.5\,kN$$

Similarly, for points O and N, the tributary areas are $0.5(3 \times 6)$ m^2 and $0.5(3.25 \times 6)$ m^2, respectively and the forces are:

point O : $1 \times 0.5(3 \times 6) = 9\,kN$

point N : $1 \times 0.5(3.25 \times 6) = 9.75\,kN$

Thus, the total loading on the frame is as illustrated in Fig. 5.23. For the required moment effect, the wind load is the principal variable load, so the factored ULS load is 1.5 times the characteristic value.

The portal method for analysing frames with lateral loads makes the following assumptions:

1. Points of contraflexure occur at the centre of all columns and beams (except pin-ended columns).
2. Following from assumption (1), a hinge is assumed at the centre of each member.
3. At each level, the frame is divided into single portal frames. The portal method assumes that the portion of the lateral load at that level which is carried by each frame is proportional to its span length. However, the authors have found that it is more accurate to divide the lateral load in proportion to the stiffness of each bay of the frame.

Fig. 5.24 illustrates the application of assumptions (1) and (2) to the frame of Fig. 5.23. In this figure also, the wind loads have been factored to their ultimate (ULS) values. At roof level, the frame can be divided into two separate portal frames, as illustrated in Fig. 5.25.

By assumption (3), $Q_1 + Q_2 = 6.75$ kN where the magnitudes of Q_1 and Q_2 are determined using the relative stiffness of each bay of the frame:

$$Q_1 = \frac{\text{stiffness of bay 1}}{\text{stiffness of bay 1} + \text{stiffness of bay 2}} \times 6.75$$

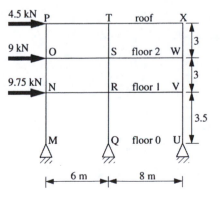

Figure 5.23 Frame loading

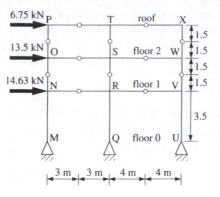

Figure 5.24 Frame showing ULS loading and assumed hinges

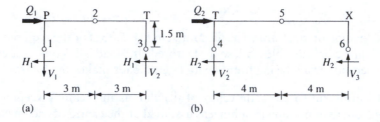

Figure 5.25 Roof level frames: (a) left portion; (b) right portion

For each bay, the stiffness of shared columns is assumed to be $\frac{1}{2}I_c$, as illustrated in Fig. 5.26. From Appendix A, the stiffness of the frames illustrated in Fig. 5.26(b) (with $I_c = \frac{1}{2}I_b$) is:

$$K = \frac{12EI_b}{h^2(6h + l)}$$

where h is the height and l is the span length of the frame. Therefore:

$$Q_1 = \frac{\dfrac{12EI_b}{h^2(6h + l_1)}}{\dfrac{12EI_b}{h^2(6h + l_1)} + \dfrac{12EI_b}{h^2(6h + l_2)}} \times 6.75 = \frac{\dfrac{1}{(6h + l_1)}}{\dfrac{1}{(6h + l_1)} + \dfrac{1}{(6h + l_2)}} \times 6.75$$

$$= \frac{1/15}{1/15 + 1/17} \times 675 = 3.59 \text{ kN}$$

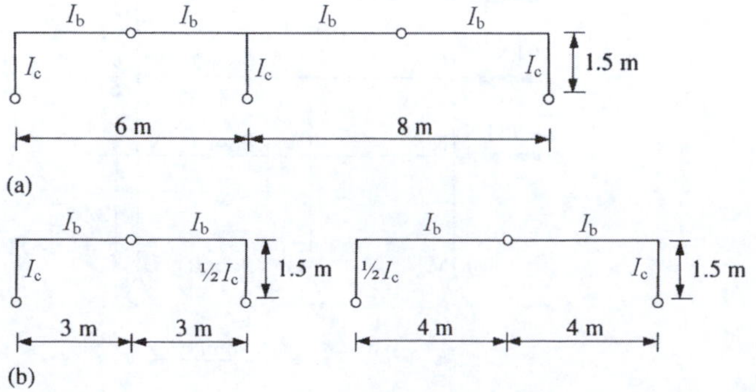

Figure 5.26 Implementation of assumption (3) for analysis of sway frames

Similarly:

$$Q_2 = 3.16 \, \text{kN}$$

By symmetry of Fig. 5.25(a–b), we have:

$$H_1 = \frac{1}{2}Q_1 = \frac{1}{2}(3.59) = 1.8 \, \text{kN}$$

and:

$$H_2 = \frac{1}{2}Q_2 = \frac{1}{2}(3.16) = 1.58 \, \text{kN}$$

Taking moments about hinge 2 of Fig. 5.25(a) gives:

$$3V_1 = 1.5H_1$$
$$\Rightarrow V_1 = 0.9 \, \text{kN}$$

Taking moments about hinge 5 of Fig. 5.25(b) gives:

$$4V_3 = 1.5H_2$$
$$\Rightarrow V_3 = 0.59 \, \text{kN}$$

From these results, the free-body diagrams for the two portal frames of Fig. 5.25 can be redrawn as illustrated in Fig. 5.27. The bending moment diagram of Fig. 5.28 can then be deduced from Fig. 5.27.

The situation at the next level is illustrated in Fig. 5.29. As for the sub-frame representing the top level, the sub-frame for floor 2, illustrated in Fig. 5.29(a), can be separated into the two portal frames of Fig. 5.29(b–c). As before:

$$Q_3 = \frac{1/15}{1/15 + 1/17} \times 13.5 = 7.17 \, \text{kN}$$

and:

$$Q_4 = 6.33 \, \text{kN}$$

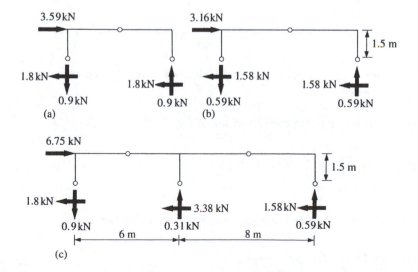

Figure 5.27 Roof level frames with reactions: (a) left portion; (b) right portion; (c) complete frame

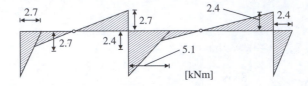

Figure 5.28 Bending moment diagram for roof level frame

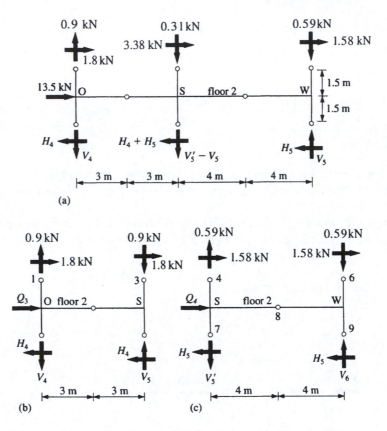

Figure 5.29 Floor 2 frames: (a) complete frame; (b) left portion; (c) right portion

Taking moments about hinge 7 in the sub-frame of Fig. 5.29(c):

$$2 \times 1.58(3) + Q_4(1.5) = (V_6 - 0.59)(8)$$
$$\Rightarrow V_6 = 2.96 \text{ kN}$$

Taking moments about hinge 8 in the right-hand portion of that sub-frame:

$$1.58(1.5) + 0.59(4) + H_5(1.5) = V_6(4)$$
$$\Rightarrow H_5 = 4.74 \text{ kN}$$

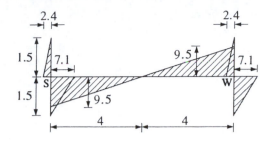

Figure 5.30 Bending moment diagram for sub-frame of Fig. 5.29(c)

Hence the moment in the column just below W is $(H_5 \times 1.5 =) 7.1$ kNm. Similarly the moment in the column just above W is $(1.58 \times 1.5 =) 2.4$ kNm.

By equilibrium of moments at W, the moment in the beam at this point is $(2.4 + 7.1 =) 9.5$ kNm. Hence the bending moment diagram for this sub-frame is as illustrated in Fig. 5.30 and the maximum moment due to wind in member SW is 9.5 kNm. (Alternatively, summing the vertical forces in the right-hand portion of the sub-frame gives a shear force of 2.37 kN at hinge 8. Hence the moment in the beam left of W is $4 \times 2.37 = 9.5$ kNm.)

5.6   Analysis of slabs

Slabs are the structural systems that actually create the living/working floor space in a structure by spanning the voids between framing members. Most permanent and variable gravity loads are applied directly to the floor slabs which then transfer the load, primarily by bending action, to the supporting beams and/or columns. It was shown in Section 2.2 that there are many different forms of construction for slabs. The precise system used depends on a number of factors including the required span distances and the magnitude of the applied vertical loads.

While beams extend only in one direction at a time, slabs extend in two directions, thus filling a complete two-dimensional surface. Hence, slabs can be thought of as the two-dimensional equivalent of beams since bending occurs about two perpendicular axes, as illustrated in Fig. 5.31. Using this analogy, the moment-curvature relationship for a beam is replaced by a two-dimensional equivalent in the case of a slab.

The distributions of stress in a small segment of slab are illustrated in Fig. 5.31(c). Direct stress on the face perpendicular to the X-axis, known as the X-face, is denoted σ_x. As in a beam, the distribution of direct stress is the result of an internal bending moment. On the X-face, this moment per unit breadth is denoted m_x, as illustrated in Fig. 5.31(d). Note that, for slabs, m_x is not moment per unit breadth **about** the X-axis but moment per unit breadth on the X-face or moment for which **reinforcement parallel** to the X-axis would be required. Similarly, there is a direct moment per unit breadth on the Y-face which is associated with the distribution of the direct stress, σ_y.

In addition to these direct moments, there are two twisting moments per unit breadth of equal magnitude. They are m_{xy}, on the X-face, and m_{yx}, on the Y-face, and it is these which cause the shear stresses τ_{xy} and τ_{yx}, respectively. These twisting moments tend to cause warping in the slab as illustrated in Fig. 5.32. Finally, there are vertical shear forces in slabs, V_x and V_y, on the X- and Y-faces, respectively. These shear forces result in the distribution of stresses τ_{xz} and τ_{yz}, respectively.

From a design viewpoint, slabs are hardly more complex than beams. Reinforcement parallel to the X-axis (ordinary reinforcement or prestress) is provided to resist m_x and reinforcement parallel to the Y-axis is provided to resist m_y. The concrete is assumed to be capable of resisting compressive stresses σ_x and σ_y simultaneously without any reduction in strength. Similarly, V_x and V_y are resisted

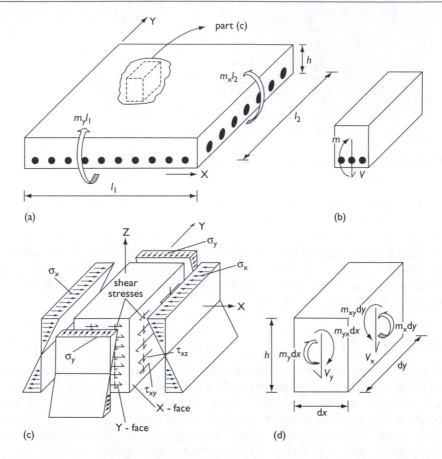

(a)   (b)

(c)   (d)

Figure 5.31 Slab with moments and stresses: (a) slab; (b) beam; (c) stresses in portion of slab; (d) moments and shear forces on portion of slab

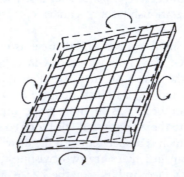

Figure 5.32 Warping of slab

in the same way as for beams. The twisting moments, m_{xy} and m_{yx}, however, are unique to slabs. Depending on their sign, these moments can act with or against m_x and m_y as can be seen from Fig. 5.31(d) (m_{yx} with/against m_x and m_{xy} with/against m_y). Formulae for the combination of these moments for design purposes are given in Appendix E. As $\tau_{xy} = \tau_{yx}$, the absolute values of m_{xy} and m_{yx} are equal. Fortunately, for most slabs, the twisting moments are relatively small and can

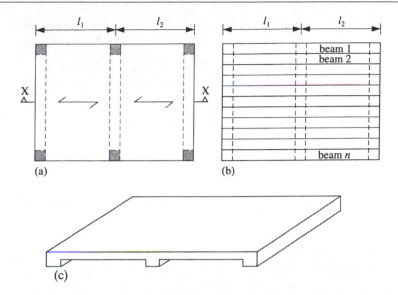

Figure 5.33 One-way spanning slab: (a) plan view of slab; (b) strips of slab treated as beams; (c) section X–X

frequently be neglected. It is only when slabs are skewed or where the corners of rectangular slabs are held down that there may be areas where the twisting moments are of significant magnitude locally.

Although all slabs extend in two directions, some forms of slab construction result in the applied loads being transferred predominantly in one direction. This may arise either if the slab is supported only along two parallel edges or if its span in one direction is much longer than its span in the other direction. In the latter case, the majority of the load is transferred across the shorter span since this has a greater stiffness. In slabs of this type, known as one-way spanning slabs, the moment about one axis is negligible compared with the moment about the other axis and the slab can be thought of as a series of adjoining beams, as illustrated in Fig. 5.33. Consequently, the analysis of one-way spanning slabs is considerably more straightforward than the analysis of two-way spanning slabs.

EC2 allows a slab to be analysed as one-way spanning if either:

(a)   it possesses two unsupported edges which are (sensibly) parallel; or
(b)   it is the central part of a (sensibly) rectangular slab supported on four edges with a ratio of the longer to shorter span greater than two.

Linear elastic methods

As described in Chapter 4, linear elastic methods of analysis are based on the assumption of linear elastic behaviour. Linear elastic analysis of one-way spanning slabs is carried out in the same way as the linear elastic analysis of beams. In effect, this means taking, say, 1 m wide strips of slab and carrying out a beam analysis for a typical strip. An example of a one-way spanning continuous slab analysis was given in Example 5.1.

Linear elastic analysis of two-way spanning slabs, both flat (beamless) slabs and slabs supported by beams, is usually carried out on computer using one of two common methods, both of which are applications of the stiffness method: grillage analysis and finite element analysis. In a grillage analysis, the slab is idealized as a mesh of beams, as illustrated in Fig. 5.34, which, being skeletal,

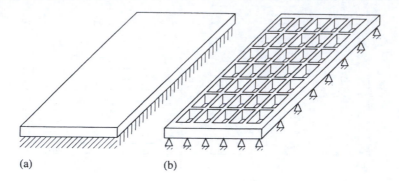

Figure 5.34 Grillage analysis of slab: (a) actual slab; (b) grillage of beams (from Hambly 1991, Figure 3.8)

can readily be analysed by the stiffness method. Each beam in the grillage mesh represents a strip of the slab and the stiffnesses of the beams in the mesh can be calculated on this basis. Grillage analysis is reasonably accurate for most problems except in areas of high concentrated loads such as near a point support. However, there are a number of special guidelines recommended for the proper application of the grillage method. Interested readers should refer to OBrien and Keogh (1999).

In a finite element analysis, the slab is idealized as a collection of discrete slab segments or elements joined only at nodes, as illustrated in Fig. 5.35. The analysis of such a model can be viewed as a special case of the stiffness method. For accurate results, a reasonably large number of elements should be used, especially in regions where moments are changing rapidly, such as over supports. Unlike the grillage method, guidelines for the application of the finite element method are relatively straightforward. Simple plate elements should not be excessively long and narrow as this may result in an accumulation of errors due to the rounding off of large numbers.

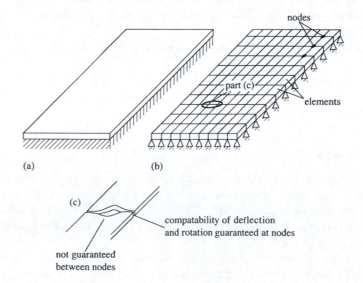

Figure 5.35 Finite element analysis of slab: (a) actual slab; (b) finite element mesh; (c) detail between two nodes

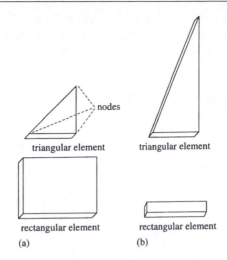

Figure 5.36 Length to breadth ratios for finite elements: (a) good; (b) poor

As a general guide, results should be reasonable provided the length/breadth ratio of elements is less than or equal to two, as illustrated in Fig. 5.36. It is also important to remember that elements are only joined at the nodes (which are sometimes only at the corners). In such cases, it is necessary to avoid 'T-junctions', as illustrated at point A in Fig. 5.37(b). At such a point, elements 2 and 3 are joined to each other but element 1, not having a node at A, is not joined to either of the other two elements. For this reason, it is possible for element 1 to have a different deflection than elements 2 and 3 at this point.

The accuracy of a finite element analysis can vary depending on the density of the mesh chosen (i.e. distribution of elements and nodes). Forces, stresses and displacements are calculated at the nodes; therefore, a model with a higher number of nodes within a certain area will produce more accurate results. However, a very dense mesh can result in singularities or very high stress peaks at points of concentrated loads or supports. It is not practical to design for such high moments as cracking and yielding would occur in these areas, and moments would be distributed to adjacent areas of the slab. As such, common practice is to smooth out local

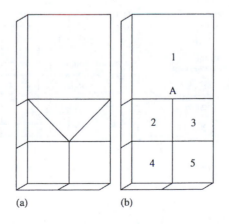

Figure 5.37 Connections between elements: (a) good; (b) poor

peaks by averaging the total moment over a portion of the slab width. For example, when designing flat slabs, the total moment in the slab is determined and distributed over column and middle strip, avoiding the need to design for high peak moments at the supports.

Example 5.7 Finite element method

Problem: For the slab of Fig. 5.38(a), use a linear elastic analysis to find the distribution of internal bending moment due to an applied uniformly distributed vertical loading of 15 kN/m².

Solution: It is often unnecessary, for the purposes of everyday design, to spend too much time obtaining a highly accurate linear elastic analysis of a minor building slab. This is particularly true considering the implications of the assumptions such as the degree of rotational fixity at the supports. Nevertheless, available software now makes the generation of a finite element mesh very easy and the solution time is negligible on typical computers.

For the mesh of elements illustrated in Fig. 5.38(b), the 1 m element width in the Y-direction has been chosen partly for convenience and partly to give a minimum of two elements across the panel ACHG. Similarly, in the X-direction, two elements are provided across this panel and the spacing is chosen so that the corner points, H and I, coincide with the nodes between elements. In a more accurate analysis, additional elements could be provided in the region of points H and I (because the moment in this region is rapidly changing) but this seems inappropriate for everyday design.

The resistance of the supporting columns and walls to rotation has been ignored and only simple (translational) supports have been provided at these points. This means that provision for the hog moment that will arise in this region will need to be made using engineering judgement. (The rotational stiffness of the columns can, in fact, be modelled by representing each column as a simple support plus springs that resist rotation. If the column is fixed at its other end, the spring stiffnesses are, from Appendix A, No. 1, $4EI_x/h$ and $4EI_y/h$ for the two coordinate directions.) Analysis of the finite element model gives the distributions of moment per unit breadth illustrated in Fig. 5.39. Reinforcement parallel to the X-axis should be provided to resist m_x and reinforcement parallel to the Y-axis should be provided to resist m_y. Some

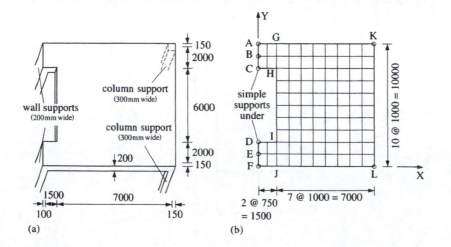

Figure 5.38 Slab of Example 5.7: (a) actual slab; (b) finite element mesh

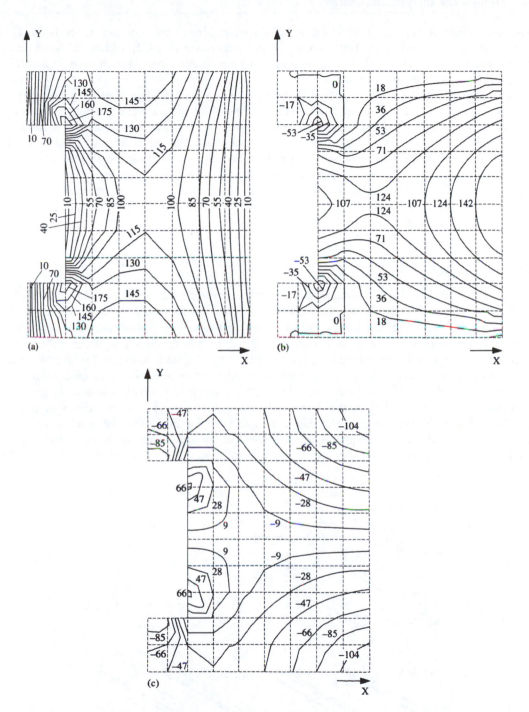

Figure 5.39 Results of finite element analysis: (a) contours for moment per unit breadth on X-face, m_x (kNm/m); (b) contours for moment per unit breadth on Y-face, m_y (kNm/m); (c) contours for twisting moment per unit breadth, m_{xy}(kNm/m)

additional reinforcement in both directions is required near the corners to resist m_{xy}. The calculation of this is given in Appendix E. (*Note*: According to EC2, additional local analyses may be necessary in areas where the assumption of a linear stress distribution is invalid, for example, in the vicinity of supports).

The grillage method is used in design offices for the analysis of larger and more complex slabs (such as bridge decks). The finite element method, being more accurate and usually easier to apply, has become an increasingly popular alternative in recent years. A distinct disadvantage of these linear elastic methods of analysis is that they do not account for the elastic-plastic behaviour of the members at the ultimate limit state (see Section 4.4). For slabs (more than beams) there is usually a great deal of ductility and hence a considerable capacity to resist load after initial yielding. This disadvantage of linear elastic analysis can be overcome to a reasonable extent by carrying out plastic moment redistribution (Section 5.3) of the results. Alternatively, a full plastic analysis is possible using the yield line method.

Yield line method

The yield line method is the two-dimensional equivalent of the plastic method of analysis described in Section 4.4. As stated above, slabs generally possess great ductility and hence a capacity to redistribute the moments obtained from a linear elastic analysis. It is important to understand that yielding of the reinforcement at any particular point in a slab does not necessarily imply an impending collapse. When a plastic hinge occurs in a region of high moment, any further increase in moment is redistributed from the yielded section to adjacent sections. For example, yielding of the member of Fig. 5.40(a) first occurs over the supports where the elastic moment is greatest, as illustrated in Fig. 5.40(c). Any further increase in moment at the supports is redistributed into the span. Note that, unlike in beams, plastic hinges in a slab develop along lines of equal moment which are known as yield lines. Collapse of the member occurs only when sufficient yield lines have developed to transform the member into a mechanism. For the member of Fig. 5.40(a), collapse

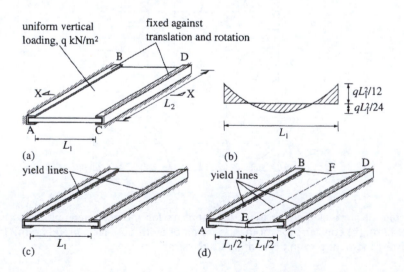

Figure 5.40 Yield line analysis of slab: (a) geometry; (b) linear elastic bending moment diagram at section X–X on one metre strip (typical of all sections); (c) first yield lines; (d) collapse mechanism

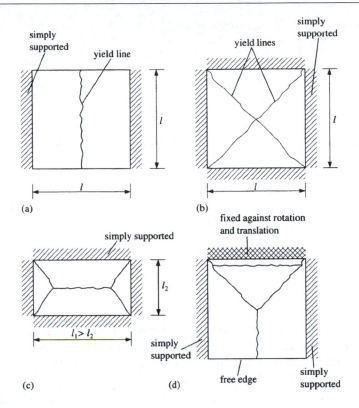

Figure 5.41 Typical yield line patterns in slabs

occurs only when the moment is increased sufficiently to cause a yield line to form at mid-span, as illustrated in Fig. 5.40(d).

In the application of the yield line method, the regions of slab between the yield lines are assumed to remain elastic so that all plastic hinge rotation of the slab occurs along the yield lines. Since the regions of slab between the yield lines remain plane, the yield lines must always be straight. Fig. 5.41 illustrates the yield line pattern in a number of slabs subject to uniform load acting perpendicular to the plane of the slabs.

The yield line method of analysis for slabs can be used to determine the magnitude of the ultimate loads which cause full plastic (mechanism) failure of the member. The following two examples illustrate the application of the method.

Example 5.8 Yield line method applied to one-way spanning slab

Problem: For the slab of Fig. 5.40(a) with known reinforcement, the moment capacities per unit breadth in sag and hog, m_x and m'_x respectively, have been calculated as 50 kNm per metre breadth. Find the intensity of the uniformly distributed load, q, that causes collapse of the slab if $L_1 = 8$ m and $L_2 = 6$ m. Assume an infinite capacity for plastic hinge rotation.

Solution: For the slab of Fig. 5.40(a) to collapse, the applied moment must be sufficiently large to develop yield lines at three locations in the member, as illustrated in Fig. 5.40(d). The kinematic approach is based on the conditions in the slab *during* collapse. During collapse:

$$\left.\begin{array}{l}\text{External rate of work}\\\text{done by the applied}\\\text{loads}\end{array}\right\}=\left\{\begin{array}{l}\text{Internal rate of work}\\\text{dissipated along the}\\\text{yield lines}\end{array}\right.$$

that is:

$$\dot{W}_e = \dot{W}_i \tag{5.2}$$

Consider a section through the slab during collapse as illustrated in Fig. 5.42. Let the rate of deflection at the centre during collapse be $\dot{\delta}$(metres per unit time). Therefore, the rate of rotation at the supports is equal to $2\dot{\delta}/L_1$. Now, the rate of work done during collapse is the product of the external forces and their average rate of deflection.

Note: Lines AB and CD of Fig. 5.40(d) do not move at all while line EF moves through $\dot{\delta}$. Hence, the average deflection of panels ABEF and CDEF is $\dot{\delta}/2$.

For this example, the external load is qL_1L_2. Therefore, the external rate of work done by the applied loads is:

$$\dot{W}_e = qL_1L_2\frac{\dot{\delta}}{2}$$

The internal rate of work dissipated along the yield lines is the product of their ultimate moment capacities and their rates of rotation. The capacity to resist hog moment per metre width is m'_x. Therefore, the total ultimate moment capacity for hog, M'_x, is given by:

$$M'_x = m'_x L_2$$

The rate of rotation along the yield lines at the edges is $2\dot{\delta}/L_1$. Hence, the internal rate of work dissipated along each fixed support is:

$$m'_x L_2 \left(2\dot{\delta}/L_1\right) = 2m'_x L_2\dot{\delta}/L_1$$

Similarly, along the mid-span yield line the ultimate moment capacity for sag is M_x and the rate of rotation is, from Fig. 5.42, $4\dot{\delta}/L_1$. Hence, the internal rate of work dissipated at mid-span is:

$$m_x L_2 \left(4\dot{\delta}/L_1\right) = 4m_x L_2\dot{\delta}/L_1$$

The total internal rate of work done, $\dot{W}_i$, is the sum of the rates of work done along each yield line, that is:

$$\dot{W}_i = 2(2m'_x L_2\dot{\delta}/L_1) + 4m_x L_2\dot{\delta}/L_1$$

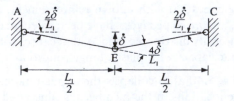

Figure 5.42 Section X–X through slab of Fig. 5.40(a) during collapse

By Equation (5.2), we then have:

$$\dot{W}_e = \dot{W}_i$$

$$\Rightarrow qL_1L_2\frac{\dot{\delta}}{2} = 2(2m'_xL_2\dot{\delta}/L_1) + 4m_xL_2\dot{\delta}/L_1$$

$$\Rightarrow q = \frac{8(m_x + m'_x)}{(L_1)^2} = \frac{8(50+50)}{(8)^2}$$

$$\Rightarrow q = 12.5 \, \text{kN/m}^2$$

Example 5.9 Yield line method applied to two-way spanning slab

Problem: The slab of Fig. 5.43 is isotropic, that is, it has the same moment capacity per unit breadth in all directions, *m*. Given that *m* is 60 kNm per unit width, and that the slab is simply supported, find the intensity of the uniformly distributed load, *q*, that causes collapse of the slab. Assume infinite capacity for plastic hinge rotation and take $L = 7$ m.

Solution: The only way in which the slab of Fig. 5.43(a) can fail is with a yield line along the middle as illustrated in Fig. 5.43(b) (symmetry requires D to be mid-way between A and C). Since the yield line is not parallel to either of the coordinate axes the solution of this example proves to be more complex than the previous example.

Panel ABD must rotate about the Y-axis as it is supported at A and B. Let the rate of deflection at D equal $\dot{\delta}$. Then, on a section through D, say S–S (Fig. 5.44(a–b)), the rate of rotation of ABD about the Y-axis is $\dot{\theta}_y = \dot{\delta}/(L/2) = 2\dot{\delta}/L$. Similarly, panel BCD will rotate about the X-axis and again, point D deflects at a rate of, $\dot{\delta}$. From Fig. 5.44(c), the rate of rotation of this panel about the X-axis is $\dot{\theta}_x = \dot{\delta}/(L/2) = 2\dot{\delta}/L$. Hence the rates of rotation about the two axes are equal:

$$\dot{\theta}_y = \dot{\theta}_x = 2\dot{\delta}/L$$

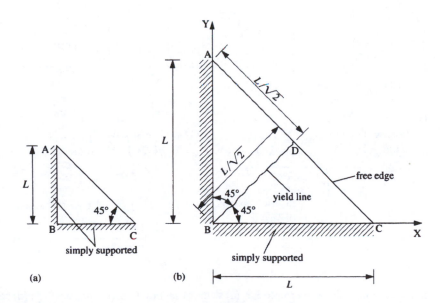

Figure 5.43 Slab of Example 5.9: (a) geometry; (b) yield pattern

Now, the average rate of deflection of panel ABD is the rate of deflection of the centroid of the triangle, which occurs at one-third of the distance from AB to D. Hence the average rate of deflection is $\dot{\delta}/3$.

The average rate of deflection of panel BCD is the same. Hence, the total external rate of work done is:

$$\dot{W}_e = 2\left(\frac{1}{2}qL\frac{L}{2}\frac{\dot{\delta}}{3}\right) = \frac{qL^2\dot{\delta}}{6}$$

The internal rate of work done, $\dot{W}_i$, is found by determining the rate of rotation of the yield line. To calculate this rate of rotation, it is necessary to take a cross-section perpendicular to the yield line, such as U–U (Fig. 5.44(a) and (d)). At section U–U, the rate of rotation at D is found to be $2\sqrt{2}\dot{\delta}/L$ from Fig. 5.44(d). Section V–V is half-way between B and D (Fig. 5.44(a) and (e)). Although the length and the rate of deflection at the yield line are halved, the rate of rotation at the yield line remains the same as that at U–U. Hence, the rate of rotation is constant throughout the length of the yield line and the internal rate of work done is:

$$\dot{W}_i = m\left(\frac{L}{\sqrt{2}}\right)\left(\frac{2\sqrt{2}\dot{\delta}}{L}\right) = 2m\dot{\delta}$$

By Equation (5.2), we have:

$$\dot{W}_e = \dot{W}_i \Rightarrow \frac{qL^2\dot{\delta}}{6} = 2m\dot{\delta}$$

$$\Rightarrow q = \frac{12m}{L^2} = \frac{12(60)}{49} = 14.7\,\text{kN/m}^2$$

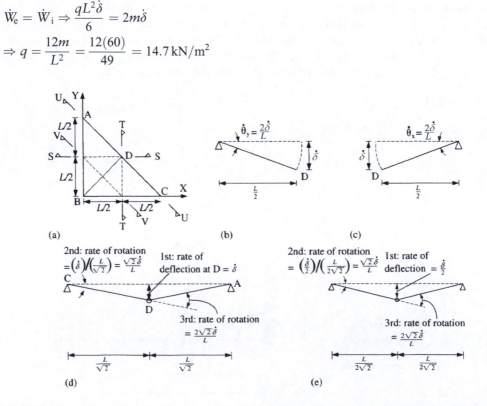

Figure 5.44 Sections through slab of Example 5.9: (a) slab; (b) section S–S; (c) section T–T; (d) section U–U; (e) section V–V

A disadvantage of the yield line method is that it gives no information on the response of slabs under service loads (for instance, it gives no information on elastic deflections). In addition, the method provides no check on the extent of the plastic hinge rotations required to achieve the critical yield pattern. EC2 allows full use of the yield line method provided certain ductility requirements are satisfied. A concrete section fails when the concrete crushes, i.e. when the strain in the concrete reaches a critical level. The higher the strain in the steel when this occurs, the more ductile the section is. Hence, for example, Case A in Fig. 5.45 is more ductile than Case B. According to EC2, when plastic analysis for beams, frames or slabs is being carried out, sufficient ductility can be assumed if the following conditions are fulfilled:

(a) $x_u/d \leq 0.25$ for concrete grades $\leq$ C50/60
     $x_u/d \leq 0.15$ for concrete grades $\geq$ C55/67
     where x_u is the distance from the extreme compressive fibre in the section to the neutral axis, and d is the effective depth to the tension reinforcement.
(b) reinforcing steel is Class B or C.
(c) the ratio of the moments at internal supports to the moments in the span should be between 0.5 and 2.

Lower-bound approaches and Hillerborg strip method

Consider the rectangular, simply supported slab of Fig. 5.46(a) which is subjected to a uniform load of intensity q kN/m^2. Suppose we were to assume that all load on this slab spans in one direction across the shorter span, that is, in an E–W direction (the arrows in the figure indicate the assumed direction of span). Under this assumption, the maximum bending moment in this member is $ql_1^2/8$ per metre width (from Appendix B) as illustrated in Fig. 5.46(b). As pointed out in Section 5.4, if we provide reinforcement with a moment capacity of no less than $ql_1^2/8$ in the E–W direction and there is infinite ductility, then the member will not collapse.

Such lower-bound methods are widely used in practice and are a safe method of analysis for slabs under ultimate loads provided the slab is sufficiently ductile. However, the results of a lower-bound method do not guarantee a satisfactory design at SLS as cracking may occur if the assumed load path is not reasonable. Furthermore, the slab may not necessarily be safe if it has insufficient ductility to redistribute the moments from the linear elastic values to the final values assumed. Slabs will

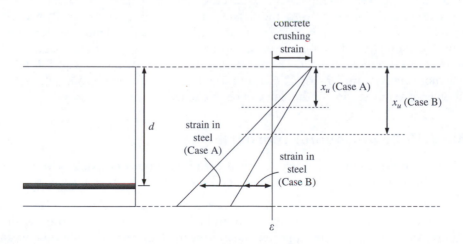

Figure 5.45 Ductility of steel in reinforced concrete beam, where Case A is more ductile than Case B

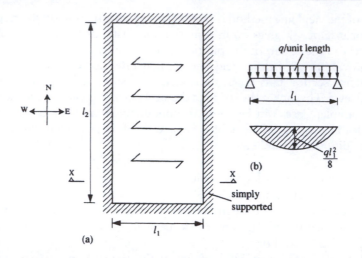

Figure 5.46 Rectangular slab: (a) plan view; (b) section X–X and bending moment diagram

generally have sufficient ductility to do this provided the moments assumed do not greatly differ from those of a linear elastic analysis.

One particular lower-bound approach that can be very useful for the analysis of slabs is the Hillerborg strip method. Its main advantage is that it is easy to implement and can be done without the aid of a computer. Like the example described above, the Hillerborg strip method is based on the assumption that load tends to be carried to the nearest support. However, the strip method extends this approach by assuming that load is always carried to the nearest support.

To illustrate the Hillerborg strip method, consider the slab of Fig. 5.46(a). The slab is first divided into regions, shown in Fig. 5.47(a), which indicate to which supports the load in that portion of the slab is transferred. The lines defining the regions are plotted at angles of 45° from each corner. The slab is then divided into a number of strips spanning E–W and N–S as illustrated in Fig. 5.47(b). Then, load on segment S (Fig. 5.47(b)), say, is deemed to be spanning E–W by strip B–B as it is nearer to the east support. Load on segment T, on the other hand, is nearer to the south support and is deemed to be spanning N–S by strip C–C. The bending moment diagrams for the various strips can readily be determined. For strip A–A, the bending moment diagram (from Appendix B) is as illustrated in Fig. 5.47(c). For strip B–B, the vertical reactions at each end must equal half the applied load of $2qy$. The moment is then found from a consideration of the free-body diagram taken from the left support to a distance y from it. The resulting diagram is illustrated in Fig. 5.47(d). Similarly, the bending moment diagram for strip C–C is found and is illustrated in Fig. 5.47(e). In this way, the bending moments at any section of the slab can be calculated.

Example 5.10 Lower-bound approach

Problem: Using a lower-bound approach, specify the required moment capacities for the slab shown in Fig. 5.48 if the uniformly distributed ultimate load is 20 kN/m².

Solution: At the western side of the slab, the regions that are supported by each edge are illustrated using arrows in Fig. 5.49. At the eastern end, two strong bands are provided to transfer the reaction from the slab to the column and the supported edges. In general, the moment differs between strips of slab and hence the reinforcement required to resist them also differs. For example, in the slab of

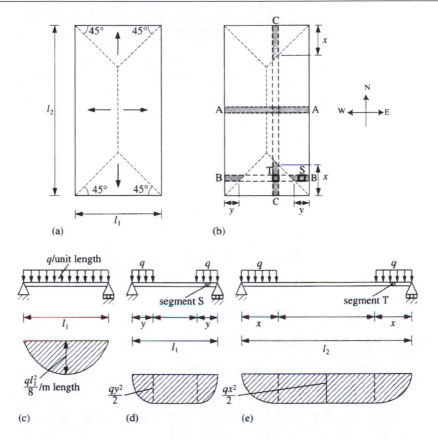

Figure 5.47 (a) Plan geometry showing regions; (b) strips through slab; (c) strip A–A and bending moment diagram; (d) strip B–B and bending moment diagram; (e) strip C–C and bending moment diagram

Fig. 5.47, the maximum moments in strips A–A and B–B were different. For convenience, therefore, it is common to divide the slab into rectangular regions in which the strips are provided with the same quantity of reinforcement. The western end of the slab of Fig. 5.48 has been divided into the regions illustrated in Fig. 5.50, for which there are two bands which have different levels of reinforcement in each direction. The selection of the dimensions of these bands is somewhat arbitrary but should, where possible, be reasonably similar to Hillerborg's suggested 45° line (shown in Fig. 5.49).

Consider the strips of slab spanning E–W. From Fig. 5.50, two strips need to be considered, namely the strips at sections A–A and B–B. Analysis of section A–A is trivial as no load is assumed to be carried by it in the E–W direction (as is spans N–S) and so no reinforcement is required there (however, nominal reinforcement is always provided throughout all slabs). The eastern end of the slab is unsupported so the right-hand side of strip B–B is supported by strong band No. 2. Since the strong bands are assumed to be 1 m wide, this provides a reaction over a width of 1 m to the right-hand side of strip B–B, r_{B2}. The loads carried by a 1 m wide strip of slab through section B–B are illustrated in Fig. 5.51 along with the resulting bending moment diagram. Reinforcement with a moment capacity equal to the maximum moment in this diagram, namely 40.8 kNm/m, must be

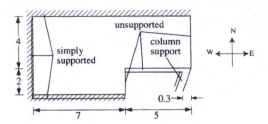

Figure 5.48 Slab of Example 5.10

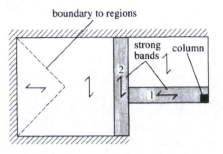

Figure 5.49 Slab showing regions and strong bands

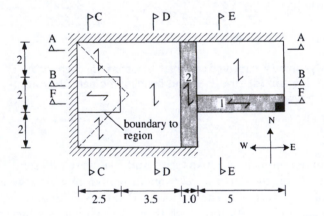

Figure 5.50 Slab showing revised regions

provided running E–W, as illustrated in Fig. 5.52. By equilibrium, the reaction r_{B2} is 9.6 kN/m. This will later be applied as a load to band 2.

The two strips of slab running N–S that need to be analysed are those at sections C–C and D–D of Fig. 5.50. The loads and resulting bending moment diagrams for 1 m wide strips through each section are illustrated in Fig. 5.53. The corresponding requirements for N–S reinforcement are illustrated in the diagram of Fig. 5.52.

In addition to the reinforcement requirements calculated above, reinforcement is required for strips at section E–E and in the strong bands, 1 and 2. First, consider a 1 m wide strip at section E–E (Fig. 5.50) which is illustrated in Fig. 5.54(a). The strong band (No. 1) running E–W effectively acts as a 1 m wide support to this strip.

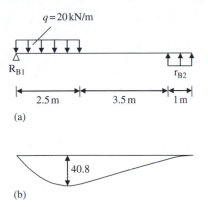

(a)

(b)

Figure 5.51 Section B–B; (a) loading; (b) bending moment diagram

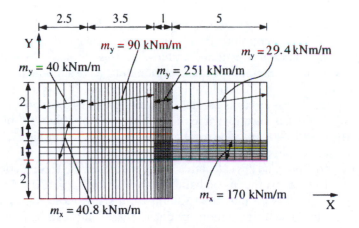

Figure 5.52 Reinforcement requirements as calculated using lower-bound approach

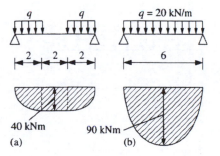

(a)   (b)

Figure 5.53 Loading and bending moment diagrams: (a) section C–C; (b) section D–D

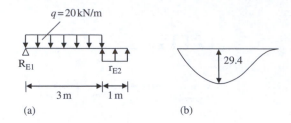

(a)   (b)

Figure 5.54 Section E–E: (a) geometry and loading (b) bending moment diagram

From Fig. 5.54(a), by taking moments about the left-hand support, the reaction, r_{E2}, on strong band 1 from this 1 m strip is given by:

$$r_{E2} \times 1 \times 3.5 = 20 \times 3 \times 1.5$$
$$\Rightarrow r_{E2} = \frac{20 \times 3 \times 1.5}{1 \times 3.5} = 25.7\,\text{kN/m}$$

Since strong band 1 is 1 m wide, this reaction will later be applied to band 1 as a UDL downward of 25.7 kN/m. From Fig. 5.54(a), the point of zero shear in the beam is 1.7 m from the left-hand support and the corresponding maximum bending moment at this point is 29.4 kNm (Fig. 5.54(b)). Therefore, the reinforcement running N–S in this region must have a moment capacity of 29.4 kNm.

As the E–W span of the slab is considerably longer than the N–S span, it has been decided to assume that band 1 only spans as far as band 2 (i.e. band 2 acts as a support for band 1). The total load on the 1m wide band 1 is the uniform loading on the slab of 20 kN/m² plus the reaction from strips along E–E (r_{E2} = 25.7 kN/m). The total loading is illustrated in Fig. 5.55, and the resulting maximum bending moment in the strip is 170 kNm.

The reaction of band 1 on band 2 is calculated as r_{B3} = 104 kN/m. This reaction is applied to band 2 over a 1 m length. The total load acting on band 2 is the slab loading of 20 kN/m², the reaction from strong band 1 (r_{B3} = 104 kN/m) plus the reaction from strips along B–B (r_{B2} = 9.6 kN/m), as illustrated in Fig. 5.56(a). The bending moment due to these loads can be found by superimposing the bending moments due to each of the loads. The bending moment diagram is illustrated in Fig. 5.56(b). From this, it can be seen that the reinforcement in band 2 must have a moment capacity not less than 251 kNm.

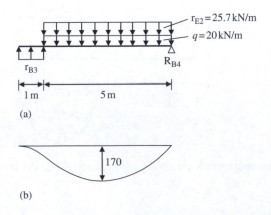

(a)

(b)

Figure 5.55 Band 1: (a) geometry and loading (b) bending moment diagram

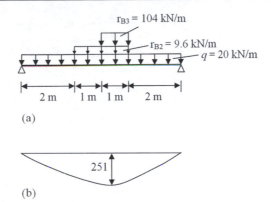

(a)

(b)

Figure 5.56 Band 2: (a) geometry and loading; (b) bending moment diagram

All of the reinforcement requirements are illustrated in the diagram of Fig. 5.52. The linear elastic bending moment diagram for this slab has been found using finite element analysis and, for comparison, the reinforcement requirements for the corresponding parts of the slab have been calculated and are illustrated in Fig. 5.57. Importantly, this example illustrates that a lower-bound approach can give considerably different results to linear elastic analysis. For this example, the lower-bound approach results in larger moments and reinforcing requirements in the strong bands than the finite element analysis. In the finite element analysis, simplifications associated with the lower-bound approach are not assumed (e.g. 1 m wide strong bands) and bending moments tend to be distributed over larger areas, resulting in different reinforcing requirements. This illustrates that these two methods assume different load paths, with the finite element analysis giving more accurate results. Using the lower-bound approach, a lot of cracking would occur before the load was redistributed to the strong bands.

Analysis of flat slabs

In flat slab construction, the slab is supported directly on columns, as illustrated in Fig. 5.58, rather than on beams. The strip of slab, ABC, in the figure could reasonably be modelled as a continuous

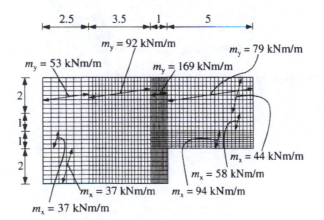

Figure 5.57 Reinforcement requirements as calculated using finite element analysis (without plastic moment redistribution)

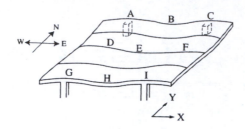

Figure 5.58 Flat slab construction

beam supported by the columns at A and C. The strip, DEF, could also be modelled as a continuous beam as it is supported at D and F by strips of slab running N–S, that is, D provides some support to DEF by virtue of its proximity to the columns at A and G. However, the support at D and F is less effective than that provided directly by a column – D and F are in effect spring supports capable of significant deflection when load is applied to them.

The analysis of a continuous beam on spring supports gives a significantly different bending moment diagram than a continuous beam on rigid supports. In general, when supports are 'springy', hog moments reduce and sag moments become correspondingly larger. The bending moment diagram for the moment on the X-face in the flat slab of Fig. 5.58 has been found by linear elastic (finite element) analysis and is illustrated in Fig. 5.59.

Flat slabs can be analysed by linear elastic, yield line, grillage or equivalent frame methods. Of these, linear elastic methods coupled with plastic moment redistribution are most effective but require the use of a computer. The yield line method is less satisfactory as a greater deal of plastic moment redistribution is required to achieve the yield pattern (Fig. 5.60). In fact, the yield pattern for an interior panel is identical to that for a slab supported on all sides by beams, even though the elastic

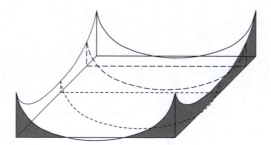

Figure 5.59 Bending moment diagram for m_x for flat slab of Fig. 5.58

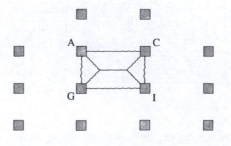

Figure 5.60 Yield pattern for interior panel in flat slab construction

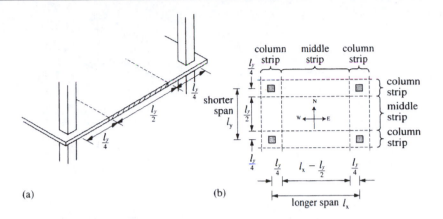

Figure 5.61 Column and middle strips in flat slab using equivalent frame method: (a) N–S section; (b) plan

bending moment diagrams for the two types of slab are significantly different. Thus, in many countries, the equivalent frame (strip) method has traditionally been favoured over the yield line method. In EC2, guidelines on the analysis of flat slabs using the equivalent frame method are given, although it does allow the use of all the above-mentioned methods.

For the analysis of flat slabs using the equivalent frame method, EC2 recommends that the slab be divided into sections which are separated by the centre lines of the columns. These sections are then analysed as equivalent frames in each direction to find the distribution of bending moments. According to EC2, for vertical loading, the stiffness of the slab is calculated using the full width of the panel, whereas for horizontal loading, the stiffness of the slab is assumed to be 40 per cent of the total stiffness of the panel. For each direction, the total load on one panel of the slab should be used for analysis. The bending moments obtained from analysis are then distributed across the width of the slab to allow for deviations in the bending moment as illustrated in Fig. 5.59. EC2 does this by using the concept of column strips and middle strips of slab (see Fig. 5.61). These are strips within the slab which are reinforced to resist different portions of the total moment.

The column strip, although not as wide, is resting directly on the columns and, as a result, it takes more of the total moment than the middle strip. However, the effect is less pronounced for the sag moment as the flexible support to the middle strip causes the hog to reduce and the sag to increase. EC2 recommends that the column strip be assigned 60–80 per cent of the total hog moment and 50–70 per cent of the sag moment. The middle strip is designed to resist the remaining 40–20 per cent and 50–30 per cent of the hog and sag moments, respectively. When distributing the moments between the column and middle strips, it should be noted that the total hogging moment should sum to 100 per cent and the total sagging moment should sum to 100 per cent. This process is carried out for both m_x and m_y. The reinforcement running E–W can be viewed as transferring the load towards the supports in this direction while the reinforcement running N–S transfers the load towards the supports in the N–S direction. Both are necessary in order to transfer the load to the column supports. However, according to EC2, the moments transferred to edge or corner columns should be limited, unless perimeter beams that are adequately designed for torsion are included in design.

5.7   Analysis of shear wall systems

Shear wall systems are commonly incorporated into a multi-storey building to provide lateral stability. Shear walls, which are made of concrete or masonry walls, act as vertical cantilevers in resisting horizontal forces (see Fig. 5.62(a)). Their great cross-sectional depth, d, gives them an

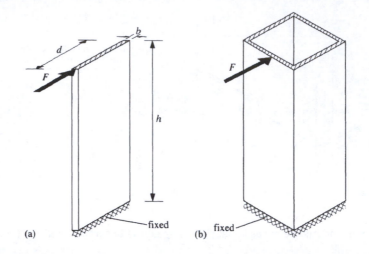

Figure 5.62 (a) shear wall; (b) core

extremely high second moment of area ($I = bd^3/12$) and hence a great stiffness against horizontal deflection. As discussed in Chapter 1, when the depth, d, exceeds 1.5 times the height, h, the mechanism by which shear walls resist load changes from flexure to membrane action. This, however, does not reduce the effectiveness of shear walls in any way.

Another effective way of resisting horizontal forces is the vertical cantilever of box section known as a core such as illustrated in Fig. 5.62(b). Cores are commonly incorporated into multi-storey buildings where they can serve the additional function of providing fire resistance around elevators and stairwells (holes for doorways, if sufficiently small, do not greatly reduce the effectiveness of the core in resisting horizontal loads). Cores are more efficient than shear walls at resisting horizontal loads because they have a much higher second moment of area with a significant portion of the cross-sectional area concentrated at the ends of the member. Non-symmetrical shear wall assemblies, such as that illustrated in Fig. 5.63(b), can have a twisting effect on the building and consequently are more complex to analyse than symmetrical shear wall assemblies. For this reason, shear walls and cores placed symmetrically within a building are generally favoured whenever possible.

The horizontal forces acting on a building are usually transferred by the floor slabs and screed to the shear wall members. For this reason, it is normal to assume that the forces acting on shear walls do so at floor levels. The following two examples illustrate a method for determining the proportion of force carried by each member in both symmetrical and non-symmetrical shear wall assemblies. However, in the non-symmetrical shear wall example, EC2 recommends that this method should only be used where the plan layout of the walls is reasonably symmetrical.

Example 5.11 Symmetrical shear wall assembly

Problem: Fig. 5.63(a) illustrates the plan of a typical multi-storey building of height, h. The structure is braced against wind load in the N–S direction by the symmetrical arrangement of the core and shear walls illustrated in the figure. Determine the portion of the wind load, F, carried by each shear wall and hence derive an expression for the moment at the base of the external walls due to horizontal load. Assume all walls have a thickness of 200 mm.

Solution: Each of the four shear walls and the core of Fig. 5.63(a) acts as a cantilever in transferring the wind load to the foundation. Although all the cantilevers are of the same height, they have differing stiffnesses since each has a different second moment of area. Thus, if the individual

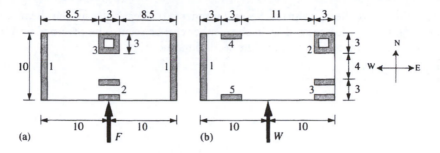

Figure 5.63 Shear wall/core systems for resisting horizontal load (plan views): (a) symmetrical shear wall system for N–S load; (b) non-symmetrical system

cantilevers were to be subjected to an applied horizontal force, P, each would deflect by a different amount. However, in most shear wall assemblies, the cantilevers are connected to the floor slabs at each level. Since the floor slabs are much stiffer against horizontal force than the shear walls, it can be assumed that each cantilever is forced to deflect by the same amount, as illustrated in Fig. 5.64.

Ignoring the resistance of the floor slab to rotation, the horizontal force, P, required to cause a deflection, δ, in a cantilever is, from Appendix A, No, 4:

$$P = \frac{3EI\delta}{h^3} \tag{5.3}$$

where h is the height of the cantilever and I is the second moment of area of the cantilever section about the relevant axis. Let F_1, be the force carried by each of the two outer shear walls, F_2 be the force carried by each of the two internal shear walls and F_3 be the force carried by the core. Then the total force required is:

$$F = 2F_1 + 2F_2 + F_3$$

Each shear wall deflects by the same amount, δ say, at the level of the floor slab. Thus, as Equation (5.3) holds for each, we have:

$$F_1 = \frac{3EI_{x1}\delta}{h^3} \quad F_2 = \frac{3EI_{x2}\delta}{h^3} \quad and \quad F_3 = \frac{3EI_{x3}\delta}{h^3}$$

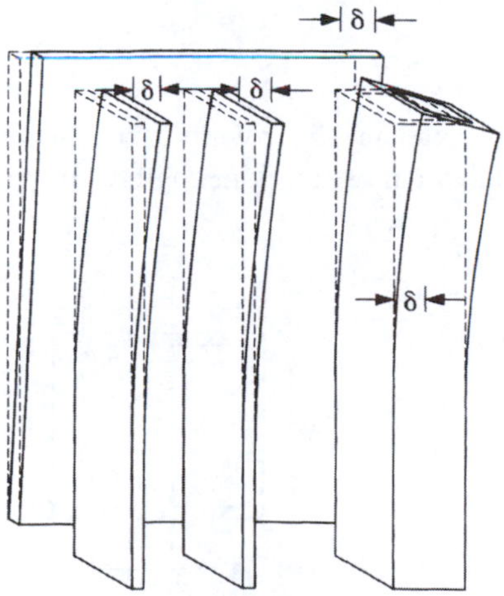

Figure 5.64 Horizontal deflection in part of a symmetrical shear wall/core system

where I_{x1}, I_{x2} and I_{x3} are the appropriate second moments of area. Since all the shear wall members are the same height, the force resisted by each member can be seen to be proportional to its second moment of area. Referring to Fig. 5.63(a), the second moments of area are:

$$I_{x1} = \frac{bd^3}{12} = \frac{0.2(10)^3}{12} = 16.67\,\text{m}^4$$

Similarly:

$$I_{x2} = \frac{3(0.2)^3}{12} = 0.002\,\text{m}^4$$

With reference to Fig. 5.65, the second moment of area for the core is:

$$I_{x3} = \left(\frac{3(3)^3}{12} - \frac{2.6(2.6)^3}{12} \right) = 2.94\,\text{m}^4$$

Hence:

$$\sum I_x = 2I_{x1} + 2I_{x2} + I_{x3}$$
$$= 2(16.67) + 2(0.002) + 2.94$$
$$= 36.28\,\text{m}^4$$

The portion of the total wind load, F, taken to member i, is then given by:

$$F_i = F\left(\frac{I_{xi}}{\sum I_x} \right) \tag{5.4}$$

Hence, the portion taken by each external shear wall is:

$$F_1 = \frac{F(16.67)}{36.28} = 0.46F$$

In a similar manner, the portion of the load carried by the internal shear walls and the core is:

$$F_2 = \frac{F(0.002)}{36.28} = 0.00006F$$
$$F_3 = \frac{F(2.94)}{36.28} = 0.081F$$

Thus, it can be seen that the two external shear walls carry the majority of the load (92 per cent) and the contribution of the internal shear walls is negligible. In practice, it is common to ignore the

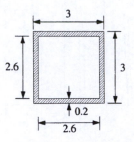

Figure 5.65 Cross-section through core

resistance of members, such as the internal shear walls of Fig. 5.63(a), which are bending about their weak axes. The moment at the base of each external wall is simply:

$$F_1h = 0.46Fh$$

Example 5.12 Non-symmetrical shear wall assembly

Problem: The structure illustrated in Fig. 5.63(b) is braced against wind load in the N–S direction by a non-symmetrical arrangement of shear walls. Determine the portion of the wind load, W to be carried by the west shear wall and the moment at the base of wall 4. Assume all walls have a thickness of 200 mm.

Solution: As mentioned previously, when shear walls/cores are non-symmetrical, we must allow for the resulting twist in the building as a whole. In this example, twisting results from the fact that the external west shear wall is much stiffer than the core (the stiffness of the other shear walls for N–S movement is negligible as can be seen from the previous example). This twisting is then resisted by the E–W stiffness of all the shear walls and the core.

The analysis procedure for a non-symmetrical shear wall system is similar to that for a group of bolts in a structural steel connection. The rotation is assumed to act about a stationary point known as the shear centre, S. If a force is applied in line with the shear centre, as illustrated in Fig. 5.66, no twisting will occur in the building. Thus, the first step is to determine the location of the shear centre. To do this we must calculate the stiffness of each core/wall in each direction (E–W and N–S). Since all members are the same height, the stiffness of each is proportional to its second moment of area. These are given in Fig. 5.67 where each shear wall/core is represented as springs parallel to the coordinate directions. The shear centre is then the centre of resistance of the springs.

The first step in finding the shear centre is to calculate the total stiffness in each direction, that is:

$$\sum I_x = 16.67 + 2.94 \ = 19.61\,\text{m}^4$$

and:

$$\sum I_y = 0.45 \times 2 + 2.94 + 0.9 \ = 4.74\,\text{m}^4$$

Now, to find the centre of resistance, S (see Fig. 5.68(a)), take 'moments' about any point, A, say.

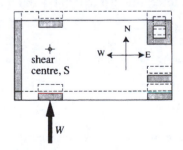

Figure 5.66 Deflection due to force at shear centre

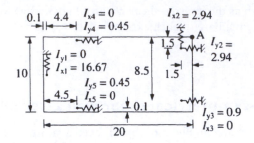

Figure 5.67 Springs representing wall/core stiffnesses

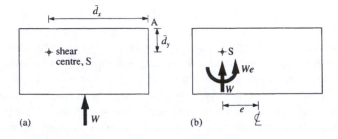

Figure 5.68 (a) applied load; (b) equivalent loading at shear centre

Hence, to find the x-coordinate of the shear centre:

$$19.9I_{x1} + 1.5I_{x2} = \bar{d}_x \sum I_x$$
$$\Rightarrow \bar{d}_x = 17.141 \text{ m}$$

and, to find the y-coordinate of the shear centre:

$$1.5I_{y2} + 8.5I_{y3} + 0.1I_{y4} + 9.9I_{y5} = \bar{d}_y \sum I_y$$
$$\Rightarrow \bar{d}_y = 3.494 \text{ m}$$

The central wind load, W, is equivalent to a force and a torque at the shear centre as illustrated in Fig. 5.68(b). The magnitude of the torque is given by:

$$T = We \tag{5.5}$$

where e is the eccentricity of the force as illustrated in the figure. In this example:

$$T = W(17.141 - 10) = 7.141W$$

This torsion generates a force on member i in the X-direction according to the expression:

$$F_{xi} = \frac{T(\bar{d}_y - d_{yi})I_{yi}}{\sum_j \left[I_{yj}(\bar{d}_y - d_{yj})^2 + I_{xj}(\bar{d}_x - d_{xj})^2\right]}$$

where $(\bar{d}_x - d_{xi})$ and $(\bar{d}_y - d_{yi})$ are the distances of member i from the shear centre in the E–W and N–S directions respectively (note that if member i is to the left of the shear centre $(\bar{d}_x - d_{xi})$ will be negative).

In the Y-direction, there is a corresponding torsional force but there is also a direct axial force which is proportional to the member stiffness. Thus, for member i:

$$F_{yi} = \frac{WI_{xi}}{\sum I_{xj}} + \frac{T(\bar{d}_x - d_{xi})I_{xi}}{\sum_j [I_{yj}(\bar{d}_y - d_{yj})^2 + I_{xj}(\bar{d}_x - d_{xj})^2]}$$

Thus, the total portion of the wind load, W, carried by the external west shear wall in the N–S direction is:

$$F_{y1} = \frac{16.67\,W}{19.61} - \frac{7.141\,W(19.9 - 17.141)(16.67)}{904.0}$$
$$= 0.487\,W$$

The portion of W carried by wall 4 in the E–W direction is:

$$F_{x4} = \frac{7.141\,W(0.45)(3.494 - 0.1)}{904.0}$$
$$= 0.0121\,W$$

and the moment at its base is:

$$M_4 = 0.0121\,Wh$$

Problems

Section 5.2

5.1   One span of a continuous beam is illustrated in Fig. 5.69(a). It is stated in Appendix D that the bending moment diagram for such a span can be constructed by superimposing the bending moment diagram associated with the support moments (Fig. 5.69(b)) on the bending moment diagram due to the applied loading on the simply supported span (Fig. 5.69(c)). Verify that this is indeed true for the beam illustrated in Fig. 5.69.

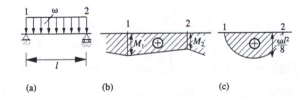

(a)   (b)   (c)

Figure 5.69 Beam of Problem 5.1: (a) geometry and loading; (b) bending moment diagram (BMD) associated with support moments (M_1 and M_2 are shown positive (sag) here); (c) BMD due to loading on simply supported beam

5.2   For the three-span beam illustrated in Fig. 5.70, find the factored ultimate bending moment envelope which gives the maximum hogging and sagging moments at all points. The unfactored permanent loading is 15 kN/m on span AB and 10 kN/m on spans BC and CD. The unfactored variable gravity loading is 20 kN/m on all spans.

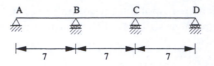

Figure 5.70 Beam of Problem 5.2

5.3 Determine the reaction at B for the beam whose geometry and loading are illustrated in Fig. 5.71.

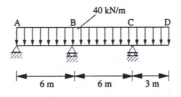

Figure 5.71 Beam of Problem 5.3

Section 5.3

5.4 The structure of Problem 4.7 is to be analysed by linear elastic analysis with up to 25 per cent plastic moment redistribution. Determine an appropriate design bending moment diagram and compare the results with those of Problem 4.7.

Section 5.4

5.5 In Example 5.4, the initial bending moment diagram is as illustrated in Fig. 5.10(b), but a change in the beam depth results in this being revised to the bending moment diagram illustrated in Fig. 5.10(c). Moment capacities have been provided based on Fig. 5.10(b), i.e. without reanalysis, and are illustrated in Fig. 5.72. Determine if this design is safe given that the frame is sufficiently ductile to allow 20 per cent plastic moment redistribution at all points.

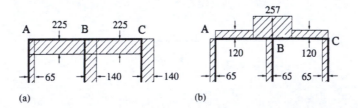

Figure 5.72 Problem 5.5: (a) capacities to resist positive moments; (b) capacities to resist negative moments

Section 5.5

5.6 A typical storey in a multi-storey two-dimensional braced frame is illustrated in Fig. 5.73. Determine the maximum contribution of the load on a typical storey to the axial force in the interior column and the corresponding moment. The total ULS loading varies from 15 kN/m to 40 kN/m.

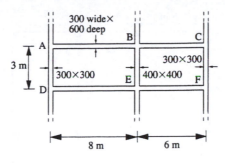

Figure 5.73 Frame of Problem 5.6

Section 5.6

5.7 For the slab illustrated in Fig. 5.74, the ratio of the larger to the smaller span dimension is 2:1 and hence EC2 allows the central part to be designed as if the slab were one-way spanning. Use the finite element method to analyse this slab for an applied vertical loading of 8 kN/m² given that the slab is 200 mm thick.
   (a) Present a contour plot of the moments per unit breadth in each direction. Compare the results to those which are obtained assuming one-way spanning behaviour.
   (b) Determine the twisting moment per unit breadth in the corners of the slab.
   (c) There are in fact 300 × 300 mm columns at the four corners of the slab which extend 3 m to the slabs above and below. Revise the computer model to allow for the rotational stiffness provided by these columns and determine the effects on m_x and m_y. The modulus of elasticity of the concrete in the columns can be assumed to be 32×10^6 kN/m².

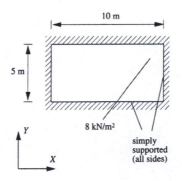

Figure 5.74 Plan view of slab of Problems 5.7 and 5.8

5.8 If the slab of Problem 5.7 is provided with moment capacities of 15 kNm/m in all directions, calculate the mechanism load factor using the yield line method.

5.9 For the flat slab illustrated in Fig. 5.75, determine the maximum sagging and hogging moments in the column and middle strips using the equivalent frame method. The

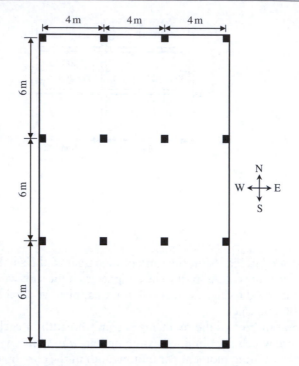

Figure 5.75 Plan view of slab of Problem 5.9

thickness of the slab is 300 mm and the columns are 300 mm × 300 mm. The characteristic floor loads are:

permanent gravity (incl. self-weight) = 9.5 kN/m²
imposed = 4.0 kN/m²

Chapter 6

Preliminary sizing of members

6.1 Introduction

The analysis and design of concrete structures is essentially an iterative process. The magnitude of permanent gravity loads, for instance, is dependent on the member sizes and hence an initial estimate of member sizes is required before the structure can be analysed. If, however, the final dimensions required to resist, say, bending moment in a member differ from the initial estimate of the member size, then the design process may be repeated with revised member dimensions until these initial assumptions are satisfied. The more accurate the initial size estimates, the less iteration that will be involved as the solution will converge more rapidly to the precise requirements.

The approximate methods and rules of thumb for the preliminary sizing of concrete members described in this chapter provide the means to make a reasonable first estimate of member sizes for the analysis and design of a viable structural scheme. The advantage of these methods is that they can be carried out on a slip of paper or a hand calculator. In addition, these methods can be used to estimate the material requirements of different schemes, thus allowing a quick assessment of the economy of each scheme.

The variables involved in the sizing of concrete members are the geometry (span and depth of beams and slabs, effective height of columns, breadth of ribs and beams), material grades (concrete and steel strengths) and areas of reinforcement. The factors that affect the choice of each variable include strength, durability and fire-resistance requirements. It is traditional to decide on the material grades at an early stage of the design and to employ the same grades for members of the same type. Thus, material grades are not normally a variable at the member design stage.

Within the overall design process, the preliminary sizing process itself can also be iterative. For example, the required depth for beams, say, is determined relatively simply but the minimum breadth can depend on the breadth required to fit the reinforcement. This causes problems as it is necessary first to know the breadth before the required reinforcement can be calculated. It is therefore usual to adopt trial geometric dimensions and to calculate the required area of reinforcement for these dimensions. If the area of reinforcement is unacceptably high for practical or performance reasons, the geometry is revised and the process is repeated.

It is important to note that the content in this chapter is to be used for the 'approximate' preliminary design of concrete structures and is not sufficient for a final design. There should always be a detailed design and analysis to check the adequacy of each structural member. For the rapid appraisal of alternative structural schemes, these approximate methods can be used in conjunction with the approximate values for permanent and variable gravity loads given in Sections 3.4 and 3.5.

6.2　Material grades

Concrete

Concrete can be designed to give a wide variety of strengths, depending on the relative proportions of its constituents. For this reason, it is usual to specify a particular grade of concrete when designing reinforced and prestressed concrete structures. Since compressive strength is one of the most important properties of concrete, the grade is measured by the characteristic compressive strength, typically at 28 days, that is the strength below which not more than 5 per cent of specimens are expected to fail when tested 28 days after casting. The characteristic strength is determined by means of standardized tests carried out on either cylindrical or cubed specimens. For a particular design mix, the characteristic cylinder strength, f_{ck}, is slightly less than the characteristic cube strength, $f_{ck,cube}$, (the ratio of f_{ck} / $f_{ck,cube}$, is approximately equal to 0.8 in most cases; see Table 6.1). EC2 design rules are based on cylinder strengths. Thus, for a design in accordance with EC2, a specimen having grade 30 concrete has a characteristic **cylinder** strength of 30 N/mm^2 (the corresponding cube strength is 37 N/mm^2). Since BS8110 rules were based on cube strengths, grades that give both cylinder and cube strengths (in that order) are commonly used to avoid ambiguity. Hence, EC2 refers to C30/37 for the above concrete grade. For most applications, concrete is only specified in multiples of five from a lower grade of about C25/30 to an upper grade of about C50/60. For preliminary design, a popular choice would be concrete of characteristic cylinder strength of 30–35 N/mm^2 (i.e. C30/37 or C35/45).

Concrete may be produced using three classes of cement, S, N and R. The different classes of cement produce concrete with different rates of early strength gain, and are classified as:

- Class S – slow early strength gain (CEM 32.5 N cement)
- Class N – normal early strength gain (CEM 32.5 R and CEM 42.5 N cement)
- Class R – rapid early strength gain (CEM 42.5 R, CEM 52.5 N and CEM 52.5 R cement)

Reinforcement

The grade of steel reinforcement is measured by its characteristic yield strength, f_{yk}. EC2 states that the characteristic yield strength of steel should be within the range 400–600 N/mm^2. For steel reinforcement in the UK a value of 500 N/mm^2 should be used. Reinforcing steel is also available in three classes of ductility, A, B and C, where class A has the lowest ductility and class C has the highest. Reinforcement can commonly be obtained only in the following standard diameters: 8, 10, 12, 16, 20, 25, 32 and 40 mm.

Table 6.1 Characteristic strengths of cylinders and cubes (from BS EN 1992-1-1)

Specimen	Characteristic strength (N/mm^2)										
Cylinder, f_{ck}	12	16	20	25	28*	30	32*	35	40	45	50
Cube, $f_{ck,cube}$	15	20	25	30	35*	37	40*	45	50	55	60

* from Concise Eurocode 2 (derived values)

6.3   Preliminary sizing of beams

Fire resistance and durability

The size of structural beams is influenced by the cover needed to ensure adequate resistance to fire and corrosion. The concrete cover is the **distance between the outermost surface of reinforcement (usually the links) and the nearest concrete surface,** as shown in Fig. 6.1. Minimum values of cover required for corrosion resistance are governed by the environmental conditions to which the beams are exposed. EC2 defines the six exposure classes of Table 6.2 corresponding to different environmental conditions. The minimum cover and strength requirements for all reinforced concrete members corresponding to exposure classes 1, 2, 3 and 4 are given in Table 6.3. The values in the table are taken from the United Kingdom National Annex (UK NA) to EC2 and correspond to members designed using Portland cement. Further recommendations are provided in the UK NA to EC2 for additional exposure classes, cement combination types, minimum cement contents and water cement ratios. To allow for construction tolerance, up to 10 mm must be added to the minimum cover values specified in Table 6.3, depending on quality assurance measures in place. This gives the **nominal cover** to reinforcement that is used for design and is specified on working drawings.

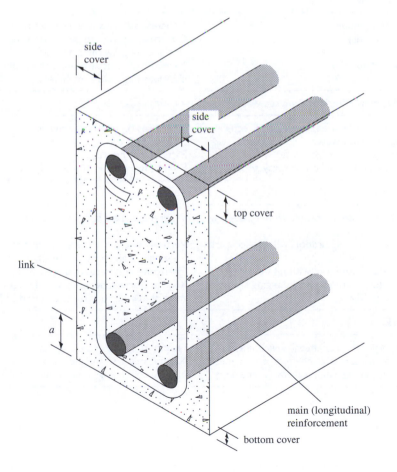

Figure 6.1   Concrete cover

Table 6.2 Exposure classes related to environmental conditions (from BS EN 1992-1-1)

Exposure class	Description of the environmental conditions	Examples of environmental conditions
1. No risk of corrosion or attack		
X0	Without reinforcement: all exposures except freeze/thaw, abrasion or chemical attack With reinforcement: where environment is very dry	Interior of building with very low air humidity
2. Corrosion induced by carbonation		
XC1	Dry or permanently wet	Interior of building with low air humidity or concrete submerged in water permanently
XC2	Wet, rarely dry	Concrete surface with long term water contact (e.g. foundations)
XC3	Moderate humidity	Interior of building with moderate or high air humidity or exterior sheltered from rain
XC4	Cyclic wet and dry	Concrete surface exposed to water but not in XC2
3. Corrosion induced by chlorides		
XD1	Moderate humidity	Concrete surfaces exposed to airborne chlorides
XD2	Wet, rarely dry	Concrete exposed to industrial water containing chlorides (e.g. swimming pools)
XD3	Cyclic wet and dry	Members of structures exposed to spray containing chlorides (e.g. bridge members)
4. Corrosion induced by chlorides from sea water		
XS1	Exposed to airborne salt but not directly in contact with sea water	Structures on or near to the coast
XS2	Permanently submerged in sea water	Parts of marine structures
XS3	Concrete in tidal, splash or spray zones	Parts of marine structures
5. Freeze/thaw attack		
XF1	Moderate water saturation, without de-icing agents	Vertical concrete surfaces exposed to rain and freezing
XF2	Moderate water saturation, with de-icing agents	Vertical concrete surfaces of road structures exposed to freezing and airborne de-icing agents
XF3	High water saturation, without de-icing agents	Horizontal concrete surfaces exposed to rain and freezing
XF4	High water saturation with de-icing agents or sea water	Road and bridge decks exposed to de-icing agents.Splash zone of marine structures exposed to freezing.
6. Chemical attack		
XA1	Slightly aggressive chemical environment	Natural soils and ground water
XA2	Moderately aggressive chemical environment	Natural soils and ground water
XA3	Highly aggressive chemical environment	Natural soils and ground water

Table 6.3 Minimum cover requirements (in mm) for durability (from UK NA to BS EN 1992-1-1)

Exposure class	Minimum strength class							
	C20/25	C25/30	C28/35	C32/40	C35/45	C40/50	C45/55	C50/60
X0	Not recommended to be applied to reinforced concrete							
XC1	15	15	15	15	15	15	15	15
XC2	–	25	25	25	25	25	25	25
XC3	–	35	30	25	25	20	20	20
XC4	–	35	30	25	25	20	20	20
XD1	–	–	35	30	30	25	25	25
XD2	–	–	40	35	35	30	30	30
XD3	–	–	–	–	50	45	40	40
XS1	–	–	–	–	40	35	35	30
XS2	–	–	40	35	35	30	30	30
XS3	–	–	–	–	–	50	45	45

* Up to 10 mm must be added to these values to give nominal cover which is used in design

For bond requirements, the minimum cover to main reinforcement must exceed the bar diameter for single bars or the equivalent diameter for bundled bars. This should be increased by 5 mm if the maximum aggregate size is greater than 32 mm. Similarly, up to 10 mm is added to the minimum cover required for bond to give the nominal cover that is used for design.

Steel reinforcement loses its strength when subjected to high temperatures. Concrete is a good insulator, with the result that it protects the steel and extends the time taken for the temperature in the steel to reach a dangerous level. Hence, greater cover means greater time for occupants to escape from a burning building before it collapses. In the part of EC2 relating to structural fire design (i.e. EN-1992-1-2), the axis distance, a (Fig. 6.1), is specified for different fire resistance times rather than the cover. The axis distance is measured from the centre of the main reinforcing bars to the surface of the concrete. Minimum beam and web breadths are also given for specified fire resistance times. The values of minimum beam breadth and **axis distance** sufficient to satisfy the requirements of fire resistance can be taken from Table 6.4. The values in the table are those recommended in the ISE manual for the **preliminary design** of a continuous beam (with a maximum of 15 per cent moment redistribution) in accordance with EC2. For prestressed concrete, two pairs of corresponding values are specified for the minimum breadth and axis distance, and either pair can be used for design. These values should be increased for simply supported beams or where a higher percentage of moment redistribution is required.

Effective depth of beams

The strength of beams in flexure is governed principally by the effective depth, that is, the depth from the extreme compression fibre of the beam to the centroid of the tension steel. A large effective depth results in a relatively small required quantity of reinforcement, as shown in Fig. 6.2(a). However, deeper beams at each floor level increase the permanent gravity load to the columns and result in a higher building overall which requires extra cladding.

Reduced structural depth on each floor can reduce heating costs and could allow an extra floor in a multi-storey building when planning controls restrict the overall height. Reducing the effective depth also has its price as it results in a need for more reinforcement to resist the bending moments due to

Table 6.4 Minimum preliminary design values of overall breadth and axis distance for continuous beams for fire requirements (from ISE manual)

Fire rating	Reinforced concrete		Prestressed concrete	
	Minimum overall breadth	Minimum axis distance	Minimum overall breadth	Minimum axis distance
(h)	(mm)	(mm)	(mm)	(mm)
1	150	25	120	40
			300	27
2	200	50	200	60
			500	45
3	240	60	–	–
4	280	75	280	90
			700	65

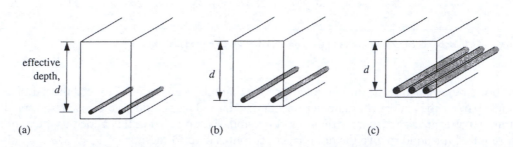

(a)   (b)   (c)

Figure 6.2 Effective depth of beams: (a) large effective depth – building higher than economical; (b) ideal effective depth – moderate quantity of reinforcement required; (c) small effective depth – excess quantity of reinforcement required

the applied loads. For extremely small effective depths, the required quantity of reinforcement often becomes impractical in the breadth available (Fig. 6.2(c)).

Engineers in different design offices have different rules of thumb by which they decide on the best effective depth for a given situation. These rules are generally represented in the form of span/depth ratios with modifications for influencing factors. The **maximum** values of span/effective depth ratios recommended in the ISE manual for deflection control are given in Table 6.5. The values in Table 6.5 are not definitive – it is for every designer to use his/her own discretion.

Table 6.5 Span/effective depth ratios for the preliminary design of reinforced and isolated prestressed concrete beams (from ISE manual). For reinforced concrete spans in excess of 7 m, multiply the value in the table by 7/(span in metres). For flanged sections with (flange width/rib width) > 3, the value in the table should be multiplied by 0.8

	Reinforced concrete	Prestressed concrete
Cantilever	6	8
Simply supported	14	18
Continuous		
Reinforced concrete interior span	20	
Reinforced concrete end span	18	
Prestressed concrete		22

Example 6.1 Effective depth of beams

Problem: Suggest preliminary depths for the beam spans illustrated in Fig. 6.3 if there are no constraints on structural depth and the fire rating is 1 hour. Assume a characteristic cylinder strength, f_{ck}, of 30 N/mm^2.

Solution: For ease of construction, the same depth will be used in all spans. The depth will be governed by conditions in span AB or span BC. In span BC, adopt the recommended span/effective depth ratio of 20:1. To allow for the fact that the beam is a flanged section (and assuming a flange to rib width ratio greater than 3), the span/effective depth ratio is multiplied by 0.8. Since the span is greater than 7 m, the span/effective depth ratio must also be multiplied by 7/9. Hence:

$$\text{(effective depth)}_{BC} = 9000/(20 \times 0.8 \times 7/9) = 723 \text{ mm}$$

Span AB is the end span of a continuous beam and also a flanged section. Hence:

$$\text{(effective depth)}_{AB} = 7000/(0.8 \times 18) = 486 \text{ mm}$$

Clearly, span BC governs and the minimum effective depth of all spans is 723 mm.

The total beam depth is the effective depth plus the axis distance, a, where a is the sum of the nominal cover to reinforcement, the thickness of link and half the diameter of the tension reinforcement bars, as illustrated in Fig. 6.4. Assuming a tension reinforcement diameter of 25 mm, and exposure class XC3, the minimum cover for bond and durability becomes the greater of 25 mm (bar diameter) and 30 mm (Table 6.3). To allow for construction tolerance, an additional 10 mm is added to these values, resulting in a nominal cover of 40 mm which is used in design. Assuming a link diameter of 10 mm, the axis distance then becomes 40 + 10 + 25/2 = 62.5 mm. For preliminary design purposes, the ISE manual specifies a minimum axis distance of 25 mm for a 1 hour fire resistance (Table 6.4). Therefore, a cover of 40 mm is adequate, resulting in a total minimum beam depth, h, of:

$$h = 723 + 62.5 = 786 \text{ mm}$$

As the ISE span/depth ratios tend to be conservative, this is rounded down to 750 mm, which corresponds to an actual effective depth of 688 mm).

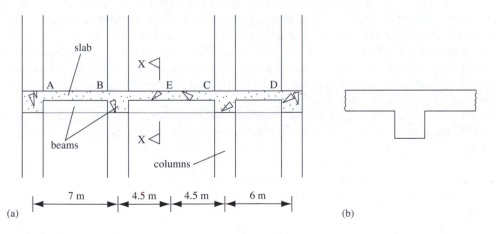

Figure 6.3 Calculation of effective depth: (a) sectional elevation of building showing beam and slab construction; (b) section X–X

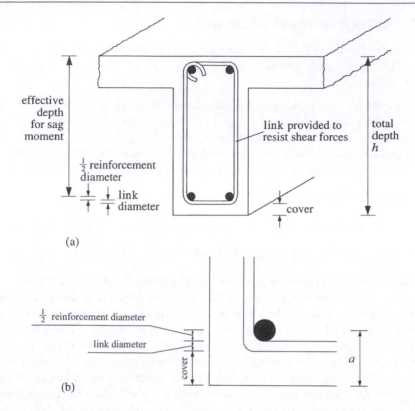

Figure 6.4 Total depth of beam

Breadth of beams

The breadth of rectangular concrete beams and webs in flanged beams (i.e. T- and L-beams) has a much lesser effect on the resistance of a beam to bending moment than effective depth does. Breadth is frequently governed by the practical consideration of simply fitting all the reinforcement into the section while avoiding congestion. The bar arrangement must allow sufficient concrete on all sides of the bar to transfer the axial forces to and from the bar. This anchorage of the reinforcement in the concrete is commonly referred to as the 'bond'. In addition, the bars must be arranged so that the fresh concrete can be placed and compacted properly. The spacing, therefore, must be great enough to allow the passage of aggregate between the bars. Also, access must be provided for a vibrator hose to get all the way to the bottom of the beam.

Reinforcement can be arranged in many ways to satisfy minimum spacing requirements, some of which are illustrated in Fig. 6.5. The minimum horizontal and vertical spacing specified by EC2 is the greater of:

- one bar diameter (or equivalent diameter for bar bundles)
- the maximum aggregate size + 5 mm, or
- 20 mm.

For the beam of Fig. 6.5(a), the minimum spacing requirements will not govern the beam breadth. However, for the beam of Fig. 6.5(b), assuming a maximum aggregate size of 20 mm, a required

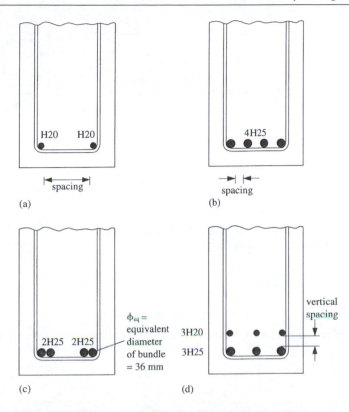

Figure 6.5 Alternative methods of placing reinforcement: (a) one bar at each corner; (b) conventional method; (c) in bundles; (d) in two layers

nominal cover of 25 mm and a link diameter of 10 mm, the minimum practical breadth for a design in accordance with EC2 is:

$$2(\text{cover}) + 2\phi_{\text{link}} + 7\phi_{\text{max}} = 2(25) + 2(10) + 7(25) = 245 \text{ mm}$$

which could be a governing criterion.

   In heavily reinforced beams, the bars can be 'bundled' to reduce the breadth required (Fig. 6.5(c)) or placed in more than one layer (Fig. 6.5(d)). To calculate the minimum spacing in the case of bundled bars, the bundle is treated as a single bar of equivalent cross-sectional area. The notional diameter of the equivalent bar must be less than 55 mm. Bars placed in more than one layer should be located vertically above each other to facilitate effective vibration.

   In long beams, continuity of reinforcement is often achieved by 'lapping'. With this technique, the bars are overlapped a specific distance, known as the 'lap length', so that the force can be transferred from one bar to the other. If the bars are placed alongside each other, as shown in Fig. 6.6(a), the breadth requirement of the beam will be increased. Alternatively, one bar may be placed below the other, as shown in Fig. 6.6(b), in which case there is no increase in the breadth requirement. However, this type of arrangement can cause complications if there is more than one layer of reinforcement.

   Breadth is a major influencing factor on the shear strength of beams. A method has been proposed in the ISE manual for the estimation of the beam breadths required to satisfy shear strength

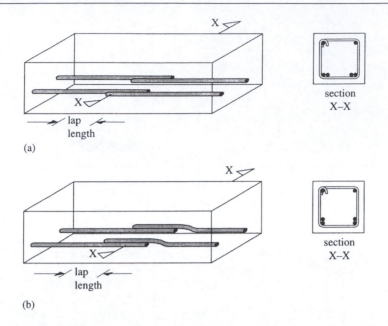

Figure 6.6 Lapping of reinforcement: (a) bars placed alongside each other; (b) bars placed one above the other

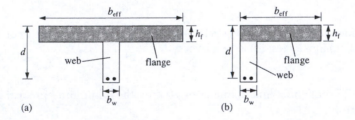

Figure 6.7 Flanged beams: (a) T-beam; (b) L-beam

requirements. The preliminary breadth is determined by limiting the shear stress in beams to 2.0 N/mm² for concrete of minimum characteristic cylinder strength of 25 N/mm². The required breadth, b, for rectangular beams and the required web width, b_w (Fig. 6.7), for flanged beams is given by:

$$b_w = \frac{V}{2.0d} \quad [\text{mm}] \tag{6.1}$$

where V is the maximum ultimate limit state shear force (in Newtons) in the member, considered as simply supported, and d is the effective depth (in millimetres). According to the ISE manual, preliminary design of prestressed concrete members should be carried out at serviceability limit state (SLS). Thus, for prestressed concrete, V is the maximum shear force at SLS.

Flanged beams occur where the beams are cast monolithically with the slab. The portion of the slab near the beams which acts with the beam in compression is known as the flange of the beam. The effective breadth of the flange depends on the web dimensions, the support conditions and the span

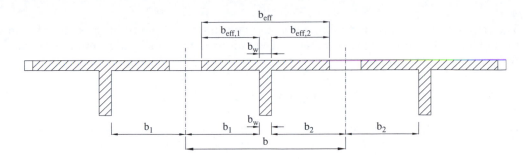

Figure 6.8 Calculation of effective width of flanged beam (where b is the actual flange width)

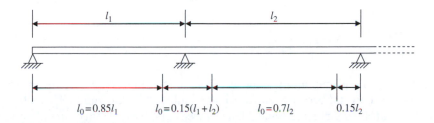

Figure 6.9 The distance between points of zero moment, l_0, in a continuous beam

lengths. The following recommendations for the calculation of effective flange breadths, b_{eff}, are given by EC2 (Fig. 6.8), provided that the ratio of adjacent spans is between 0.67 and 1.5 and/or the length of the cantilever is less than half the adjacent span:

$$b_{eff} = b_w + b_{eff,1} + b_{eff,2} \le b \qquad (6.2)$$

$$b_{eff,1} = 0.2b_1 + 0.1l_0 \le 0.2l_0 \text{ and } b_{eff,1} \le b_1 \qquad (6.3)$$

where l_0 is the distance between points of zero moment in the span and b_1 is half the clear distance between the webs of adjacent beams. If the distance between the flanges is equal for all beams then $b_{eff,2} = b_{eff,1}$, otherwise $b_{eff,2}$ is calculated in similar way to $b_{eff,1}$ in Equation 6.3. It is common practice to assume the point of contraflexure in a continuous beam to be 15 per cent of the length of the span away from the supports. Hence, the calculation of l_0 for different sections of a continuous beam is as illustrated in Fig. 6.9.

Example 6.2 Breadth of beams

Problem: For the beam of Fig. 6.3, the total factored ultimate loading, ω, is 43 kN/m and an approximate analysis has indicated a maximum moment of 220 kNm (sag) at E. Preliminary calculation (see Section 6.5) has indicated that, for the selected depth of 800 mm, an area of tensile reinforcement of 950 mm² is required at E. At this section: (a) select a preliminary web breadth and (b) determine the effective flange breadth assuming a characteristic cylinder strength, f_{ck}, of 30 N/mm².

Solution:

(a) The effective depth, d, is the total beam depth minus the axis distance or the sum of the cover, the link diameter and half the diameter of the reinforcing bars. Using the values from Example 6.1, the effective depth becomes:

$$d = 750 - (40 + 10 + 12.5) = 688 \text{ mm}$$

Assuming each span to be simply supported (as specified for this estimation), the shear force at the ends of each member is $\omega L/2$. Hence, the maximum shear occurs in span BC and is given by:

$$V_{max} = \frac{\omega L}{2} = \frac{43 \times 9}{2} = 194 \text{ kN}$$

From Equation (6.1), the minimum web breadth which satisfies the shear stress requirements is:

$$b_w = \frac{194000}{2.0 \times 688} = 141 \text{ mm}$$

which is less than the minimum value given in Table 6.4 for a fire resistance of 1 hour (i.e. 150 mm). The total area of reinforcement required is 950 mm². Three bars of 20 mm diameter have a total area of 942 mm². Hence, three bars, two 20 mm diameter, and one 25 mm diameter, will be used. It is necessary to check that the web breadth of 150 mm satisfies the minimum spacing requirements for these bars.

Assuming the maximum aggregate size is 20 mm, according to EC2, the minimum bar spacing is 25 mm. The nominal cover is $(30 + 10 =)$ 40 mm due to durability requirements (i.e. exposure class XC3, C30/35). Thus, the minimum web breadth required for all bars on one level is given by:

$$b_w = 40 + 10 + 2(20) + 3(25) + 10 + 40 = 215 \text{ mm}$$

which is more stringent than the breadth required for shear and fire. Thus, the preliminary web breadth is taken as 225 mm.

Note: It is often convenient for shuttering purposes to make the beam and column breadths equal. Accordingly, the column breadth may influence the preliminary beam breadth selected.

(b) The ratio of adjacent spans is between 0.67 and 1.5, therefore, according to EC2, the distance between points of zero moment in span BC is:

$$l_0 = 0.7(9) = 6.3 \text{ m}$$

Assuming an actual flange width spacing between drop beams of $b = 7$ m, the term b_1 in Equation (6.3) is $(7000 - 225)/2 = 3388$ mm. Hence, from Equation (6.3):

$$b_{eff,1} = b_{eff,2} = 0.2(3388) + 0.1(6300) = 1308 \text{ mm}$$
$$> 0.2l_0 = 0.2(6300) = 1260 \text{ mm}$$
$$\Rightarrow b_{eff,1} = b_{eff,2} \, 1260 \text{ mm}$$

From Equation (6.2):

$$b_{eff} = b_w + b_{eff,1} + b_{eff,2} = 225 + 1260 + 1260 = 2745 \text{ mm}$$

6.4 Preliminary sizing of slabs

Fire resistance and durability

As for beams, the dimensions of slabs may be governed by fire-resistance requirements or the cover needed to ensure adequate resistance to corrosion. The minimum cover requirements for corrosion resistance in the five exposure classes of Table 6.2 are given in Table 6.3. Again, up to 10 mm must be added to the values for minimum cover in Table 6.3 to give the nominal cover which is used for design. The fire-resistance requirements in Table 6.6 are those recommended for preliminary design in the ISE manual for slabs spanning continuously over supports where moment redistribution is limited to 15 per cent. For simply supported slabs the minimum size and axis distance should be increased. For prestressed beams and ribbed slabs, width can be traded against distance to provide the required fire resistance. Table 6.6 therefore gives two sets of data, one based on minimum width and the other based on minimum axis distance.

Effective depth of slabs

As for beams, the effective depth has a major influence on the capacity of a slab to resist bending moment. With the exception of flat slabs and ribbed slabs, the depth of slabs does not control the total depth of structures. However, a small increase in slab depth adds greatly to the self-weight of the structure which significantly increases the bending moments in the slab itself, and adds load to all beams, columns and foundations which support the slab.

In order to minimize the depth of slabs while at the same time ensuring adequate moment and shear capacity, the ISE manual recommends guidelines for the preliminary sizing of slabs. These guidelines, for the minimum span/effective depth ratios of slabs, are given in Table 6.7 for reinforced concrete and Table 6.8 for prestressed concrete. For additional checks for vibration control of prestressed concrete floors, refer to the ISE manual.

The effective depth requirements for flat slabs can also be governed by 'punching shear', the tendency of a supporting column to 'punch up' through the slab (of course it is the slab that pushes down around the column). The approximate method proposed in the ISE manual for ensuring adequate resistance to punching shear in reinforced concrete flat slabs is as follows. Where no

Table 6.6 Minimum dimensions and axis distance for the preliminary design of continuous plain (solid), ribbed slabs and flat slabs. All values are in millimetres (from the ISE manual)

Member	Minimum dimension	Reinforced concrete 1	2	3	4	Prestressed concrete 1		2		4	
Plain slab	Overall depth	80	120	150	175	80		120		175	
	Axis distance	15	20	30	40	25		35		55	
Ribbed slab	Rib width	100	160	310	450	100	200	160	300	450	700
	Axis distance	25	45	60	70	40	25	60	45	85	75
	Axis distance to side of rib	–	–	–	–	50	35	70	55	95	85
	Overall Flange depth	80	120	150	175	80		120		175	
	Axis distance to flange	–	–	–	–	25		35		55	
Flat slab	Overall depth	180	200	200	200	180		200		200	
	Axis distance	15	35	45	50	30		50		65	

Table 6.7 Span/effective depth ratios for the preliminary design of reinforced concrete slabs. For two-way spanning slabs, the span/effective depth ratio is based on the shorter span. For flat slabs that exceed 8.5 m, the ratio should be multiplied by 8.5/span. For all other slab types, for spans in excess of 7 m, the values in the table should be multiplied by 7/span. (ISE manual)

Slab	Span/effective depth ratio
Cantilever	8
Simply supported (one-way or two-way spanning)	20
End span of:	
One-way spanning continuous slab	26
Two-way spanning slab, continuous over one long side	26
Interior span of:	
One-way spanning slab	30
Two-way spanning slab	30
Flat slab (based on longer span)	24

Table 6.8 Span/effective depth ratios for the preliminary design of prestressed concrete multi-span floors, with spans between 6 and 13 m (from ISE manual). For single span floors the depth should be increased by approximately 15 per cent.

Slab	Total imposed load (kN/m)		
	2.5	5	10
One-way spanning (with narrow beam, $b \approx$ span/15)			
slab	42	38	34
beam	18	16	13
Ribbed slab	30	27	24
Solid flat slab	40	36	30

shear reinforcement is to be provided (*Note*: Shear reinforcement is not effective in slabs that are less than 200 mm thick), check the shear stress in a perimeter around the column given by:

$$\text{perimeter stress} = \frac{1250w(\text{area supported by column})}{(\text{column perimeter} + 12h)h} \tag{6.4}$$

where w is the ultimate load per unit area in kilonewtons per square metre and h is the total depth of the slab at the column in millimetres. To avoid providing shear reinforcement, this stress needs to be less than or equal to 0.6 N/mm^2. Where shear reinforcement is provided, this perimeter stress can be as high as 1 N/mm^2.

The perimeter stress at the column face should also be calculated:

$$\text{perimeter stress} = \frac{1250w(\text{area supported by column})}{(\text{column perimeter})h} \tag{6.5}$$

and should be less than 15 per cent of the characteristic cylinder strength, f_{ck}. Note that the area supported by the column is in square metres and the column perimeter is in millimetres. Note also that these formulas assume a perimeter around all four sides of the column. They need to be amended for edge and corner columns where the perimeter has three and two sides respectively.

Example 6.3 Effective depth of one-way spanning slabs

Problem: A multi-storey structure is to be constructed using a continuous one-way spanning floor system. The layout for the proposed system is illustrated in Fig. 6.10.

(a) Determine a preliminary depth for the slabs given a required fire resistance of 1 hour.
(b) Comment on the proposed scheme.

Solution:

(a) As for the beams in Example 6.1, the same depth is used in all slabs for ease of construction. From Table 6.6, a fire resistance of 1 hour requires a minimum depth of 80 mm. At a typical floor level there are two interior panels and two exterior panels, one of which is a cantilever. The recommended span/effective depth ratio for each panel, taken from Table 6.7, is given by:

$$(\text{span/eff. depth})_{\text{ext. panel}} = 26$$
$$(\text{span/eff. depth})_{\text{int. panel}} = 30$$
$$(\text{span/eff. depth})_{\text{cantilever}} = 8$$

Thus, the required effective depth for each panel is:

$$(\text{eff. depth})_{\text{ext. panel}} = 4000/26 = 154 \text{ mm}$$
$$(\text{eff. depth})_{\text{int. panel}} = 4000/30 = 133 \text{ mm}$$
$$(\text{eff. depth})_{\text{cantilever}} = 2000/8 = 250 \text{ mm}$$

Therefore, an effective depth, d, of 250 mm, as governed by the cantilever panel, is adopted throughout. The total slab depth is the sum of the effective depth, the nominal cover and half the diameter of the tension reinforcement (unlike for beams, links for shear resistance are not

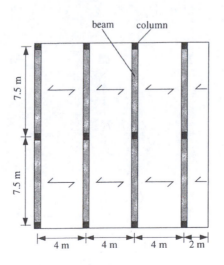

Figure 6.10 Plan view of floor layout

commonly provided in slabs). Assuming a reinforcement diameter of 12 mm, the minimum cover required for bond is 12 mm. Table 6.3 specifies a minimum cover for durability purposes of 15 mm (exposure class XC1), giving a nominal cover of 25 mm. For a fire resistance of 1 hour, Table 6.6 gives a preliminary minimum axis distance of 15 mm, which is less onerous than the cover required for durability. Thus, the total depth, h, becomes:

$$h = 250 + 25 + \frac{1}{2}(12) = 281 \text{ mm}$$

The span/effective depth ratios specified in Table 6.7 are conservative, so the total depth is rounded down from 281 mm to 275 mm.

(b) The proposed layout is not economical due to the excessively long cantilever. By reducing the length of the cantilever to, say, 1.3 m and increasing the two interior spans to 4.35 m, the required effective depth is reduced to 163 mm and a total depth of $h = 200$ mm becomes feasible.

Example 6.4 Effective depth of flat slabs

Problem: A multi-storey residential building is to be constructed using a flat slab floor system. The layout of the proposed floor system is illustrated in Fig. 6.11 (all columns are 400 mm by 400 mm).

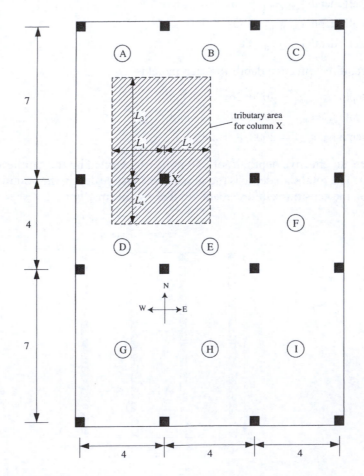

Figure 6.11 Flat slab floor of Example 6.4

Table 6.9 Panel groups for Example 6.4

Group	Type	Longer span dimension (m)	Panels
I	Exterior	7	A, B, C, G, H, I
II	Exterior	4	D, F
III	Interior	4	E

Each floor is required to sustain a total characteristic variable gravity load of 2 kN/m^2 plus the weight of non-permanent partitions. Suggest a preliminary depth for the slabs if the required fire resistance is 2 hours. Assume exposure class XC1 and a concrete cylinder strength, f_{ck}, of 30 N/mm^2.

Solution: As for the previous examples, the same depth is used in all parts of the slab. The depth may be governed by conditions in any of the nine panels. However, for convenience, the panels can be classified into three groups (see Table 6.9) depending on their type and geometry. The required depth will be governed by one of these three groups of panels.

The span/effective depth ratio given in Table 6.7 for flat slabs is adopted for all groups, that is:

$$\text{span}/\text{eff. depth} = 24$$

Thus, the required effective depth for each group is given by:

$$(\text{eff. depth})_{\text{groups I}} = 7000/24 = 292 \text{ mm}$$
$$(\text{eff. depth})_{\text{groups II}} = 4000/24 = 167 \text{mm}$$
$$(\text{eff. depth})_{\text{groups III}} = 4000/24 = 167 \text{mm}$$

Therefore, an effective depth, d, of 292 mm, as governed by group I panels, is adopted throughout. The total slab depth is equal to the effective depth plus either the axis distance or the nominal cover to the reinforcement plus half the diameter of the tension reinforcement. The nominal cover for durability for exposure class XC1 is 25 mm. From Table 6.6, the preliminary minimum axis distance for a 2 hour fire resistance is 35 mm. Assuming a tension reinforcement diameter of 16 mm, the minimum axis distance for fire resistance governs in this case, resulting in a total slab depth, h, of:

$$h = 292 + 35 = 327 \text{ mm}$$

which is rounded to 325 mm (note that this is greater than the minimum depth given in Table 6.6 and so is satisfactory). As a result of the rounding off, the effective depth is now (325–35) = 290 mm.

It should be noted that, in this example, the total depth has been checked assuming that the outermost mat of reinforcement spans in the critical direction. However, in other examples where the effective depth is governed by a square panel, the total depth is equal to the effective depth plus cover plus one and one half times the reinforcement diameter.

It remains to be checked if this slab depth satisfies the perimeter shear strength requirements (see Equations (6.4) and (6.5)). These stresses tend to be critical for the interior columns supporting the central panel where the tributary areas are greatest. For the column common to panels A, B, D and E, say, the tributary area is calculated with these panels fully loaded while the minimum loading is applied to all other panels. The tributary lengths, L_1, L_2, L_3 and L_4 (see Fig. 6.11), can be determined accurately using the formulae in Appendix C. However, since we are only interested in the preliminary size of the slab, it would be excessive to use this level of accuracy to calculate the tributary lengths. For calculating the preliminary shear capacity of the slab using Equations (6.4) and (6.5), it is reasonable to assume the following:

$$L_1 = \frac{5}{8}(4) = 2.5\,\text{m}$$

$$L_2 = \frac{1}{2}(4) = 2.0\,\text{m}$$

$$L_3 = \frac{5}{8}(7) = 4.38\,\text{m}$$

$$L_4 = \frac{1}{2}(4) = 2.0\,\text{m}$$

(the accuracy of this assumption is left to the reader to evaluate). Hence, the tributary area is:

$$(L_1 + L_2) \times (L_3 + L_4) = 4.5 \times 6.38 = 2.87\,\text{m}^2$$

Assuming a self-weight of concrete of 25 kN/m², allowing for a 75 mm floor finish of density 23 kN/m³, and assuming a weight for ceilings and services of 0.5 kN/m², gives a total permanent gravity load of:

$$\text{Permanent load} = 25 \times 0.325 + 23 \times 0.075 + 0.5$$
$$= 10.35\text{kN}/\text{m}^2$$

Applying the partial safety factors from Chapter 3 (1.35 for permanent and 1.5 for variable gravity load), the total maximum ultimate load per unit area on this floor is:

$$\text{ultimate load}, w = 1.35 \times 10.35 + 1.5 \times (2 + 1)$$
$$= 18.5\,\text{kN}/\text{m}^2$$

Thus:

$$\frac{1250w(\text{area supported by column})}{(\text{column perimeter} + 12h)h} = \frac{1250 \times 18.5 \times 28.7}{(4 \times 400 + 12 \times 325)325}$$
$$= 0.37$$
$$< 0.6\,\text{N}/\text{mm}^2 \Rightarrow \text{no shear reinforcement required}$$

and

$$\frac{1250w(\text{area supported by column})}{(\text{column perimeter})h} = \frac{1250 \times 18.5 \times 28.7}{4 \times 400 \times 325}$$
$$= 1.28\,\text{N}/\text{mm}^2$$
$$\leq 0.15 f_{ck}\ (= 4.5\,\text{N}/\text{mm}^2)$$

As both inequalities are satisfied, the flat slab depth of 325 mm is taken as the preliminary solution.

Note 1: A slab depth of 325 mm is very large and is much deeper than would be commonly found in practice. Many designers will use a greater span/depth ratio than the value of 24 given in Table 6.7.

Note 2: Such short spans in flat slab construction have today become unusual. Panels of up to 7 m × 8 m are not now uncommon. For such spans, preliminary design in accordance with the above criteria would give excessively deep slabs, much deeper than are actually found in practice. This again implies that a span/depth ratio of 24 is quite conservative.

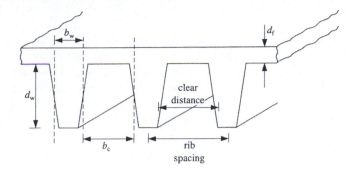

Figure 6.12 Ribbed slab

Proportions of ribbed slabs

To ensure adequate stiffness against bending and torsion and to allow ribbed slabs to be treated as solid slabs for the purposes of analysis, EC2 recommends that the following restrictions on size be satisfied (refer to Fig. 6.12):

1. The rib spacing must not exceed 1.5 m.
2. The rib depth, d_w, must not exceed four times the rib breadth, b_w.
3. The flange depth, d_f, must exceed the greater of one-tenth of the clear distance between ribs, b_c, and 50 mm. In the case of ribbed slabs incorporating permanent blocks (Fig. 2.19(d)), this lower limit of 50 mm may be reduced to 40 mm.
4. Transverse ribs must also be provided at a maximum spacing of 10 times the overall depth of the slab.

In addition to the minimum values given in Table 6.6 to satisfy fire-resistance requirements, the breadth of ribs may be governed by shear strength requirements. The method proposed in the ISE manual for the estimation of rib breadths limits the shear stress in the rib to 0.6 N/mm² for reinforced concrete and 0.8 N/mm² for prestressed concrete. This is for concrete with a characteristic cylinder strength of 25 N/mm² or more. The minimum required breadth for reinforced concrete is therefore given by:

$$b_w = \frac{V}{0.6d} \ \ [\text{mm}] \tag{6.6}$$

where V is the maximum shear force in Newtons on the rib considered as simply supported and d is the effective depth in millimetres. For characteristic cylinder strengths less than 25 N/mm², the breadth should be increased in proportion.

6.5 Reinforcement in beams and slabs

Once the section dimensions of a beam or slab are fixed and the grade of concrete has been decided upon, the steel reinforcement is the only remaining design variable. The quantity of flexural reinforcement required follows directly from the moment capacity required in the member to resist moments due to applied loads.

Preliminary analysis

At the preliminary design stage, the adequacy of the reinforcement typically only needs to be checked at mid-span and over the supports. The bending moment and shear force may be obtained using elastic analysis. Alternatively, some approximate rules of thumb are given in the ISE manual for the estimation of internal moments and shear forces at these locations due to applied loads. For continuous beams and one-way spanning slabs, bending moments and shear forces may be taken from Table 6.10 if the following conditions are satisfied:

1. The variable gravity load does not exceed the permanent gravity load.
2. There are at least three spans.
3. The spans do not differ in length by more than 15 per cent of the longest span.

For two-way spanning reinforced concrete slabs on linear supports, the average moment per metre width, m, can be taken as:

$$m = q\left(\frac{L_x L_y}{18}\right) \qquad [\text{kNm/m}] \tag{6.7}$$

where L_x and L_y are the shorter and longer span lengths, respectively, in metres and q is the ultimate (factored) load in kilonewtons (kN) per square metre. If $L_y > 1.5\, L_x$, the slab should be treated as a one-way spanning slab. Similarly, for two-way spanning prestressed concrete slabs the average moment per metre width, m, can be taken from Table 6.11.

For reinforced concrete solid flat slabs, the moments per unit width in strips of slab between columns can be taken as 1.5 times those for one-way spanning slabs, whereas for prestressed concrete, moments are determined as for one-way spanning slabs.

For one-way spanning ribbed slabs, the bending moments at mid-span can be assessed on a width equal to the rib spacing and assuming simple supports throughout. For two-way

Table 6.10 Ultimate bending moments and shear forces for continuous beams and one-way spanning slabs where L is the span length in metres (from the ISE manual)

	Uniformly distributed load, q, acting on all spans	Central point load, Q, acting on all spans
	(kN/m)	(kN)
Bending moment		
at support	$-0.10qL^2$	$-0.150QL$
at mid-span	$0.08qL^2$	$0.175QL$
Shear force	$0.65qL$	$0.65Q$

Table 6.11 Average bending moments per metre width for two-way spanning prestressed concrete slabs on linear supports with an SLS load of q_{SLS} (from the ISE manual)

	Short span (kNm/m)	Long span (kNm/m)
At continuous support	$-q_{SLS}\left(\frac{L_x L_y}{17}\right)$	$-q_{SLS}\left(\frac{L_x^2}{17}\right)$
At mid-span	$q_{SLS}\left(\frac{L_x L_y}{20}\right)$	$q_{SLS}\left(\frac{L_x^2}{20}\right)$

spanning ribbed slabs where $L_y < 1.5\ L_x$, the average rib moment in both directions can be taken as:

$$m = q\left(\frac{L_x L_y}{18}\right)c \quad [\text{kNm/m}] \tag{6.8}$$

where c is the rib spacing. If $L_y > 1.5\ L_x$, the slab should be treated as one-way spanning. Similarly, for prestressed concrete ribbed slabs, the moments in Table 6.11 are multiplied by the rib spacing.

Reinforcement in rectangular sections

When a rectangular section is subjected to sag moment as illustrated in Fig. 6.13(a), the top fibres are compressed and the bottom fibres are extended. Concrete has good compressive strength and very low tensile strength. For design purposes, zero tensile strength can be assumed. Thus, the bending causes tension only in the reinforcing bars near the bottom of the section and compression in the concrete near the top of the section (and in the top reinforcement if any is present). For the high levels of stress that exist at ULS, the distribution of compressive stress in the concrete is of the form illustrated in Fig. 6.13(b). This roughly parabolic distribution of stress is often approximated by an equivalent uniform (rectangular) distribution. By force equilibrium at the face illustrated in Fig. 6.13(b), we have:

compression force, C = tension force, T

Moment equilibrium at the face gives:

$$M = Tz \tag{6.9}$$

where z is the lever arm between centres of the tension force and the compression force as shown in Fig. 6.13(b).

An estimate of the lever arm that is sufficiently accurate for detailed design is based on an assumed uniform distribution of compressive stress in the concrete (so-called equivalent rectangular stress block). Using the equivalent rectangular stress block, EC2 gives:

$$z = d\left(0.5 + \sqrt{0.25 - 0.88K}\right) \tag{6.10}$$

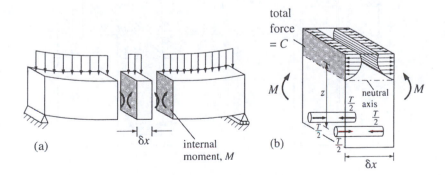

Figure 6.13 Reinforcement requirement in reinforced concrete beam: (a) beam in bending; (b) segment of beam in bending at ULS

(Equation (6.10) is derived in Chapter 8.) However, the maximum value of z should be $0.95d$. In this equation, d is the effective depth of the section and the parameter K is given by:

$$K = \frac{M}{f_{ck}bd^2} \tag{6.11}$$

where f_{ck} is the characteristic compressive cylinder strength of concrete at 28 days and b is the breadth of the section. Note that the units used in Equations (6.10) and (6.11) must be consistent. The area of tension reinforcement required to resist the internal moment, M, is then calculated using the following formula:

$$A_s = \frac{M}{0.87f_{yk}z} \tag{6.12}$$

where f_{yk} is the characteristic yield strength of the reinforcement.

For relatively high moment, specifically if $K > 0.167$, compression reinforcement is required to maintain ductility in the section. For preliminary purposes, the required area of compression reinforcement, A'_s, is calculated using the formula:

$$A'_s = \frac{M - 0.167f_{ck}bd^2}{0.87f_{yk}(d - d')}$$

where d' is the effective depth from the extreme fibre in compression to the centroid of the compression steel. The required area of tension reinforcement then becomes:

$$A_s = \frac{0.234bdf_{ck}}{f_{yk}} + A'_s \tag{6.13}$$

To ensure ductility of the section, it must be checked that the difference between provided and required steel is greater for the compression reinforcement than the tension reinforcement.

Reinforcement in flanged beams and ribbed slabs

The design procedure for flanged beams and ribbed slabs depends on whether the neutral axis lies within the flange or the web (Fig. 6.14). If the neutral axis lies within the flange, its location can be calculated using:

$$x = (d - z)/0.4 \tag{6.14}$$

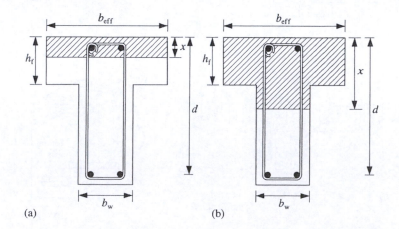

Figure 6.14 Alternative locations of neutral axis for a flanged section in sag: (a) within flange; (b) within web. Shaded area indicates zone of compression

where z is determined using Equations (6.10) and (6.11) with $b = b_{\text{eff}}$, the effective flange breadth. If the neutral axis does not lie within the flange, Equation (6.14) is inappropriate. However, this equation can still be used to determine whether x is greater or less than the flange depth, that is $x \gtrless h_{\text{f}}$.

Neutral axis within flange

If the neutral axis lies within the flange for a sagging section (i.e. $x < h_{\text{f}}$), as shown in Fig. 6.14(a), the compression zone is rectangular and the section can be designed as a rectangular section. The shape of the tensile zone of the concrete, assumed cracked, is of no relevance in this case. It can be seen in Fig. 6.15 that the equivalent stress block specified in EC2 does not extend fully to the neutral axis (the reason for this is to provide a simpler distribution in which the resultant compressive force acts at a point reasonably near to the actual resultant). It is clear from the figure that the compression zone actually remains rectangular for all $x \leq h_{\text{f}}/0.8$. For $x > h_{\text{f}}/0.8$, the assumption of rectangular section behaviour ceases to be valid since the compression zone moves into the web.

Thus, for all sections where $x \leq h_{\text{f}}/0.8$, the area of tension steel required is derived from Equations (6.10)–(6.12) using b_{eff} in place of b. If $x > h_{\text{f}}/0.8$, the neutral axis does not lie within the flange and these equations are invalid.

Neutral axis within web

If the neutral axis lies within the web for a section in sag (Fig. 6.14(b)), the compression zone is not rectangular and the section cannot be designed using Equations (6.10)–(6.12). It is possible to derive a formula similar to Equation (6.12) for such a case but for preliminary design purposes this unduly complicates the procedure. It is simpler to redesign the section by increasing the web depth so that the neutral axis lies within the flange.

For flanged sections in hog (Fig. 6.16), the neutral axis generally lies in the web. Hence, the minimum tension steel requirements are found using Equations (6.10)–(6.12) with $b = b_{\text{w}}$, the web width.

EC2 restrictions

It should be checked that the areas of tension steel and of compression steel do not exceed the recommended maximum value of $0.04A_{\text{c}}$ (except in long members where reinforcement is lapped)

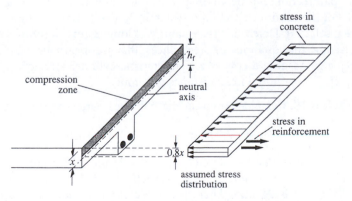

Figure 6.15 Sagging flanged section assuming equivalent rectangular stress block

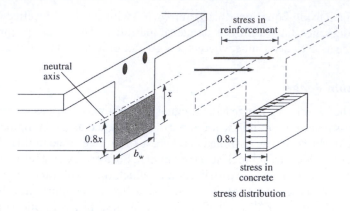

Figure 6.16 Hogging flanged section with neutral axis in web

where A_c is the total cross-sectional area of the concrete section. The area of tension steel should also not be less than:

$$A_{s,\,min} = 0.000156 f_{ck}^{2/3} b_t d \geq 0.0013 b_t d \qquad \text{(for concrete} \leq \text{C50/60)}$$

where f_{ck} is the characteristic compressive cylinder strength of concrete at 28 days, b_t is the mean breadth of the tension zone and d is the effective depth of the section. For flanged beams, EC2 recommends that at intermediate supports of continuous beams, the total area of tension reinforcement of a flanged cross-section should be spread over the effective width of flange, although, the reinforcement may be more concentrated over the web width.

According to the ISE manual, the minimum area of compression reinforcement, where it is required, should be $0.002bh$ for a rectangular beam. In addition, secondary reinforcement must be provided in one-way spanning slabs and should be at least 20 per cent of the principal reinforcement. The spacing of principal reinforcement should not exceed the lesser of $3h$ and 400 mm where h is the total depth of the slab. In secondary slab reinforcement, the spacing should not exceed the lesser of $3.5h$ and 450 mm. This should be reduced to $2h$ or 250 mm for principal reinforcement and $3h$ or 400 mm for secondary reinforcement in areas of maximum moment or concentrated loads.

Example 6.5 Reinforcement in beams

Problem: A four-span rectangular beam of span length 8 m is to carry a total ULS factored gravity load of 57 kN/m. For this section, a preliminary depth, h, of 500 mm and a preliminary breadth, b, of 275 mm have been selected. Determine the quantity of longitudinal reinforcement required to resist the applied loads. For this quantity of steel, check that the beam breadth is sufficient to allow adequate bar spacing. Assume a concrete characteristic cylinder strength, f_{ck}, of 35 N/mm^2 and exposure class XC1. The required fire resistance is 1 hour.

Solution: From Table 6.10, the mid-span moment in the beam due to the applied load of 57 kN/m is:

$$M = 0.08qL^2 = 0.08(57)(8)^2 = 292 \text{ kNm}$$

Assuming (initially) a tension reinforcement diameter of 25 mm, the nominal cover requirements for bond and durability are 35 mm and 25 mm, respectively, and the required axis distance for fire resistance from Table 6.4 is 25 mm. Therefore, the cover is governed by bond requirements and assuming a link diameter of 10 mm, the effective depth is given by:

$$d = 500 - 35 - 10 - \frac{1}{2}(25)$$
$$= 443 \text{ mm}$$

From Equation (6.11), we have:

$$K = \frac{M}{f_{ck}bd^2}$$
$$= \frac{292 \times 10^6}{35 \times 275 \times 443^2}$$
$$= 0.155$$

As $K < 0.167$, compression reinforcement will not be required. From Equation (6.10), the lever arm is:

$$z = d\left(0.5 + \sqrt{0.25 - 0.88K}\right)$$
$$= d\left(0.5 + \sqrt{0.25 - 0.88(0.155)}\right)$$
$$= 0.84d = 370 \text{ mm}$$

Thus, assuming $f_{yk} = 500 \text{ N/mm}^2$, the required area of tension reinforcement is:

$$A_s = \frac{M}{0.87f_{yk}z}$$
$$= \frac{292 \times 10^6}{0.87 \times 500 \times 370}$$
$$= 1814 \text{ mm}^2$$

One 25 mm diameter bar has a cross-sectional area of 491 mm². Thus, four 25 mm diameter bars (total area = 1963 mm²) must be provided as tension reinforcement.

The bar arrangement for this section is illustrated in Fig. 6.17. Although no compression reinforcement is required, two 10 mm diameter bars (nominal reinforcement) are provided in the top of the

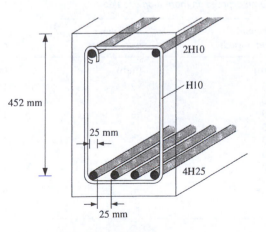

Figure 6.17 Bar arrangement for section of Example 6.5

section to act as hangers for the links. The minimum spacing of the tension bars is 25 mm. Thus, the minimum beam breadth required for this arrangement is:

$$b = 2(35) + 2(10) + 7(25) = 265 \text{ mm}$$

Hence, the assumed breadth of 275 mm is satisfactory for this bar arrangement, and the side cover can be increased to 40 mm.

6.6   Preliminary sizing of columns and walls

Columns

Columns are classified in EC2 as being:

(a) either 'non-slender (stocky)' or 'slender'; and
(b) either 'braced' or 'unbraced'.

These distinctions are made because the strength of slender and/or unbraced columns can be substantially reduced by the horizontal deflection of the column under the applied loads. The majority of structural columns are within the stocky classification and slenderness effects can be ignored. For simplicity, only 'stocky' columns should be used at the preliminary design stage.

The criteria by which columns are classified as non-slender or slender in EC2 are quite complex. However, according to the ISE manual, for preliminary design a stocky column is generally one for which the ratio of the effective height to the least lateral dimension does not exceed 15 for braced columns. Again, for preliminary design, the effective height can be taken as 0.85 times the clear column height for braced columns. For the final design, the classification of columns should be carried out in accordance with EC2.

The minimum dimensions required for the fire resistance of rectangular columns are given in Table 6.12. In EC2, different values are provided depending on the value of a reduction factor based on the design load level in a fire situation. For preliminary design, only the most conservative case is considered here. Refer to EN 1992-1-2 for more information.

Table 6.12 Minimum dimensions for the preliminary design of columns (from BS EN 1992-1-1)

	Column exposed on more than one side		Column exposed on one side	
Fire resistance	Minimum column width	Minimum axis distance	Minimum column width	Minimum axis distance
(h)	(mm)	(mm)	(mm)	(mm)
0.5	200	32	155	25
	300	27		
1	250	46	155	25
	350	40		
1.5	350	53	155	25
	450*	40		
2	350*	57	175	35
	450*	51		
3	450*	70	230	55
4	–	–	295	70

* minimum of 8 bars

A preliminary estimate of the capacity of a stocky column to resist axial force is given in the ISE manual as:

$$N = A_c\left(0.44f_{ck} + \frac{p}{100}\left(0.67f_{yk} - 0.44f_{ck}\right)\right) \qquad (6.15)$$

where A_c is the cross-sectional area of the column (mm^2)
f_{ck} is the characteristic cylinder strength (N/mm^2)
f_{yk} is the characteristic strength of the reinforcement (N/mm^2)
p is the percentage of reinforcement ($100\,A_s/A_c$)

A satisfactory preliminary column design is one which has an ultimate load capacity, calculated using Equation (6.15), greater than the total applied load, that is the sum of the ultimate loads applied at each floor level. This equation is applicable to braced frames in which axial forces, rather than moments, are the dominant criteria for column design. However, some moment occurs in all frames due to the lack of symmetry in the arrangement of loads. To allow for this in preliminary design, the ultimate applied load from the floor immediately above the column being considered should be multiplied by the following factors (ISE manual):

Columns balanced in two perpendicular directions
    (Fig. 6.18(a)):   1.25
Columns balanced in only one direction
    (Fig. 6.18(b)):   1.50
Columns unbalanced in two perpendicular directions
    (Fig. 6.18(c)):   2.00

The maximum percentage of longitudinal reinforcement allowed in columns by EC2 is 8 per cent. Allowing for laps, this limits the area in regions between laps to 4 per cent. However, it is recommended to take the percentage of reinforcement, p, as 2 per cent at the preliminary stage of design. This leaves scope either to increase or decrease the amount of reinforcement in the detailed design. The minimum area of longitudinal reinforcement is given by:

$$A_{s\,min} = \text{greater of}\left(0.1N\gamma_s/f_{yk} \text{ and } 0.002A_c\right)$$

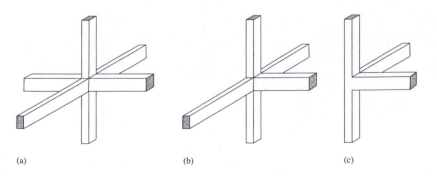

    (a)   (b)   (c)

Figure 6.18 Column types: (a) balanced in two perpendicular directions (internal columns); (b) balanced in one direction only (façade columns); (c) unbalanced in two perpendicular directions (corner columns)

Example 6.6 Preliminary sizing of columns

Problem: An internal column in a multi-storey building, illustrated in Fig. 6.19, has a clear height of 5.5 m between the foundation pad and the first floor. The column carries a total characteristic permanent gravity load of 750 kN and a total characteristic variable gravity load of 440 kN from upper levels. In addition, characteristic permanent and variable gravity loads of 170 kN and 90 kN, respectively, are transferred to the column at the first floor. If the column is to have a square cross-section, select appropriate preliminary dimensions and determine the quantity of reinforcement required. Assume that the column is braced and that $f_{ck} = 35$ N/mm^2 and $f_{yk} = 500$ N/mm^2.

Solution: Using the partial safety factors from Chapter 3, the ultimate load on the column from the upper levels is:

$$N_{\text{upper levels}} = 1.35 \times 750 + 1.5 \times 440$$
$$= 1673 \text{ kN}$$

Similarly, the ultimate load from the first floor is:

$$N_{\text{first floor}} = 1.35 \times 170 + 1.5 \times 90$$
$$= 365 \text{ kN}$$

The applied loads from the first floor are factored by 1.25 (i.e. column balanced in two perpendicular directions) to compensate for bending effects. Thus, the total ultimate load on the column is:

$$N_{\text{total}} = N_{\text{upper levels}} + 1.25 \times N_{\text{firstfloor}}$$
$$= 1673 + 1.25 \times 365$$
$$= 2129 \text{ kN}$$

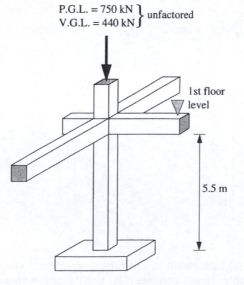

Figure 6.19 Internal ground floor column in multi-storey structure

For a reinforcement percentage of 2, Equation (6.15) gives:

$$N = A_c \left(0.44 f_{ck} + \frac{p}{100} (0.67 f_{yk} - 0.44 f_{ck}) \right)$$

$$\Rightarrow \quad 2129 \times 10^3 = A_c \left(0.44(35) + \frac{2}{100} (0.67 \times 500 - 0.44 \times 35) \right)$$

$$\Rightarrow \quad A_c = \frac{2129 \times 10^3}{(0.44(35) + 2/100(0.67 \times 500 - 0.44 \times 35))}$$

$$= 97696 \text{ mm}^2$$

Assuming the column is square, the side length, h, is given by:

$$h = \sqrt{97696} = 313 \text{ mm}$$

This is rounded up to give a preliminary side length of 325 mm.

For the calculated dimensions of the column, it is necessary to confirm the initial assumption that the column is 'stocky'. The effective height of the column (assuming it is braced) is:

$$l_0 = 0.85l = 0.85(5.5) = 4.675 \text{ m}$$

Thus:

$$\frac{l_0}{h} = \frac{4.675}{0.325} = 14.4 < 15$$

Hence, the column is indeed 'stocky'. As the dimensions have been rounded up, it is worth recalculating the assumed percentage of reinforcement:

$$N = A_c \left(0.44 f_{ck} + \frac{p}{100} (0.67 f_{yk} - 0.44 f_{ck}) \right)$$

$$\Rightarrow \quad \frac{p}{100} = \frac{N - 0.44 f_{ck} A_c}{A_c (0.67 f_{yk} - 0.44 f_{ck})} = \frac{2129 \times 10^3 - 0.44(35)(325)^2}{(325)^2 (0.67 \times 500 - 0.44 \times 35))}$$

$$\Rightarrow \quad \frac{p}{100} = 0.0149$$

which gives an area of reinforcement of 1,456 mm^2. It must be checked if this is greater than the minimum area of longitudinal reinforcement of:

$$A_{s,min} = \text{greater of} \left(0.1 N \gamma_s / f_{yk} \text{ and } 0.002 A_c \right)$$

$$= \text{greater of} (0.1 \times 2129 \times 10^3 \times 1.15/500 \text{ and } 0.002 \times 325^2)$$

$$= \text{greater of} (490 \text{ and } 211)$$

$$\Rightarrow A_{s,min} = 490 \text{mm}^2$$

Therefore, 1,456 mm^2 of reinforcement is sufficient and one possible solution is with the provision of eight 16 mm diameter bars as illustrated in Fig. 6.20 (total area of 1,608 mm^2). In addition, links are provided to give restraint against buckling of the main reinforcement.

EC2 specifies that the spacing of links in columns should not exceed the minimum of:

(a) 20 times the minimum diameter of the longitudinal bars
(b) the least dimension of the column
(c) 400 mm

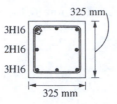

3H16

2H16

3H16

325 mm

325 mm

Figure 6.20 Preliminary solution for column of Example 6.6

However, the spacing should be reduced by a factor of 0.6 in certain areas (e.g. near beams/slabs and lapped joints). Refer to EC2 for more details.

Walls

Shear walls are designed to resist horizontal wind loads. However, they also provide stability to the structure and, for this reason, they are designed to resist a small percentage of the vertical loads. According to the ISE manual, walls carrying vertical loads can be designed as columns for preliminary design. Hence, as for columns, only non-slender walls should be used at the preliminary design stage. Shear walls should be designed as vertical cantilevers. When the dimensions are such that shear walls are not 'deep', the quantity of reinforcement required is determined in the same manner as for beams. Where a shear wall has a return at the compression end such as in Fig. 6.21, it should be designed as a flanged beam.

In general, walls should have a thickness of no less than 150 mm, but 180 mm may be more practical. This allows sufficient access for compaction hoses and, in most cases, ensures adequate strength and stiffness. EC2 specifies that the area of vertical reinforcement in walls be kept between 0.2 and 4 per cent and when the minimum area of reinforcement controls design, half the bars should be located on each face. Horizontal reinforcement should also be provided at each surface and it should not be less than the greater of 25 per cent of the vertical reinforcement and 0.1 per cent of the cross-sectional area. Transverse reinforcement (i.e. links) should be provided if the vertical reinforcement exceeds 2 per cent or if the vertical bars form the outer layer of reinforcement.

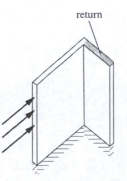

return

Figure 6.21 Wall with a return at the compression end

Problems

Sections 6.3 and 6.5

6.1   Suggest preliminary dimensions and areas of reinforcement for a continuous rectangular beam with two 7 m spans. The total factored ULS loading (deemed to include self-weight) is 15 kN/m and the fire rating is 1 hour. Assume exposure class XC3 and $f_{ck} = 30$ N/mm^2.

6.2   The four-span T-beam illustrated in Fig. 6.22 is subjected to unfactored permanent and variable gravity loadings of 25 kN/m (excluding self-weight of drop beam) and 20 kN/m respectively. Assuming exposure class XC1, determine preliminary values for the maximum areas of top and bottom reinforcement given a fire rating of 1 hour. The beam is integral with 325×325 mm columns.

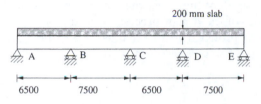

Figure 6.22 Beam of Problem 6.2

Sections 6.4 and 6.5

6.3   For the slab of Example 6.3, make a preliminary estimate of the maximum required areas of reinforcement. The characteristic variable gravity loading of 5 kN/m^2 includes an allowance for non-permanent partitions.

Section 6.6

6.4   Determine preliminary dimensions and quantities of reinforcement for a reinforced concrete perimeter wall subjected to a total factored ULS load of 5,000 kN uniformly distributed over its 5 m length. Of this load, 1,250 kN is applied from the floor immediately above. The wall is part of a braced structure and is 4 m high.

Chapter 7

Case studies

7.1   Introduction

This chapter brings together all of the topics that have been presented in the previous chapters and develops an understanding of the design process up to the point of detailed design. All the basic building blocks have been provided in the chapters up to this point and this chapter provides examples showing how it all fits together. A range of different examples are considered, including:

- a simple industrial building
- an office building
- an office building with an unusual shape
- a residential building with basement car park
- a hotel
- a grandstand.

Some examples are worked through in detail in this chapter, while for others, only an outline of the shape, size and/or layout is provided, for which the students are expected to propose arrangements of beams, slabs and columns, i.e. the structural system with which to resist the applied load. The authors have consulted widely with engineers in industry in compiling these model solutions. However, it should be borne in mind that design is subjective and other engineers may prefer a different solution.

This complete design process begins with a scheme design for the structure of the building to determine how the loads are transferred to the foundations. At this stage of the process, it is also necessary to consider how the horizontal loads are resisted and if expansion joints are necessary. Materials other than concrete are considered as a good engineer should be open to all structural materials. The process, essentially, is to propose a stable and robust scheme of beams, columns and/ or other structural members that satisfies the brief. For concrete solutions, preliminary sizing of the slabs/beams/columns is carried out using the procedures outlined in Chapter 6. Next, the loads are calculated and the structure is analysed to find the stresses in the main members. Finally, preliminary areas of reinforcement are calculated and the critical sections are checked to determine if the design is reasonable.

7.2   Case Study 1 – Simple Industrial Building

Problem: Design a single-storey industrial building approximately 9 m high with plan dimensions 42 m × 70 m. The maximum roof pitch should be 6 degrees, and for structural purposes it can be assumed to be flat. To facilitate the movement of goods around the building, there should be a minimum of internal obstacles. The client also requires two 6 m wide × 6.6 m high doors on the north face of the building and anticipates possible future expansion to the south.

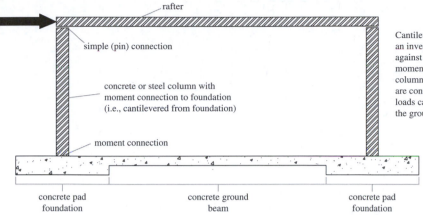

Cantilevered columns behave similarly to an inverted portal frame. Stability against lateral loads is provided by moment connections between the columns and large pad foundations that are connected by ground beams. Lateral loads cause bending in the columns and the ground beam.

Figure 7.1 Arrangement of cantilevered columns and unbraced rafters

Solution: A single-storey industrial building of this size can be constructed using a variety of structural schemes. The central issue is how to resist horizontal (wind) loading. One of the simplest means of taking horizontal force is to have cantilever columns fixed rigidly to their foundations. While ground conditions have not been specified and are assumed to be unknown, rigid foundations tend to be expensive, especially for such a high building. A solution would be to tie foundation pads with ground beams to resist their rotation – Fig. 7.1. While feasible, this would be an unusual solution and the relative height of the building (9 m) would tend to rule it out.

An alternative and generally efficient means of resisting horizontal load is a portal frame. A frame such as that illustrated in Fig. 7.2 is rigidly connected at the joints and resists horizontal load by the bending of all its members. Haunches (local thickenings) are used at the joints to transfer moment between members. Single bay portal frames are economical up to about 20 m so, for this 42 m × 70 m building, a 2 × 21 m bay portal frame would be a good solution for the 42 m dimension. It is difficult to manufacture portal frames for two directions so the 70 m dimension requires an alternative solution.

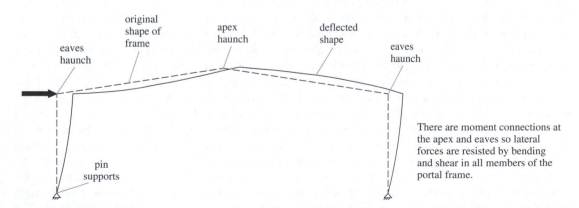

There are moment connections at the apex and eaves so lateral forces are resisted by bending and shear in all members of the portal frame.

Figure 7.2 Deflection of portal frame under E–W lateral load

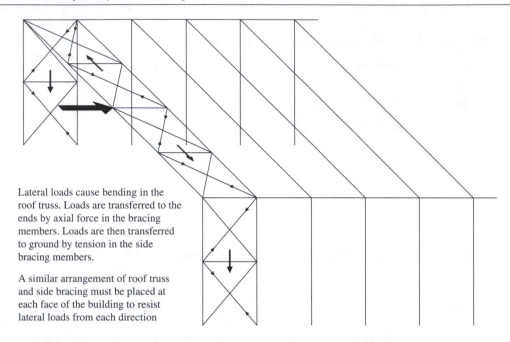

Lateral loads cause bending in the roof truss. Loads are transferred to the ends by axial force in the bracing members. Loads are then transferred to ground by tension in the side bracing members.

A similar arrangement of roof truss and side bracing must be placed at each face of the building to resist lateral loads from each direction

Figure 7.3 System of roof truss and side bracing to resist N–S lateral loading

Roof trusses with side-wall bracing is the other principal means of resisting horizontal load. Fig. 7.3 shows how a horizontal load at roof level can be transferred to the sides using a truss and how the reactions are taken down to the foundations using cross-bracing in the side walls. In reality, the roof is sloped but this does not change the effectiveness of this scheme. The span/depth ratio for a truss is about 15, so a truss spanning 42 m would be about 3 m deep in the plane of the roof. If a roof truss is only provided at one end of the building, wind loads would have to be transferred to it by compression in several roof members. As this might lead to buckling, it is common practice to provide one roof truss at each end of the building.

With cross-bracing, one diagonal member in tension resists wind in each direction so compression forces do not need to be considered, hence the term X-bracing. Therefore, an angle section is adequate for cross-bracing as only tensile forces need to be resisted. An alternative option would be to use one diagonal hollow section that is designed to resist wind loads in both directions (i.e. to resist tension and compression), as illustrated in Fig. 7.4. For aesthetic reasons, circular hollow sections are often used in roof trusses where the bracing is visible. However, in walls, where bracing members must fit within the cavity, more slender equal angle members are generally favoured.

This building requires systems to resist horizontal load in both N–S and E–W directions. Roof trusses with side wall-bracing could be used for both directions but, given that the 42 m dimension fits so well with a 2-bay portal frame solution, a combination is chosen here. This is why a portal frame is adopted for one direction and roof trusses for the other. Fig. 7.4 illustrates the solution in plan. The roof truss is made up of diagonal circular hollow sections, which are designed to resist tensile and compressive forces. While precast concrete portal frames are feasible, steel is chosen here. The portal frames are repeated at 7 m centres. Apex and eaves haunches, as illustrated in Fig. 7.5, are included to provide moment continuity across the joints and to increase the stiffness and strength of the frame. On the south face, where future expansion is possible, a portal frame is used as elsewhere in the building. On the north face however, where there are no plans for future expansion,

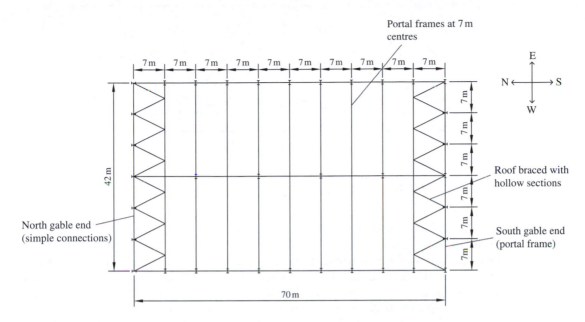

Figure 7.4 Plan view of industrial building

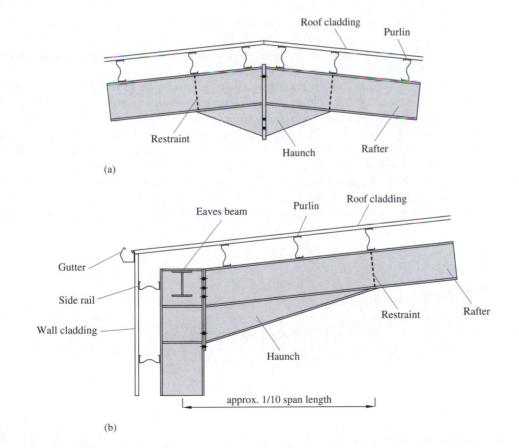

Figure 7.5 Joints of industrial building: (a) apex detail; (b) eaves detail

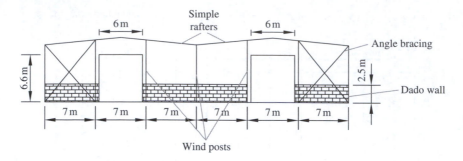

Figure 7.6 North elevation with simple connections and vertical bracing

bracing is used to resist E–W load (Fig. 7.6) as it is more economical and does not cause any obstruction to future expansion.

N–S wind load is taken by the roof trusses to wall bracing in the end bays of the east and west walls. When N–S wind load acts on the north (or south) face of the building, the cladding spans vertically between the horizontal side rails. This load is carried by the side rails to vertical beams, known as wind posts, that span from the ground to the roof truss (Fig. 7.6).

The design brief requires the number of internal obstructions to be minimized. One line of columns down the centre of the building is a good solution. However, the scheme can be improved further by removing half of these central columns. This is achieved by introducing 'jumper' valley beams (Fig. 7.7) spanning 14 m (N–S) between every second portal frame. Hence, cross-sections in alternate bays are as illustrated in Fig. 7.8(a) or (b). Where there is a jumper valley beam, it is assumed to act as a simple support and the frame resists horizontal load by bending in the remaining members of the frame.

The portal frames are placed at 7 m centres as this is an economical span for purlins (roof) and side rails (walls). Purlins and side rails are light cold rolled steel sections that support the cladding. They become uneconomical at spans of around 8 m (depending on local market conditions). Side rails span the 7 m between portal frames and are placed at 1.2 m centres. For purlins, the spacing is reduced near the valleys on the roof to allow for snow drift loading. A masonry wall is specified for the first 2.5 m to resist accidental impact from cars or forklifts and to improve the aesthetics.

The *ISE Manual for the Design of Steelwork Building Structures* recommends that movement joints are provided at approximately 60 m centres, although this can be increased for single-storey sheeted buildings. Since this single-storey sheeted building is just 70 m in length, a decision is made to

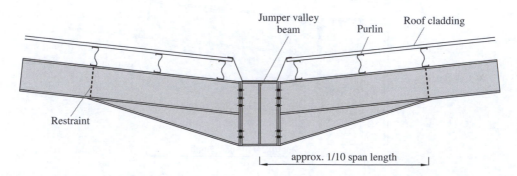

Figure 7.7 Jumper valley beam detail

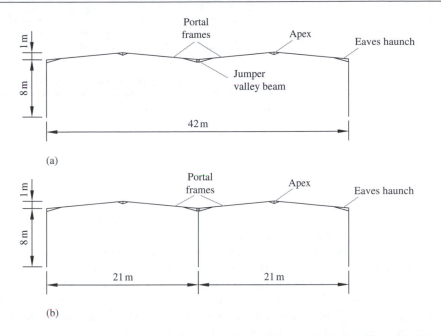

(a)

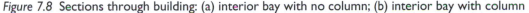

(b)

Figure 7.8 Sections through building: (a) interior bay with no column; (b) interior bay with column

provide no movement joints. The anticipated thermal movement can be calculated quite easily to confirm that this decision meets with the allowable tolerances of the building finishes.

In a simply supported structure, columns are generally designed using Universal Column sections (UCs) as they are more efficient and economical for members that are primarily subjected to axial compression. Universal Beam sections (UBs) are normally chosen for beams as they are more efficient in bending. However, the external columns of the portal frame are constructed using UBs, as bending is much more dominant than axial compression in the portal frame. The internal columns of the two span portal frames are UCs, as bending moments are less significant.

Load paths

To ensure that all loads can be adequately supported by the structure, the load paths for vertical and lateral loads should be identified. As well as the magnitude of the loads, the mechanism(s) by which loads are transferred through members to ground (i.e. bending, shear, axial force) can dictate how they are designed and must be identified.

First, let us consider the load paths for vertical loads acting on this building. Vertical loads act on the roof cladding. The cladding spans onto the purlins, which span onto the rafters (Fig. 7.5). Loads are transferred from the cladding to the purlins and from the purlins to the rafters by bending and shear.

For the portal frames, the vertical loads on the rafters are transferred by bending and shear to the exterior and internal columns or to the jumper valley beam. The loads on the columns are then transferred to foundation level by axial compression and bending. As the frames are pinned at the base, only axial forces and shear forces are transferred to the foundations. The loads on the jumper valley beam are transferred to the adjacent internal columns by bending and shear and then to ground by axial compression in the columns.

For the steel frame at the north end (Fig. 7.6), the vertical loads on the external rafters are transferred by bending and shear to the wind posts which are at 7 m centres. Since there are no

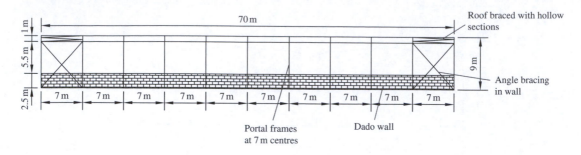

Figure 7.9 West elevation showing vertical X-bracing and roof truss

moment connections between the rafters and columns at this end, the loads on the columns are transferred to ground by axial compression and shear.

The structure must also be designed to resist horizontal loads in each direction (from wind). First, considering loads in the E–W direction, the horizontal wind loads acting on the west face (or east face) of the building are transferred from the cladding to the side rails which span N–S to the exterior columns of the portal frames. These horizontal loads are transferred to the foundations by moment and shear throughout the frame. As it is not a portal frame, the E–W loads on the north face are transferred through equal angle cross-bracing to ground (Fig. 7.6).

N–S wind loads acting on the northern face (or southern face) of the building are transferred from the cladding to the side rails which span E–W to the wind posts, which in turn span between the foundations and the roof truss. The horizontal loads at roof level are transferred to the sides using the truss, and are then transferred to ground through equal angle cross-bracing in the side walls (Fig. 7.9).

7.3 Case Study 2 – Office Building

Problem: Design a five-storey office building with the footprint illustrated in Fig. 7.10. Car parking has been provided elsewhere and is not needed in the structure.

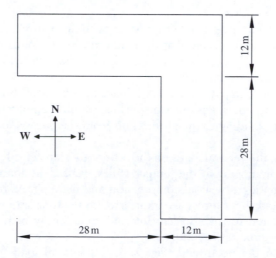

Figure 7.10 Footprint of office building

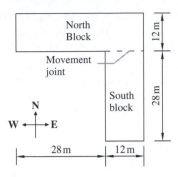

Figure 7.11 Location of movement joint in office building

Solution: When designing an office building, it should be noted that the function and layout may change over time. As such, it is desirable to have open spans where possible with non-structural internal partitions so that the building can be versatile to suit the changing needs of the client. Therefore, for this building, no more than one line of internal columns is considered acceptable.

For this problem, there is more than one feasible solution, so different options are considered before the optimum design is chosen. Due to the significant change in geometry in this building (the L-shape), a movement joint is positioned as shown in Fig. 7.11 for all proposed designs.

In the absence of detailed architectural drawings, it is assumed that the external façade is a mix of glazing and solid panels, with a high proportion of glazing. A flat roof is also assumed.

Scheme options

Concrete is chosen as a structural material as it is low maintenance and versatile, providing excellent fire resistance. Using concrete, three scheme options are considered:

(a) Option 1 – Flat slab with central and external columns, as illustrated in Fig. 7.12. When constructing a building using flat slabs there are no downstand beams so the clear floor to ceiling height can be greater, leaving more space for services and/or reducing the overall height of the building. Formwork is also simpler and can be erected quickly. As with all scheme options, since there is a movement joint in this building, the lateral stability of each part of the building must be considered separately. For this option, three stair cores are provided for vertical circulation and lateral stability, with one at the centre and one at each end of the building (stairs are generally needed near the ends of buildings for fire escape purposes). Lateral wind loads acting on the side of the building are transferred from the façade to the slabs, which transfer the load to the cores by membrane action. The cores behave as rigid hollow boxes cantilevered from the foundations, and loads in the cores are transferred to ground by bending and shear, as discussed in Section 2.3 and illustrated in Fig 2.37. A dowelled joint is adopted at the movement joint in order to transfer E–W loads from the southern block into the central core in the northern block. The dowels are debonded on one side of the joint to allow the southern block to move in the N–S direction due to thermal expansion/contraction. The two part assembly of a proprietary shear connector is illustrated in Fig. 7.13. The sleeve debonds one side of the shear connector from the surrounding concrete, allowing movement to occur in the axial direction only.

(b) Option 2 – Precast floor slab units on external edge beams and columns, as illustrated in Fig. 7.14. At 12 m, this is a very long span. However, the advantages of having no internal columns are considerable, both for construction and in use. The precast 'hollowcore' units are pretensioned prestressed concrete and are widely available. They come in 1.2 m strips and only span one way. Corbels (shelves) have to be provided in the cores when the slabs do not span all the way to the beam. Fig. 7.15 illustrates a cross-section of an edge beam, column and hollowcore unit, detailing how the beam and slab are tied together using a reinforcing bar which is bent down into the structural screed. Lateral stability is provided using two cores in each wing of the building, with one stair and one lift core at the centre and one stair core at each end of the building.

(c)   Option 3 – One-way spanning slab that spans onto cross beams supported on three columns, as illustrated in Fig. 7.16, with edge beams offset a distance of 1.5 m from the edges to minimize beam depth. A typical cross-section of a column and downstand beam is illustrated in Fig. 7.17. To allow for quicker and simpler erection of shuttering during construction, the breadth of the downstand beam is designed to be the same as the breadth of the corresponding column. As with Option 1, three stair cores are provided, with one at the centre and one at each end of the building. A dowelled joint is adopted at the movement joint to transfer lateral loads from the southern block into the central core in the northern block.

Depending on the layout of services within the building, it may be more convenient to have the downstand beams running along the length of the building rather than across the width. In this

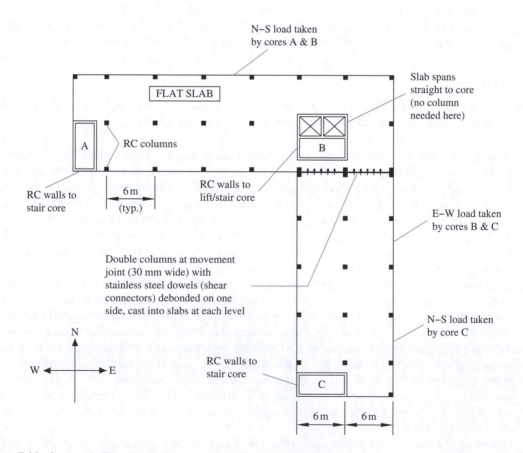

Figure 7.12 Option 1: Flat slab with external and central columns

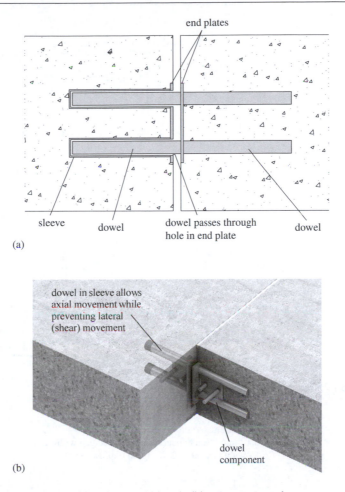

Figure 7.13 Shear connector detail: (a) schematic detail; (b) proprietary shear connector (courtesy of Ancon)

case, an alternative option would be to run the beams in the opposite direction, with edge beams on either side of the building and one beam down the centre, with the slab spanning 6 m between beams.

Scheme selection

Of the three viable schemes considered, Option 1 is selected for a variety of reasons, including the following:

1. No downstand beams within the building means greater flexibility for services and/or reduced overall building height
2. No downstand beams on perimeter allows for full height external glazing
3. Simple shutter arrangements and temporary propping requirements reduce construction time
4. Flat slabs are fast and easy to construct.

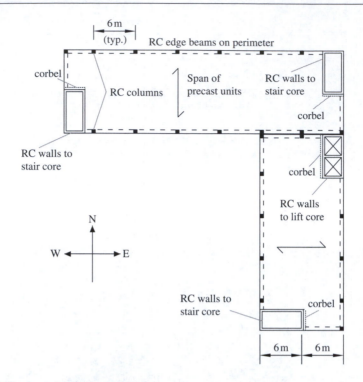

Figure 7.14 Option 2: Precast slabs on external beams

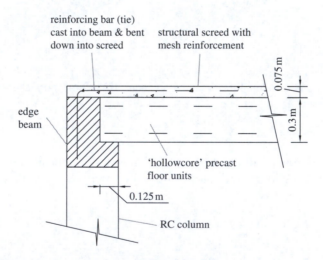

Figure 7.15 Detail for precast floor units supported by an edge beam

Loading

In EC1, the recommended imposed load for an office area is 3.0 kN/m². An additional load of 1.0 kN/m² for moveable partitions is added, giving a total imposed office load of $3.0 + 1.0 = 4.0$ kN/m². Since it is unclear how the building will be used, and to allow for future flexibility in change of use, the building is designed using a total imposed load allowance of 5.0 kN/m².

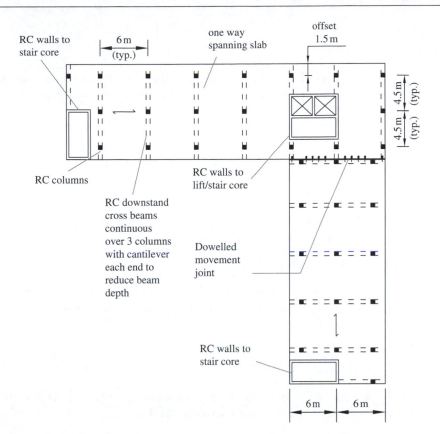

Figure 7.16 Option 3: One-way spanning slabs

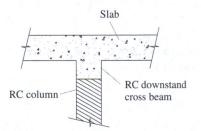

Figure 7.17 Detail of RC one-way spanning slab and downstand cross beam supported by RC column

Preliminary design for Option I

For preliminary design of the slab, C30/37 concrete and 16 mm diameter reinforcing bars are assumed. Exposure class XC1 is appropriate for the interior of the building. The minimum depth of the slab is initially dictated by the fire rating. From Table 6.6, assuming a 1 hour fire rating, the minimum depth is 180 mm with a minimum axis distance of 15 mm. The depth of the slab is sized using the span/effective depth ratio for flat slabs, which is given in Table 6.7.

span/eff.depth $= 24$

$\Rightarrow d = 6000/24 = 250\,\text{mm}$

The minimum cover depth for durability (exposure class XC1) is 15 mm. However, assuming 16 mm diameter reinforcing bars, the minimum required cover for bond is 16 mm, resulting in a nominal cover, c_{nom}, of:

$c_{\text{nom}} = 16 + 10 = 26\,\text{mm}$

which gives an axis distance greater than 15 mm. Thus, the total depth of the slab is:

$h = 250 + 26 + 16/2 = 284\,\text{mm}$

Since the span/effective depth ratios in Table 6.7 tend to be conservative, the height of the slab is rounded down to 275 mm to give a more economical design.

$\Rightarrow d = 275 - 26 - 16/2 = 241\,\text{mm}$

Assuming a self-weight of concrete of 25 kN/m^2, and assuming a weight for floor finishes/ceilings/services of 0.5 kN/m^2 (assuming client does not require screed), gives a total permanent gravity load of:

$$\text{Permanent load} = 25 \times 0.275 + 0.5$$
$$= 7.4\,\text{kN/m}^2$$

Applying the partial safety factors from Chapter 3 (1.35 for permanent and 1.5 for variable gravity load), the total maximum ultimate load per unit area on this floor is:

$$\text{Ultimate load}, w = 1.35 \times 7.4 + 1.5 \times 5$$
$$= 17.5\,\text{kN/m}^2$$

For flat slabs, it is necessary to check if the slab depth satisfies the punching shear requirements specified in Equations (6.4) and (6.5). These equations will be critical for the interior columns supporting the central panel since these columns have larger tributary areas per perimeter length. Taking the column on gridline C2 in the north block of the building (Fig. 7.18) the tributary area of the column for preliminary design is:

tributary area $= (L1 + L2) \times (L3 + L4)$

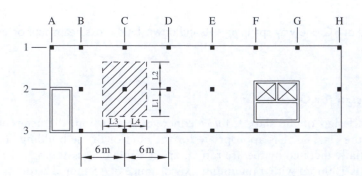

Figure 7.18 Plan of north block, illustrating tributary area of column C2

where L1 and L2 are $\frac{5}{8}(6) = 3.75$ m and L3 and L4 are each 3.0 m. Hence the tributary area is:

tributary area $= (3.75 + 3.75) \times (3.0 + 3.0) = 45 \text{ m}^2$

For flat slabs, it is necessary to check if shear reinforcement is required. Assuming a 400 mm × 400 mm 400 mm square column:

$$\frac{1250w(\text{area supported by column})}{(\text{column perimeter} + 12h)h} = \frac{1250 \times 17.5 \times 45}{(4 \times 400 + 12 \times 275)275}$$

$$= 0.73 \text{ N/mm}^2$$

$$> 0.6 \text{ N/mm}^2 \Rightarrow \text{shear reinforcement required}$$

and

$$\frac{1250w(\text{area supported by column})}{(\text{column perimeter })h} = \frac{1250 \times 17.5 \times 45}{4 \times 400 \times 275}$$

$$= 2.23 \text{ N/mm}^2$$

$$\leq 0.15 f_{ck} \; (= 4.5 \text{ N/mm}^2)$$

Note: Even if the total depth of the slab is increased to 300 mm, shear reinforcement is still required.

As the second inequality is satisfied, the flat slab depth of 275 mm with shear reinforcement is taken as the preliminary solution. When designing a slab, critical design situations need to be established and the Eurocode recommends simplified loading arrangements that can be used for buildings. For this example, two loading arrangements are considered: all spans loaded and alternate spans loaded. To determine the maximum hogging moment at the supports, all spans are loaded with the total design load (i.e. factored permanent and variable loads). To determine the maximum sagging moment at mid span, alternate spans are loaded with the total design load, while the remaining spans are loaded only with the design permanent gravity load. For this example, the total design load at ULS is 17.5 kN/m^2 and the design permanent gravity load at ULS is 1.35 $(7.4) = 10$ kN/m^2.

The flat slab is analysed using the equivalent frame method outlined in Section 5.6. It is split into column and middle strips, with a typical bay illustrated in Fig. 7.19. The width of the column strip is one-quarter of the width of the shorter side of the bay (i.e. column strips are 1.5 m wide on each side). The slab must be analysed in both directions and the resulting bending moment is distributed over the column and middle strips. According to the guidelines in EC2, when analysing flat slabs using the equivalent frame method, the relative stiffness of the slab and columns should be included in a frame model, and the full width of a panel should be used to determine the stiffness of the slab. In this example, the width of each panel of the slab is 6 m and the relative stiffness of the columns is negligible. Thus, the slab is analysed as an equivalent continuous beam in each direction. For final design, detailing guidelines should be followed to ensure sufficient reinforcement is provided to resist any bending moment in the columns due to rotation of the slab.

Analysis of N–S gridline

First, the bending moments in the N–S direction are determined with all spans loaded (e.g. along gridline C in Fig. 7.18). The total loading on a 6 m wide panel is:

$$\omega = (17.5)(6) = 105 \text{ kN/m}$$

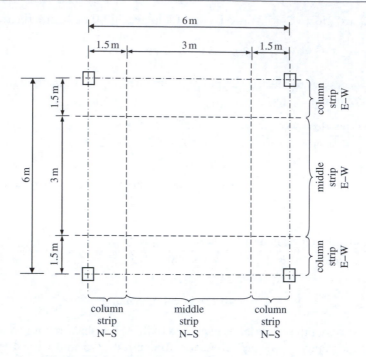

Figure 7.19 Column and middle strips of a typical bay

The slab is simplified as a continuous beam with pinned ended supports (Fig. 7.20). The maximum hogging moment at the support is $\omega l^2/8$ (since there is no rotation at the central support, each span behaves like a propped cantilever, as illustrated in Appendix B, No. 6). The bending moment diagram is illustrated in Fig. 7.20, with a maximum hogging moment of:

$$M_{\text{hog}} = \omega l^2/8 = (105)(6)^2/8 = 473 \text{ kNm}$$

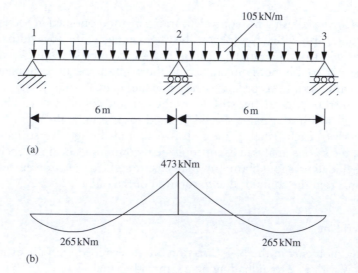

Figure 7.20 (a) loading arrangement 1 for N–S continuous beam; (b) corresponding bending moment diagram

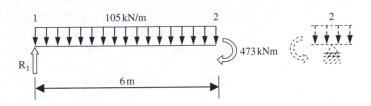

Figure 7.21 Free-body diagram between gridlines 1 and 2

To find the maximum sagging moment for all spans loaded, the point of zero shear is found. The free-body diagram for the section between gridline 1 and 2 is illustrated in Fig. 7.21. The reaction at 1 is determined by taking moments about point 2.

$$6R_1 + 473 = 105 \times 6 \times \frac{6}{2}$$
$$\Rightarrow R_1 = 236 \text{ kN}$$

The maximum sag moment occurs where shear force is zero. This distance x is found using the free-body diagram of Fig. 7.22. By equilibrium of the vertical forces:

$$V_x = R_1 - \omega x = 0$$
$$\Rightarrow x = \frac{R_1}{\omega} = \frac{236}{105}$$
$$= 2.25 \text{ m}$$

Thus, the maximum sagging moment, M_x, is given by:

$$M_x = R_1 x - \omega x^2/2$$
$$= (236)(2.25) - (105)(2.25)^2/2$$
$$= 265 \text{ kNm}$$

Therefore, the maximum hogging moment is 473 kNm and the maximum sagging moment is 265 kNm. This procedure is repeated with span 1–2 loaded with the total design load ($17.5 \times 6 = 105$ kN/m) and span 2–3 loaded with the design permanent gravity load ($10 \times 6 = 60$ kN/m). This results in a maximum hogging moment of 371 kNm and a maximum sagging moment of 305 kNm. Considering both load cases, the maximum hogging moment in the slab is 473 kNm and the maximum sagging moment is 305 kNm.

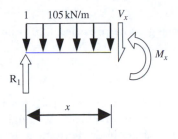

Figure 7.22 Free-body diagram between 1 and x

The maximum moments are distributed across the width of the slab, which is divided into column and middle strips, as illustrated in Fig. 7.19. The combined width of the column strips is 3 m and the width of the middle strip is also 3 m. EC2 specifies that the column strips take 60–80 per cent of the total hog moment and the middle strips take 40–20 per cent. The values recommended by the ISE manual are within these ranges and are adopted here. Hence, the column strips are assigned 75 per cent of the hogging moment ($473 \times 0.75 = 354$ kNm) and the middle strip is assigned 25 per cent ($473 \times 0.25 = 118$ kNm). This results in moments per unit width of:

$$m_{x\_hog}(\text{column}) = 354/3 = 118 \,\text{kNm/m}$$
$$m_{x\_hog}(\text{middle}) = 118/3 = 39 \,\text{kNm/m}$$

For the sagging moment, the column strips are assigned the percentage recommended in the ISE manual, which is 55 per cent of the moment ($305 \times 0.55 = 168$ kNm) and the middle strip is assigned 45 per cent ($305 \times 0.45 = 137$ kNm). This results in moments per unit width of:

$$m_{x\_sag}(\text{column}) = 168/3 = 56 \,\text{kNm/m}$$
$$m_{x\_sag}(\text{middle}) = 137/3 = 46 \,\text{kNm/m}$$

The slab must be designed to resist the moment in each strip in the N–S direction. The required area of hogging reinforcing steel for the column and middle strips is calculated below. The required area of sagging reinforcement can be calculated in a similar manner.

Middle strip hogging moment

From Equation (6.11):

$$K = \frac{M}{f_{ck}bd_{N-S}^2} = \frac{39 \times 10^6}{30 \times 1000 \times 241^2} = 0.022$$

As $K < 0.167$, compression reinforcement will not be required. From Equation (6.10), the lever arm ratio is:

$$\frac{z}{d_{N-S}} = \left(0.5 + \sqrt{0.25 - 0.88K}\right) = \left(0.5 + \sqrt{0.25 - 0.88(0.022)}\right) = 0.98 > 0.95$$
$$\Rightarrow z = 0.95d_{N-S} = 229 \,\text{mm}$$

Thus, the required area of tension reinforcement is:

$$A_s = \frac{M}{0.87f_{yk}z} = \frac{39 \times 10^6}{0.87 \times 500 \times 229} = 392 \,\text{mm}^2/\text{m}$$

⇒H16 top bars at 250 mm spacing to be provided (250 mm is maximum spacing for principal reinforcement, see Section 6.5)

$$\Rightarrow A_s = 804 \,\text{mm}^2/\text{m}$$

Note: It is recommended that the same bar diameter is used for the column and middle strips, varying the spacing to give the required area of reinforcement. Although the required area of reinforcement is small for the middle strip, there is a larger hogging moment in the column strip. Therefore, 16 mm diameter reinforcing bars are chosen.

Column strip hogging moment

From Equation (6.11):

$$K = \frac{118 \times 10^6}{30 \times 1000 \times 241^2} = 0.0677$$

As $K < 0.167$, compression reinforcement will not be required. From Equation (6.10), the lever arm ratio is:

$$\frac{z}{d_{N-S}} = \left(0.5 + \sqrt{0.25 - 0.88(0.0677)}\right) = 0.936$$
$$\Rightarrow z = 0.936 d_{N-S} = 226 \text{ mm}$$

Thus, the required area of top reinforcement is:

$$A_s = \frac{M}{0.87 f_{yk} z} = \frac{118 \times 10^6}{0.87 \times 500 \times 226} = 1200 \text{ mm}^2/\text{m}$$

EC2 defines A_t as the total area of reinforcement required to resist the full negative moment from the half panels on each side of the column. In this example since each bay is 6 m wide, A_t is the total area of steel required to resist the hogging moment in 1 bay (i.e. total area of top reinforcement required in column and middle strips). EC2 then recommends that $0.5 A_t$ be placed within an area of 0.125 times the panel width on either side of the column.

$$\Rightarrow A_t \text{ for 1 bay (6 m wide)} = (392 \times 3) + (1200 \times 3) = 4776 \text{ mm}^2$$

Reinforcement of area $0.5 A_t$ ($= 2388 \text{ mm}^2$) is distributed over an area of $2(0.125 \times 6) = 1.5$ m. Therefore, $1592 \text{ mm}^2/\text{m}$ is required over an area of 1.5 m, which is 0.75 m on either side of the centre line of the column.

$$\Rightarrow \text{H16 top bars at 125 mm spacing to be provided}$$

$$\Rightarrow A_s = 1608 \text{ mm}^2/\text{m}$$

In the column strip outside the middle 1.5 m, the area of reinforcement required is the area of reinforcement required in the column strip ($1,200 \text{ mm}^2/\text{m}$) minus the area of reinforcement provided in the middle 1.5 m:

$$A_s = 1200 \times 3 - (1.5)(1608) = 1188 \text{ mm}^2/\text{m}$$

$$\Rightarrow \text{H16 top bars at 150 mm spacing to be provided}$$

$$\Rightarrow A_s = 1340 \text{ mm}^2/\text{m}$$

Analysis of E–W Gridline

For the E–W direction, the slab is analysed along gridline 2 between A and F (Fig. 7.18). Firstly, the bending moments are determined with all spans loaded. As above, the total loading on a 6 m wide panel is:

$$\omega = (17.5)(6) = 105 \text{ kN/m}$$

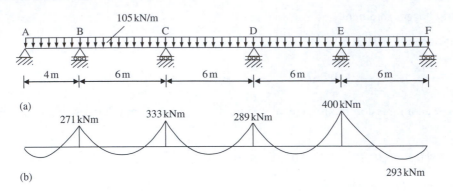

(a)

(b)

Figure 7.23 (a) loading arrangement I for E–W continuous beam; (b) corresponding bending moment diagram

The slab, simplified as a continuous beam (illustrated in Fig. 7.23), is analysed using moment distribution. The bending moment diagram is illustrated in Fig. 7.23, with a maximum hogging moment of 400 kNm.

The maximum sagging moment for all spans loaded is in span E–F. To find the maximum sagging moment, the point of zero shear is found. The reaction at F is determined by taking moments about point E.

$$6R_F + 400 = 105 \times 6 \times \frac{6}{2}$$
$$\Rightarrow R_F = 248 \text{ kN}$$

The shear force at a distance x from support F is zero. By equilibrium of the vertical forces:

$$V_x = R_F - \omega x = 0$$
$$\Rightarrow x = \frac{R_F}{\omega} = \frac{248}{105}$$
$$= 2.36 \text{ m}$$

Thus, the maximum sagging moment, M_x, is given by:

$$M_x = R_F x - \omega x^2/2$$
$$= (248)(2.36) - (105)(2.36)^2/2$$
$$= 293 \text{ kNm}$$

Therefore, for all spans loaded, the maximum hogging moment is 400 kNm and the maximum sagging moment is 293 kNm. This procedure is repeated with alternate spans loaded, resulting in a maximum hogging moment of 315 kNm and a maximum sagging moment of 328 kNm. Considering both load cases, the maximum hogging moment in the slab is 400 kNm and the maximum sagging moment is 328 kNm.

Note: Both load cases have been considered systematically in this example. Generally, the load case with all spans loaded will give the maximum hogging moment and the load case with alternate spans loaded will give the maximum sagging moment for design.

Again, the maximum moments are distributed across the width of the slab, into column and middle strips, as illustrated in Fig. 7.19. For the E–W direction, the total width of the column strips is 3 m and the width of the middle strip is 3 m. The column strips are assigned 75% of the hogging moment ($400 \times 0.75 = 300$ kNm) and the middle strip is assigned 25% ($400 \times 0.25 = 100$ kNm). This results in moments per unit width of:

$$m_{x\_hog}(\text{column}) = 300/3 = 100 \text{ kNm/m}$$
$$m_{x\_hog}(\text{middle}) = 100/3 = 33 \text{ kNm/m}$$

For the sagging moment, the column strips are assigned 55% of the moment ($328 \times 0.55 = 180$ kNm) and the middle strip is assigned 45% ($328 \times 0.45 = 148$ kNm). This results in moments per unit width of:

$$m_{x\_sag}(\text{column}) = 180/3 = 60 \text{ kNm/m}$$
$$m_{x\_sag}(\text{middle}) = 148/3 = 49 \text{ kNm/m}$$

The slab is then designed to resist the moment in each strip in the E–W direction. Considering each strip separately, the required area of reinforcing steel can be calculated using the method demonstrated above for the N–S direction.

Note: For design, it is assumed that the outer steel of the slab is in the N–S direction, with $d_{N-S} = 241$ mm. Therefore, the steel running in the E–W direction is the inner layer of reinforcement with an effective depth of $d_{E-W} = 241 - 16 = 225$ mm.

Column design (e.g. gridline C2)

Taking a critical column, on gridline C2 say, the tributary area is illustrated in Fig. 7.18 and is calculated as:

$$(\text{L1} + \text{L2}) \times (\text{L3} + \text{L4}) = 7.5 \times 6.0 = 45 \text{ m}^2$$

The total design load of the slab is 17.5 kN/m², resulting in a total axial load on the column from each floor of:

$$N_{\text{floor}} = 45 \times 17.5$$
$$= 787.5 \text{ kN}$$

Bending moments can also be transferred from the slab into the column due to the lack of symmetry in the arrangement of loads. For detailed design, this moment can be calculated using sub-frame analysis, as discussed in Section 5.5. For preliminary design, the ultimate applied load from the floor immediately above the column being considered should be multiplied by a factor, as discussed in Section 6.6. Thus, the applied loads from the first floor are factored by 1.25 (i.e. column balanced in two perpendicular directions, Fig. 6.18(a)) to allow for bending effects. The total ultimate load on the column is:

$$N_{\text{total}} = (3 \times 787.5) + (1.25 \times 787.5)$$
$$= 3347 \text{kN}$$

Assuming 2 per cent reinforcement and $f_{ck} = 35$ kN/m², Equation (6.15) gives:

$$N = A_c \left(0.44 f_{ck} + \frac{p}{100} \left(0.67 f_{yk} - 0.44 f_{ck} \right) \right)$$
$$\Rightarrow \quad 3347 \times 10^3 = A_c \left(0.44(35) + \frac{2}{100} \left(0.67 \times 500 - 0.44 \times 35 \right) \right)$$
$$\Rightarrow \quad A_c = \frac{3347 \times 10^3}{(0.44(35) + 2/100(0.67 \times 500 - 0.44 \times 35))}$$
$$= 153,588 \text{ mm}^2$$

Assuming the column is square, the side length, h, is given by:

$$h = \sqrt{153,588} = 391.9 \, \text{mm}$$

This is rounded up to give a preliminary side length of 400 mm, which was the assumed column size for the punching shear check above.

For the calculated depth of the column, it is necessary to confirm the initial assumption that the column is 'stocky'. Assuming a clear height between the foundation pad and the first floor of 4 m, the effective height of the column is:

$$l_0 = 085l = 085(4) = 3.4 \, \text{m}$$

Thus:

$$\frac{l_0}{h} = \frac{3.4}{0.4} = 8.5 < 15$$

Hence, the column is indeed 'stocky'. As the dimensions of the column have been rounded up to 400 mm, the required percentage of reinforcement is recalculated.

$$N = A_c \left(0.44 f_{ck} + \frac{p}{100} \left(0.67 f_{yk} - 0.44 f_{ck} \right) \right)$$

$$\Rightarrow \quad \frac{p}{100} = \frac{N - 0.44 f_{ck} A_c}{A_c (0.67 f_{yk} - 0.44 f_{ck})} = \frac{3347 \times 10^3 - 0.44(35)(400)^2}{(400)^2 (0.67 \times 500 - 0.44 \times 35)}$$

$$\Rightarrow \quad \frac{p}{100} = 0.0173$$

which gives an area of reinforcement of 2,768 mm². It must be checked if this is greater than the minimum area of longitudinal reinforcement of:

$$A_{s,\min} = \text{greater of } \left(0.1 N \gamma_s / f_{yk} \text{ and } 0.002 A_c \right)$$

$$= \text{greater of } (0.1 \times 3347 \times 10^3 \times 1.15/500 \text{ and } 0.002 \times 400^2)$$

$$= \text{greater of } (770 \text{ and } 320)$$

$$\Rightarrow A_{s,\min} = 770 \text{mm}^2$$

Therefore, 2,768 mm² of reinforcement is sufficient and one possible solution is with the provision of four 20 mm diameter bars and four 25 mm diameter bars as illustrated in Fig. 7.24 (total area of

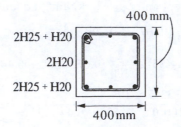

Figure 7.24 Preliminary solution for column

3,220 mm^2). In addition, links are provided to give restraint against buckling of the main reinforcement.

7.4   Case Study 3 – Doughnut Shaped Office Building

Propose an outline structural scheme for a four-storey office building with the footprint illustrated in Fig. 7.25. Include the choice of material, grid layout, how lateral stability is provided and some preliminary calculations for the slab.

Solution: As with Case Study 2, for this building it is desirable to have large open spans with the minimum number of internal columns to allow the layout to be adapted as its function changes. Despite the unusual geometry of this building there is more than one suitable scheme design, so at preliminary design stage, two alternatives are considered, and the advantages and disadvantages of each are noted.

   For vertical circulation, cores are provided, with each including a stair core and two lifts. One core is provided at each side of the building, with one just inside the lobby area. In the absence of detailed architectural details, it is assumed that the external façade is a mix of glazing and solid panels, with a high proportion of glazing. A flat roof is also assumed. The maximum dimension of the building is equal to 50 m, and as there is no significant change in geometry, a movement joint is not considered necessary.

Scheme options

Due to the curved shape of the building, *in situ* concrete is considered a suitable structural material. Concrete is versatile and formwork can be shaped to give the required outline of the building. Concrete is also low maintenance and provides excellent fire resistance. Using *in situ* concrete, two scheme options are proposed:

(a)   Option 1 – a flat slab with columns around the inner and outer perimeters of the building – see Fig. 7.26. When constructing a building using flat slabs there are no downstand beams so the clear floor to ceiling height can be greater, leaving more space for services and/or reducing the overall height of the building. This also simplifies the formwork arrangement, which is desirable given the complex shape of the building. Generally, for a flat slab to be economic, more than one bay is required. Since there is just one bay radially in this example, the columns are offset from the outer perimeter by 1 m, reducing the span and increasing efficiency. Between the columns, there is a large radial span of 12 m, which would result in a very deep (and heavy) reinforced concrete flat slab. Therefore, the slab is prestressed (post-tensioned) radially to minimize the structural depth and to reduce deflections. The outer circumference length is divided into 20 bays (each bay 18°). Circumferentially, the spans between the columns are smaller (maximum of 7.51 m for typical bay), so ordinary reinforcing steel is sufficient in this direction. The radial prestressing will have the effect of stressing the concrete tangentially too, but this effect is neglected here. It is noted that a portion of the slab around the cores may have to be reinforced rather than prestressed radially, as prestressing the slab in this area would lead to stresses developing due to the restraint provided by the cores. This scheme has the advantage of large open spans with no internal columns, making the building more versatile. For aesthetic reasons, columns are designed with a circular cross-section.

(b)   Option 2 – One-way spanning slab spanning radially onto circumferential cross beams. The beams in turn are supported on three rows of columns running circumferentially around the

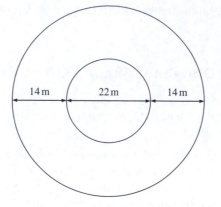

Figure 7.25 Footprint of office building for Case Study 3

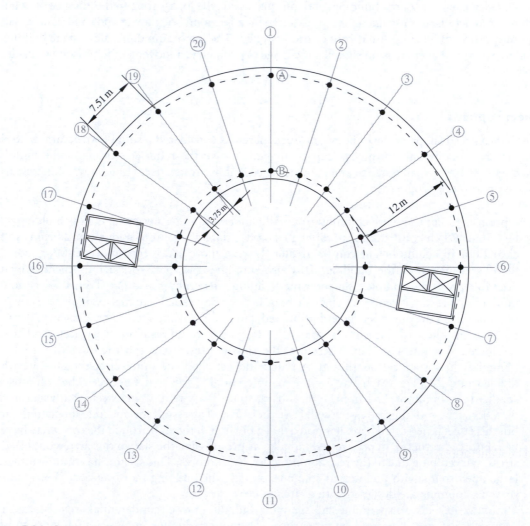

Figure 7.26 Option 1: Flat slab with external columns

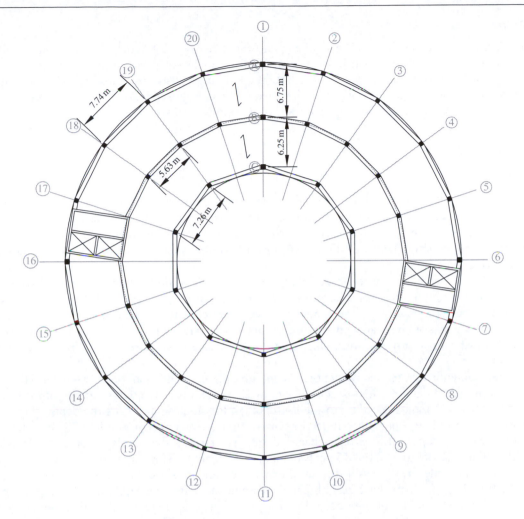

Figure 7.27 Option 2: One way spanning slabs

building, as illustrated in Fig. 7.27. As with Option 1, the building is divided into 20 bays and the columns are positioned on gridlines 18° apart. Around the inner circumference, columns are placed on every second gridline, resulting in a more economical span. The inner edge beams are offset a distance of 500 mm so that the straight beam does not affect the outline of the inner circumference of the building. To allow for quicker and simpler erection of shuttering during construction, the breadth of the downstand beam is designed to be the same as the breadth of the corresponding column.

Lateral stability

As discussed previously, for vertical circulation there are two cores, one at each side of the structure. Lateral loads acting on any side of the building are transferred from the façade to the slabs on each floor by bending and shear. These loads are then transferred to the cores by membrane action in the slab. The cores behave as rigid hollow boxes cantilevered from the foundations, and provide stability in all directions. Lateral loads in these cores are transferred to ground by bending and shear, as discussed in Section 2.3 and illustrated in Fig 2.37.

Loading

In EC1, the recommended imposed load for an office area is 3.0 kN/m^2. An additional load of 1.0 kN/m^2 for moveable partitions is also added, giving a total imposed office load of $3.0 + 1.0 = 4.0 \text{ kN/m}^2$. Since it is unclear how the building will be used, and to allow for future flexibility in change of use, the building is designed using an imposed load allowance of 5.0 kN/m^2. Preliminary sizing of the slab is carried out for both options discussed.

Preliminary design

Option I

The solution is a flat slab that is prestressed in one direction and reinforced in the other. Even with only one span radially, the slab is a flat slab and column strips can be assumed between columns and middle strips elsewhere. Radially, the column strips take most of the moment but are post-tensioned to resist it. Tangentially, there are column strips over both the inner and outer rings of columns and middle strips in-between.

Because of the geometry of the structure, the radial prestress inadvertently generates stresses in the tangential direction. Prestressing causes a reduction in the circumference of the outer perimeter and an increase in the circumference of the inner one. The effect is to generate tangential compression on the outer parts of the slab and tension in the inner parts. As the stresses due to prestress are used mostly to improve serviceability, the slabs is treated as unprestressed in the tangential direction.

The recommended span/effective depth ratio in Table 6.8 for multi-span prestressed flat slabs is 36 for a total imposed load of 5 kN/m^2. For a span of 12 m, this gives an effective depth of the slab of $(12,000/36 =)$ 333 mm. According to the ISE manual, for single span floors the depth of the slab should be increased by approximately 15 per cent, resulting in an effective depth of 383 mm. For this building, there is a 1 m cantilever on each end of the single span. If the cantilever were 15 per cent of the main span (i.e. 1.8 m) then this 15 per cent increase in depth for single span floors wouldn't apply. Given the 1m cantilever, it is assumed that the increase in depth should be approximately equal to $(1.8 - 1.0)/1.8 \times 15 = 7$ per cent, resulting in a recommended effective depth of 356 mm. Assuming exposure class XC1 and a 30 mm deep duct for the prestressing tendons, the minimum cover for durability is 15 mm and the minimum cover for bond is 30 mm. This results in a nominal cover of 40 mm. A 1 hour fire rating for a prestressed concrete flat slab requires a minimum axis distance of 30 mm and a minimum overall depth of 180 mm (ISE manual), neither of which are governing criteria for this problem. Thus, the overall required depth is:

$$h = 356 + 30/2 + 40 = 411 \text{ mm}$$

Since recommendations in the ISE manual tend to be conservative, this is rounded down to 400 mm.

The depth of the slab may also be governed by shear capacity. For this check, assuming a weight for floor finishes/ceilings/services of 0.5 kN/m^2, gives a total permanent gravity load of:

$$\text{Permanent load} = 25 \times 0.4 + 0.5$$
$$= 10.5 \text{ kN/m}^2$$

Applying the partial safety factors from Chapter 3 (1.35 for permanent and 1.5 for variable gravity load), the total maximum ultimate load per unit area on this floor is:

$$\text{Ultimate load}, w = 1.35 \times 10.5 + 1.5 \times 5$$
$$= 21.7 \text{ kN/m}^2$$

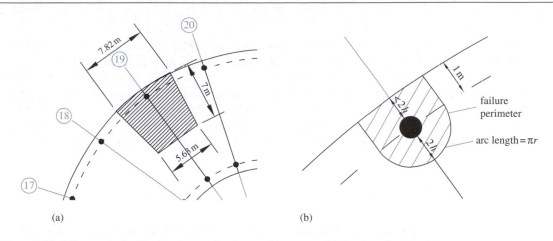

Figure 7.28 (a) assumed tributary area of column; (b) assumed failure perimeter

For flat slabs, it is necessary to check if the slab depth satisfies the punching shear strength requirements given by Equations (6.4) and (6.5). These equations are used as a punching shear check for the preliminary design of slabs. These equations will be critical for the columns around the outer circumference of the building, since these columns have the largest tributary areas. For example, taking the column on gridline A19, Fig. 7.28(a), the tributary area of the column for preliminary design is assumed to be:

$$A = (7.82 + 5.63)/2 \times 7 = 47 \, \text{m}^2$$

It is assumed for preliminary design that the columns are circular with a 500 mm diameter. The perimeter stress is given by Equation (6.4) below:

$$\text{perimeter stress} = \frac{1250w(\text{area supported by column})}{(\text{column perimeter} + 12h)h}$$

For a square column, the expression '(column perimeter + 12h)' corresponds approximately to the failure perimeter assumed to be 2h from the edge of the column, with rounded corners. For this example the failure perimeter is assumed to be 2h from the column perimeter, extended straight to the edge of the slab, as illustrated in Fig. 7.28(b). Thus, the failure perimeter is calculated as:

$$\text{failure perimeter} = \pi r + (2 \times 1000) = \pi(500/2 + 2 \times 400) + (2 \times 1000) = 5300 \, \text{mm}$$

The perimeter stress can now be checked to determine if shear reinforcement is required.

$$\frac{1250w(\text{area supported by column})}{(\text{failure perimeter})h} = \frac{1250 \times 21.7 \times 47}{(5300)400}$$

$$= 0.6$$

$$\leq 0.6 \, \text{N/mm}^2 \ \Rightarrow \ \text{shear reinforcement not required}$$

and assuming class C40/50 concrete:

$$\frac{1250w(\text{area supported by column})}{(\text{column perimeter})h} = \frac{1250 \times 21.7 \times 47}{2 \times 250 \times \pi \times 400}$$

$$= 2.03 \, \text{N/mm}^2$$

$$\leq 0.15 f_{ck} \ (= 6 \, \text{N/mm}^2)$$

Option 2

For Option 2, the recommended span/effective depth ratio in Table 6.7 (from the ISE manual) for the end span of a one-way spanning slab is 26. For a span of 6.75 m, this gives an effective depth of the slab of (6750/26 =) 260 mm. Assuming class C30/35 concrete, exposure class XC1 and 16 mm diameter reinforcing bars, the minimum cover for durability is 15 mm and the minimum cover for bond is 16 mm. This results in a nominal cover of 26 mm. A 1 hour fire rating for a solid reinforced concrete slab requires a minimum axis distance of 15 mm and a minimum overall depth of 80 mm (Table 6.6), neither of which are governing criteria for this problem. This results in an overall depth of:

$$h = 260 + 16/2 + 26 = 294 \, \text{mm}$$

This is reduced to 275 mm as the recommendations in the ISE manual tend to be conservative. Thus, the effective depth is $275 - 26 - 16/2 = 241$ mm. Assuming a weight for floor finishes/ceilings/services of 0.5 kN/m^2, gives a total permanent gravity load of:

$$\text{Permanent load} = 25 \times 0.275 + 0.5$$
$$= 7.4 \, \text{kN/m}^2$$

Applying the partial safety factors, the total maximum ultimate load per unit area on this floor is:

$$\text{Ultimate load}, w = 1.35 \times 7.4 + 1.5 \times 5$$
$$= 17.5 \, \text{kN/m}^2$$

Preliminary design of slab

For preliminary design, analysis is carried out based on the 'all spans loaded' loading arrangement to find the maximum hogging moment in the slab. Taking a strip of the slab along gridline 20 for example, the load per metre width on the slab is 17.5 kN/m. The slab is simplified as a continuous two-span beam with pinned end supports (Fig. 7.29(a)). Along gridline 20, the slab spans 6.75 m between columns on gridline A and B and 6.75 m between the column on gridline B and the beam on gridline C (Fig. 7.27). Since the length and load on each span is equal, there is no rotation over the central support, and each span behaves as a propped cantilever. Thus, the maximum hogging moment over the central support is (Appendix B, No. 6):

$$M_B = \frac{wl^2}{8} = \frac{17.5 (6.75)^2}{8}$$
$$\Rightarrow M_B = 100 \, \text{kNm}$$

For final design, the moment would be redistributed to account for the ductile behaviour of reinforced concrete in bending, as discussed in Section 5.3. However, for preliminary design, the slab is sized based on the linear elastic bending moment. The bending moment diagram is illustrated in Fig. 7.29(b). From Appendix B, No. 6, the reaction at A and C is:

$$R_A = R_C = \frac{3wl}{8} = \frac{3(17.5)(6.75)}{8} = 44.3 \, \text{kN}$$

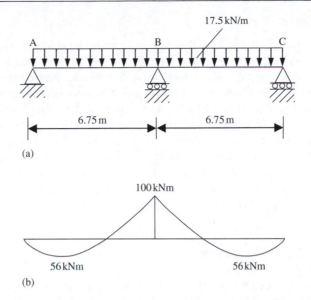

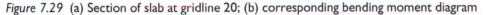

Figure 7.29 (a) Section of slab at gridline 20; (b) corresponding bending moment diagram

The reaction at B is:

$$R_B = 2 \times \frac{5wl}{8} = 2 \times \frac{5(17.5)(6.75)}{8} = 147.7 \text{ kN}$$

To calculate the required area of reinforcement for hogging sections, from Equation (6.11), we have:

$$K = \frac{M}{f_{ck}bd^2}$$
$$= \frac{100 \times 10^6}{30 \times 1000 \times 241^2}$$
$$= 0.0572$$

As $K < 0.167$, compression reinforcement will not be required. From Equation (6.10), the lever arm is:

$$z = d\left(0.5 + \sqrt{0.25 - 0.88K}\right)$$
$$= d\left(0.5 + \sqrt{0.25 - 0.88(0.0572)}\right)$$
$$= 0.947d = 228 \text{ mm}$$

Thus, the required area of tension reinforcement is:

$$A_s = \frac{M}{0.87f_{yk}z}$$
$$= \frac{100 \times 10^6}{0.87 \times 500 \times 228}$$
$$= 1005 \text{ mm}^2$$

One 16 mm diameter bar has a cross-sectional area of 201 mm². Thus, H16 top bars at 175 mm spacing are provided (total area = 1,149 mm²) as tension reinforcement.

Preliminary design of beam

The uniformly distributed loads on the beams are equal to the reactions from the slab (i.e. $R_A = R_C = 44.3$ kN/m, $R_B = 147.7$ kN/m). However, the self-weight of the beam itself must also be determined. The recommended span/effective depth ratio for an interior span of a continuous reinforced concrete beam is 20 (Table 6.5). For flanged sections with (flange width/rib width) >3, this value should be multiplied by 0.8. Initially, it is assumed that the (flange width/rib width) >3, resulting in a span/effective depth ratio of 16. Gridline B is critical, with a load from the slab of $\omega = 147.7$ kN/m and span of 5.63 m. This gives a recommended effective depth for the beam of (5630/16 =) 352 mm. Assuming 12 mm shear links, 25 mm diameter reinforcing bars and exposure class XC1, the minimum cover for durability is 15 mm and the minimum cover for bond is 25 mm. This results in a nominal cover of 35 mm. A 1 hour fire rating requires a minimum axis distance of 25 mm and a minimum breadth of 150 mm (Table 6.4), neither of which govern. This results in an overall depth of:

$$h = 352 + 35 + 12 + 25/2 = 412 \text{ mm}$$

However, the depth of the beam may also be governed by shear capacity. The cross-section of the columns are assumed to be 500 mm × 500 mm. To simplify the fabrication and erection of formwork, the breadth of the beams is designed to be the same as the columns. Rearranging Equation (6.1), the required depth of the beam can be calculated using:

$$d = \frac{V}{2.0b_w}$$

where V is the maximum shear force in the beam, considered as simply supported:

$$V = \frac{\omega l}{2} = \frac{147.7 \times 5.63}{2} = 416 \text{ kN}$$

Thus, to provide adequate shear capacity, the effective depth of the beam must be greater than:

$$d = \frac{416 \times 10^3}{2.0(500)} = 416 \text{ mm}$$

This results in an overall depth of:

$$h = 416 + 35 + 12 + 25/2 = 476 \text{ mm}$$

This is rounded to the nearest 25 mm to give an overall depth of 475 mm and an effective depth of 415.5 mm (say 416 mm). The additional factored permanent load on the beam due to self-weight is:

$$\text{Additional factored permanent load} = 1.35 \times 25 \times 0.5 \times (0.475 - 0.275)$$
$$= 3.4 \text{ kN/m}^2$$

The total maximum ultimate load on the beam is therefore:

$$\text{Ultimate load, } \omega = 147.7 + 3.4$$
$$= 151.1 \text{ kN/m}$$

Since the beams are continuous with equal spans, with all spans loaded there is no rotation over the supports and each beam behaves as though it has fixed-fixed supports (Appendix B, No. 8), with a maximum hogging moment over the supports of:

$$M_{\text{hog}} = \frac{\omega l^2}{12} = \frac{151.1(5.63)^2}{12} = 399 \text{ kNm}$$

To calculate the required area of reinforcement for hogging sections, from Equation (6.11), we have:

$$K = \frac{M}{f_{\text{ck}} b_{\text{w}} d^2}$$
$$= \frac{399 \times 10^6}{30 \times 500 \times 416^2}$$
$$= 0.154$$

As $K <0.167$, compression reinforcement will not be required. From Equation (6.10), the lever arm is:

$$z = d\left(0.5 + \sqrt{0.25 - 0.88K}\right)$$
$$= d\left(0.5 + \sqrt{0.25 - 0.88(0.154)}\right)$$
$$= 416 \times 0.84 = 349 \text{ mm}$$

Thus, the required area of tension reinforcement is:

$$A_{\text{s}} = \frac{M}{0.87 f_{\text{yk}} z}$$
$$= \frac{399 \times 10^6}{0.87 \times 500 \times 349}$$
$$= 2628 \text{ mm}^2$$

One 25 mm diameter bar has a cross-sectional area of 491 mm². Thus, six H25 reinforcing bars are provided (total area = 2,945 mm²) spread over the effective width of the flange but concentrated over the web width. The 500 mm wide beam is sufficient to fit this many bars with adequate space between them.

Scheme selection

Of the two viable schemes detailed above, Option 1 is selected for a variety of reasons, including the following:

1. No internal columns provides a space with optimum flexibility.
2. No upstand or downstand beams within the building means greater flexibility for services and/or reduced overall building height.
3. No downstand beams on perimeter allows for full height external glazing.
4. Simple shutter arrangements and temporary propping requirements reduce construction time.
5. Flat slabs are fast and easy to construct.
6. Flat slabs are economical (given current practices at time of publication).

7.5   Case Study 4 – Residential Building with Underground Car Park

Problem: Propose two alternatives designs for a three-storey residential building within the site boundaries illustrated in Fig. 7.30 with a minimum of 32 one-bedroom apartments of about 50 m^2 each. An underground car park is required, with a minimum of one 2.4 m × 4.8 m space per apartment.

Solution: It is common for multi-storey residential buildings to be constructed using precast concrete floor slabs (e.g. wideslab, hollowcore or similar type reinforced or prestressed units) supported on masonry (blockwork) loadbearing walls. This form of construction is quick to construct, is durable and has good thermal and sound insulation properties. In many cases, the internal partition walls between apartments are also constructed from blockwork to provide additional sound insulation and fire resistance.

The optimum structural grid for a car park is generally not the same as the optimum grid for a residential building, so consideration must be given to this when choosing the structural layout. When designing a residential building with an underground car park, a decision therefore has to be made on whether to (attempt to) line up the structural grid of the building and the car park in order to transfer vertical loads directly to the foundations, or to design a 'transfer' structure (e.g. a deep 'podium' flat slab or beams) at ground floor level that transfers the loads from the grid of the building above to the grid of the car park below. This decision may determine the location of the building on the site. The degree of complexity of the structure and the relative (mis)alignment of the two structural grids will often determine the appropriate transfer solution. For either solution, the stairs and lift core for the building should continue down to basement level to provide structural stability and vertical circulation from the car park to the apartments. Consideration must also be given to the location of the ramp for vehicles to access the underground car park. It is assumed for this case study that landscaping is provided on the remaining area of the site.

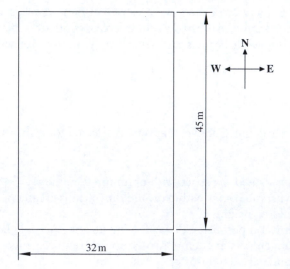

Figure 7.30 Footprint of site boundaries

Scheme options

Blockwork is chosen as the structural material for the walls and precast concrete slabs are chosen for the floors. Using this structural arrangement and the design flexibility permitted in terms of the layout of the apartments within the allowable site, two alternative scheme options are considered:

(a) Option 1 – *Transfer beams*

The proposed layout plan of the residential building for this solution is illustrated in Fig. 7.31. There are six apartments on each floor, with a total of 18 apartments in one building. Two identical buildings are constructed on the site to provide a total of 36 apartments, with floor area of approximately 54 m² each.

Precast concrete floors span between the external walls and the internal corridor walls as indicated. A basement car park is provided throughout the entire site, with the structural grid illustrated in Fig. 7.32. The footprints of the residential buildings overhead are shown shaded. Parking is provided along the east and west sides and in the central area. A 6 m wide aisle provides circulation space. As indicated, columns are offset from the corners of the 4.8 m long parking bays to allow adequate space for cars parked along the east and west walls to manoeuvre around them. Since spaces are 2.4 m wide, E–W gridlines 1–7 are designed at 7.5 m centres, to allow for three spaces between columns and a 0.3 m column width. This arrangement provides 37 spaces, which is above the minimum of one per apartment. Additional spaces around cores (which are not large enough for car parking spaces) are allocated to bicycle parking, bin storage and stores.

The layout of the two buildings was chosen to line up with the grid of the car park. One building is located at each end of the site, with the ramp to the car park between the buildings. The exterior structural walls of each building line up with the retaining wall surrounding the exterior of the car park or columns on an E–W gridline, designed at 7.5 m centres. The interior

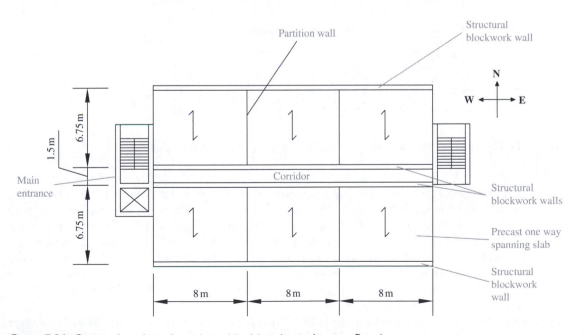

Figure 7.31 Option 1 – plan of residential building (typical upper floor)

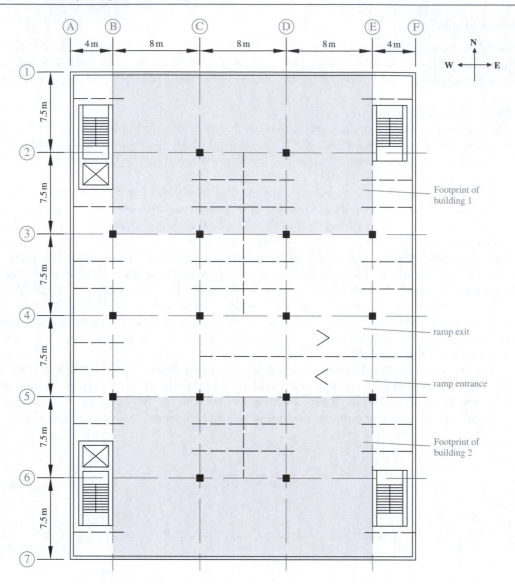

Figure 7.32 Option 1 – Basement plan showing layout of underground car park

structural walls of the building do not line up with the 7.5 m grid of the car park below, so a wide transfer beam (running E–W) is designed to transfer the loads from the walls of the building to the columns, as illustrated in Fig. 7.33. A partial cross-section along gridline C at ground floor level is given in Fig. 7.34. The weight of the blockwork walls is taken by the wide transfer beams through local bending transversely though the beam. *Note*: This arrangement reduces the floor to ceiling height of the car park. Horizontal wind loads are transferred through diaphragm action in the precast floors to the cantilever cores. Two cores per building resist E–W and N–S forces.

(b) Option 2 – *Transfer 'podium' slab*
A similar structural solution for the upper floors is adopted for Option 2, with precast concrete slabs spanning onto masonry walls and non-structural partitions between apartments. The

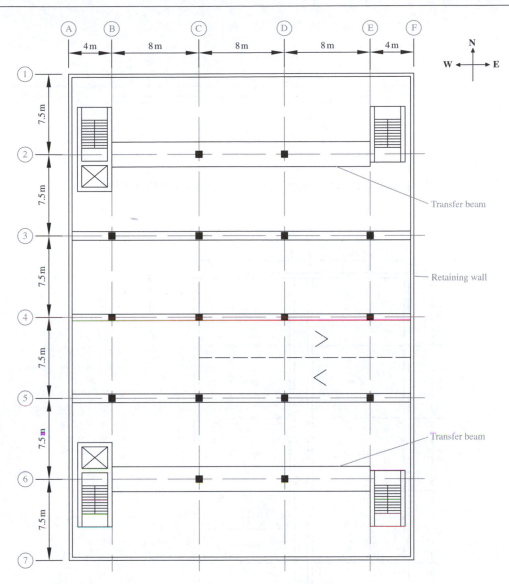

Figure 7.33 Option I – Ground floor plan

building is 17.5 m × 36 m, with eleven apartments of area 48 m² on each floor and a corridor down the centre, as illustrated in Fig. 7.35. Precast concrete hollowcore slabs span 8 m (across apartments) and 1.5 m (across corridors) between masonry walls. Stairs and a lift core are located at the centre of the building to provide some structural stability and vertical circulation. Additional stability against lateral loads is provided by a shear wall which continues down to basement level on the south face of the building (Fig. 7.36). The normal masonry wall on that face, together with the basement retaining wall, constitute the shear wall. It may be necessary to specify no damp proof course so that there is consistency between the concrete and the masonry. E–W wind load on the north end of the building is transferred by diaphragm action in the slab back to the core and shear wall.

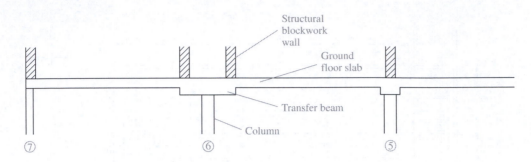

Figure 7.34 Option 1 – partial cross-section along gridline C at ground floor level

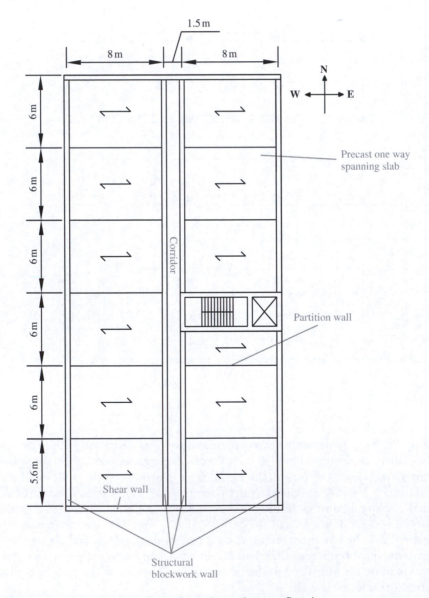

Figure 7.35 Option 2 – plan of residential building (typical upper floor)

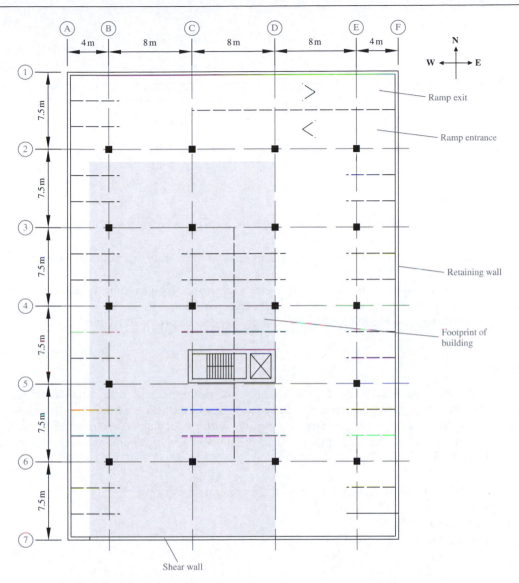

Figure 7.36 Option 2 – Basement plan showing layout of underground car park

A similar grid as before is chosen for the car park (Fig. 7.36) but the changed location of the core and ramps provides 46 spaces. For this scheme, a podium flat slab is designed to transfer the loads from the structural grid of the upper floors of the building to the structural grid of the car park below. Since the podium slab at ground floor level is an *in-situ* concrete flat slab, this requires less complicated shuttering, speeding up construction time. The location of the building on the site is chosen to minimize the number of car parking spaces taken by the stairs and lift core (i.e. the east wall of the building is located along gridline D, as illustrated in Fig. 7.36). The car park ramp is situated at the top end of the site (see Fig. 7.36). A cross-section along gridline 3 at ground floor level is given in Fig. 7.37.

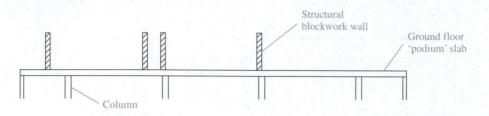

Figure 7.37 Option 2 – cross-section along gridline 3 at ground floor level

Scheme selection

Of the two schemes considered, Option 2 is selected for a variety of reasons, including the following:

1. No downstand beams in the car park means greater flexibility for services and/or reduced building height.
2. For the car park, simple shutter arrangements and temporary propping requirements reduce construction time as flat slabs are fast and easy to construct.
3. Option 2 requires less cladding and services only need to be provided to one building.

Loading

In EC1, the recommended imposed load for areas of domestic and residential activities (not including balconies) is 2.0 kN/m^2 and 3–5 kN/m^2 for staircases and corridors (i.e. imposed load category C3 – areas where people may congregate, without obstacles for moving people). As well as the self-weight of the slab, the following permanent gravity loads are added (according to the ISE manual):

Permanent blockwork partitions = 2.5 kN/m^2 (on plan over floor area)
Ceilings and services = 0.5 kN/m^2
Floor finishes = 1.8 kN/m^2
Landscaping (external areas of deck) = 5 kN/m^2

Preliminary sizing for Option 2

The precast concrete units are often sized based on manufacturers' specifications. For example, for an 8 m clear span, manufacturers' tables indicate that 200 mm deep hollowcore floor units with a 75 mm structural screed (reinforced with steel mesh) can carry an unfactored imposed load of 7.9 kN/m^2. The structural screed is reinforced with mesh to effectively 'tie' the precast units together which allows diaphragm action in the floor plates and acts against disproportionate or progressive collapse.

Podium slabs may be sized for preliminary purposes using the rule of thumb of 100 mm per upper floor plus 100 mm, giving in this case a preliminary minimum podium slab depth of 400 mm. It has been found that provision of an additional 100 mm depth above this minimum will significantly reduce punching shear and bending reinforcement requirements. Therefore, a preliminary depth of 500 mm is adopted for the design. A detailed finite element or grillage analysis of the podium slab is required for final design and detailing of reinforcement.

7.6 Case Study 5 – Hotel

Problem: Propose two alternative designs for a five-storey hotel building with the footprint illustrated in Fig. 7.38. A minimum of 80 hotel rooms is required, about 36 m² (4 m × 9 m) each, including bathroom. The ground floor is to accommodate a reception lobby and conference/dining facilities, so a minimum number of internal columns is required.

Solution: Hotel buildings are commonly constructed using concrete due to its excellent fire resistance and sound insulating properties. *In situ* or precast concrete, or a combination of both, can be used. Due to the cellular repetitive nature of hotel structures, innovative construction methods have been developed, such as prefabricated bathrooms modules (or pods). They are constructed in a factory and are fully assembled on delivery, thus reducing construction time and the number of trades on site.

The ground floor of the hotel requires large open spans for the lobby, meeting rooms and dining areas and the floors above accommodate the hotel rooms, with a regular repeatable layout. A structural scheme must be provided to transfer the loads from the upper floors to ground, minimizing the number of internal columns at ground floor level. The stair and lift cores for the building should continue down to ground level to provide structural stability and vertical circulation from the reception lobby to the bedrooms.

Scheme options

Concrete is chosen as the structural material for the two alternative scheme designs.

(a)   Option 1 – *in situ* reinforced concrete frame with infill precast floors

To provide large open spans at ground floor level, the central columns in the N–S direction are spaced at 11 m centres, with columns also on the outer perimeter of the building (5 m from the central columns). In the E–W direction, the central columns are spaced at 8 m centres, as illustrated in Fig. 7.39. This provides an area for the reception lobby in the southern part of the building and an area for kitchens and services in the northern part. The large span in the central area provides adequate open space for conference and dining facilities.

This grid is repeated for the upper floors, forming a reinforced concrete frame structure. The proposed structure for the upper floors, providing a total of 84 rooms, is illustrated in Fig. 7.40. The slabs span 11 m and 5 m in the N–S direction between reinforced concrete beams. Precast hollowcore slabs approximately 375 mm deep (including structural screed) or Double-Tee

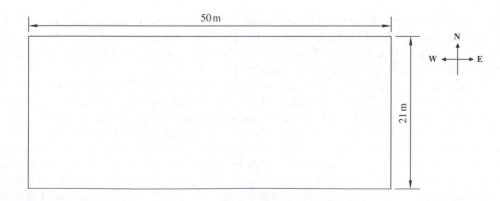

Figure 7.38 Footprint of building

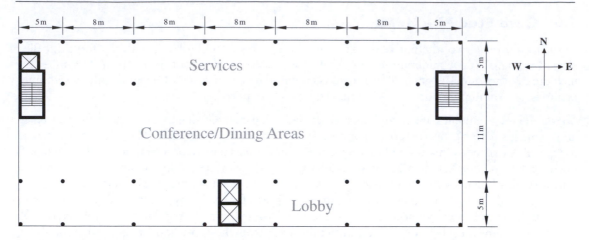

Figure 7.39 Option 1 – layout of ground floor

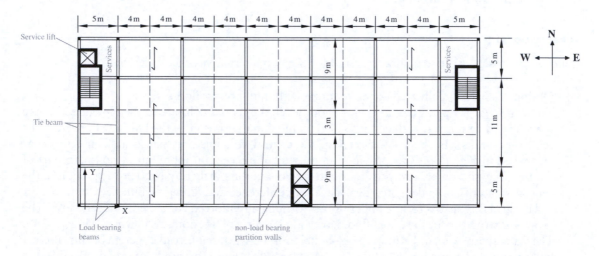

Figure 7.40 Option 1 – layout of upper hotel floors

slabs approximately 550 mm deep (including structural screed) can be used to span 11 m between beams. Double-Tee slabs, illustrated in Fig. 7.41, are precast prestressed slabs and are suitable for long spans. Although Double-Tee slabs are deeper than hollowcore slabs, they are structurally more efficient as the relative self-weight is less.

Continuous reinforced concrete beams span 8 m in the E–W direction, transferring the loads from the slab to the columns by bending and shear. There are moment connections between the beams and columns, forming a reinforced concrete frame, so loads in the columns are transferred to ground by axial compression and bending in the columns. Tie beams spanning in the N–S direction between each frame are there to provide stability during and after construction. Non-structural partitions are used to subdivide the area between adjacent walls into individual rooms. Lightweight sound proof partitions are chosen to reduce the dead load on the frame.

A cross-section of the building is illustrated in Fig. 7.42. The precast floor spans 11 m and 5 m between the reinforced concrete beams, which span in the opposite direction. The beam and slab

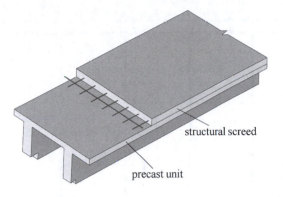

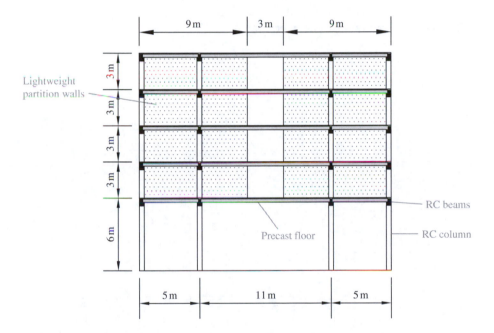

Figure 7.41 Double-Tee precast slabs (courtesy of *The Concrete Centre*)

Figure 7.42 Option 1 – cross-section through building (N–S section)

are tied together using a reinforcing bar which is bent down into the structural screed, similar to the detail illustrated in Fig. 7.15.

Horizontal wind load are transferred by diaphragm action in the structural screed to the lift/stair cores. These loads are then transferred to ground by cantilever bending and shear.

(b) Option 2 – *In situ* concrete 'Tunnel form' or 'Twin wall' precast construction

For Option 2, a similar layout to Option 1 is adopted for the ground floor level, with a large 11 m span at the centre of the building in the N–S direction and columns at 8 m centres in the E–W direction (see Fig. 7.39). For the upper floors, a cellular structure is chosen with reinforced concrete walls constructed at 8 m centres and slabs spanning between walls, as illustrated in Fig. 7.43. Non-structural partitions are used to subdivide the area between walls into two individual rooms. The structural walls act as deep beams which transfer load by membrane action to the columns

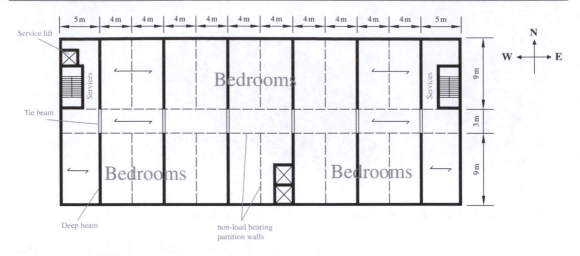

Figure 7.43 Option 2 – layout of upper hotel floors

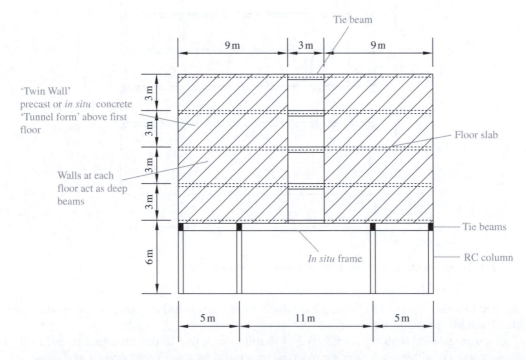

Figure 7.44 Option 2 – cross-section through building (N–S section)

below (Fig. 7.44). Tie beams (spanning into the page) are also included to provide stability. Once the upper floors are erected, the walls behave as two deep cantilever beams, tied at each level by corridor beams and supported by the columns at first floor level. This design will result in significant compression forces in the corridor beams. During construction, the *in situ* frame in the ground floor supports the walls and construction loading until the concrete has set.

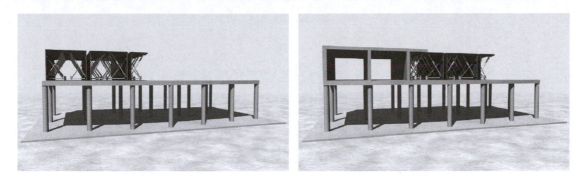

Figure 7.45 'Tunnel form' process (courtesy of *The Concrete Centre*)

Either 'Tunnel form' or 'Twin wall' construction could be chosen for this scheme design. Working on a 24-hour cycle, 'Tunnel form' construction uses formwork to pour the wall and slab together before removing the formwork and moving onto the next cell, as illustrated in Fig. 7.45. This is a fast and economical method of constructing hotel buildings. The 'Twin wall' system consists of two thin concrete plates (about 65 mm thick) with a cavity between them, as illustrated in Fig. 7.46. During the offsite construction phase, the plates are connected with cast-in steel lattice girders (Fig. 7.46(b)). The precast plates are erected on site and then filled with *in situ* concrete. Precast hollowcore or wideslab floors can be used with 'Twin wall' construction. The members are tied together with reinforcing steel and *in situ* concrete, as illustrated in Fig. 7.46, resulting in a monolithic structure. This allows diaphragm action in the floor plates and acts against disproportionate or progressive collapse. This permanent formwork system is a quick construction method as formwork does not have to be removed, and the combination of precast and *in situ* provides excellent fire and sound resistance. The quality of the finish on the exterior of the plates means that plastering is often not required.

Scheme selection

Of the two alternative schemes considered, Option 2 using 'Twin wall' construction is selected for a variety of reasons, including the following:

1. Using deep beam action solves the issue of very long (11 m) spans over the conference/dining areas.
2. The 'Twin wall' plates have a smooth external finish so no plastering is required.
3. Precasting the walls reduces construction time and the number of trades required on site.

Loading

In EC1, the recommended imposed load for bedrooms in hotels (not including balconies) is 2.0 kN/m^2 and 3–5 kN/m^2 for access areas such as staircases and corridors (i.e. imposed load category C3 – areas where people may congregate, without obstacles for moving people). An additional imposed load of 1 kN/m^2 should be added to this to account for lightweight partitions. As well as the self-weight of the slab, the following permanent gravity loads are added (according to the ISE manual):

Ceilings and services = 0.5 kN/m^2
Floor finishes = 1.8 kN/m^2

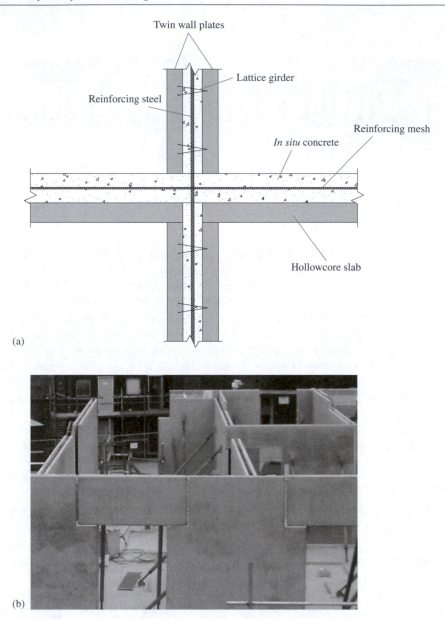

Figure 7.46 Option 2 – 'Twin wall': (a) detail; (b) 'Twin wall' construction (courtesy of Keegan Quarries Ltd)

7.7 Case Study 6 – Grandstand

Problem: New grandstands are required for the east and west sides of the rugby pitch illustrated in Fig. 7.47. Together, the new stands are to cater for 12,000 seats. Existing stands at the north and south ends are to be retained. Where possible, no seat in either stand should be more than 90 m from the centre of the pitch and no seat should be more than 150 m from any point on the pitch. To minimize disruption to the playing season, the construction time on site should be as short as possible. Vomitrys (exit stairs) are to be provided at 15 m intervals. Assume the slope is a uniform

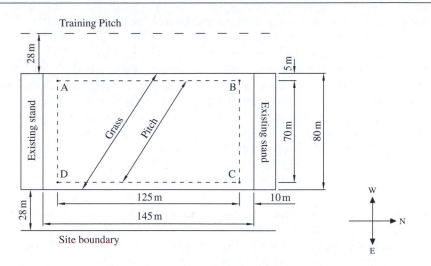

Figure 7.47 Layout of site

32 degrees. Assume that the average plan area of stand per seat is 0.5 m² (on plan), including walkways and stairs.

Solution: Feasible areas for seating are identified by:

(a) drawing a circle of radius 90 m from the centre of the pitch; and
(b) drawing part-circles of radius 150 m from the four corners of the pitch, A, B, C and D. This is illustrated in Fig. 7.48.

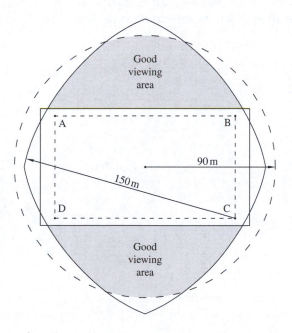

Figure 7.48 Optimum location of seating areas

The areas that satisfy both of these constraints are shown shaded in the figure.

(a) Option 1 – Cantilever roof structure
The simplest solution is to design two rectangular stands, as illustrated in Fig. 7.49. To minimize the time of construction while retaining the advantages of durability and fire resistance, precast concrete is chosen for the main structure. For simplicity, a cantilever roof structure is selected, as illustrated in Fig. 7.50. The stands are 145 m long so a movement joint is included at the centre. As with all scheme options with movement joints, the lateral stability of each part of the structure must be considered separately. For this option, stair cores at 15 m centres provide adequate stability to the stands. For maximum strength to weight ratio, structural steel trusses are selected for this – concrete would make for a very heavy cantilever roof. There will be a large bending moment at the base of the cantilever roof. To support this, a pair of point supports is provided, as illustrated in the figure. For space efficiency, an overhang of 4 m at the back of the stands is included (Fig. 7.50).

Three 7 m spans in the E–W direction keep the spans modest while providing reasonable clear space for the many services needed in modern stands. In the N–S direction, a grid spacing of 7.5 m keeps the spans modest and is consistent with the 15 m vomitry spacing. Each rectangular stand gives (145 m $\times$ 21 m =) 3,045 m^2 of seating area which provides approximately 6,090 seats, some of which would be more than 90 m from the centre of the pitch and more than 150 m from its furthest end. Together, the two stands give a total of 12,180 seats, in excess of the 12,000 target.

Wind load is transferred by diaphragm action in the inclined seating units/horizontal floor slabs at each level to the stair cores. These loads are then transferred to ground by cantilever bending and shear in the cores.

(b) Option 2 – 'Goalpost' roof structure
To maximize the number of seats that are less than 150 m from the corners of the pitch, a stand curved on plan is preferable, as illustrated in Fig. 7.51. In order for the curved stand to have the same capacity as the rectangular equivalent, its depth at the centre must be increased to 28 m. This increase in span significantly impacts upon the size of cantilever truss required, and also on

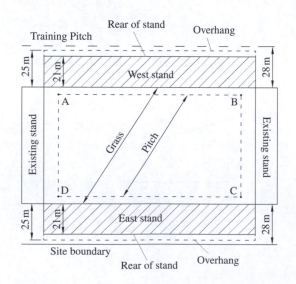

Figure 7.49 Option 1 – plan view of rectangular stands

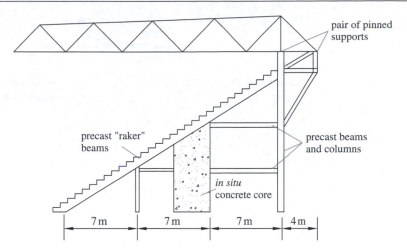

Figure 7.50 Option 1 – cross-section

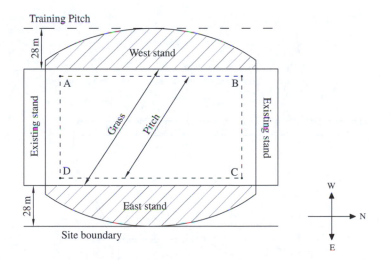

Figure 7.51 Option 2 – plan view of curved stands

the main concrete support frame. Hence, an alternative 'goalpost' roof solution is adopted here. This involves a goalpost shaped external structure near the front of the stand which provides support to a series of secondary roof trusses – Fig. 7.52.

An added advantage of the goalpost structure is that the secondary roof trusses are supported towards the front, thereby removing the moment reaction at the rear. The seating area extends to the back of the structure, providing a total area of 3,033 m² of seating all the way in each stand. Together, the east and west curved stands in this solution provide approximately 12,132 seats. A simplified cross-section of the proposed solution is given in Fig. 7.53. As with the previous option, a movement joint is included at the centre of the stands and stability is provided by stair cores at 15 m centres.

Beam, arch and portalized options were all considered for the goalpost structure. An arch would have been structurally efficient but would have required extensive foundations to

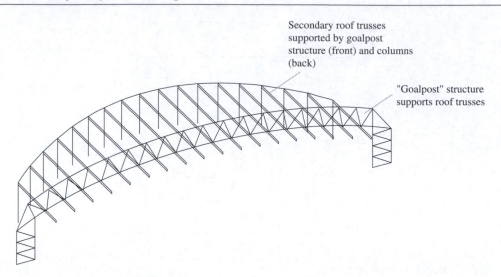

Figure 7.52 Option 2 – example of a goalpost structure

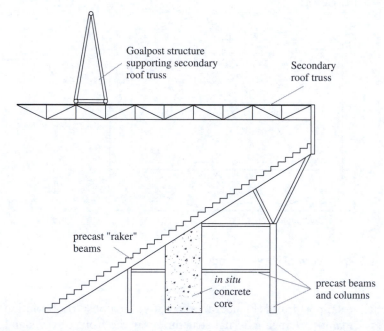

Figure 7.53 Option 2 – cross-section

take the horizontal thrust. A trussed portal frame is therefore adopted. A pinned joint connecting the base of the frame to the foundations prevents moment from being transmitted through this point. However, there is still a horizontal reaction which does require substantial foundations.

As for Option 1, wind load is transferred by diaphragm action in the inclined seating units/ horizontal floor slabs at each level to the stair cores. These loads are then transferred to ground by cantilever bending and shear in the cores.

Scheme selection

Of the two alternative schemes considered, Option 2 is selected for a variety of reasons, including the following:

1. This solution provides more seats within the optimum viewing areas.
2. For aesthetic reasons, the curved stand with the goalpost roof structure is a more attractive solution.
3. Reduced member sizes in the main concrete frame as no moment support to truss is required at the rear of the stand.

Loading

In EC1, the recommended imposed load for areas susceptible to large crowds, including grandstands, is 5–7.5 kN/m². For circulation and service areas inside the stand, the recommended imposed load is 3.0–5.0 kN/m² (i.e. imposed load category C3 – areas where people may congregate, without obstacles for moving people).

Note: Since grandstands are often used for concerts, they need to be checked for dynamics. One method involves ensuring that the natural frequency of the structure is outside the range that can be excited by crowds acting in unison.

Part III

Detailed member design

Introduction

Part III of this book deals with the topics traditionally covered in concrete design courses. While the 'new' topics such as qualitative design are most important, an understanding of the detailed requirements of codes of practice is also necessary.

Perhaps the most important mechanism of load transfer is flexure. This is dealt with in two chapters. Chapter 8 deals with the design of ordinary reinforced concrete to resist internal bending moment. Design for bending in prestressed concrete is dealt with in Chapter 9. Members subject to axial force are treated in Chapter 10. This chapter includes all members which transmit load through axial force mechanisms. Thus, in addition to columns and walls, this chapter considers tension members and deep beams. The internal shear force, which usually accompanies bending, is dealt with in Chapter 11 for both ordinary reinforced and prestressed concrete members. Design to resist torsional moment is also treated in this chapter as there are many parallels with design for shear. As the analysis of slabs was dealt with in Part II, the design of slabs for flexure and shear is identical to that of beams as can be seen from the examples in these chapters. An exception to this rule is the design of slabs for punching shear. Punching shear in slabs is treated in a separate section in Chapter 11.

This part of the book is intended to complete the education of the engineer in concrete design. It also reflects the final stage in the complete design process. Thus, the designer, having gone through the stages of conceptual design and preliminary analysis and design, must now carry out a detailed analysis and, using the information in this part of the book, must complete a detailed design for all structural members.

Chapter 8

Design of reinforced concrete me
for bending

8.1   Introduction

Beams and slabs, by definition, transmit load by bending and must be designed to resist the resulting moment. Two limit states are generally considered. At the serviceability limit state (SLS), cracking occurs which must be limited to prevent air and moisture from reaching the reinforcement and causing corrosion. Also at SLS, deflections occur which must be kept within acceptable levels. From the design viewpoint, SLS checks are characterized by relatively low stresses. It follows that it can be reasonably assumed that the stress/strain relationship is linear, that is, **stress is proportional to strain at SLS**. It is a fundamental assumption of bending theory that plane sections remain plane. Hence, the distribution of **strain through a cross-section is always linear**, as illustrated in Fig. 8.1(b). It then follows from the linear stress/strain relationship that the **stress distribution is also linear at SLS**, as illustrated in Fig. 8.1(c).

At the ultimate limit state (ULS), the safety of the beam or slab is considered. Thus, at the critical cross-sections, the capacity to resist moment is compared with the moment due to applied loads. From the design viewpoint, ULS checks are characterized by high stresses and by non-linear stress/strain relationships. The assumption that plane sections remain plane is still valid. However, **distributions of stress are generally non-linear at ULS**.

8.2   Second moments of area (linear elastic)

An internal bending moment, M, due to applied loads, generates a distribution of axial stresses within the section as illustrated in Fig. 8.1. Recall that, for **linear elastic** behaviour, the relationship between the applied moment, M, and the stress at a distance y above the centroid of the section, $\sigma(y)$, is given by the flexure formula:

$$\frac{M}{I} = \frac{\sigma(y)}{y} = \frac{E\varepsilon(y)}{y} \tag{8.1}$$

where $\varepsilon(y)$ is the strain at a distance y above the centroid and E is the modulus of elasticity. The term I is known as the second moment of area. It is a constant which relates the applied moment at a section to the stress or strain which it causes.

For a given cross-section of area A, as illustrated in Fig. 8.2, the second moment of area about the axis containing the centroid, C, is defined by the integral:

$$I = \int_A y^2 \, dA \tag{8.2}$$

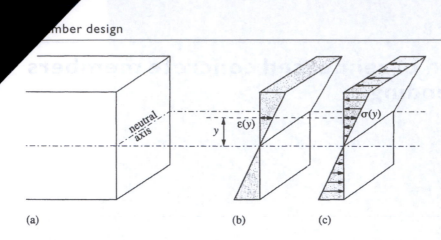

Figure 8.1 Linear distributions of stress and strain at SLS (a) section; (b) strain distribution; (c) stress distribution

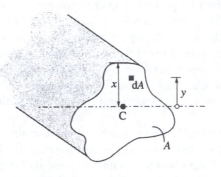

Figure 8.2 Cross-section

The centroid of an area is the geometric centre of the area. For a section in pure bending, the axis containing the centroid is the neutral axis of the section. **The neutral axis is defined as the axis at which strains are zero.** For the linear elastic case, the location of the neutral axis is found by taking first moments of area about any point. For this purpose, moments are usually taken about the extreme fibre in compression which, for sag moment, is the top fibre. The calculation of neutral axis location and of the second moment of area, I, is illustrated in the following examples.

Example 8.1 Rectangular section

Problem: The rectangular section of Fig. 8.3 is subjected to a sag moment of 500 kNm.

(a) Determine the neutral axis location and the second moment of area.
(b) Assuming that stress is everywhere proportional to strain, find the maximum tensile strain in the section given that $b = 300$ and $h = 500$.

Solution: (a) Taking first moments of area about P in Fig. 8.3:

$$\sum (\text{area} \times \text{distance from P}) = (\text{total area}) \times x$$

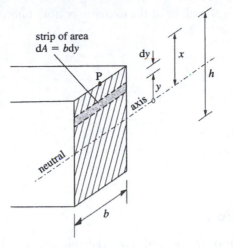

strip of area
$dA = bdy$

Figure 8.3 Section of Example 8.1

where x is the distance from the neutral axis to the extreme fibre in compression, which in this case is P. Hence, for a rectangular section:

$$(bh)(h/2) = (bh)x$$
$$\Rightarrow x = h/2$$

The second moment of area is calculated using Equation (8.2) where y is measured from the neutral axis:

$$I = \int_{y=-h/2}^{y=h/2} y^2 dA$$

Referring to the strip of area, dA, in Fig. 8.3:

$$I = \int_{y=-h/2}^{y=h/2} y^2 (b\, dy)$$

$$= b \int_{y=-h/2}^{y=h/2} y^2\, dy$$

$$= b \frac{y^3}{3} \Big|_{-h/2}^{h/2}$$

$$= \frac{b}{3} \left[\left(\frac{h}{2} \right)^3 - \left(-\frac{h}{2} \right)^3 \right]$$

$$= \frac{b}{3} \left[\frac{h^3}{8} - \left(-\frac{h^3}{8} \right) \right]$$

$$= \frac{b}{3} \frac{h^3}{4} = \frac{bh^3}{12}$$

(b) The maximum tensile stress will be at the extreme bottom fibre, that is $y = -250$ mm. Applying the flexure formula (Equation (8.1)):

$$\frac{M}{I} = \frac{\sigma(y)}{y}$$

$$\Rightarrow \sigma(y) = \frac{My}{I}$$

$$= \frac{(500 \times 10^6)(-250)}{(300 \times 500^3/12)}$$

$$= -40 \text{ N/mm}^2$$

Example 8.2 T-section

Problem: Find the location of the neutral axis (centroid) for the T-section illustrated in Fig. 8.4. Hence calculate the second-moment of area of the section.

Solution: Taking moments of area about point Q in Fig. 8.4:

$$(A_1 + A_2)x = A_1(100) + A_2(200 + 600/2)$$

where $A_1 = 200 \times 1000 = 200000$ mm^2 and $A_2 = 600 \times 300 = 180000$ mm^2. Hence:

$$x = \frac{200000(100) + 180000(500)}{200000 + 180000}$$

$$= 290 \text{ mm}$$

That is, the neutral axis is 290 mm from the top fibre or 90 mm from the bottom of the flange. Hence, the second moment of area is:

$$I = \int_A y^2 dA$$

$$= \int_{-(600-90)}^{+90} y^2 (300dy) + \int_{+90}^{+290} y^2 (1000dy)$$

$$= 300 \frac{y^3}{3} \Big|_{-510}^{90} + 1000 \frac{y^3}{3} \Big|_{90}^{290}$$

$$= 100[90^3 - (-510)^3] + 333.3[290^3 - 90^3]$$

$$= 21.23 \times 10^9 \text{ mm}^4$$

Alternatively, the second moment of area can be evaluated by summing, for all parts of the section, the second moment of area of that part plus the product of area and the distance to the section neutral axis squared:

$$I = \frac{1000 \times 200^3}{12} + A_1(x - 100)^2 + \frac{300 \times 600^3}{12} + A_2(600/2 - 90)^2$$

$$= (667 + 7220 + 5400 + 7938) \times 10^6$$

$$= 21.23 \times 10^9 \text{ mm}^4$$

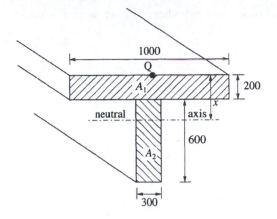

Figure 8.4 Section of Example 8.2

The neutral axis can be located from first principles using equilibrium of the forces parallel to the X-axis of the beam. For pure bending there are no axial forces, that is:

$$F = \int_A \sigma(y)\mathrm{d}A = 0 \tag{8.3}$$

Similarly Equations (8.1) and (8.2) can be derived by summing moments about the neutral axis. Example 8.3 illustrates how stresses can be calculated from first principles.

Example 8.3 Stresses from first principles

Problem:

(a)   Using the equilibrium of force method (Equation (8.3)), locate the neutral axis of the section of Fig. 8.5(a).
(b)   Derive an expression for the stress $\sigma(y)$ at a distance y above the neutral axis.

Solution: (a) As plane sections remain plane, the strain distribution for this section is as illustrated in Fig. 8.5(b). If the maximum compressive strain is ε_{cu} as illustrated, the strain at a distance y from the neutral axis is (by similar triangles):

$$\frac{\varepsilon(y)}{y} = \frac{\varepsilon_{cu}}{x}$$

$$\Rightarrow \varepsilon(y) = \frac{\varepsilon_{cu}y}{x}$$

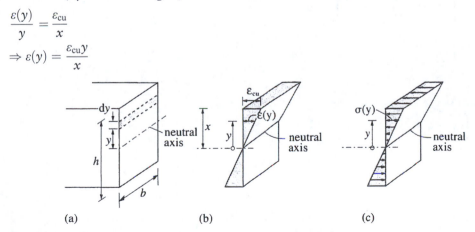

(a)   (b)   (c)

Figure 8.5 Section of Example 8.3: (a) section; (b) strain distribution; (c) stress distribution

As stress is proportional to strain everywhere, the stress at this level is:

$$\sigma(y) = E\varepsilon(y)$$
$$= \frac{E\varepsilon_{cu}y}{x}$$

Thus, Equation (8.3) becomes:

$$F = \int_A \sigma(y)\mathrm{d}A = 0$$
$$\Rightarrow \int_A \frac{E\varepsilon_{cu}y}{x}(b\,\mathrm{d}y) = 0$$
$$\Rightarrow \frac{E\varepsilon_{cu}b}{x}\int_{-(b-x)}^{x} y\,\mathrm{d}y = 0$$
$$\Rightarrow \frac{E\varepsilon_{cu}b}{x}\frac{y^2}{2}\bigg|_{-(b-x)}^{x} = 0$$

which reduces to:

$$\frac{E\varepsilon_{cu}b}{2x}(-b^2 + 2bx) = 0$$

As none of the constants is zero, this implies:

$$-b^2 + 2bx = 0$$
$$\Rightarrow x = b/2$$

which is the centroid of the section.

(b) The sum of all moments about the neutral axis must equal the applied moment, M, that is:

$$M = \int_A \sigma(y)y\,\mathrm{d}A$$
$$= \int_A \left(\frac{E\varepsilon_{cu}y}{x}\right)y\,\mathrm{d}A$$
$$= \frac{E\varepsilon_{cu}}{x}\int_A y^2\,\mathrm{d}A = E\left(\frac{\varepsilon(y)}{y}\right)\int_A y^2\,\mathrm{d}A$$
$$= \frac{\sigma(y)}{y}\int_A y^2\,\mathrm{d}A$$

Hence:

$$\sigma(y) = \frac{My}{\displaystyle\int_A y^2\,\mathrm{d}A}$$
$$= \frac{My}{I}$$

where:

$$I = \int_A y^2\,\mathrm{d}A$$

Uncracked sections

For concrete sections, the second moment of area considered in Examples 8.1–8.3 is known as the 'gross' second moment of area, I_g. In the calculation of I_g, the presence of the reinforcement is ignored and it is assumed that the entire section is available to resist bending. In the calculation of internal bending stress, however, it is sometimes useful to take account of the composite nature of reinforced concrete, that is to account for the presence of the reinforcement, which is considerably stiffer than concrete. The flexure formula (Equation (8.1)) is directly applicable only to homogeneous sections in bending. Therefore, in order to use it, the composite steel and concrete section is transformed into an equivalent homogeneous concrete section.

Transformed sections

Consider the two-material composite beam of Fig. 8.6(a). The two materials are bonded together so that, in bending, no slip can occur between them. The top material has an elastic modulus of E_1 and the bottom material has an elastic modulus of E_2 such that $E_1 < E_2$. If a bending moment, M, is applied to the section, the assumption that plane sections remain plane is still valid, and the strain varies linearly as illustrated in Fig. 8.6(b). The stress at any level in the top material is equal to $E_1\varepsilon$ and the stress at any level in the bottom material is given by $E_2\varepsilon$. The stress distribution corresponding to these conditions is illustrated in Fig. 8.6(c). To transform this into an equivalent homogeneous section having one elastic modulus $E = E_1$, the breadth of the bottom material is increased, as

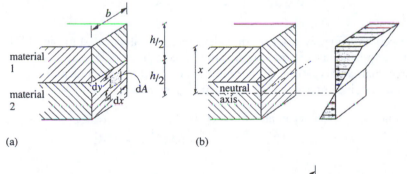

(a)   (b)

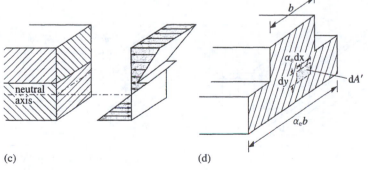

(c)   (d)

Figure 8.6 Composite section; (b) strain distribution; (c) stress distribution; (d) equivalent section with elastic modulus E_1

illustrated in Fig. 8.6(d). The equivalent breadth is determined by considering a force dF acting on an area $dA = dx\,dy$ of the section of Fig. 8.6(a). Thus:

$$dF = \sigma dA = (E_2\varepsilon)dxdy$$

The equivalent area, dA', of height dy in the section of Fig. 8.6(d) is:

$$dA' = \alpha_e dxdy$$

where $\alpha_e\,dx$ is the equivalent breadth. In the equivalent section the force acting on this area is calculated using the equivalent area and the new modulus E_1. Hence it is:

$$dF' = (E_1\varepsilon)dA' = (E_1\varepsilon)\alpha_e dxdy$$

Equating the forces dF and dF' gives:

$$(E_2\varepsilon)dxdy = (E_1\varepsilon)\alpha_e dxdy$$
$$\Rightarrow \alpha_e = E_2/E_1 \tag{8.4}$$

Thus, for a composite section (two materials such that $E_1 \neq E_2$) of breadth b, the breadth of the section is to be increased to $\alpha_e b$ over the depth of the stiffer material, where α_e, known as the effective modular ratio, is given by Equation (8.4).

In the case of reinforced concrete, the equivalent transformed section is obtained by replacing the areas of tension and compression reinforcement, A_s and A'_s; respectively, by their equivalent areas of concrete, $\alpha_e A_s$ and $\alpha_e A'_s$, where α_e is the ratio of the elastic modulus for steel to that for concrete:

$$\alpha_e = E_s/E_c$$

Thus, for the section of Fig. 8.7(a), the equivalent section is illustrated in Fig. 8.7(b). The depth of each strip is not important as it is only the area that affects the calculations. For convenience, a depth equal to the steel diameter is assumed. The compression reinforcement displaces an area of concrete, A'_s. Thus the total equivalent area of the top strip is given by:

$$\text{(area of concrete)} + \text{(equivalent area of steel)} = (b\phi' - A'_s) + (\alpha_e A'_s)$$
$$= b\phi' + (\alpha_e - 1)A'_s$$

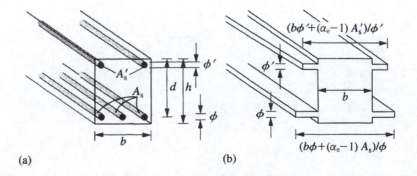

Figure 8.7 (a) Reinforced concrete section; (b) equivalent concrete section

where b is the breadth of the section and ϕ' is the diameter of the compression reinforcement. Similarly, the total equivalent area of the bottom strip is given by the area of the rectangular section plus $(\alpha_e-1)A_s$.

Once the equivalent section is established, the calculation of I is identical to the calculation of I_g. In this case, however, I is known as the 'uncracked' second moment of area, I_u, as the possibility of the concrete cracking has not yet been considered. It is important to remember that stresses that are calculated using the transformed section refer to the less stiff material, that is, concrete.

Cracked sections

In calculating I_g and I_u, the concrete is assumed to be uncracked. However, reinforced concrete sections in pure bending develop tensile cracks under very small loads. Thus, for the calculation of the stresses in a section under applied loads and for the calculation of crack widths, the second moment of area of the section must account for the cracking which has occurred. One estimate is to assume that the concrete has fully cracked; that is, that cracks have penetrated the tension region as far as the neutral axis. Thus, no tensile stress exists in the concrete in this region and this concrete is effectively absent as illustrated in Fig. 8.8. There will, of course, continue to be tension in the bottom reinforcement.

As in the case of an uncracked section, we use the concept of equivalent areas of concrete when calculating the second moment of area for a cracked section. Thus, for the cracked section of Fig. 8.8, the equivalent section is illustrated in Fig. 8.8(d). Like the uncracked section, the total equivalent area of the top strip is given by:

$$b\phi' + (\alpha_e - 1)A'_s$$

(i.e. the original area of the gross section at this level plus $(\alpha_e-1)A'_s$). In the bottom strip, however, there is no usable concrete to displace and so the total equivalent area of the strip is given by:

$$\alpha_e A_s$$

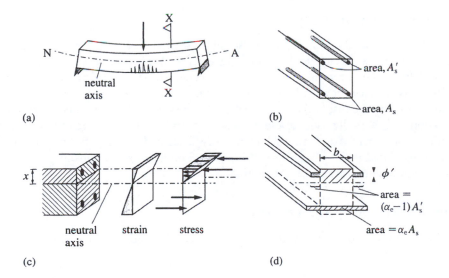

(a)   (b)   (c)   (d)

Figure 8.8 Cracked section: (a) beam in bending; (b) section X–X; (c) stress and strain distributions; (d) equivalent concrete section

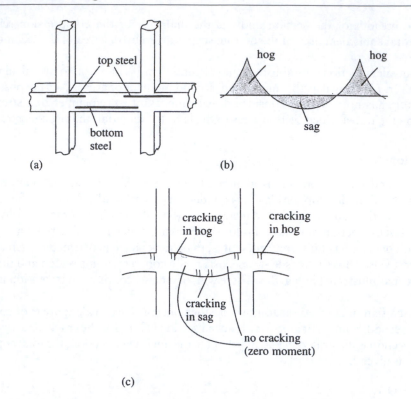

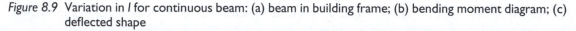

Figure 8.9 Variation in *I* for continuous beam: (a) beam in building frame; (b) bending moment diagram; (c) deflected shape

Once the equivalent section is established, the calculation of *I* is identical to the calculation of I_g or I_u. In this case, *I* is known as the 'cracked' second moment of area, I_c.

Choice of second moment of area

To use I_g or I_u in the calculation of the stresses at a cracked section would give poor results for reinforced concrete. For such calculations, only I_c gives reasonably accurate and conservative results, and it can be as little as half the value of I_g or I_u. I_c is generally used to calculate crack widths under service loads. However, for the calculation of deflection of reinforced concrete members, the use of I_c will give only approximate results as not all sections in the span will be cracked. Furthermore, parts of a beam in hog will generally have different cracked second moments of area than parts in sag. The variation in *I* for one span of a continuous member is illustrated in Fig. 8.9. For accurate deflection calculations, a weighted mean of I_u, I_c for the top steel and I_c for the bottom steel, known as the effective second moment of area, I_{eff}, can be used.

Example 8.4 Rectangular section

Problem: The rectangular beam illustrated in Fig. 8.10 is subjected to a variable uniform loading of 15 kN/m.

(a) Calculate the equivalent 'uncracked' and 'cracked' second moments of area of the section for the dimensions given. Assume $E_c = 30{,}500$ N/mm^2 and $E_s = 200{,}000$ N/mm^2.

(b)   Determine the bending moment at the centre, B, due to the variable loading. Assuming cracked conditions and no long-term effects, calculate the stress in the concrete at the extreme top fibre at B and the stress in the tension reinforcement.

Solution: (a) The label, '2H16', as illustrated in Fig. 8.10(b), indicates two 16mm diameter reinforcing bars ($f_{yk} = 500$ N/mm^2). Hence the area of compression reinforcement is:

$$A'_s = 2(\pi 16^2/4) = 402 \text{ mm}^2$$

Similarly:

$$A_s = 3(\pi 25^2/4) = 1473 \text{ mm}^2$$

The modular ratio is:

$$\alpha_e = \frac{E_s}{E_c} = \frac{200000}{30500} = 6.56$$

Hence, the equivalent uncracked concrete section is as illustrated in Fig. 8.11(a). As in Example 8.2, the location of the neutral axis is found by first calculating the total equivalent area and then taking moments about P in Fig. 8.11(a):

$$\text{total (equivalent) area} = (300)(500) + 2235 + 8190$$
$$= 160,425 \text{mm}^2$$

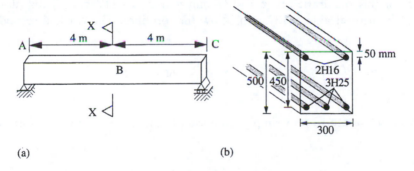

(a)   (b)

Figure 8.10 Beam of Examples 8.4 and 8.5: (a) elevation; (b) section X–X

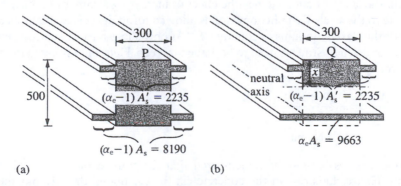

(a)   (b)

Figure 8.11 Equivalent concrete section of Example 8.4: (a) uncracked; (b) cracked

Taking moments about P:

$$(160425)x = (300)(500)(250) + (2235)(50) + (8190)(450)$$
$$\Rightarrow x = 257\text{mm}$$

Using the alternative method demonstrated in Example 8.2, the second moment of area is:

$$I_u = \frac{(300)(500)^3}{12} + (300)(500)(x - 250)^2 + (2235)(x - 50)^2 + (8190)(450 - x)^2$$

Note: The second moment of area of the steel reinforcement about its own axis is small and is neglected here. Substituting for x above gives:

$$I_u = 3.533 \times 10^9 \text{mm}^4$$

The equivalent cracked section is illustrated in Fig. 8.11(b). The equivalent total area is:

$$\text{total(equivalent)area} = (300)(x) + 2235 + 9663$$
$$= 300x + 11898\text{mm}^2$$

Taking moments about Q results in a quadratic equation in x:

$$(300x + 11898)x = 300(x)(x/2) + (2235)(50) + (9663)(450)$$
$$\Rightarrow 150x^2 + 11898x - 4460100 = 0$$

The two roots of this quadratic are $x = 137$ mm and $x = -217$ mm. Thus (discounting the negative root), the neutral axis is 137 mm below the top fibre. The cracked second moment of area, I_c, is:

$$I_c = \frac{300x^3}{12} + (300x)\left(\frac{x}{2}\right)^2 + (2235)(x - 50)^2 + (9663)(450 - x)^2$$
$$\Rightarrow Ic = 1.221 \times 10^9 \text{ mm}^4$$

(b) The moment at the centre of a simply supported beam subjected to a uniformly distributed loading, ω, is:

$$M_B = \frac{\omega L^2}{8} = \frac{15(8)^2}{8} = 120 \text{ kNm}$$

In the long term, the creep of concrete has the effect of increasing strain and redistributing stresses from the concrete to the steel. This phenomenon is allowed for in design calculations by the use of a reduced value for the elastic modulus of concrete, E_c. As only short-term effects are being considered in this example, no such adjustments need to be made for creep. Hence the stress in the concrete at the extreme top fibre, Q, is:

$$\sigma(y = x) = \frac{Mx}{I_c} = \frac{(120 \times 10^6)(137)}{(1.221 \times 10^9)}$$
$$= 13.5 \text{ N/mm}^2$$

When calculating stresses it must be remembered that the transformed section is an equivalent **concrete** section. Hence the stress in the concrete can be calculated directly but the stress in the steel cannot. At the level of the tension reinforcement:

$$y = -(450 - x)$$
$$= -313 \, \text{mm}$$

Hence the stress that would be in the concrete at this level (were it not cracked) is:

$$\sigma(y = -313) = \frac{M(-313)}{I_c} = \frac{(120 \times 10^6)(-313)}{(1.221 \times 10^9)}$$
$$= -30.8 \, \text{N/mm}^2$$

The strain at this level is then $-30.8/E_c$. This strain is the same in both the steel and concrete. The steel stress is thus calculated as:

$$f_s = E_s(-30.8)/E_c$$
$$= \alpha_e(-30.8)$$
$$= -202 \, \text{N/mm}^2$$

that is, a tensile stress of 202 N/mm^2.

It can be seen from Example 8.4 that the stress in reinforcement at a distance y above the neutral axis is:

$$f_s = \frac{\alpha_e M y}{I_c} \tag{8.5}$$

8.3   Elastic deflections and crack widths

For a member in bending (Fig. 8.12), the flexure formula gives us:

$$\frac{M}{I} = \frac{E}{R}$$
$$\Rightarrow M = EI \times \text{curvature}$$

where the curvature is defined as the reciprocal of the radius of curvature, R. The moment-curvature relationship for the reinforced concrete member of Fig. 8.12(a) is illustrated in Fig. 8.12(c). It can be seen that the segment remains uncracked and has a large stiffness, EI_u, until the moment reaches the **cracking moment**, M_{cr} (point A). When this happens, the member cracks and the stiffness at the cracked section reduces to EI_c. As the load (and hence the moment) is increased further, more cracks occur and existing cracks increase in size. Eventually, the reinforcement yields at the point of maximum moment, corresponding to point C on the diagram. Above this point, the member displays large increases in deflection for small increases in moment. The service load range is between the origin and point C on the diagram and it is in this range that deflections are checked and stresses calculated.

Consider a point B within the service load range. This curvature represents the instantaneous (short-term) curvature under an applied moment, M. If the moment is sustained, however, the curvature increases with time to point D due to the creep of the concrete. The curvature at this point is known as the long-term or sustained curvature. As deflection results from curvature, there are both instantaneous and sustained deflections which must be considered in the design of members with bending.

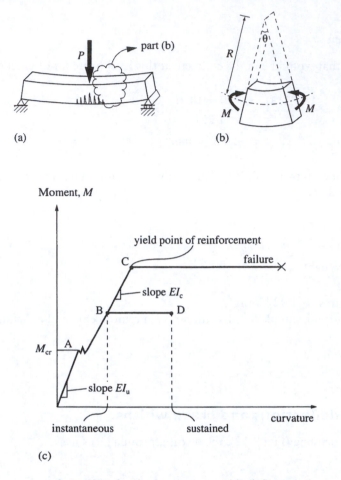

Figure 8.12 Moment/curvature relationship for beam segment: (a) deflected shape; (b) curvature of segment of beam, curvature $= 1/R$; (c) moment/curvature plot for segment of part (b)

Limits on deflection

The deformations (or deflections) which result from bending must be limited such that they do not adversely affect the function and appearance of the member or the entire structure. EC2 limits deflections to span/250 under the quasi-permanent load combination of Table 3.13, that is, the sustained loads. Excessive deflections may affect the aesthetics but also the function of the structure. For instance, sagging roof slabs can result in the ponding of rainwater, which can result in higher applied loads and possibly leakage.

In addition to causing detriment to the appearance and function of a member, deflections can cause damage to non-structural elements (partitions, windows, doors, etc.) which are supported by the structural members in bending. To prevent such damage, EC2 recommends an upper deflection limit of span/500 for normal circumstances.

Calculation of deflection

The deflection of reinforced concrete members can be calculated using the structural analysis methods described in Chapter 4. The instantaneous deflection, δ_i, is calculated using the characteristic combination of applied loads given in Table 3.13. The sustained deflection, δ_s, is calculated using the quasi-permanent load combination of Table 3.12. However, deflection calculations are not necessary on a 'deemed to satisfy' basis if the preliminary span/depth ratios recommended in Table 6.5 are not exceeded.

Where a calculation of deflection is considered necessary, the method of calculation must take account of a phenomenon known as 'tension stiffening' and, for sustained deflections, the effects of creep. In a beam subjected to bending, all parts of the beam will not be in a cracked state at any one time as cracks tend to occur at discrete intervals. The extra stiffness in a beam due to the uncracked portions (between cracks) is known as tension stiffening. In accounting for tension stiffening, EC2 allows the use of the following approximate formula:

$$\delta = \zeta \delta_c + (1 - \zeta)\delta_u \tag{8.6}$$

where δ_u = deflection assuming an uncracked section (i.e. use I_u)
$\quad\;\; \delta_c$ = deflection assuming a fully cracked section (i.e. use I_c)
$\quad\;\; \zeta$ = distribution factor given by Equation (8.7)

Note: According to EC2, the deflection parameter in this formula can be replaced by other deformation parameters (e.g. strain, curvature or rotation).

For uncracked sections $\zeta = 0$. For the case where cracking has occurred:

$$\zeta = 1 - \beta (f_{sr}/f_s)^2 \tag{8.7}$$

where β = coefficient which accounts for the duration of the loading or of repeated loading: $\beta = 1.0$ for single short-term loading; $\beta = 0.5$ for sustained loading or repeated loading
$\quad\;\; f_s$ = stress in tension steel assuming a cracked section
$\quad\;\; f_{sr}$ = stress in tension steel assuming a cracked section due to loading which causes initial cracking

The calculation of long-term deflection, δ_s, must take account of creep. The effect of creep is to increase the strain in the concrete with time, as illustrated in Fig. 8.13. Deflection calculations can allow for this increase in stress by reducing the value of the elastic modulus for concrete, E_c, to an effective elastic modulus, $E_{c,eff}$. The effective modulus recommended by EC2 is given by:

$$E_{c,eff} = \frac{E_{cm}}{1 + \varphi} \tag{8.8}$$

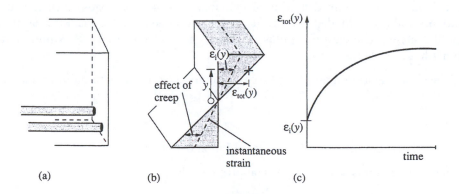

Figure 8.13 Increase of strain with time: (a) section; (b) strain distribution; (c) strain/time relationship

Table 8.1 Creep coefficient, φ, for concrete using type N cement (normal strength gain). A_c is the cross-sectional area of the concrete and u is the length of the perimeter of that area which is exposed to drying (values obtained from equations in BS EN 1992-1-1, Annex B)

		Notional size, $2A_c/u$ (mm)							
		100	200	300	400	100	200	300	400
Concrete class	Age of concrete at loading (days)	Dry atmospheric conditions, i.e. inside (RH = 50 per cent)				Humid atmospheric conditions, i.e. outside (RH = 80 per cent)			
C25/30	$t_0 = 7$	3.8	3.4	3.2	3.1	2.6	2.5	2.4	2.3
	$t_0 = 28$	3.0	2.6	2.5	2.4	2.0	1.9	1.8	1.8
C30/37	$t_0 = 7$	3.4	3.1	2.9	2.8	2.4	2.2	2.2	2.1
	$t_0 = 28$	2.6	2.4	2.2	2.1	1.8	1.7	1.7	1.6
C40/50	$t_0 = 7$	2.7	2.4	2.3	2.2	1.9	1.8	1.8	1.7
	$t_0 = 28$	2.1	1.9	1.8	1.7	1.5	1.4	1.4	1.3

Table 8.2 Secant modulus of elasticity for concrete, E_{cm}, and mean value of axial tensile strength of concrete, f_{ctm}, for a range of strength classes (from BS EN 1992-1-1)

	Strength class						
	C20/25	C25/30	C30/37	C35/45	C40/50	C45/55	C50/60
E_{cm} (kN/mm^2)	30	31	33	34	35	36	37
f_{ctm} (N/mm^2)	2.2	2.6	2.9	3.2	3.5	3.8	4.1

where φ is a creep coefficient taken from Table 8.1 and E_{cm} is the secant modulus of elasticity for concrete which can be taken from Table 8.2. It can be seen from this table that the creep coefficient is largest in young, heavily loaded concrete where the atmospheric conditions are dry. This is because creep is related to the degree of moisture present in concrete.

Example 8.5 Deflection calculation

Problem: The beam of Fig. 8.10 is subjected to a total permanent gravity loading (including self-weight) of 15 kN/m. Of the total variable gravity loading, 8 kN/m has been judged to be quasi-permanent. Find the long-term deflection at mid-span due to these loads given that the beam is located indoors and is first loaded 7 days after casting. Assume type N cement (see Section 6.2) and class C25/30 concrete. The secant modulus of elasticity for concrete is 31,000 N/mm^2 and the mean tensile strength is 2.6 N/mm^2.

Solution: The elastic deflection at the centre of a beam of length L, due to uniform loading ω, is:

$$\delta = \frac{5}{384} \frac{\omega L^4}{EI}$$

where E is the elastic modulus and I is the second moment of area.

The area of concrete in this cross-section is (approximately):

$$A_c \approx (300)(500) = 150000 \text{ mm}^2$$

Hence the notional size, as defined in Table 8.1, is:

$$\text{notional size} = 2\frac{A_c}{u} = \frac{2(150000)}{2(300+500)}$$
$$= 188 \text{ mm}$$

As the beam is first loaded at an age of 7 days, the value of φ is found by interpolating between the values of 3.4 and 3.8 in Table 8.1:

$$\varphi = 3.4 + \frac{(3.8-3.4)(200-188)}{(200-100)}$$
$$= 3.45$$

Hence the effective modulus for concrete is, from Equation (8.8):

$$E_{c,\text{eff}} = \frac{E_{cm}}{1+\varphi} = \frac{31000}{4.45}$$
$$= 6966 \text{ N/mm}^2$$

The uncracked second moment of area is generally similar in magnitude to the gross value and it is common practice to use I_g in lieu of I_u. However, in this example, the effective concrete modulus is so small that the presence of reinforcement will have a significant stiffening effect. The equivalent transformed section for uncracked conditions is calculated using $\alpha = E_s/E_{c,\text{eff}} = 200,000/6,966 = 28.7$ and is illustrated in Fig. 8.14(a).

Taking moments about the top fibre in the uncracked section gives $x = 279$ mm. Hence:

$$I_u = \frac{(300)(500)^3}{12} + (300)(500)(x-250)^2 + (11135)(x-50)^2 + (40802)(450-x)^2$$
$$= 5.028 \times 10^9 \text{ mm}^4$$

Thus the deflection, assuming an uncracked section, is:

$$\delta_u = \frac{5}{384}\frac{\omega L^4}{E_{c,\text{eff}}I_u} = \frac{5}{384}\left(\frac{(15+8)(8000)^4}{(6966)(5.028 \times 10^9)}\right)$$
$$= 35 \text{ mm}$$

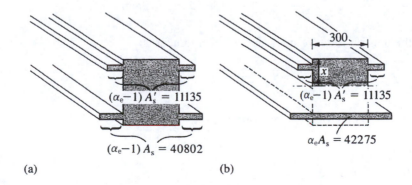

(a)   (b)

Figure 8.14 Equivalent transformed sections: (a) uncracked; (b) cracked

The equivalent transformed section for cracked conditions is illustrated in Fig. 8.14(b). Taking moments about the top fibre gives a quadratic in x:

$$150x^2 + 53410x - 19,580,500 = 0$$

The two roots of this quadratic equation are $x = 225$ mm and $x = -581$ mm of which the only feasible root is $x = 225$ mm. Using this value for x, the cracked second moment of area is calculated as:

$$I_c = \frac{300x^3}{12} + (300x)\left(\frac{x}{2}\right)^2 + (11135)(x - 50)^2 + (42275)(450 - x)^2$$
$$= 3.62 \times 10^9 \text{ mm}^4$$

Hence the deflection, assuming a cracked section, is:

$$\delta_c = \frac{5}{384} \frac{\omega L^4}{E_{c,eff}I_c} = \frac{5}{384}\left(\frac{(15 + 8)(8000)^4}{(6966)(3.62 \times 10^9)}\right)$$
$$= 49 \text{ mm}$$

The maximum moment in the beam is:

$$M = \frac{\omega L^2}{8} = \frac{(15 + 8)(8)^2}{8} = 184 \text{ kNm}$$

The stress in the tension steel due to this moment is given by Equation (8.5):

$$f_s = \alpha_e My/I_c$$

where $y = -(450 - x) = -225$ mm. Hence:

$$f_s = \frac{(28.7)(184 \times 10^6)(-225)}{3.62 \times 10^9} = -328 \text{ N/mm}^2$$

The section first cracks when the maximum tensile stress in the concrete equals the tensile strength, f_{ctm}, which is given in Table 8.2. Hence:

$$M_{cr}y/I_u = -2.6 \text{ N/mm}^2$$
$$\Rightarrow M_{cr} = \frac{(-26)I_u}{y} = \frac{-26(5.028 \times 10^9)}{-(500 - 279)} = 59.15 \times 10^6 \text{ Nmm}$$

The stress in the tension reinforcement due to this moment is:

$$f_{sr} = \frac{\alpha_e M_{cr}y}{I_c} = \frac{(28.7)(59.15 \times 10^6)(-225)}{3.62 \times 10^9} = -106 \text{ N/mm}^2$$

Hence, the distribution factor given by Equation (8.7) is:

$$\zeta = 1 - \beta(^{f_{sr}/f_s)2}$$
$$= 1 - (0.5)(-106/-328)^2$$
$$= 0.948$$

The long-term deflection, from Equation (8.6), is:

$$\delta = \zeta\delta_c + (1 - \zeta)\delta_u$$
$$= (0.948)(49) + (1 - 0.948)(35)$$
$$= 48 \text{ mm}$$

Limits on cracking

The formation of cracks in reinforced concrete members with bending is inevitable. This is undesirable because flexural cracking leads to corrosion of the reinforcement, with the rate of corrosion directly related to the width of the cracks. For this reason, plus the fact that cracks are unsightly and can cause public concern, crack widths are strictly controlled in many structures. Since the environment in the interior of buildings is usually non-severe, corrosion does not generally pose a problem and limits on crack widths will be governed by their appearance. On the other hand, for reinforced concrete structures in aggressive environments, corrosion is a problem and stringent limits are imposed.

The limits recommended by EC2 take account of durability (corrosion) requirements for different environmental conditions and requirements for appearance. The six exposure classes relating to environmental conditions defined in EC2 are given in Table 6.2. In the United Kingdom National Annex (UK NA) to EC2, for exposure classes 1–4, the recommended limit on crack widths is 0.3 mm under the quasi-permanent combination of service load. For reinforced concrete members in exposure classes X0 and XC1, crack width has no influence on durability and the limit is set for appearance only. Where appearance is not an issue, this limit may be relaxed. No specific limit is proposed for exposure classes 5 and 6.

Calculation of crack widths

Crack widths are calculated using the quasi-permanent service load combination of Table 3.13. In general, however, crack width calculations, like deflection calculations, are not necessary on a 'deemed to satisfy' basis if certain detailing rules specified in EC2 are satisfied. If crack control is required, a minimum amount of bonded reinforcement, (specified by an equation in EC2) must be provided in areas where cracking is expected to occur.

In specific cases where a crack width calculation is considered necessary, the following formula is specified in EC2:

$$w_k = s_{r,max}(\varepsilon_{sm} - \varepsilon_{cm}) \tag{8.9}$$

where w_k = design crack width
$s_{r,max}$ = maximum crack spacing (Equation (8.11))
ε_{sm} = mean strain in the tension steel allowing for tension stiffening (Equation (8.10))
ε_{cm} = mean strain in the concrete between cracks

The value of $(\varepsilon_{sm} - \varepsilon_{cm})$ may be calculated using Equation (8.7):

$$(\varepsilon_{sm} - \varepsilon_{cm}) = \frac{f_s - k_t \dfrac{f_{ct,eff}}{\rho_{p,eff}}\left(1 + \alpha_e\rho_{p,eff}\right)}{E_s} \geq 0.6\frac{f_s}{E_s} \tag{8.10}$$

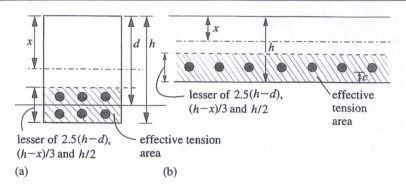

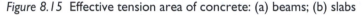

Figure 8.15 Effective tension area of concrete: (a) beams; (b) slabs

where f_s = stress in the tension reinforcement

k_t = factor depending on duration of load ($k_t = 0.6$ for short-term loading; $k_t = 0.4$ for long-term loading)

$f_{ct,eff}$ = mean value of tensile strength of concrete, effective at the time when first cracks are expected to occur (may be taken as f_{ctm}, Table 8.2)

$\rho_{p,eff}$ is the effective reinforcement ratio = $A_s/A_{c,eff}$ (for reinforced concrete), where $A_{c,eff}$ is the effective tension area of the concrete, as illustrated in Fig. 8.15

$\alpha_e = E_s/E_{cm}$

E_s = elastic modulus for steel

The maximum crack spacing in millimetres is calculated using the equation:

$$s_{r,max} = 3.4c + 0.425k_1k_2\frac{\phi}{\rho_{p,eff}} \tag{8.11}$$

where c = cover to longitudinal reinforcement

k_1 = coefficient which accounts for the bond properties of the reinforcement: $k_1 = 0.8$ for high-bond bars; $k_1 = 1.6$ for plain bars (prestressing tendons)

k_2 = coefficient which takes account of the form of strain distribution: for bending, $k_2 = 0.5$

φ = bar diameter

Alternatively, the maximum crack width can be calculated using a maximum crack spacing of $s_{r,max} = 1.3(h - x)$ when the spacing of the bonded reinforcement is greater than $5(c + \varphi/2)$.

The application of these formulae is illustrated in Example 8.6. These equations do not appear to take account of the increase in crack size that will occur with time due to creep. While this is not stated in EC2, it would seem prudent for sustained loading to use a reduced value for the elastic modulus of concrete such as that given in Equation (8.8).

Example 8.6 Crack width calculation

Problem: Determine the crack width due to a characteristic variable gravity loading of 3 kN/m² for the strip of slab illustrated in Fig. 8.16, which is located indoors. The concrete class is C30/37, $E_{cm} = 33,000$ N/mm² and $f_{ctm} = 2.9$ N/mm². Assume long-term loading with $\varphi = 2.4$. The elastic modulus for the reinforcement is 200,000 N/mm².

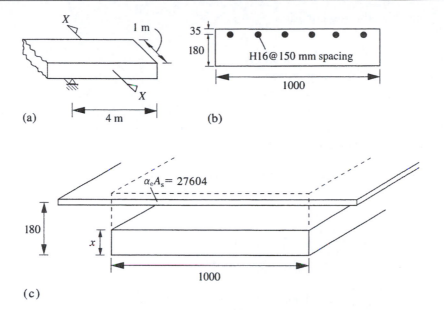

Figure 8.16 Strip of slab of Example 8.6: (a) elevation; (b) section X–X; (c) transformed section

Solution: Since the slab is subjected to long-term loading, the effects of creep are considered and $E_{c,eff}$ is used in the elastic analysis of the section to find f_s.

$$E_{c,eff} = \frac{E_{cm}}{1 + \varphi} = \frac{33000}{3.4}$$
$$= 9706 \, N/mm^2$$
$$\Rightarrow \alpha_e = 20.6 \, (\text{for the calculation of } f_s)$$

As the bars are spaced at 150 mm centres, there will, on average, be 1,000/150 bars in a 1 m strip, giving the area of tension reinforcement, A_s, as:

$$A_s = \left(\frac{1000}{150}\right)\left(\frac{\pi 16^2}{4}\right) = 1340 \, mm^2$$

As the applied moment is hogging, the top of the slab is in tension as can be seen in the transformed section of Fig. 8.16(c). The neutral axis location and second moment of area can be shown to be:

$$x = 76 \, mm$$
$$I_c = 445 \times 10^6 \, mm^4$$

The maximum moment in a cantilever due to a uniform load is:

$$M = \frac{\omega L^2}{2} = \frac{(3)(4)^2}{2} = 24 \, kNm/m$$

The stress in the tension reinforcement due to this moment is:

$$f_s = \frac{\alpha_e M[-(180 - x)]}{I_c} = \frac{(20.6)(24 \times 10^6)[-(180 - 76)]}{445 \times 10^6} = -116 \, N/mm^2$$

$(\varepsilon_{sm} - \varepsilon_{cm})$ is calculated using Equation (8.10):

$$(\varepsilon_{sm} - \varepsilon_{cm}) = \frac{f_s - k_t \dfrac{f_{ct,eff}}{\rho_{p,eff}}\left(1 + \alpha_e \rho_{p,eff}\right)}{E_s} \geq 0.6\frac{f_s}{E_s}$$

where $k_t = 0.4$ for long-term loading, $f_{ct,eff} = f_{ctm} = 2.9\,\text{N/mm}^2$, $\alpha_e = 20.6$ and $E_s = 200{,}000\,\text{N/mm}^2$. The height of the effective tension area is the lesser of $2.5(h-d) = 88$, $(h-x)/3 = 46$ and $h/2 = 105$. Therefore, the effective reinforcement ratio is:

$$\rho_{p,eff} = A_s/A_{c,eff} = 1340/(1000 \times 46) = 0.0291$$

Therefore:

$$(\varepsilon_{sm} - \varepsilon_{cm}) = \frac{116 - 0.4\left(\dfrac{2.9}{0.0291}\right)\left[1 + (20.6)(0.0291)\right]}{200,000}$$

$$= \frac{116 - 63.8}{200000} = 0.000261 \left(< 0.6\frac{f_s}{E_s} = 0.000348 \right)$$

$$\Rightarrow (\varepsilon_{sm} - \varepsilon_{cm}) = 0.000348$$

The spacing of the bonded reinforcement does not exceed $5(c + \phi/2)$, therefore, the maximum crack spacing is calculated using Equation (8.11):

$$s_{r,max} = 3.4c + 0.425k_1 k_2 \frac{\phi}{\rho_{p,eff}}$$

where $c = 27\,\text{mm}$, $k_1 = 0.8$ for high bond bars, $k_2 = 0.5$ for bending.

$$\Rightarrow s_{r,max} = 3.4(27) + 0.425(0.8)(0.5)\frac{16}{0.0291} = 185\,\text{mm}$$

from which the design crack width is:

$$w_k = s_{r,max}(\varepsilon_{sm} - \varepsilon_{cm}) = 185(0.000348) = 0.064\,\text{mm}$$

This is less than the recommended limit of 0.3 mm.

8.4 Stress/strain relationships and modes of failure

The short-term stress/strain relationship for a concrete test cylinder in compression is illustrated in Fig. 8.17. The stress/strain relationship is approximately linear (i.e. $f_c \approx E_{cm}\varepsilon_c$) for low values of strain. Thus, the slope of the initial part of the diagram is reasonably constant. As the stress is increased, the stress/strain relationship becomes substantially non-linear until the point is reached where very little extra stress causes a large increase in strain. The peak in stress is known as the compressive strength. Beyond this point of maximum strength, the stress must be decreased if failure is to be prevented. The mean value of concrete cylinder compressive strength is denoted f_{cm} and the corresponding compressive strain is denoted ε_{c1}. Finally, at a certain ultimate strain, ε_{cu1}, the concrete crushes regardless of the level of stress. Much research has been carried out to predict ε_{c1}

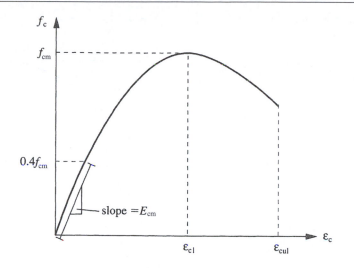

Figure 8.17 Stress/strain relationship for concrete

and ε_{cu1}, for different grades of concrete. The design values recommended by EC2 for all concrete grades are as follows:

$$\varepsilon_{c1} = 0.002$$
$$\varepsilon_{cu1} = 0.0035$$

As the stress/strain relationship for concrete is non-linear, an average slope, known as a secant modulus, is used in lieu of the regular concrete modulus. This is defined as the slope of the line joining the origin to the point corresponding to 40 per cent of the characteristic strength (as illustrated in Fig. 8.17). Where great accuracy is not required, EC2 allows the estimation of the secant modulus, E_{cm}, with the formula:

$$E_{cm} = 22(f_{cm}/10)^{0.3} \qquad (8.12)$$

where $f_{cm} = f_{ck} + 8$, E_{cm} is in Newtons per square millimetre (N/mm^2) and f_{ck} is the characteristic cylinder strength (subscript k implies 'characteristic' throughout EC2), also in Newtons per square millimetre.

The stress/strain relationship for reinforcement is reasonably linear for a considerable portion of the diagram. The elastic modulus, E_s, is equal to the slope of the linear portion of the diagram. EC2 allows a mean value of 200,000 N/mm^2 to be assumed for E_s. For cold worked steel, the 0.2 per cent proof stress, $f_{0.2k}$, at which yielding is said to occur, is defined as the stress at which strain has exceeded the value predicted by the linear relationship by 0.002. Hot rolled steel has a definite yield point and characteristic yield stress, f_{yk}, at which yielding occurs. Typical stress/strain diagrams for cold worked and hot rolled steel reinforcement in compression or tension are illustrated in Fig. 8.18. The yield stress is of equal magnitude in tension and compression. Beyond this point, increases in stress result in progressively larger increases in strain. The ultimate tensile and compressive strengths are also equal in magnitude and are denoted by f_{tk}.

Under service loads (SLS), strains in reinforced concrete members in bending are small and both the steel and concrete stress/strain relationships remain linear. Initially, segments of a

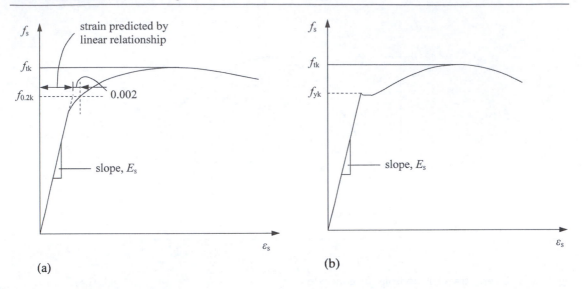

Figure 8.18 Stress/strain relationship; (a) cold worked steel reinforcement, (b) hot rolled steel reinforcement

member such as illustrated in Fig. 8.12(b) are uncracked and the stress distribution is as illustrated in Fig. 8.19(b) with a maximum compressive stress of Mx_u/I_u, where x_u is the depth to the uncracked neutral axis. As the moment is increased, the section cracks and the stresses increase in magnitude while the distribution remains linear (see Fig. 8.19(c)). As loads are increased further, the strain distribution remains linear but the stress distribution becomes non-linear as illustrated in Fig. 8.19(d). At the ultimate limit state, under maximum load, strains in the concrete at the extreme fibres in compression reach ε_{cu1}, and the distribution of stress is as illustrated in Fig. 8.19(e). The stress distribution in the concrete in compression has approximately the same shape as the stress/strain relationship of Fig. 8.17. However, the two do not correspond directly as the situation for a fibre of concrete in bending is different from that for concrete in a standard test cylinder. Considerable research effort has been expended in attempting to determine the exact shape of the distribution in Fig. 8.19(e). Fortunately, the calculation of the moment required to cause this ultimate distribution of stress is insensitive to the exact shape of the distribution and crude approximations of the shape are quite acceptable for design purposes.

The cracking moment, M_{cr}, was introduced in the previous section as the moment at which the concrete member first cracks. After members have cracked, the ultimate capacity to resist moment, M_{ult}, is greatly affected by the area of tension reinforcement, A_s. If this area is particularly small, the moment capacity after first cracking will be less than M_{cr} which will result in sudden failure of the member when the applied moment reaches M_{cr} (see Fig. 8.20(a)). To prevent such sudden failure, sufficient reinforcement must be provided so that:

$$M_{ult} > M_{cr} \tag{8.13}$$

To calculate a conservatively high value for M_{cr}, for use in this inequality, the tensile strength with a 95 per cent probability of not being exceeded can be used:

$$f_{ctk,0.95} = 0.39f_{ck}^{2/3} \tag{8.14}$$

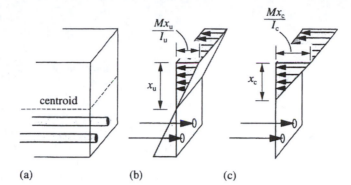

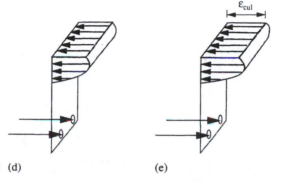

Figure 8.19 Stress distributions for increasing moment: (a) section; (b) uncracked; (c) cracked; (d) non-linear; (e) ultimate

The corresponding cracking moment is defined by:

$$-f_{ctk,0.95} = \frac{M_{cr}[-(h-x)]}{I_u} \tag{8.15}$$

where h is the section depth and x is the distance from the neutral axis to the extreme fibre in compression. Substituting for $-f_{ctk,0.95}$ in Equation (8.15) and rearranging gives:

$$M_{cr} = \frac{0.39f_{ck}^{2/3}I_u}{(h-x)} \tag{8.16}$$

To ensure that M_{ult} exceeds M_{cr}, EC2 specifies a minimum value for A_s which results in an ultimate moment in excess of M_{cr}.

When the area of reinforcement, A_s, is increased so that $M_{ult} > M_{cr}$, the beam does not fail when it first cracks, but continues to deform elastically as illustrated in Fig. 8.20(b). However, as A_s is still small, the reinforcement still yields before the maximum compressive strain in the concrete reaches ε_{cu1}. The plastic deformation which occurs after yielding of the reinforcement results in considerable deformation of the beam before, finally, the reinforcement fails completely. The more normal mode of failure, in practice, occurs when a larger area of tension reinforcement is present. In this case, the reinforcement yields but the maximum compressive strain in the concrete reaches ε_{cu1}, before it fails. Thus, failure of the beam is due to crushing of the concrete in compression, as illustrated in Fig. 8.20(c). Because the amount of reinforcement is small, the beam is termed **under-reinforced** in both these cases (Fig. 8.20 (b–c)). They are referred to as **ductile failures** because of the large plastic

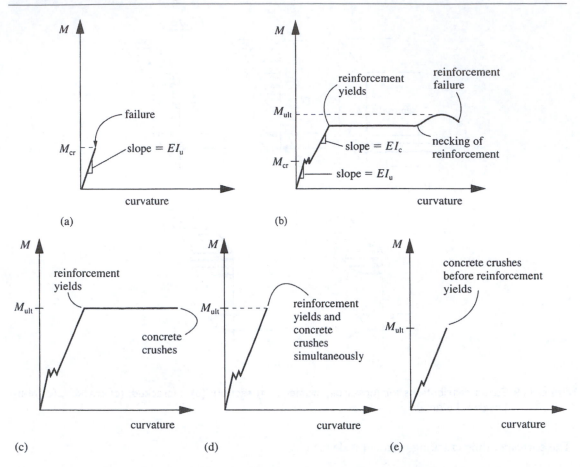

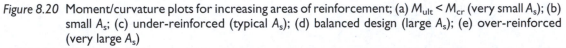

Figure 8.20 Moment/curvature plots for increasing areas of reinforcement; (a) $M_{ult} < M_{cr}$ (very small A_s); (b) small A_s; (c) under-reinforced (typical A_s); (d) balanced design (large A_s); (e) over-reinforced (very large A_s)

deformations which occur prior to collapse. Since there is ample warning of imminent failure this is highly desirable in design.

If A_s is increased further to a level where the concrete crushes at the same time as the reinforcement first yields, as illustrated in Fig. 8.20(d), the mode of failure is known as balanced. Since there is no plastic deformation of the reinforcement prior to collapse the beam acts as a brittle material giving little warning of failure. If A_s is increased still further, the concrete crushes before any yielding of the reinforcement, as illustrated in Fig. 8.20(e). In this case there is too much reinforcement and the beam is termed over-reinforced. Since failure is again sudden and brittle, this is an unacceptable form of failure. To prevent brittle failure, EC2 specifies maximum values for A_s.

8.5   Ultimate moment capacity

Simplified stress/strain diagrams for member design

Rather than using the actual stress/strain diagram for concrete given in Fig. 8.19(e), other idealized diagrams can be used in order to simplify member design. The idealized stress/strain diagrams for

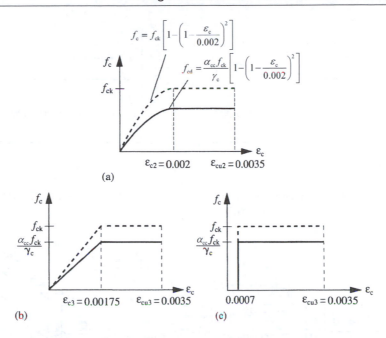

$$f_c = f_{ck}\left[1-\left(1-\frac{\varepsilon_c}{0.002}\right)^2\right]$$

$$f_{cd} = \frac{\alpha_{cc}f_{ck}}{\gamma_c}\left[1-\left(1-\frac{\varepsilon_c}{0.002}\right)^2\right]$$

Figure 8.21 Simplified stress/strain diagrams for concrete (idealized shown dashed, design stress shown solid): (a) parabolic-rectangular; (b) bi-linear; (c) rectangular

concrete specified in EC2 are the parabolic-rectangular, the bi-linear and the equivalent rectangular diagrams illustrated in Fig. 8.21. The diagrams to be applied in design (shown solid) are derived from the idealized diagrams (shown dotted) by means of a reduction of all stresses in the idealized diagrams by the factor α_{cc}/γ_c, where:

α_{cc} = coefficient which takes account of long-term effects on the compressive strength and of the unfavourable effects resulting from the way in which the load is applied: $\alpha_{cc} = 0.85$ for compression in flexure and axial loading, $\alpha_{cc} = 1.0$ for other phenomena (conservatively, α_{cc} may be taken as 0.85 for all phenomena).

γ_c = partial factor of safety for concrete strength equal to 1.5.

For reinforcement, the simplified stress/strain diagram adopted in EC2 is the bi-linear diagram illustrated in Fig. 8.22(a) in which f_{yk} is the characteristic yield strength and f_{tk} is the characteristic ultimate strength. The limiting strain in the reinforcement depends on the ductility class of the reinforcing steel. For further simplicity, the diagram can be altered so that the top branch is horizontal, as illustrated in Fig. 8.22(b). In this case, no limit on the steel strain is necessary. In both cases, the idealized diagrams are shown dotted. The corresponding design diagrams are obtained by factoring the stresses of the idealized diagrams by $1/\gamma_s$ where γ_s is the partial factor of safety for the strength of steel. At ULS, $\gamma_s = 1.15$. The design value for the elastic modulus, E_s, is 200,000 N/mm².

General procedure for calculating ultimate moment capacity

Given the assumptions stated above, the ultimate moment capacity, M_{ult}, can be calculated for a section of specified dimensions and areas of reinforcement such as that illustrated in Fig. 8.23. In accordance with EC2, any combination of the design stress/strain diagrams for steel and concrete can be used.

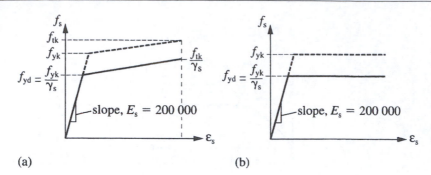

Figure 8.22 Idealized stress/strain relationships for reinforcement (from BS EN 1992-1-1): (a) inclined top branch; (b) horizontal top branch

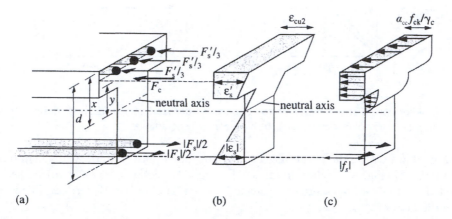

Figure 8.23 Distribution of stress and strain in typical section: (a) section; (b) linear strain distribution; (c) parabolic-rectangular stress distribution

For all modes of failure (with the exception of the very under-reinforced section, a case which can be ignored), ultimate moment failure occurs when the strain in the extreme concrete fibre in compression reaches $\varepsilon_{cu2} = 0.0035$, at which stage the concrete stress distribution is assumed to equal one of the distributions illustrated in Fig. 8.21 (see Fig. 8.23(c)). As in the calculation of second moments of area (Example 8.3), equilibrium of forces can then be used to determine the ultimate moment capacity, M_{ult}, of a section in bending. The different forces that act on a reinforced concrete section in sag, illustrated in Fig. 8.24, are as follows:

(a) concrete in compression above the neutral axis, F_c (gross section)
(b) reinforcement in compression above the neutral axis, F'_s
(c) concrete displaced by compression reinforcement that would have been in compression, F_{disp}
(d) reinforcement in tension below the neutral axis, F_s. Using the sign convention of positive compression, F_s, being tensile, is always negative.

A similar set of forces acts on a reinforced concrete section in hog. If the total force, F, acting on an area, A, is given by:

$$F = \int_A \sigma dA \tag{8.17}$$

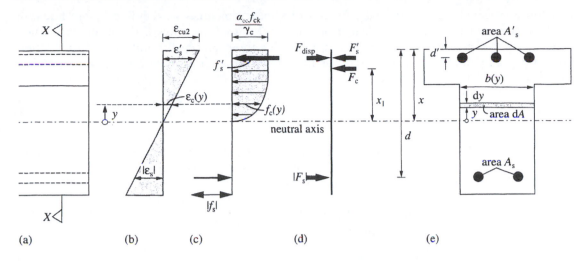

Figure 8.24 Components of axial force in typical section: (a) elevation; (b) strain; (c) parabolic-rectangular stress distribution; (d) axial forces on section; (e) section X–X

then the compressive force in the concrete, F_c, is given by:

$$F_c = \int_0^x f_c(y)\,dA = \int_0^x f_c(y)b(y)\,dy \tag{8.18}$$

where $f_c(y)$ is the concrete stress at height y and $b(y)$ is the breadth of the section at that height, as illustrated in Fig. 8.24(e). By allowing later for the force that would have been in the concrete displaced by compression reinforcement, this integration can be over the gross section. The formula for $f_c(y)$ depends on which idealized stress/strain diagram for concrete is used.

The compression force in the top steel, F'_s, is equal to $A'_s f'_s$ where f'_s is the stress there. Similarly, the force in the bottom steel, F_s, is equal to the product of area and stress, $A_s f_s$ (both f_s and F_s, being tensile, will be negative). The concrete displaced by the compression reinforcement has an area A'_s and would, if present, have had a stress equal to the stress in the concrete at that level. This depends on the point at which the compression reinforcement intersects the stress distribution, but the situation illustrated in Fig. 8.24 is typical, where the stress in the displaced concrete is $\alpha_{cc}f_{ck}/\gamma_c$. Hence the force is typically:

$$F_{disp} = A'_s \left(\frac{\alpha_{cc}f_{ck}}{\gamma_c} \right) \tag{8.19}$$

By equilibrium, the sum of the compressive forces must equal zero, that is:

$$F_c - F_{disp} + F'_s + F_s = 0$$
$$\Rightarrow \int_0^x f_c(y)b(y)\,dy - F_{disp} + A'_s f'_s + A_s f_s = 0 \tag{8.20}$$

The only unknown in Equation (8.20) is x. In practice, a trial and error approach is sometimes used to find the correct value. An initial value for x is assumed and the forces on the section are determined. If Equation (8.20) is not satisfied, the value for x is adjusted and the procedure repeated. Several refined estimates of x may sometimes be required before Equation (8.20) is satisfied to a sufficient degree of accuracy. Once x is known, the ultimate moment capacity of the section is found by taking moments about any horizontal axis in the section. For instance, by taking moments about the neutral axis, the ultimate moment capacity, M_{ult}, is found to be (refer to Fig. 8.24(d)):

$$M_{ult} = F_c x_1 + F'_s(x - d') - F_{disp}(x - d') + |F_s|(d - x) \qquad (8.21)$$

where x_1 is the distance from the compressive force to the neutral axis. Note that the tension reinforcement component is **additive** to the components due to the compressive force. This is because, while the forces act in opposite directions, the components of moment in all cases (except the displaced concrete) act anticlockwise. The depth from the extreme fibre in compression to the tension reinforcement is known as the **effective depth** and is traditionally denoted as d (see Fig. 8.24 (e)). Substituting for the forces in Equation (8.21) gives:

$$M_{ult} = (x_1) \int_0^x f_c(y) b(y) \, \mathrm{d}y + A'_s f'_s(x - d') - F_{disp}(x - d') + A_s |f_s|(d - x) \qquad (8.22)$$

Ultimate moment capacity of rectangular sections

For a rectangular section with specified total depth h and having constant breadth b, as illustrated in Fig. 8.25(a), Equation (8.18) becomes:

$$F_c = b \int_0^x f_c(y) \, \mathrm{d}y \qquad (8.23)$$

The term $\int_0^x f_c(y)$ is equal to the total area under the compressive stress block for the concrete. Clearly, the value of the total area is dependent on which of the distributions of Fig. 8.21 is adopted. Fortunately, while the distributions are quite different in form, the areas under each are in fact quite similar in magnitude. The simplest of these is the equivalent rectangular stress block of Fig. 8.21(c). In this case, the stress is constant for all strains in excess of 0.0007 and is zero elsewhere. When a section is on the point of collapse, the strain at the extreme fibre is ε_{cu3}, as illustrated in Fig. 8.25(b).

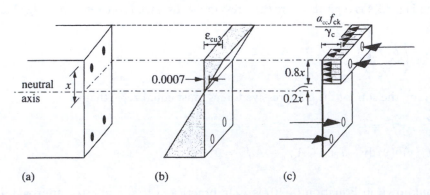

Figure 8.25 Stress and strain distributions at ULS: (a) elevation and section; (b) strain; (c) equivalent rectangular stress block

Thus, the strain increases linearly from zero at the neutral axis to a maximum of ε_{cu3}, at the extreme fibre in compression. The strain is 0.0007 at a distance from the neutral axis of:

$$\left(\frac{0.0007}{\varepsilon_{cu3}}\right)x = \frac{0.0007x}{0.0035} = 0.2x$$

Hence the stress distribution is as illustrated in Fig. 8.25(c). The total compressive force in the concrete is simply the product of stress and area. Taking the gross section:

$$F_c = \left(\frac{\alpha_{cc}f_{ck}}{\gamma_c}\right)(08xb) \tag{8.24}$$

Taking the recommended values, $\alpha_{cc} = 0.85$ and $\gamma_c = 1.5$, Equation (8.24) becomes:

$$F_c = 0.453xbf_{ck} \tag{8.25}$$

The force, F_c, acts at the centroid of the compressive stress block, that is, at a distance of $0.4x$ from the top of the section.

The depth to the neutral axis, x, is calculated from equilibrium of forces, i.e. using Equation (8.20). Substitution of Equation (8.25) into Equation (8.20) gives:

$$0.453xbf_{ck} - F_{disp} + A'_sf'_s + A_sf_s = 0 \tag{8.26}$$

In using Equation (8.26) to calculate x, assume initially, say, that both the tension and compression reinforcements have yielded, that is $-f_s = f'_s = f_{yk}/\gamma_s$. Once x is found, these assumptions can be checked from the strain diagram of the section. If they are incorrect (that is, if either or both of the reinforcements has not yielded), the procedure is repeated using the revised assumption. Having found x, the ultimate moment capacity is calculated using Equation (8.22). The method is illustrated in the following examples.

Example 8.7 Ultimate moment capacity of rectangular beam

Problem: For the beam of Fig. 8.26, calculate the value for the applied load, P, which will cause collapse (ignoring the self-weight of the beam). Use the rectangular stress distribution for concrete and the horizontal top branch in the stress distribution for the reinforcement. Assume the concrete has $f_{ck} = 35$ N/mm^2 and the reinforcement has $f_{yk} = 500$ N/mm^2.

Solution: The moment capacity of the beam is found from a consideration of the stress distribution in the section. Assuming that the reinforcement has yielded, the stress distribution is as illustrated in Fig. 8.27(c). Using Equation (8.24), the total compressive force in the concrete is:

$$F_c = \left(\frac{\alpha_{cc}f_{ck}}{\gamma_c}\right)(0.8xb) = \frac{(0.85)(35)(0.8x)(250)}{1.5} = 3967x \text{ N}$$

Assuming that the reinforcement has yielded, its total force is:

$$F_s = -A_s\left(\frac{f_{yk}}{\gamma_s}\right) = -3\left(\frac{\pi 32^2}{4}\right)\left(\frac{500}{1.15}\right)$$
$$= -1,049,020 \text{ N}$$

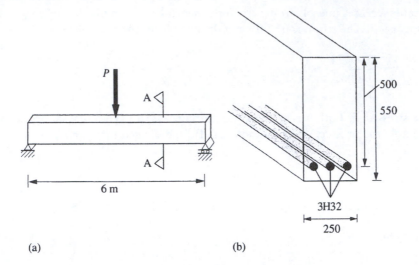

(a)   (b)

Figure 8.26 Beam of Examples 8.7 and 8.10: (a) elevation; (b) section A–A

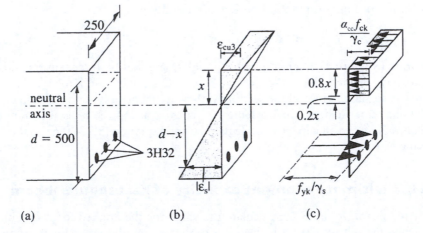

(a)   (b)   (c)

Figure 8.27 Assumed distributions of stress and strain: (a) elevation and section; (b) strain; (c) stress

As there is no net axial force in the member, $F_c + F_s = 0$. Hence:

$$3967x - 1049020 = 0$$
$$\Rightarrow x = 264 \text{ mm}$$

The assumption that the reinforcement has yielded is now checked. The strain, ε_s, in the reinforcement is calculated using similar triangles (see Fig. 8.27(b)):

$$\frac{|\varepsilon_s|}{(d - x)} = \frac{\varepsilon_{cu3}}{x}$$
$$\Rightarrow |\varepsilon_s| = \frac{(500 - 264)(0.0035)}{264} = 0.00313$$

The yield strain (from Fig. 8.22) is:

$$\frac{f_{yk}/\gamma_s}{E_s} = \frac{500/1.15}{200000} = 0.00217$$

Therefore the assumption that the steel has yielded is a correct one. The moment capacity is found by taking moments about the neutral axis:

$$M_{ult} = F_c(0.6x) + |F_s|(d - x)$$

but:

$$|F_s| = F_c$$
$$\Rightarrow M_{ult} = |F_s|(0.6x + d - x)$$
$$= |F_s|(d - 0.4x)$$
$$= (1049020)[500 - 0.4(264)]$$
$$= 414 \text{ kNm}$$

The bending moment diagram for this beam is illustrated in Appendix B, No. 2. The maximum applied moment can be seen to be $Pl/4$. Thus, the beam will collapse when:

$$\frac{P_{ult}l}{4} = M_{ult}$$
$$\Rightarrow P_{ult} = \frac{4M_{ult}}{l} = \frac{4(414)}{6}$$
$$= 276 \text{ kN}$$

Example 8.8 Parabolic/rectangular stress block

Problem: For the slab illustrated in Fig. 8.28 calculate the ultimate moment capacity (per unit breadth). Use the parabolic-rectangular stress block for the concrete and the horizontal top branch in the stress/strain relationship for reinforcement. Assume that $f_{ck} = 40$ N/mm^2 and $f_{yk} = 500$ N/mm^2.

Solution: The total compressive force on the concrete is:

$$F_c = F_c^1 + F_c^2$$

where F_c^1 is the compressive force in the rectangular portion of the stress distribution and F_c^2 is the compressive force in the parabolic portion (see Fig. 8.28(c)). The point on the stress distribution where the stress changes from being constant to being parabolic occurs when the strain is 0.002 (see Fig. 8.21 (b)). Hence, referring to Fig. 8.28(b), the change point occurs at a distance from the neutral axis given by:

$$\text{change point} = \left(\frac{0.002}{0.0035}\right)x = 0.571x$$

Hence:

$$F_c^1 = \left(\frac{\alpha_{cc}f_{ck}}{\gamma_c}\right)(x - 0.571x)1000$$
$$= \left(\frac{(0.85)(40)}{1.5}\right)(x - 0.571x)1000$$
$$= 9724x \text{ N}$$

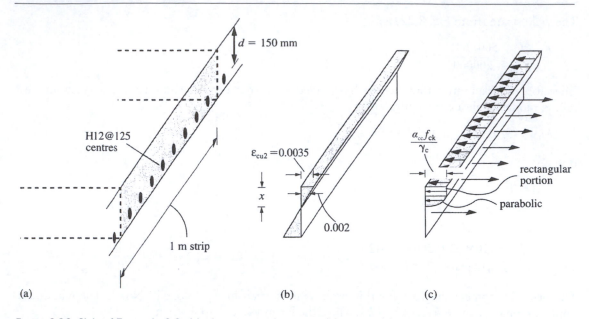

(a)   (b)   (c)

Figure 8.28 Slab of Example 8.8: (a) elevation and section; (b) strain; (c) stress

The strain in the concrete at any point, $\varepsilon_c(y)$, is found from Fig. 8.28(b) (by similar triangles):

$$\frac{\varepsilon_c(y)}{y} = \frac{\varepsilon_{cu2}}{x}$$

$$\Rightarrow \varepsilon_c(y) = \frac{0.0035y}{x}$$

In the parabolic region (refer to Fig. 8.21(a)):

$$f_c(y) = \frac{\alpha_{cc}f_{ck}}{\gamma_c}\left[1 - \left(1 - \frac{\varepsilon_c}{0.002}\right)^2\right]$$

$$= \frac{(0.85)(40)}{1.5}\left[1 - \left(1 - \frac{0.0035y}{0.002x}\right)^2\right]$$

$$= \frac{(0.85)(40)}{1.5}\left[3.5\frac{y}{x} - 3.062\frac{y^2}{x^2}\right]$$

$$= \frac{79.33y}{x} - \frac{69.41y^2}{x^2}$$

Hence the force in this region is (stress × area):

$$F_c^2 = \int_A f_c(y)\mathrm{d}A = b\int_0^{y=0.571x} f_c(y)\,\mathrm{d}y$$

$$= b\int_0^{y=0.571x}\left(\frac{79.33y}{x} - \frac{69.41y^2}{x^2}\right)\mathrm{d}y$$

which (remembering that x is constant) reduces to:

$$F_c^2 = 8625x \text{ N}$$

Assuming that the reinforcement has yielded gives:

$$F_s = -A_s \left(\frac{f_{yk}}{\gamma_s} \right) = -\left(\frac{1000}{125} \right) \left(\frac{\pi 12^2}{4} \right) \left(\frac{500}{1.15} \right) = -393382 \text{ N}$$

Then equilibrium of axial forces gives:

$$F_c^1 + F_c^2 + F_s = 0$$
$$\Rightarrow 9724x + 8625x - 393382 = 0$$
$$\Rightarrow x = 21.4 \text{ mm}$$

The corresponding steel strain is well in excess of the yield strain. To get the moment capacity, M_{ult}, moments are taken about the neutral axis. The contribution of the parabolic portion of the stress distribution to the moment capacity is found by integrating the product of force and lever arm:

$$M_c^2 = \int_A f_c(y) y \, dA$$
$$= b \int_0^{y=0.571x} f_c(y) y \, dy$$
$$= b \int_0^{y=0.571x} \left(\frac{79.33y^2}{x} - \frac{69.41y^3}{x^2} \right) dy$$

where $x = 21.4$ mm. This reduces to:

$$M_c^2 = 1.41 \text{ kNm}$$

The total moment capacity is the sum of the contributions from the parabolic and rectangular portions of the compressive block and that of the reinforcement:

$$M_{ult} = F_c^1 \left(\frac{0.571x + x}{2} \right) + M_c^2 + |F_s|(d - x) = 0$$

which reduces to 55.5 kNm for the 1 m strip.

Example 8.9 Reinforcement not yielded

Problem: Calculate the ultimate hogging moment capacity for the section illustrated in Fig. 8.29. Use the rectangular stress distribution of Fig. 8.21(c) with $f_{ck} = 35$ N/mm². For the reinforcement, use the simplified stress/strain distribution of Fig. 8.22(b) with $f_{yk} = 500$ N/mm².

Solution: As the moment is hogging, the bottom of the section is in compression and the stress and strain distributions are as illustrated in Fig. 8.30. Provided $0.8x < 150$, this section will behave exactly as a rectangular section because the shape of the compression zone will be rectangular. Assuming therefore that $0.8x < 150$, the breadth is 2×200 mm and the compressive force in the concrete is:

$$F_c = \left(\frac{\alpha_{cc}f_{ck}}{\gamma_c}\right)(400)(0.8x) = \left(\frac{(0.85)(35)}{(1.5)}\right)(400)(0.8x) = 6347x \text{ N}$$

Assuming that the tension reinforcement has yielded implies:

$$|F_s| = A_s\left(\frac{f_{yk}}{\gamma_s}\right) = (1608)\left(\frac{500}{1.15}\right) = 699,130 \text{ N}$$

Assuming that the compression reinforcement has yielded implies:

$$|F'_s| = A'_s\left(\frac{f_{yk}}{\gamma_s}\right) = (452)\left(\frac{500}{1.15}\right) = 196,522 \text{ N}$$

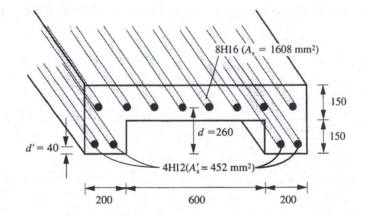

Figure 8.29 Section of Example 8.9

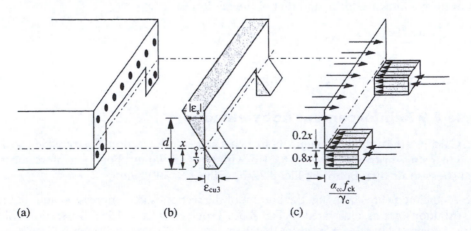

Figure 8.30 Strain and stress distributions for section of Example 8.9: (a) elevation and section; (b) strain; (c) stress

The force that would have been present were the concrete not displaced by the compression reinforcement is given by:

$$F_{\text{disp}} = \left(\frac{\alpha_{cc}f_{ck}}{\gamma_c}\right)A_s' = \left(\frac{(0.85)(35)}{(1.5)}\right)(452) = 8965\,\text{N}$$

By equilibrium of the forces on the section:

$$F_c - F_{\text{disp}} + F_s' = |F_s|$$
$$\Rightarrow 6347x - 8965 + 196,522 = 699,130$$
$$\Rightarrow x = 80.6\,\text{mm}$$

The assumptions made earlier are now checked. By similar triangles in Fig. 8.30(b):

$$\frac{|\varepsilon_s|}{(d-x)} = \frac{\varepsilon_{cu3}}{x}$$
$$\Rightarrow |\varepsilon_s| = \frac{(260 - 80.6)(0.0035)}{80.6} = 0.0080 > 0.00217$$

which implies that the tension reinforcement has yielded and the initial assumption is correct. Similarly:

$$\varepsilon_s' = \frac{(80.6 - 40)(0.0035)}{80.6} = 0.00176$$

This time, it emerges that the initial assumption that the compression reinforcement had yielded is incorrect. Thus, the value for x must be recalculated using a revised expression for F_s':

$$\varepsilon_s' = \frac{\varepsilon_{cu3}(x - 40)}{x}$$
$$\Rightarrow f_s' = \frac{E_s\varepsilon_{cu3}(x - 40)}{x}$$
$$\Rightarrow F_s' = \frac{A_s'E_s\varepsilon_{cu3}(x - 40)}{x} = \frac{(452)(200,000)(0.0035)(x - 40)}{x}$$
$$= \frac{(316400)(x - 40)}{x}$$

Then equilibrium gives:

$$F_c - F_{\text{disp}} + F_s' = |F_s|$$
$$\Rightarrow 6347x - 8965 + \frac{(316,400)(x - 40)}{x} = 699,130$$

which reduces to:

$$6.347x^2 - 392x - 12,656 = 0$$

Solving this equation for x, the only positive root is $x = 85.1$ mm. This time, on checking, all of the assumptions are found to be correct; that is, the tension reinforcement has yielded, the compression reinforcement has not and $0.8x < 150$ mm.

The ultimate moment capacity is now calculated by taking moments about the neutral axis:

$$M_{\text{ult}} = (6347x)(0.6x) - (8965)(x - 40) + \frac{(316,400)(x - 40)}{x}(x - 40)$$
$$+ (699,130)(260 - x)$$
$$\Rightarrow M_{\text{ult}} = 150 \text{ kNm}$$

Reinforcement required to resist moment

Up to now, all of the examples considered have involved the calculation of the moment capacity for cross-sections in which the areas of reinforcement present were known. However, the more usual problem is that the applied moment is known and the area of reinforcement required to resist it is sought, that is, we usually wish to determine the area of reinforcement necessary to provide a specified moment capacity. When there is no compression reinforcement, the procedure for calculating the required area of tension reinforcement, A_s, is as follows:

(a)   Determine an expression for F_c as a function of x.
(b)   Taking moments about the neutral axis gives Equation (8.21) (with $F'_s = F_{\text{disp}} = 0$). Use equilibrium of axial forces (i.e. $F_c = |F_s|$) to remove F_s from this equation to give an equation relating M_{ult} to x. Solve this (by iteration if necessary) to find x.
(c)   Back-substitute into Equation (8.21) to find F_s and hence find A_s.

Using this procedure the formula for a rectangular section is derived below. Formulae for other sections can be derived in a similar manner.

Formula for A_s required in a rectangular section

It is assumed that the reinforcement has yielded. Steps will be taken later to ensure that the beam is under-reinforced, as illustrated in Fig. 8.20(c), and hence that this assumption is valid.

(a) If there is no compression reinforcement, then, for the rectangular section of arbitrary dimensions illustrated in Fig. 8.31:

$$F_c = \left(\frac{\alpha_{cc}f_{ck}}{\gamma_c}\right)(0.8xb)$$

(b) Summing moments about the neutral axis gives the ultimate moment capacity:

$$M_{\text{ult}} = F_c(0.6x) + |F_s|(d - x)$$

But $F_c = |F_s|$ due to equilibrium of the horizontal forces, from which:

$$M_{\text{ult}} = F_c(d - 0.4x)$$
$$\Rightarrow M_{\text{ult}} = F_c z \tag{8.27}$$

where z is the lever arm equal to $(d - 0.4x)$. Hence:

$$x = \frac{d - z}{0.4} \tag{8.28}$$

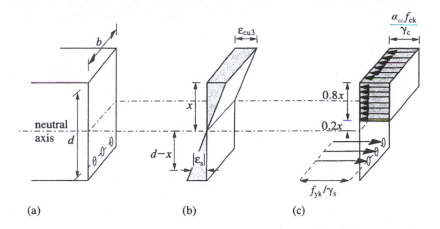

Figure 8.31 Rectangular section of arbitrary dimensions: (a) elevation and section; (b) strain distribution; (c) stress distribution

It is, in this case, more convenient to solve Equation (8.27) to find z and subsequently to use Equation (8.28) to find x. Thus:

$$M_{ult} = F_c z = \left(\frac{\alpha_{cc} f_{ck}}{\gamma_c} \right) (0.8xb)z$$

$$= \left(\frac{\alpha_{cc} f_{ck}}{\gamma_c} \right) (0.8) \left(\frac{d-z}{0.4} \right) bz$$

which reduces to:

$$\frac{M_{ult}}{bd^2 f_{ck}} = \frac{2\alpha_{cc}}{\gamma_c} \left(1 - \frac{z}{d} \right) \left(\frac{z}{d} \right)$$

It is convenient to define a non-dimensional constant, K, as:

$$K = \frac{M_{ult}}{bd^2 f_{ck}} \tag{8.29}$$

Then, taking $\alpha_{cc} = 0.85$ and $\gamma_c = 1.5$ gives:

$$\left(\frac{z}{d} \right)^2 - \left(\frac{z}{d} \right) + 0.88K = 0$$

$$\Rightarrow \frac{z}{d} = \frac{1 \pm \sqrt{1 - 4(0.88K)}}{2}$$

$$= 0.5 \pm \sqrt{0.25 - 0.88K}$$

Of the two roots, it is generally only the positive one that is valid. Hence:

$$z = d(0.5 + \sqrt{0.25 - 0.88K}) \tag{8.30}$$

(c) Having calculated z, it is a simple matter to calculate the area of reinforcement required. From Equation (8.27):

$$M_{ult} = F_c z = |F_s|z$$

$$\Rightarrow M_{ult} = A_s \left(\frac{f_{yk}}{\gamma_s} \right) z$$

$$\Rightarrow A_s = \frac{M_{ult}}{(f_{yk}/\gamma_s)z}$$

(8.31)

Thus, for a rectangular section, Equations (8.30) and (8.31) can be used to find the area of reinforcement required to provide a moment capacity of M_{ult}.

Example 8.10 Reinforcement in rectangular beam

Problem: The ultimate applied sag moment on the section of Fig. 8.31 is 250 kNm. If the breadth of the beam is 300 mm and the overall depth of the beam is 600 mm, use the equivalent rectangular stress block to calculate the area of tension reinforcement required to resist the applied moment. Assume a cover of 50 mm and $f_{ck} = 35$ N/mm^2.

Solution: Assuming a 10 mm diameter stirrup and a longitudinal tension reinforcement bar diameter of 25 mm, the effective depth is:

$$d = 600 - 50 - 10 - 25/2$$
$$= 527mm$$

From Equation (8.29):

$$K = \frac{M_{ult}}{bd^2 f_{ck}} = \frac{250 \times 10^6}{(300)(527)^2(35)} = 0.0857$$

Equation (8.30) then gives:

$$z = d(0.5 + \sqrt{0.25 - 0.88K})$$
$$= (527)(0.5 + \sqrt{0.25 - 0.88(0.0857)})$$
$$= 484 \, mm$$

The area of reinforcement required to resist M_{ult} is found from Equation (8.31):

$$A_s = \frac{M_{ult}}{(f_{yk}/\gamma_s)z} = \frac{250 \times 10^6}{(500/1.15)484} = 1188 \, mm^2$$

3H25 gives an area of $3(\pi 25^2/4) = 1473$ mm^2, which exceeds the required area and is thus sufficient to resist the applied moment.

Flanged beams and ribbed slabs

It was briefly stated in Chapter 6 that the design of reinforced flanged beams (i.e. T-beams and L-beams) and ribbed slabs depends on the location of the neutral axis. For light or moderately loaded beams, the neutral axis lies within the flange for a section in sag as illustrated in Fig. 8.32(b). In such circumstances, the compression zone is rectangular (assuming that the flange is rectangular) and the

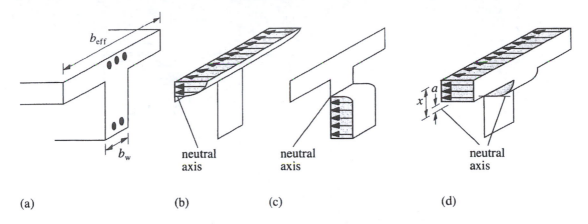

Figure 8.32 Flanged sections: (a) elevation and section; (b) rectangular compression zone (sag); (c) rectangular compression zone (hog); (d) T-shaped compression zone (sag)

section can be designed as a rectangular section having a breadth equal to the effective breadth of the flange, b_{eff}. Similarly, if the moment is small, the neutral axis lies within the web for a section in hog, as illustrated in Fig. 8.32(c). Assuming a constant web breadth, the section can then be designed as a rectangular section having a breadth equal to the web breadth, b_w. (This was true of Example 8.9.) In certain cases, however, the neutral axis may not lie within the flange for sections in sag or in the web for sections in hog. In such cases, the design procedure is not as straightforward since the compression zone is no longer rectangular as illustrated in Fig. 8.32(d) and so the member cannot be treated as a simple rectangular section. From Equation (8.18), the total compressive force in the concrete for the section of Fig. 8.32(d) is given by:

$$F_c = b_w \int_0^a f_c(y)\, dy + b_{eff} \int_a^x f_c(y)\, dy \tag{8.32}$$

Using Equation (8.32) and an appropriate idealized stress block, the ultimate moment capacity for such a section can be calculated.

Example 8.11 Flanged beam

Problem: Determine the ultimate moment capacity in sag of the flanged beam of Fig. 8.33 using the equivalent rectangular stress block. Take $f_{ck} = 30$ N/mm^2.

Solution: Assuming the tension reinforcement to have yielded and the compressive stress block to be in the top 175 mm of the section:

$$|F_s| = \left(\frac{A_s f_{yk}}{\gamma_s}\right) = \left(\frac{(804)(500)}{1.15}\right) = 349,565 \text{ N}$$

$$F_c = \left(\frac{\alpha_{cc} f_{ck}}{\gamma_c}\right)[(100)(0.8)x]$$

$$= \left(\frac{(0.85)(30)}{1.5}\right)[(100)(0.8)x]$$

$$= 1360x \text{ N}$$

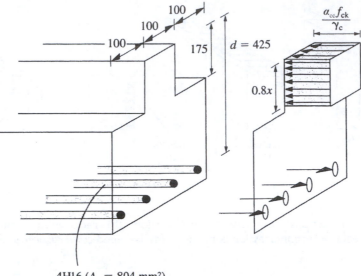

4H16 (A_s = 804 mm²)

(a)   (b)

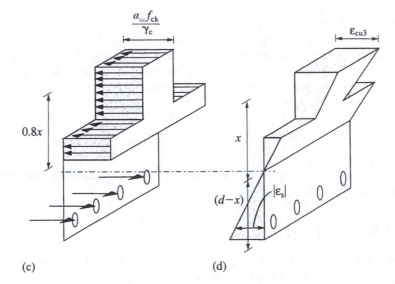

(c)   (d)

Figure 8.33 Beam of Example 8.11: (a) elevation and section; (b) stress distribution assuming 0.8x < 175 mm; (c) stress distribution assuming 0.8x > 175 mm; (d) strain distribution assuming 0.8x > 175 mm

Equating $|F_s|$ and F_c gives:

$$349,565 = 1306x$$
$$\Rightarrow x = 257 \text{ mm}$$

As 0.8x (= 206 mm) exceeds 175 mm, the assumption is clearly not valid. It is next assumed that the compressive stress block extends below the top 175 mm of the section as illustrated in Fig. 8.33(c). Hence:

$$F_c = \left(\frac{\alpha_{cc} f_{ck}}{\gamma_c}\right)[(100)(175) + (300)(0.8x - 175)]$$

$$= \left(\frac{(0.85)(30)}{1.5}\right)[(100)(175) + (300)(0.8x - 175)]$$

$$= 4080x - 595,000$$

This time, equating $|F_s|$ and F_c gives:

$$349,565 = 4080x - 595,000$$
$$\Rightarrow x = 232\text{mm}$$

which verifies the assumption regarding the extent of the compressive stress block. Now the assumption that the steel has yielded is checked. From similar triangles in Fig. 8.33(d):

$$|\varepsilon_s| = \frac{\varepsilon_{cu3}(d - x)}{x} = \frac{(0.0035)(425 - 232)}{232} = 0.00291 > 0.00217$$

Finally the ultimate moment capacity, M_{ult}, can be calculated by taking moments about the neutral axis:

$$M_{ult} = \left(\frac{(0.85)(30)}{1.5}\right)\left[(100)(175)(x - 175/2) + (300)(0.8x - 175)\left(0.2x + \frac{0.8x - 175}{2}\right)\right]$$
$$+ (349565)(425 - x)$$
$$\Rightarrow M_{ult} = 113\text{ kNm}$$

Non-uniform sections

For non-rectangular beams which have a varying breadth, such as those illustrated in Fig. 8.34, the calculation of the ultimate moment capacity is somewhat more complex than for rectangular sections. The reason behind this is that, since the breadth, $b(y)$, varies with depth, the total compressive force in the concrete must be calculated directly from Equation (8.18) by integration.

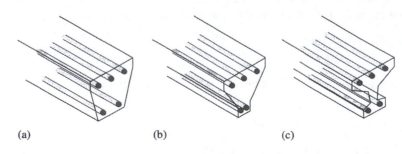

(a)   (b)   (c)

Figure 8.34 Non-uniform sections

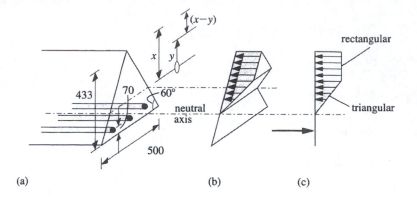

Figure 8.35 Beam of Example 8.12: (a) elevation and section; (b) stress distribution; (c) elevation of stress distribution

The complexity of the problem is further increased if the bi-linear or parabolic-rectangular stress block is used, in which case the stress is not constant over the depth. The procedure is illustrated in Example 8.12.

Example 8.12 Non-uniform section

Problem: For the section illustrated in Fig. 8.35, determine the area of tension reinforcement required to resist an applied ultimate moment of 62 kNm. Assume the bi-linear stress distribution for concrete, as illustrated in Fig. 8.21(b), with $f_{ck} = 30$ N/mm². Assume the simplified stress distribution for reinforcement with $f_{yk} = 500$ N/mm².

Solution: The section breadth varies from 500 mm at the base to zero at the top. Thus at a distance y above the neutral axis (i.e. $x - y$ below the top fibre), the breadth is:

$$b(y) = \frac{(x - y)500}{433}$$

Referring to Fig. 8.21(b), the triangular portion of the stress distribution ends at a strain of 0.00175. This occurs at a distance y_0 above the neutral axis where:

$$y_0 = \frac{0.00175x}{\varepsilon_{cu3}} = \frac{0.00175x}{0.0035} = 0.5x$$

In this region, the stress in the concrete varies linearly with strain. From Fig. 8.21(b), the stress at a distance y above the neutral axis is related to the strain at this point by:

$$f_c(y) = \left(\frac{\alpha_{cc}f_{ck}}{\gamma_c}\right)\left(\frac{\varepsilon_c(y)}{0.00175}\right)$$

Hence the compressive force in the first portion of the stress distribution is:

$$F_c^1 = \int_A f_c(y)dA = \int_{y=0}^{y=y_0} f_c(y)b(y)\,dy$$

$$= \int_{y=0}^{y=0.5x}\left(\frac{\alpha_{cc}f_{ck}}{\gamma_c}\frac{\varepsilon_c(y)}{0.00175}\right)\left(\frac{(x-y)500}{433}\right)dy$$

where $\varepsilon_c(y)$ is given by:

$$\varepsilon_c(y) = \frac{\varepsilon_{cu3}y}{x}$$

Hence:

$$F_c^1 = \int_{y=0}^{y=0.5x} \left(\frac{\alpha_{cc}f_{ck}}{\gamma_c} \frac{\varepsilon_{cu3}y}{0.00175x} \right) \left(\frac{(x-y)500}{433} \right) dy$$

which reduces to:

$$F_c^1 = 3.27x^2$$

The force in the upper part of the section is a simple product of stress and area:

$$F_c^2 = \left(\frac{\alpha_{cc}f_{ck}}{\gamma_c} \right) \times \frac{1}{2}(x-y_0)\left(\frac{(x-y_0)500}{433} \right)$$

which reduces to:

$$F_c^2 = 2.45x^2$$

The ultimate moment capacity is found by taking moments about the neutral axis:

$$\begin{aligned}
M_{ult} &= \int_{y=0}^{y=y_0} f_c(y)b(y)y\,dy + \int_{y=y_0}^{y=x} f_c(y)b(y)y\,dy + |F_s|(d-x) \\
&= \int_{y=0}^{y=y_0} f_c(y)b(y)y\,dy + \int_{y=y_0}^{y=x} f_c(y)b(y)y\,dy + (F_c^1 + F_c^2)(d-x) \\
&= \int_{y=0}^{y=y_0} \left(\frac{\alpha_{cc}f_{ck}}{\gamma_c} \right) \left(\frac{\varepsilon_{cu3}y}{0.00175x} \right) \left(\frac{(x-y)500}{433} \right) y\,dy \\
&\quad + \int_{y=y_0}^{y=x} \left(\frac{\alpha_{cc}f_{ck}}{\gamma_c} \right) \left(\frac{(x-y)500}{433} \right) y\,dy \\
&\quad + (3.27x^2 + 2.45x^2)(d-x)
\end{aligned}$$

This reduces to:

$$M_{ult} = 2076x^2 - 3.06x^3$$

This equation may be solved iteratively to find a value of x for which $M_{ult} = 62 \times 10^6$ Nmm. By trial and error it is found that $x = 207$ mm. For this value of x, the force in the reinforcement is:

$$\begin{aligned}
|F_s| &= F_c^1 + F_c^2 \\
&= 3.27(207)^2 + 2.45(207)^2 \\
&= 245 \text{ kN}
\end{aligned}$$

Equating this to $A_s f_{yk}/\gamma_s$ gives:

$$A_s = \frac{245 \times 10^3}{(500/1.15)} = 564 \text{ mm}^2$$

2H20 bars give an area of 628 mm².

8.6   Balanced design and section ductility

Balanced design

It was stated in Section 8.4 that a balanced design is one in which the concrete crushes at the same instant as the reinforcement first yields under the ultimate applied loads. In other words, the strain in the reinforcement reaches its yield value at the same time as the strain in the extreme concrete fibre in compression reaches ε_{cu2} as illustrated in Fig. 8.36. Then, by similar triangles on the strain diagram:

$$\frac{\varepsilon_{cu2}}{x} = \frac{(f_{yk}/\gamma_s)E_s}{d-x} \tag{8.33}$$

Assuming $f_{yk} = 500$ N/mm^2, $\varepsilon_{cu2} = 0.0035$ and $E_s = 200,000$ N/mm^2, Equation (8.33) becomes:

$$\frac{0.0035}{x} = \frac{(500/1.15)(200000)}{d-x}$$
$$\Rightarrow \frac{0.0035}{x} = \frac{0.00217}{d-x}$$

$$\Rightarrow x/d = 0.617 \tag{8.34}$$

This is true of all sections regardless of their shape. The precise amount of reinforcement which gives a balanced design (i.e. satisfies Equation (8.34)) can be readily calculated by equilibrium of the axial forces acting on the section. Unlike Equation (8.34), however, the area of reinforcement for which a design is balanced depends in the cross-section being considered. Assuming an equivalent rectangular stress block for the rectangular section of Fig. 8.37, equilibrium gives:

$$(\alpha_{cc}f_{ck}/\gamma_c)(0.8xb) = A_s f_{yk}$$
$$\Rightarrow (0.85f_{ck}/1.5)[0.8(0.617d)b] = 500A_s \tag{8.35}$$
$$\Rightarrow A_s = \frac{bdf_{ck}}{1788} \text{ mm}^2$$

If the area of reinforcement provided is greater than that given by Equation (8.35), the beam is over-reinforced, that is the reinforcement does not yield before the concrete crushes.

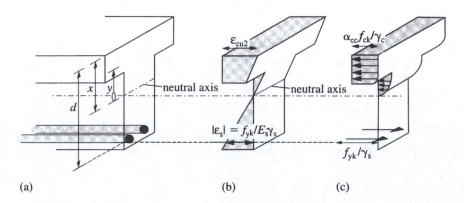

(a)   (b)   (c)

Figure 8.36 Balanced design for typical section: (a) elevation and section; (b) strain distribution; (c) stress distribution

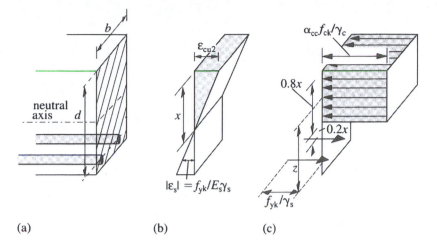

Figure 8.37 Balanced design for singly reinforced rectangular section: (a) elevation and section; (b) strain; (c) equivalent rectangular stress block

Table 8.3 Maximum values for x/d and K recommended in UK NA to BS EN 1992-1-1 for singly reinforced rectangular sections (for $f_{ck} \leq 50$ N/mm^2)

	% plastic moment redistribution						
	0	5	10	15	20	25	30
Maximum x/d	0.6*	0.55*	0.5*	0.45	0.4	0.35	0.3
Maximum K	0.207	0.194	0.181	0.167	0.152	0.136	0.120

* *Note: ISE manual recommends x/d is limited to 0.45, thus limiting K to 0.167*

Similarly, if the area of reinforcement provided is less than that given by Equation (8.35), the beam is under-reinforced and the reinforcement will yield before the concrete crushes. It was seen in section 8.4 that, for safety reasons, only under-reinforced beams are acceptable. Where plastic moment redistribution is used (refer to Section 5.3), even more ductility is needed and the limits given by the UK NA to EC2 are specified in Table 8.3. To ensure that designs are under-reinforced and substantially more ductile than the balanced case, the ISE manual recommends that x/d does not exceed 0.45 (a small x implies large strain in the steel). As discussed in Chapter 5, redistribution is limited to 30 per cent for Class B and C reinforcing steel and 20 per cent for Class A reinforcing steel.

Singly reinforced sections

The rectangular section illustrated in Fig. 8.37 has tension reinforcement only and is termed a singly reinforced rectangular section. For designs with $x/d < 0.617$, the tension reinforcement yields under ultimate loads. Assuming the simplified rectangular stress block, the forces acting on the section are as illustrated in Fig. 8.37(c). The moment capacity of the section is given by the compressive force (or tensile force) multiplied by the lever arm, z:

$$M_{ult} = F_c z$$

$$= \left(\frac{\alpha_{cc} f_{ck}}{\gamma_c} \right)(0.8x)(b)(d - 0.4x)$$

$$\Rightarrow M_{ult} = 0.453xbf_{ck}(d - 0.4x) \tag{8.36}$$

Using the non-dimensional factor defined by Equation (8.29), this gives:

$$K = \frac{M_{ult}}{bd^2 f_{ck}} = 0.453 \frac{x}{d}\left(1 - 0.4\frac{x}{d}\right) \tag{8.37}$$

Recall that, for a balanced design, $x/d = 0.617$. Thus, Equation (8.37) becomes:

$$K = 0.453(0.617)[1 - 0.4(0.617)]$$
$$= 0.21$$

Hence, for a balanced design, the ultimate moment capacity for the singly reinforced rectangular section is given by:

$$M_{ult} = Kbd^2 f_{ck} = 0.21bd^2 f_{ck} \tag{8.38}$$

To ensure ductile failure, the allowable values for K are calculated using the limits on x/d specified in the UK NA to EC2 (Table 8.3). Substituting for x/d in Equation (8.37) results in the maximum values for K, which are also given in Table 8.3. Thus, for singly reinforced rectangular sections, the safe upper limit for ultimate moment capacity, regardless of reinforcement provided, is given by:

$$M_{ult} = Kbd^2 f_{ck} \tag{8.39}$$

where the appropriate value for K is taken from Table 8.3 (but limited to 0.167 as recommended by the ISE manual).

Example 8.13 Minimum depth ductile section

Problem: Design a rectangular singly reinforced section of breadth 350 mm and of minimum depth to resist a factored ultimate moment of 400 kNm. This moment has been reduced from an applied elastic moment of 470 kNm through plastic moment redistribution. $f_{ck} = 35$ N/mm^2, $f_{yk} = 500$ N/mm^2.

Solution: The ductility of sections increases with effective depth as a higher depth allows a lower area of reinforcement. Hence this design is governed by the requirements of ductility. A redistribution of 70 kNm, or 15 per cent, has been performed. Using Table 8.3:

$$K \leq 0.167$$
$$\Rightarrow \frac{M_{ult}}{bd^2 f_{ck}} \leq 0.167$$
$$\Rightarrow d \geq \sqrt{\frac{M_{ult}}{0.167bf_{ck}}} = \sqrt{\frac{400 \times 10^6}{(0.167)(350)(35)}} = 442 \text{ mm}$$

Allowing for cover of 35 mm, a 12 mm link and a 25 mm bar diameter gives a total depth of:

$$h = 442 + 35 + 12 + 25/2 = 502 \, \text{mm}$$

This is rounded up to $h = 525$ mm, giving $d = 466$ mm. Hence:

$$K = \frac{M_{\text{ult}}}{bd^2 f_{\text{ck}}} = \frac{400 \times 10^6}{(350)(466)^2(35)} = 0.150$$

The corresponding area of reinforcement is calculated using Equations (8.30) and (8.31):

$$z = d(0.5 + \sqrt{0.25 - 0.88K})$$
$$= (466)(0.5 + \sqrt{0.25 - 0.88(0.150)})$$
$$= 393 \, \text{mm}$$

$$\Rightarrow A_{\text{s}} = \frac{M_{\text{ult}}}{(f_{\text{yk}}/\gamma_{\text{s}})z} = \frac{400 \times 10^6}{(500/1.15)393} = 2341 \, \text{mm}^2$$

5H25 gives an area of 2454 mm^2, which exceeds the required area and is sufficient to resist the ultimate moment.

Doubly reinforced rectangular sections

The upper limits on moment capacity presented above are only valid for singly reinforced sections. By providing compression reinforcement, greater capacities can be achieved while maintaining a ductile section. Consider first the case where the ultimate moment capacity, M_{ult}, equals the maximum value allowed for a singly reinforced section. If the area of tension reinforcement is A_{s1}, then equilibrium of axial forces dictates that:

$$F_{\text{c}} = \frac{A_{\text{s1}} f_{\text{yk}}}{\gamma_{\text{s}}} \tag{8.40}$$

Further, Equations (8.29) to (8.31), derived using moment equilibrium, can be used to calculate A_{s1}:

$$A_{\text{s1}} = \frac{M_{\text{ult}}}{(f_{\text{yk}}/\gamma_{\text{s}})z} \tag{8.41}$$

where:

$$z = d(0.5 + \sqrt{0.25 - 0.88K}) \tag{8.42}$$

and:

$$K = \frac{M_{\text{ult}}}{bd^2 f_{\text{ck}}} \tag{8.43}$$

Now, if additional tension reinforcement is provided of area A_{s2} and, at the same time, compression reinforcement of area A_{s}' such that:

$$A_{\text{s}}' f_{\text{s}}' = \frac{A_{\text{s2}} f_{\text{yk}}}{\gamma_{\text{s}}} \tag{8.44}$$

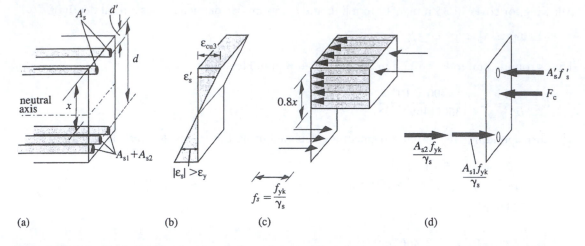

Figure 8.38 Ductile design for doubly reinforced rectangular section: (a) elevation and section; (b) strain distribution; (c) stress distribution; (d) axial forces

then, as the additional forces are self-equilibrating, equilibrium of axial forces would be satisfied without any change in x. Further, all of the tension reinforcement would have yielded as is evident from Equations (8.40) and (8.44), thus maintaining the ductility of the section. The forces on this strengthened section are illustrated in Fig. 8.38(d). The new areas of reinforcement result in an increase in moment capacity of:

$$\Delta M_{\mathrm{ult}} = A'_s f'_s (x - d') + \frac{A_{s2} f_{yk}(d - x)}{\gamma_s} \tag{8.45}$$

Substitution from Equation (8.44) gives:

$$\Delta M_{\mathrm{ult}} = A'_s f'_s (d - d') \tag{8.46}$$

Hence, the area of compression reinforcement required to provide an ultimate moment capacity of $M_{\mathrm{ult}} + \Delta M_{\mathrm{ult}}$ is:

$$A'_s = \frac{\Delta M_{\mathrm{ult}}}{f'_s (d - d')} \tag{8.47}$$

The total area of tension reinforcement required is:

$$A_s = A_{s1} + A_{s2} \tag{8.48}$$

where A_{s1} is given by Equation (8.41) and, from Equation (8.44):

$$A_{s2} = \frac{A'_s f'_s}{(f_{yk}/\gamma_s)} \tag{8.49}$$

Compression reinforcement may or may not have yielded. The strain in the compression reinforcement is:

$$\varepsilon'_s = \frac{\varepsilon_{cu3}(x - d')}{x} \tag{8.50}$$

Yield occurs when:

$$\varepsilon'_s = \left(\frac{f_{yk}}{\gamma_s}\right)\frac{1}{E_s}$$

$$\Rightarrow \frac{\varepsilon_{cu3}(x - d')}{x} = \frac{f_{yk}}{\gamma_s E_s}$$

Rearranging gives:

$$\frac{x}{d'} = \frac{E_s\varepsilon_{cu3}}{(E_s\varepsilon_{cu3} - f_{yk}/\gamma_s)} \tag{8.51}$$

Taking $E_s = 200,000$, $\varepsilon_{cu3} = 0.0035$ and $f_{yk} = 500$ gives:

$$\frac{x}{d'} = 2.639$$

Therefore, if x/d' is less than 2.639, then $f'_s = E_s\varepsilon'_s$ where ε'_s is given by Equation (8.50). Otherwise, $f'_s = f_{yk}/\gamma_s$.

The design formulae derived above are only applicable to rectangular sections and flanged sections where the compression zone is rectangular (such as where the neutral axis lies within the flange for a section in sag). In addition, the formulae have been derived using the equivalent rectangular stress block of Fig. 8.21(c). Formulae can be derived for flanged sections where the compression zone is non-rectangular and for other stress/strain diagrams.

Example 8.14 Section with compression reinforcement

Problem: For the beam of Example 8.13, determine the reinforcement requirements if the effective depth is limited to 400 mm.

Solution: As for Example 8.13, the parameter K is limited to 0.167. With the reduced effective depth, the corresponding ultimate moment capacity (the maximum possible without compression reinforcement) is:

$$M_{ult} = 0.167bd^2f_{ck} = 0.167(350)(400)^2(35)$$
$$= 327\,kNm$$

The corresponding area of tension reinforcement is found from Equations (8.30) and (8.31):

$$z = d(0.5 + \sqrt{0.25 - 0.88K})$$
$$= (400)(0.5 + \sqrt{0.25 - 0.88(0.167)})$$
$$= 328\,mm$$

$$\Rightarrow A_{s1} = \frac{M_{ult}}{(f_{yk}/\gamma_s)z} = \frac{327 \times 10^6}{(500/1.15)328} = 2293\,mm^2$$

From Equation (8.28), the corresponding value for x is:

$$x = \frac{d - z}{0.4} = \frac{400 - 328}{0.4} = 180\,mm$$

Assuming a value for d' of 60 mm, the ratio x/d' then becomes 3.00. As this exceeds 2.639, compression reinforcement will yield. Hence, from Equation (8.47), the required area of compression reinforcement is:

$$A'_s = \frac{\Delta M_{ult}}{f'_s(d-d')} = \frac{400 \times 10^6 - 327 \times 10^6}{(500/1.15)(400-60)} = 494 \text{ mm}^2$$

Equation (8.49) gives the required additional area of tension reinforcement:

$$A_{s2} = \frac{A'_s f'_s}{(f_{yk}/\gamma_s)} = A'_s = 494 \text{ mm}^2$$

giving a total required area of tension reinforcement of $2293 + 494 = 2787 \text{ mm}^2$.

The formula given in Chapter 6 for the preliminary design of reinforced concrete beams and slabs, namely Equation (6.13), is derived from Equations (8.41) and (8.47) assuming no moment redistribution has been carried out. Taking a value for K of 0.167 from Table 8.3 and substituting in Equation (8.42) gives:

$$z = 0.82d$$

Then Equation (8.41) gives:

$$A_{s1} = \frac{M_{ult}}{(f_{yk}/\gamma_s)(0.82d)} = \frac{0.167bd^2 f_{ck}}{(f_{yk}/\gamma_s)(0.82d)}$$
$$\Rightarrow A_{s1} = \frac{0.234bd f_{ck}}{f_{yk}}$$

Taking $A_{s2} = A'_s$ gives the formula for A_s presented in Chapter 6.

8.7   Anchorage length

In reinforced concrete members, the flexural strength relies on the transfer of tensile force, commonly known as bond, between the longitudinal reinforcement and the surrounding concrete. This is important near the ends of bars and where bars are lapped and force needs to be transferred from one bar, through the concrete, to another. The quality of the bond depends on the surface pattern of the bar, the dimensions of the member and the position and inclination of the bars. Bond is achieved by the combination of adhesion and a small amount of friction between the two materials. In the case of deformed bars, a significant proportion of the bond is provided by the bearing of the concrete between the ribs. This is illustrated in Fig. 8.39 where a moment due to the tensile force F_s is being transferred by the bond into the concrete. Clearly, the greater the length that the bars are embedded in the concrete, the greater the bond between the two materials.

The application of loads to a reinforced concrete member leads to bending of the member which, in turn, results in tensile forces being developed in the reinforcement. If the anchorage bond between the bars and the concrete is sufficient, the full strength of the reinforcement can be utilized. If, however, the bond is insufficient, the bar will pull out of the concrete, the tensile force will drop to zero and the member will fail. The **anchorage length**, l_b, is the length of reinforcement required to develop sufficient anchorage bond so that the full strength of the reinforcement can be used.

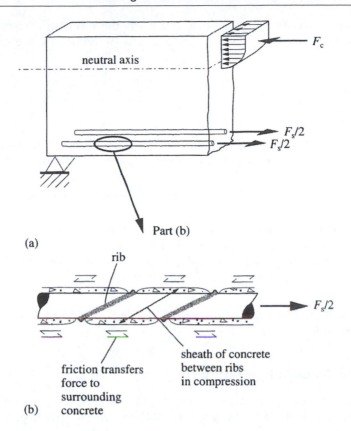

neutral axis

F_c

$F_s/2$
$F_s/2$

Part (b)

(a)

rib

friction transfers
force to
surrounding
concrete

sheath of concrete
between ribs
in compression

$F_s/2$

(b)

Figure 8.39 Bond in deformed bars: (a) portion of beam (elevation and section); (b) detail around bar

The design anchorage length, l_{bd}, recommended for design in EC2 is given by:

$$l_{bd} = \text{greater of} \left(\alpha l_{b,rqd} \frac{A_{s,req}}{A_{s,prov}} \right) \text{ and } l_{b,min} \tag{8.52}$$

where
   α = coefficient which can conservatively be taken as 1.0 (see EC2 for more details)
   $l_{b,rqd}$ = basic required anchorage length (Equation (8.54))
   $A_{s,req}$ = calculated area of reinforcement required for design
   $A_{s,prov}$ = area of reinforcement provided in design (the required area is rounded up to an integer number of bars with the result that $A_{s,prov} \geq A_{s,req}$)
   $l_{b,min}$ = minimum anchorage length: for bars in tension, $l_{b,min}$ = greater of $0.3l_{b,rqd}$, 10ϕ and 100 mm, where ϕ is the bar diameter; for bars in compression, $l_{b,min}$ = greater of $0.6l_{b,rqd}$, 10ϕ and 100 mm.
   The basic required anchorage length, $l_{b,rqd}$, is the straight length of bar required to anchor the force $A_s f_{yk}/\gamma_s$. For a bar of diameter ϕ, this force must equal the shear force developed between the bar surface and the surrounding concrete:

$$A_s f_{yk}/\gamma_s = (\pi \phi l_{b,rqd}) f_{bd} \tag{8.53}$$

where $(\pi\phi l_{b,rqd})$ is the contact surface area and f_{bd} is the ultimate design bond strength, that is, the maximum shear stress that can act at the interface of the two materials. This simplifies to:

$$l_{b,rqd} = \frac{\phi}{4} \frac{f_{yk}}{\gamma_s f_{bd}} \qquad \text{for } (\phi \leq 32 \text{ mm}) \tag{8.54}$$

The bond strength is dependent on the quality of the bond, with EC2 distinguishing between 'good' and 'poor' bond conditions. Good bond conditions are defined as follows:

1. All bars inclined at an angle of between 45°' and 90° to the horizontal during casting of the member.
2. All bars inclined at an angle of between 0° and 45° to the horizontal that are either (a) placed in members whose depth does not exceed 250 mm or (b) placed in the lower 250 mm where the depth of the member is greater than 250 mm or at least 300 mm from the top of the member where the depth of the member is greater than 600 mm.

All other bond conditions are defined as poor. The bond strength, f_{bd}, for good and poor bond conditions is as given in Table 8.4.

Example 8.15 Anchorage length

Problem: Determine the design anchorage length for the beam of Fig. 8.40. The beam must resist an ultimate sag moment of 190 kNm. The breadth is 300 mm, the effective depth, d, is 550 mm and the total depth is 600 mm. Assume $f_{ck} = 35$ N/mm², $f_{yk} = 500$ N/mm².

Solution: With $f_{ck} = 35$ N/mm², Table 8.4 gives the ultimate bond stress, f_{bd}, as 3.4 N/mm² (bond is good as the bars are placed in the lower 250 mm of the beam). The area of steel required is calculated using Equations (8.29), (8.30) and (8.31). From Equation (8.29):

$$K = \frac{M_{ult}}{bd^2 f_{ck}} = \frac{190 \times 10^6}{(300)(550)^2(35)} = 0.060$$

Table 8.4 Design values of ultimate bond strength, f_{bd}.

Characteristic cylinder strength, f_{ck}	Design value for the ultimate bond strength (N/mm²)	
	Good bond conditions	Poor bond conditions
12	1.7	1.2
16	2.0	1.4
20	2.3	1.6
25	2.7	1.9
30	3.0	2.1
35	3.4	2.4
40	3.7	2.6
45	4.0	2.8
50	4.3	3.0

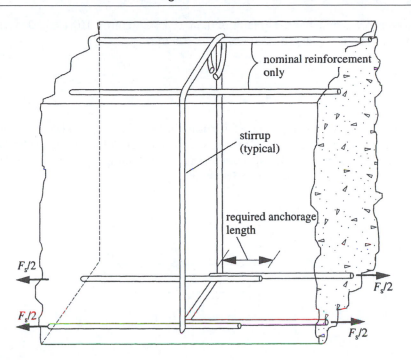

nominal reinforcement
only

stirrup
(typical)

required anchorage
length

$F_s/2$

$F_s/2$

$F_s/2$

$F_s/2$

Figure 8.40 Beam of Example 8.15

From Equation (8.30):

$$z = d(0.5 + \sqrt{0.25 - 0.88K})$$
$$= (550)(0.5 + \sqrt{0.25 - 0.88(0.060)})$$
$$= 519 \text{ mm}$$

Hence using Equation (8.31):

$$A_s = \frac{M_{ult}}{(f_{yk}/\gamma_s)z} = \frac{190 \times 10^6}{(500/1.15)519} = 842 \text{ mm}^2$$
$$\Rightarrow A_{s,req} = A_s = 842 \text{ mm}^2$$

2H25 provides an area of reinforcement of 982 mm².

$$\Rightarrow A_{s,prov} = 982 \text{ mm}^2$$

From Equation (8.54), the basic required anchorage length, $l_{b,rqd}$, is:

$$l_{b,rqd} = \frac{\phi}{4} \frac{f_{yk}}{\gamma_s f_{bd}} = \frac{25}{4} \frac{500}{(11.5)(3.4)} = 799 \text{ mm}$$

The design anchorage length, l_{bd}, is calculated from Equation (8.52):

$$l_{bd} = \alpha l_{b,rqd} \frac{A_{s,req}}{A_{s,prov}} = (1.0)(7.99)\left(\frac{842}{982}\right) = 685 \text{ mm}$$

This is greater than $l_{b,min}$ which is the greater of $0.3l_{b,rqd}$ ($= 240$ mm), 10ϕ ($= 250$ mm) and 100 mm. Rounding up, the anchorage length becomes $l_{bd} = 700$ mm.

Problems

Section 8.2

8.1 For the uncracked and homogeneous T-section illustrated in Fig. 8.41, derive formulae, (a) for the location of the centroid and (b) for the second moment of area. Hence find the maximum tensile stress in a T-section with $b_{eff} = 900$ mm, $b_w = 300$ mm, $h_f = 200$ mm and $h_w = 500$ mm, due to an applied sagging moment of 500 kNm.

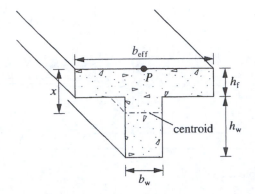

Figure 8.41 Section of Problem 8.1

8.2 For the uncracked and homogeneous concrete beam illustrated in Fig. 8.42, verify from first principles that $x = 183$ mm. Hence calculate the gross second moment of area. Find the distribution of stress at the centre of the beam due to its self-weight. Assume a density for concrete of 25 kN/m^3.

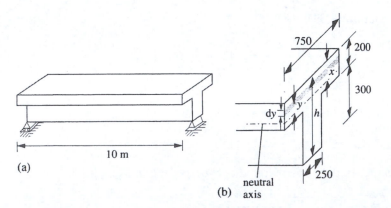

Figure 8.42 Beam of Problem 8.2: (a) span; (b) cross-sectional dimensions

8.3 Prove that the second moment of area for an uncracked reinforced section is given by:

$$I_u = \int y^2 dA + (\alpha_e - 1)A_s(d - x)^2$$

8.4   Find, by rule, the stresses due to self-weight at A for the beam illustrated in Fig. 8.43:

   (a)   at the top fibre in the concrete;
   (b)   in the reinforcement.
      Assume the section to be cracked. Let $E_s = 200{,}000$ N/mm^2 and $E_c = 30{,}000$ N/mm^2.

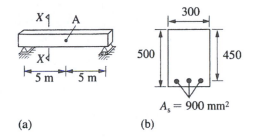

        (a)   (b)

Figure 8.43 Beam of Problem 8.4: (a) elevation; (b) section X–X

8.5   For the beam whose section is illustrated in Fig. 8.44, it can be assumed that the section is cracked and that the neutral axis is in the flange.

   (a)   Sketch the equivalent concrete section and calculate x.
   (b)   Calculate the cracked second moment of area, I_c.
   (c)   Calculate the stress in the tension steel due to an applied moment of 150 kNm.
      Take $E_s = 200{,}000$ N/mm^2 and $E_c = 32{,}000$ N/mm^2.

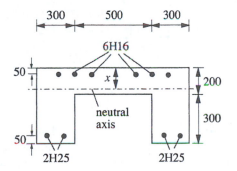

Figure 8.44 Section of Problem 8.5

Section 8.5

8.6   The section illustrated in Fig. 8.45 is on the point of rupture (i.e. strain in concrete has just reached 0.0035). Assuming initially that $x < 400$ mm:

   (a)   determine if the steel has yielded;
   (b)   find x and verify that $x < 400$ mm;
   (c)   calculate the ultimate moment capacity.
      The concrete has a cylinder strength, $f_{ck} = 35$ N/mm^2 and the reinforcement has a yield strength, $f_{yk} = 500$ N/mm^2.

8.7   Calculate the ultimate moment capacity of the section illustrated in Fig. 8.46 using the simplified stress–strain relationships of Fig. 8.21(c) and Fig. 8.22(b).

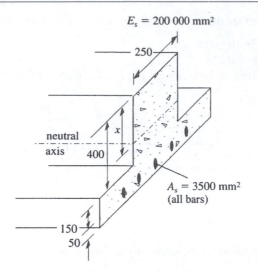

Figure 8.45 Section of Problem 8.6

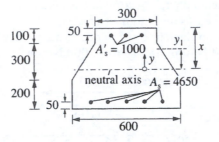

Figure 8.46 Section of Problem 8.7

8.8 (a) For the section illustrated in Fig. 8.47 derive, from first principles, a formula giving x/d in terms of $M/bd^2 f_{ck}$. Use the parabolic rectangular stress block for the concrete and the horizontal top branch in the stress–strain relationship for the reinforcement. Hence derive a formula for the area of reinforcement required to resist the applied moment.

   (b) Determine the area of reinforcement required to resist a moment of 300 kNm in a rectangular section of breadth, 300 mm, effective depth to reinforcement, 500 mm, $f_{ck} = 40$ N/mm^2 and $f_{yk} = 500$ N/mm^2.

8.9 For the section of Problem 8.6, determine from first principles the area of reinforcement that would be required to resist an applied ultimate moment of 300 kNm given that $f_{ck} = 35$ N/mm^2 and $f_{yk} = 500$ N/mm^2.

Section 8.6

8.10 (a) For the singly reinforced T-section illustrated in Fig. 8.48, determine the value of x at which the design is balanced. Calculate the corresponding ultimate moment capacity of the section. Assume $f_{ck} = 40$ N/mm^2 and $f_{yk} = 500$ N/mm^2.

(b) For a moment 2 per cent in excess of the balanced value, calculate x and show that the reinforcement has not yielded.
(c) Calculate the area of compression reinforcement required to render the section ductile and the area of tension reinforcement required to resist the higher applied moment, and verify that the section is now ductile.

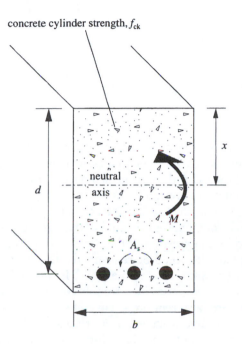

Figure 8.47 Section of Problem 8.8

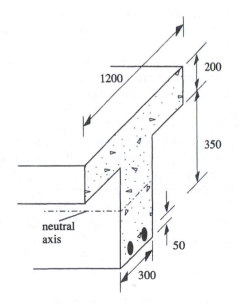

Figure 8.48 Section of Problem 8.10

Chapter 9

Design of prestressed concrete members for bending

9.1   Introduction

Loads carried in a member by bending are effectively transferred through it as compressive and tensile stresses, as illustrated in Fig. 9.1. Due to the low tensile strength of concrete, steel reinforcement must be provided in all structural concrete members subject to such bending stresses to control tensile cracking and, ultimately, to prevent failure.

In ordinary reinforced concrete, steel bars are placed within the tension zones of concrete members to carry the internal tensile forces across flexural cracks to the supports. In such a member, the applied moment is resisted by compression of the uncracked portion of the concrete section and by tension in the reinforcing bars. It should be remembered that this form of reinforcement does not prevent the development of tensile cracks in members. It is only by limiting the magnitude of the strain in the bars that the cracks are prevented from becoming excessively large.

One problem with such ordinary reinforced concrete is that the presence of cracks can lead to corrosion of the reinforcement due to its exposure to water, air and sometimes to chemical contaminants. Corrosion is generally only a problem for structures in aggressive exterior environments (bridges, marine structures, etc.) and is not critical in the majority of buildings. A further effect of cracking of ordinary reinforced members is the substantial loss in stiffness which occurs after cracking. The second moment of area of the cracked section, I_c, is far less than the second moment of area before cracking, I_u (see Section 8.2). This reduced second moment of area can result in greater deflections under service loads.

Prestressed concrete is an alternative form of reinforced concrete. In prestressed concrete, compressive stresses are introduced into a member to reduce or nullify the tensile stresses and hence, cracks. The compressive stresses are generated in a member by tensioned steel anchored at the ends of the members and/or bonded to the concrete.

Consider the simply supported member of Fig. 9.2(a) of cross-sectional area, A. The linear elastic bending stress due to the applied load at a given section of the member is illustrated in the figure. When the maximum tensile stress, $\sigma_{app,t}$, exceeds the tensile strength of the concrete, a crack forms in the section. If there is no reinforcement, the crack will propagate until it causes failure. The introduction of a steel tendon, running along the centroid of the member, tensioned to a force P and anchored at its ends (Fig. 9.2(b)), creates the stress distribution illustrated in that figure which, when combined with the stress due to applied loading (Fig. 9.2(a)), gives the distribution illustrated in Fig. 9.2(c). The maximum tensile stress in the concrete is reduced by an amount P/A and the maximum compressive stress is increased by the same amount. By applying a sufficiently large prestress force, P, the tensile stresses can be reduced below the tensile strength of the concrete to eliminate cracking.

In ways similar to this, the development of cracks in all prestressed concrete members can be controlled or prevented completely. Therefore, prestressed members are stiffer and more durable

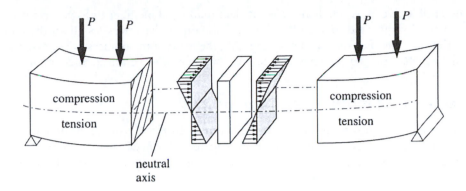

Figure 9.1 Stress due to bending in a beam

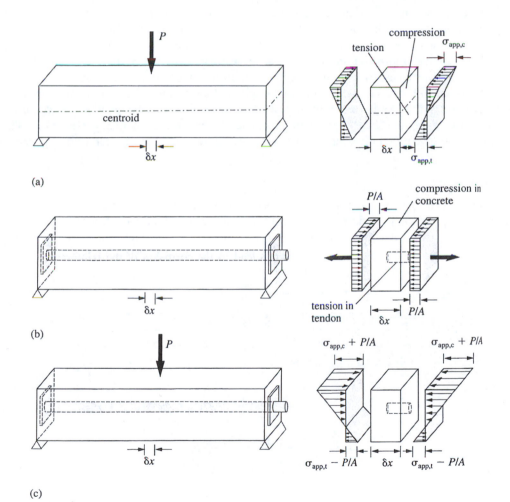

(a)

(b)

(c)

Figure 9.2 Stress distribution in prestressed concrete beam: (a) stress due to applied load; (b) stress due to prestress; (c) stress due to applied load and prestress

(i.e. less susceptible to corrosion) than the equivalent reinforced members. However, the magnitude of P is restricted by the need to keep the maximum compressive stress less than the compressive strength of the concrete. For the member of Fig. 9.2, the maximum compressive stress, $\sigma_{app,c} + P/A$, must remain less than the maximum allowable compressive stress of the concrete at all sections.

9.2 Prestressing methods and equipment

Prestress is normally applied to members by steel strands tensioned using hydraulic jacks at one or both ends of the member. The tensioning operation can be performed either: (a) before the concrete is cast, in which case the member is classed as pre-tensioned; or (b) after the concrete is cast, in which case the member is classed as post-tensioned.

Pre-tensioning

The pre-tensioning process involves three basic stages, each of which is illustrated in Fig. 9.3. In the first stage, the steel strands are placed in a casting bed, stressed to the required level and anchored between two supports (Fig. 9.3(a)). The concrete is then cast around the strands and allowed to set (Fig. 9.3(b)). During this curing stage, the strands bond to the surrounding concrete. When the concrete has developed sufficient compressive strength, the strands are released from the supports (Fig. 9.3(c)). Immediately after the release, the strands attempt to contract. Due to their bond with the concrete, this prestress contraction force is transferred to the concrete, thus forcing the concrete into compression.

Pre-tensioning is most commonly employed where many similar precast members are required. It is generally only carried out offsite at precasting factories which have permanent casting beds. The size and weight of pre-tensioned members can therefore be limited by the transportation requirements.

Post-tensioning

The stages involved in the post-tensioning process are illustrated in Fig. 9.4 for a simple beam. In the first stage of the process, the concrete is cast around a hollow duct (Fig. 9.4(a)). After the concrete has set, a tendon, consisting of a number of strands, is pushed through the duct. Alternatively, the tendon

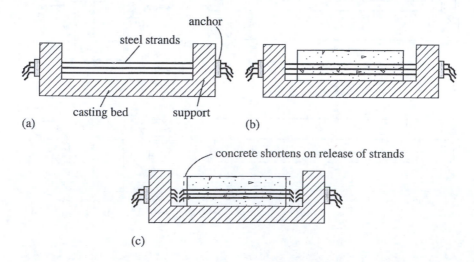

Figure 9.3 Pre-tensioning: (a) stage 1, steel strands are tensioned; (b) stage 2, concrete is cast; (c) stage 3, strands are cut

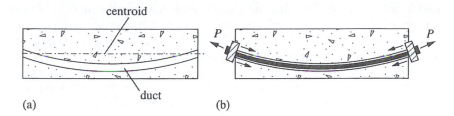

Figure 9.4 Post-tensioning: (a) beam is cast; (b) beam is prestressed

can be placed in the duct before casting. Unlike in pre-tensioned members, the tendon in post-tensioned members can be fixed in any desired linear or curved profile. By varying the eccentricity of the tendon from the centroid, the prestressing force can be utilized more effectively by applying it only where it is required (more details in Section 9.4). Once the concrete has achieved sufficient compressive strength, the tendon is jacked from one or both ends using hydraulic jacks, putting the concrete into compression (Fig. 9.4(b)). When the required level of prestress is achieved, the tendon is anchored at the ends of the member. After anchorage, the ducts are usually filled with grout (a fine cement paste) under pressure. The grout is provided mainly to prevent corrosion of the tendon but it also forms a bond between the tendon and the concrete, which reduces the dependence of the beam on the integrity of the anchor and hence improves robustness. This is known as 'bonded' construction.

Post-tensioning is the most common method of prestressing *in situ* because it does not require a casting bed. However, the technique is also used offsite to make large purpose-built individual precast units.

Prestressing steel

The steel used for prestressed concrete comes in the form of either cold-drawn high-strength wire or high-strength alloy steel bars. The use of high-yield bars, however, is generally limited as they do not have the flexibility to be profiled along the length of the member. High tensile steel wire is by far the more widely used material for both pre-tensioning and post-tensioning.

The short-term stress–strain relationship for a typical wire specimen is illustrated in Fig. 9.5. Failure generally occurs at a strain somewhere between 0.04 and 0.06. For design, EC2 specifies a

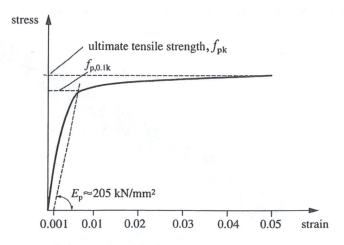

Figure 9.5 Stress–strain relationship for prestressing wire

design strain limit of $\varepsilon_{ud} = 0.02$. The **ultimate characteristic tensile strength**, f_{pk}, of most manufactured wires is approximately 1,800 N/mm² (compared with 500 N/mm² for ordinary reinforcement) and the elastic modulus may be taken to be about 205,000 N/mm² for wires and bars. It can be seen from the figure that there is no definite elastic yield point. For this reason, the concept of proof stress is used as an equivalent yield stress. The most frequently used proof stress is the characteristic 0.1 per cent proof stress, $f_{p,0.1k}$, which is the point on the stress–strain curve which intersects with a straight line drawn at a slope equal to the elastic modulus starting from 0.1 per cent strain (see Fig. 9.5). To ensure adequate ductility in tension, $f_{pk}/f_{p,0.1k} \geq 1.1$.

High-strength steel wire, which comes in a range of diameters from 3 mm to 7 mm, does not generally have sufficient strength to be used singly for prestressing purposes. Thus, for most prestressing applications, several wires are twisted together to form a strand. The wires in a strand are spun in a helical form around a central straight wire, as illustrated in Fig. 9.6. Most manufacturers supply strands made up of seven spun wires, as illustrated in the figure, although five-wire and 19-wire strands are also supplied by some manufacturers. The performance characteristics of strands differ slightly to that of the wire from which they are made, due to the straightening of the spun wires when in tension. The elastic modulus for strands may be taken as 195,000 N/mm².

In post-tensioned concrete, it is common to group many strands together to form a cable or tendon. Fig. 9.7 illustrates a tendon with multiple strands. A complete prestressing tendon can be made up of as many strands as are needed to carry the required tension, with all the strands enclosed in a single duct. In addition, large structures may have many individual tendons running parallel to each other along the length of the member.

Most codes of practice place restrictions on the maximum stress which can be applied to strands during the prestressing process. EC2 recommends that the stress in the strands during jacking should not exceed the lesser of $0.8f_{pk}$ and $0.9f_{p,0.1k}$ where f_{pk} is the characteristic value of the ultimate strength and $f_{p,0.1k}$ is the characteristic value of the 0.1 per cent proof stress. In addition, EC2 restricts the stress in the tendons immediately after they have been anchored in the concrete (not all the stress applied during jacking remains after anchorage due to losses – see Section 9.5). At this stage, the stress in the tendons should not exceed the lesser of $0.75f_{pk}$ and $0.85f_{p,0.1k}$.

Prestressing equipment

For both pre-tensioning and post-tensioning of concrete members, specialist equipment is required for stressing the steel and/or anchoring the stressed steel to the concrete. A wide variety of systems have been developed for these purposes, many of which are patented by their manufacturers.

Figure 9.6 Prestressing seven-wire strand (courtesy of CCL)

Figure 9.7 Prestressing tendon (© Board of Trinity College Dublin)

A detailed knowledge of each system is not generally necessary for design purposes since all systems achieve the same end result. Having said that, however, a general knowledge is needed so that sufficient space is allowed for anchoring and jacking equipment. For this reason, some of the different systems are presented below.

The tensioning of the steel is usually achieved by mechanical jacking using hydraulic jacks. In pre-tensioning, the jacks pull the steel against the supports of the casting beds. The strands in pre-tensioned members are often stressed individually using small jacks, such as that illustrated in Fig. 9.8(a). In post-tensioning, the jacks pull the steel against the hardened concrete member itself. As the strands are usually grouped in tendons, large multi-strand jacks, such as that illustrated in Fig. 9.8(b), are often used to tension all the strands in the tendon simultaneously.

In pre-tensioning, anchorage of the strand is provided by the bond between the strand and the concrete. However, before the prestress is transferred to the concrete, temporary anchors are required to hold the ends of the strands while they are being tensioned. One of the most popular methods of anchoring the ends of the strands in the casting bed is the wedge grip (Fig. 9.9). Wedge grips are also used to grip each strand during post-tensioning of a tendon and to hold the strands permanently in the tendon anchor afterwards. Fig. 9.10 illustrates one such anchor. The bearing plate on this anchor transmits the force in the strands to the main body of the assembly which in turn transmits the force to the surrounding concrete.

9.3   Basis of design

The design of prestressed members is generally governed by limits on tensile and compressive stress in service rather than by their strength at the ultimate limit state. Normal practice, therefore, is first to produce an initial design (i.e. to choose an appropriate prestress force and tendon location) which

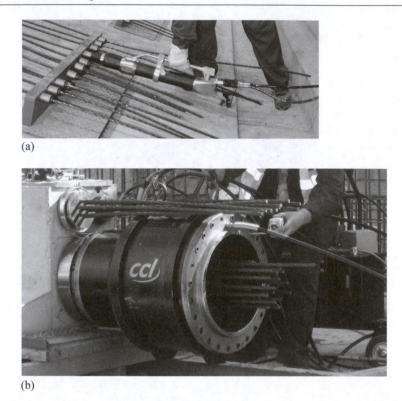

(a)

(b)

Figure 9.8 Hydraulic jacks: (a) single-strand jack (courtesy of CCL and Spiroll); (b) multi-strand jack (courtesy of CCL)

satisfies the design criteria for the serviceability limit state (SLS). This design is then checked at the ultimate limit state to ensure that it satisfies strength requirements. If it is found that the member has insufficient strength against bending, ordinary reinforcing bars can be provided to increase its moment capacity.

According to EC2, serviceability criteria must be checked at all stages in the process, e.g. at the transfer of prestress, during transportation and in service. However, as a result of the techniques employed in the construction of prestressed concrete members, in general there are two critical loading conditions that arise and are considered in this chapter. For these two critical loading conditions, the stresses, crack widths and deflections in the concrete must be checked against specified permissible values. The first condition, known as the **transfer condition**, occurs immediately on transfer of the prestress force to the concrete. At this stage, the concrete is still relatively young and its compressive strength has not reached its full design value. The stresses acting on a member during the transfer condition are prestress and stress due to the moment, M_0, induced by the applied loads present at that time. Often, the only load present at transfer is the self-weight of the member. Hence, the induced transfer moment, M_0, is often equal to the moment due to member self-weight. The second condition that must be checked is the **service condition**. This condition is reached when the concrete has matured to its full strength and the full service loads are being applied. At this stage, the applied prestress force has been reduced from its initial magnitude, P_0, due to losses which have taken place in the concrete and the steel. Prestress losses are discussed in detail in Section 9.5. The total loss of prestress force between transfer and service is generally in the region

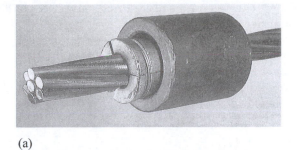

(a)

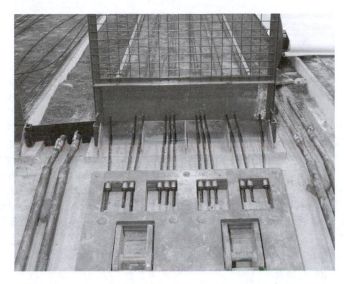

(b)

Figure 9.9 Wedge grip anchorage: (a) wedge grip assembly (© Board of Trinity College Dublin); (b) wedge
grips as end anchors in pre-tensioning facility (courtesy of CCL and Spiroll)

Figure 9.10 Anchorage assembly after casting (courtesy of CCL).

of 10–25 per cent. The stresses acting on the member during the service condition are prestress and stress due to the applied moment.

Design criteria at SLS

The design of prestressed concrete members for SLS according to EC2 should consider three design criteria:

- stress limitations
- crack width and
- deflection control.

Stress limits

Stress limits are imposed to control microcracking and to prevent excessive loss of prestress due to creep. Excessive compression causes tension perpendicular to the compressive force. Hence, to avoid longitudinal cracking of the concrete in service, EC2 recommends that the characteristic combination of compressive stress be limited to $0.6f_{ck}$ in areas of exposure classes XD, XF and XS (see Table 6.2). An additional constraint comes from creep issues under long-term load. If stress under quasi-permanent loads exceeds $0.45f_{ck}$, then the creep effect becomes large and is a non-linear function of stress. To prevent this from happening, stress should be kept below this level.

At transfer, the compressive stress in the concrete at the time of tensioning should not exceed $0.6f_{ck,0}$, where $f_{ck,0}$ is the characteristic compressive cylinder strength of the concrete at transfer. This may be increased to $0.7f_{ck,0}$ if experience or tests can validate that longitudinal cracking will not occur.

No limits are specified for the tensile stress in the concrete, but maximum crack widths are identified for different design situations (Table 9.1). Once certain conditions are met regarding bar size and stresses in the strands, crack widths do not have to be checked (see EC2 for more information). There are two major potential benefits of allowing limited cracks to develop in prestressed concrete:

1. Cost savings, because less prestressing steel is required or a smaller concrete section can be used. In the former case, ordinary reinforcing steel may be needed to supplement the prestressing steel in order to limit crack widths or to obtain the required ultimate moment capacity.
2. The upward deflection (camber) caused by prestressing can be significantly reduced by the reduction in prestress force, as illustrated in Fig. 9.11(a).

Table 9.1 Recommended maximum crack widths in mm (from UK NA to BS EN 1992-1-1)

Exposure class	Prestressed members without bonded tendons (quasi-permanent load combination)	Prestressed members with bonded tendons (frequent load combination)
X0, XC1	0.3 (for appearance only)	0.2
XC2, XC3, XC4	0.3	0.2[*]
XD1, XD2, XD3, XS1, XS2, XS3	0.3	0.2 and decompression[†]

[*]decompression should also be checked under quasi-permanent combination of loads
[†]crack width should be checked in areas where the decompression limit does not apply

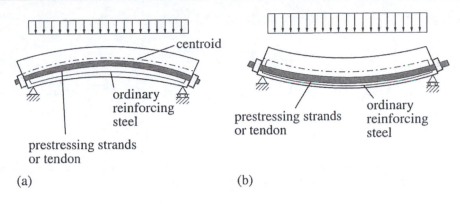

Figure 9.11 Deflections in simply supported member: (a) upward camber due to prestress and permanent gravity load; (b) total deflection due to prestress and permanent and variable gravity loads at SLS

As well as specifying maximum allowable crack widths, an additional decompression check is also specified in EC2 for particular design situations. This limit specifies that all parts of the tendon must be at least 25 mm (for buildings) or 100 mm (for bridges) inside the compression zone of the concrete. In this chapter, when designing a member to comply with this decompression limit, upper and lower decompression depths are determined, which form the minimum required compression zone around the tendon (Fig. 9.12). Generally, the lower decompression depth, y_{dl} is critical for sections in sag and the upper decompression depth, y_{du} is critical for sections in hog.

The method for designing members to comply with stress limits at the extreme fibres (top and bottom) and decompression limits, is outlined in this chapter.

Crack control

Since EC2 allows cracks to develop in prestressed concrete members, as for reinforced concrete, crack widths need to be kept within desired limits to inhibit the corrosion of the reinforcement (limits specified in Table 9.1). At a crack, the full tensile force is taken by the reinforcement. Between cracks, the bond between the reinforcement and the concrete gradually transfers the tension into the concrete as illustrated in Fig. 9.13(b). This transfer of stress builds up with distance from the crack as illustrated in Fig. 9.13(c). If the load is increased, a new crack occurs about half-way between

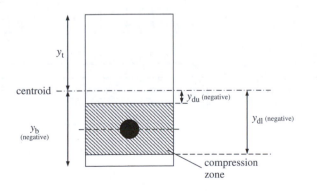

Figure 9.12 Compression zone and upper and lower decompression depths

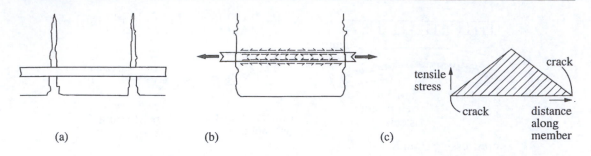

Figure 9.13 Crack spacing: (a) reinforcement crossing cracks; (b) transfer of tensile stress into concrete; (c) plot of tensile stress in concrete versus distance from crack

existing cracks when the tensile stress reaches the tensile strength of the concrete. If the bond is poor, the build-up of tensile stress is slow and the distance between cracks is large. This results in a small number of large cracks. Thus, crack widths can be kept smaller by increasing bond through the use of a larger number of smaller diameter bars (and hence a larger perimeter length per unit area).

In practice, concrete crack width is controlled in prestressed concrete by the area, spacing and level of stress in the ordinary reinforcement and by the level of prestressing. EC2 specifies an equation for the minimum area of ordinary reinforcement required to control cracking. It is noted that minimum reinforcement to limit crack growth does not have to be provided if tensile stress in the concrete does not exceed the mean tensile strength, f_{ctm}, under the characteristic combination of loads. Calculation of crack widths is not covered in this chapter. However, the procedure is the same as for reinforced concrete (Chapter 8). For prestressed concrete members, an additional term is included in the calculation of the effective reinforcement ratio to take account of prestressing tendons. The effect of prestress must also be considered when calculating the stress in the tension reinforcement (see EC2 for more details).

Example 9.1 Simple beam with tendon at centroid

Problem: The simply supported beam of Fig. 9.14 is post-tensioned by applying a prestress force of 1900 kN at the centre of the section. The tendon is bonded and has a diameter of 75 mm.

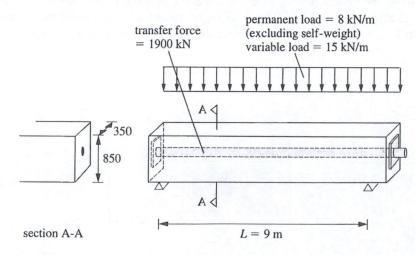

Figure 9.14 Beam of Example 9.1

The concrete is designed to achieve a compressive cylinder strength of 35 N/mm^2 at transfer and 50 N/mm^2 at service. The beam forms part of a structural scheme for a gym (i.e. imposed load category C, Table 3.3) and the permanent gravity load (excluding self-weight) is 8 kN/m of which 4 kN/m is present at transfer. In addition, there is a variable gravity load of 15 kN/m. Assuming a 20 per cent loss in prestress between transfer and service and exposure class XC3, check if the level of prestress applied satisfies the concrete compressive stress limits and the decompression limit for buildings at mid-span.

Solution: At transfer, EC2 states that the compressive stress should not be greater than $0.6f_{ck,0}$, and to assume linear creep, the compressive stress under quasi-permanent loads should not exceed $0.45f_{ck}$. As the beam is in exposure class XC3, the decompression limit should be checked under the quasi-permanent combination of loads (i.e. the tendon must lie at least 25 mm within the compression zone). Since this section is sagging, the stresses at the lower decompression depth will be critical for decompression and the stresses at the extreme top fibre will be critical for compression. To prevent decompression, the minimum stress at the lower decompression depth, y_{dl}, (25 mm below the tendon) is 0 N/mm^2, taking tension as negative. As a sign convention, the distance below the centroid is taken as negative. The diameter of the duct is 75 mm so $y_{dl} = -(75/2 + 25) = -63$ mm, see Fig. 9.15.

Transfer condition check

At transfer, self-weight and 4 kN/m of permanent gravity load are present. The cross-sectional area of the beam is given by:

$$A = 350 \times 850 = 297500 \text{ mm}^2$$

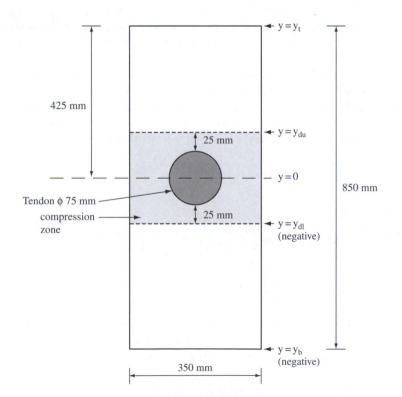

Figure 9.15 Illustration of decompression limit for Example 9.1

Hence, assuming concrete density of 25 kN/m^3 the applied load at transfer is:

$$q_0 = 25(0.298) + 4 = 11.44 \text{kN/m}$$

Since only these permanent gravity loads are acting on the beam at transfer, each SLS load combination (i.e. characteristic, frequent and quasi-permanent) will give the same result. Thus, the transfer moment is:

$$M_0 = q_0 L^2/8 = 11.44 \, (9)^2/8 = 116 \text{ kNm} \quad \text{(sag moments are considered positive)}$$

The elastic bending stress, σ, caused by an applied moment, M, is given by:

$$\sigma = My/I \tag{9.1}$$

where y is the distance from the centroid to the point where the stress is required and I is the second moment of area of the section. The stresses at the extreme top fibre ($y = y_t$) and at the lower decompression depth ($y = y_{dl}$) are given by:

$$\sigma(\text{top}) = \frac{M_0 y_t}{I} = \frac{M_0}{Z_t} \tag{9.2}$$

and

$$\sigma(y_{dl}) = \frac{M_0 y_{dl}}{I} = \frac{M_0}{Z_{dl}} \tag{9.3}$$

where Z_t ($=I/y_t$) is the section modulus for the top of the section and Z_{dl} ($=I/y_{dl}$) is the section modulus for the lower decompression depth of the section. Note that Z_{dl} is negative since y_{dl} is negative (below the centroid). Hence, the stress due to bending below the centroid is negative indicating that it is tensile as expected.

   Initially it is assumed that the section is uncracked. EC2 states that a section can be assumed to be uncracked if the flexural tensile stress does not exceed f_{ctm} (Table 8.2). This assumption will be checked later. For uncracked prestressed concrete members it has been found that sufficient accuracy in calculating bending stresses is achieved using the gross second moment of area, I_g, for I in Equation (9.1). The gross second moment of area of the member of Fig. 9.14 is:

$$I_g = \frac{bh^3}{12} = \frac{350 \times (850)^3}{12} = 17.91 \times 10^9 \text{ mm}^4$$

and the section moduli are:

$$Z_t = \frac{17.91 \times 10^9}{425} = 42.15 \times 10^6 \text{ mm}^3$$

$$Z_{dl} = \frac{17.91 \times 10^9}{-63} = -284 \times 10^6 \text{ mm}^3$$

The stresses in the concrete at transfer are due to the prestress force plus the transfer moment, M_0. In this example, the prestress force at transfer exerts a uniform axial stress of P/A. Hence, at the extreme top fibre and at the lower decompression depth, the stress **due to prestress** is:

$$\sigma_t = \sigma_{dl} = \frac{P}{A} = \frac{1900 \times 10^3}{297500} = 6.39 \text{ N/mm}^2$$

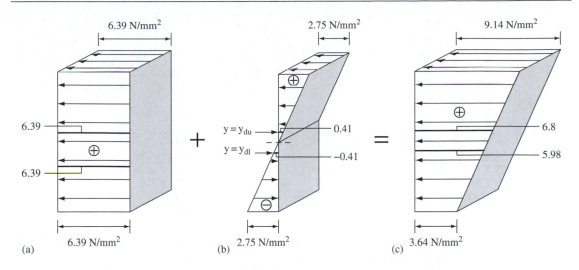

Figure 9.16 Distributions of stress at transfer: (a) prestress; (b) applied loads; (c) total

At the extreme top fibre, there is a compressive stress due to M_0 of:

$$\frac{M_0}{Z_t} = \frac{116 \times 10^6}{42.15 \times 10^6} = 2.75 \, \text{N/mm}^2$$

Similarly, the tensile stress due to M_0 at y_{dl} is:

$$\frac{M_0}{Z_{dl}} = \frac{116 \times 10^6}{-284 \times 10^6} = -0.41 \, \text{N/mm}^2$$

The stress distributions due to prestress and applied transfer moment are summarized in Fig. 9.16. Similarly, the stresses at the bottom fibre and the upper decompression depth can also be calculated. It can be seen from the figure that the total stress is compressive throughout the section at transfer with a maximum value of 9.14 N/mm² at the top fibre. The maximum permissible compressive stress at transfer is $0.6f_{ck,0} = 0.6(35) = 21$ N/mm². Hence, the amount of prestress applied satisfies the stress requirements at transfer. Since the whole section is in compression, the decompression limit is satisfied and the assumption of an uncracked section is valid at transfer.

Service condition check

The stresses in the concrete at service are due to the prestress force (after all losses have occurred) plus the service moment. For the quasi-permanent load combination at service, the full permanent gravity loads and a fraction of the variable gravity loads apply. From Table 3.11 (imposed load category C) and Table 3.13, the applied service loading for the quasi-permanent combination is:

$$q_{qp} = 25(0.298) + 8 + (0.6)15 = 24.45 \, \text{kN/m}$$

from which the quasi-permanent service moment at mid-span is:

$$M_{qp} = q_{qp}L^2/8 = 24.45(9)^2/8 = 248 \, \text{kNm}$$

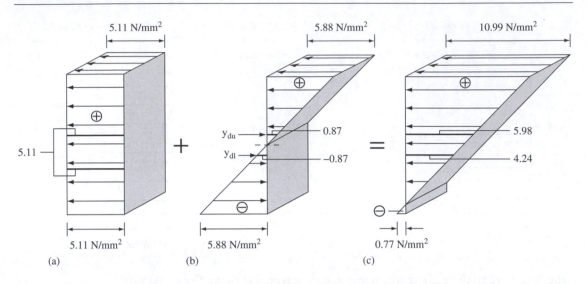

Figure 9.17 Distributions of stress at service: (a) prestress; (b) applied loads; (c) total

To allow for losses, the stresses due to prestress are reduced from their magnitudes at transfer by 20 per cent. Hence, the loss ratio is:

$$\rho = \frac{\text{stress at service}}{\text{stress at transfer}} = 0.8$$

and the stress due to prestress at service is:

$$\rho\sigma_t = \rho\sigma_{dl} = 0.8\frac{P}{A} = 0.8\frac{1900 \times 10^3}{297500} = 5.11 \text{ N/mm}^2$$

Hence, the total stress at the extreme top fibre is:

$$\rho\sigma_t + \frac{M_{qp}}{Z_t} = 5.11 + \frac{248 \times 10^6}{42.15 \times 10^6} = 10.99 \text{ N/mm}^2$$

and the total stress at the lower decompression depth is:

$$\rho\sigma_{dl} + \frac{M_{qp}}{Z_{dl}} = 5.11 + \frac{248 \times 10^6}{-284 \times 10^6} = 4.24 \text{ N/mm}^2$$

The total stress distribution due to the service loads and prestress is summarized in Fig. 9.17. The maximum permissible compressive stress at service for the quasi-permanent load combination (to assume linear creep) is $0.45f_{ck} = 0.45(50) = 22.5 \text{ N/mm}^2$ which is satisfactory. There is also compressive stress at 25 mm above and below the tendon, satisfying the decompression limit. Although there is a tensile stress of 0.77 N/mm^2 at the bottom fibre, this is less than the mean tensile strength of the concrete ($f_{ctm} = 4.1 \text{ N/mm}^2$, Table 8.2), so the initial assumption of an uncracked section is valid and there is no cracking there.

Sign conventions

The sign conventions used in this and subsequent sections are as follows:

(a)   Applied sag moments are considered positive and applied hog moments are negative.
(b)   The eccentricity, e, of prestress tendons above the centroid of the member is positive and below the centroid of the member is negative.

(c) Compressive stresses are considered positive and tensile stresses are considered negative.
(d) The distance from the centroid is positive upwards. Hence, the numerical value for the section modulus for the bottom fibre, Z_b ($= I/y_b$) is always negative. Z_{du} ($= I/y_{du}$) and Z_{dl} ($= I/y_{dl}$) are also negative if the respective decompression limits (y_{du} and y_{dl}) are below the centroid.

Note: In the following examples, it is assumed that the section is uncracked (i.e. tensile stress $< f_{ctm}$) and, as such, the gross second moment of area is used to calculate the section modulus.

Effect of tendon location

It was seen in Example 9.1 that a tendon placed at the centroid of a member exerting a force, P, on the concrete creates a uniform distribution of stress, equal to P/A, across the section. If, instead, a straight tendon is located at an eccentricity e above the centroid of the member, as illustrated in Fig. 9.18, an eccentric force is applied to the concrete. The application of this eccentric force, P, is equivalent to applying a concentric axial force, P, and a sag bending moment, M_p, at any given section as illustrated in Fig. 9.19.

For a determinate beam such as this, the moment due to prestress is simply the product of prestress force and eccentricity, that is $M_p = Pe$. For indeterminate beams, refer to Section 9.6. The axial force component creates a uniform axial stress distribution, of magnitude P/A, at any given section, as

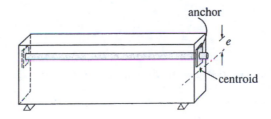

Figure 9.18 Eccentric straight tendon

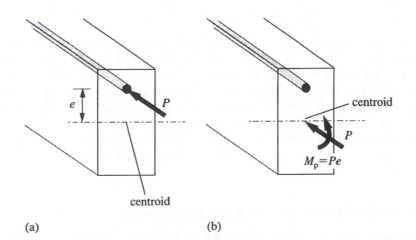

(a)   (b)

Figure 9.19 Eccentric prestress: (a) actual prestress force; (b) equivalent force and moment at centroid

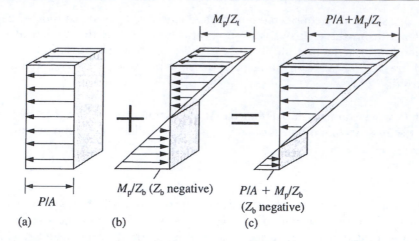

Figure 9.20 Stress distributions due to eccentric prestress: (a) axial component; (b) bending component; (c) total

illustrated in Fig. 9.20(a). The bending component, however, creates a triangular stress distribution at any given section (Fig. 9.20(b)). Hence, the total stress due to prestress at the extreme fibres is:

$$\sigma_t = \frac{P}{A} + \frac{M_p}{Z_t} \tag{9.4}$$

$$\sigma_b = \frac{P}{A} + \frac{M_p}{Z_b} \tag{9.5}$$

Similarly, the stress due to prestress at the upper and lower decompression depths can be determined:

$$\sigma_{du} = \frac{P}{A} + \frac{M_p}{Z_{du}} \tag{9.6}$$

$$\sigma_{dl} = \frac{P}{A} + \frac{M_p}{Z_{dl}} \tag{9.7}$$

The overall stress distribution due to the axial and bending components of the prestress is illustrated in Fig. 9.20(c). In this case, there are no service loads applied to the beam, and the upper portion of the beam containing the tendon is well within the compression zone, satisfying the decompression limit. However, service loads generally counteract the moments induced by eccentricity of the prestress. Due to this change in the overall stress distribution at service, there is a risk of decompression around the tendon or excessive compressive stresses developing in the top or bottom fibres. Therefore, compression limits at both extreme fibres and decompression limits at the upper and lower decompression depths must be checked under service loads. Crack widths may also need to be checked in accordance with Table 9.1.

Example 9.2 Simple beam with eccentric tendon

Problem: The simply supported member of Fig. 9.14 is now post-tensioned by applying an eccentric prestress force of 1,500 kN at 150 mm below the centroid of the member (i.e. $e = -150$). The prestress loss, the design concrete strengths and the applied loading are the same as for Example 9.1. Check if the stresses at mid-span are within the limits specified in Example 9.1.

Solution: As before, the adequacy of the member must be checked at both the transfer condition and at the service condition.

Transfer condition check

As the beam is determinate, $M_p = Pe$, and the stress at the extreme top fibre due to prestress is:

$$\sigma_t = \frac{P}{A} + \frac{Pe}{Z_t}$$
$$= \frac{1500 \times 10^3}{297500} + \frac{(1500 \times 10^3)(-150)}{42.15 \times 10^6}$$
$$= 5.04 - 5.34$$
$$= -0.3 \, \text{N/mm}^2$$

Since the tendon is located below the centroid, the stresses at the lower decompression depth will again be critical. The tendon is located 150 mm below the centroid of the member (rather than at the centroid, as with the previous example). Thus the lower decompression depth is now, $y_{dl} = -(150 + 75/2 + 25) = -213$ mm. This results in a section modulus, Z_{dl}, of:

$$Z_{dl} = \frac{I}{y_{dl}} = \frac{17.91 \times 10^9}{-213} = -84 \times 10^6 \, \text{mm}^3$$

Therefore, at the decompression depth, y_{dl}, the stress due to prestress is:

$$\sigma_{dl} = \frac{P}{A} + \frac{Pe}{Z_{dl}}$$
$$= \frac{1500 \times 10^3}{297500} + \frac{(1500 \times 10^3)(-150)}{-84 \times 10^6}$$
$$= 5.04 + 2.68$$
$$= 7.72 \, \text{N/mm}^2$$

From Example 9.1, the stress due to the applied loads at transfer at the extreme top fibre is 2.75 N/mm². The stress due to the applied loads at the lower decompression depth is:

$$\frac{M_0}{Z_{dl}} = \frac{116 \times 10^6}{-84 \times 10^6} = -1.38 \, \text{N/mm}^2$$

The total stress distribution at mid-span is illustrated in Fig. 9.21. It can be seen that this distribution complies with the specified stress limits.

Service condition check

As before (Fig. 9.17), the stress due to the applied quasi-permanent combination of service loads at the extreme top fibre is 5.88 N/mm². At the lower decompression depth, the stress due to the applied loads is:

$$\frac{M_{qp}}{Z_{dl}} = \frac{248 \times 10^6}{-84 \times 10^6} = -2.95 \, \text{N/mm}^2$$

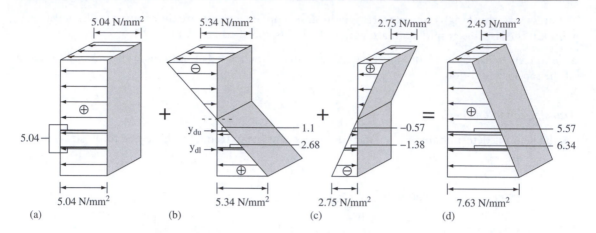

Figure 9.21 Transfer stress distributions at mid-span: (a) axial component of prestress; (b) bending component of prestress; (c) applied loads; (d) total

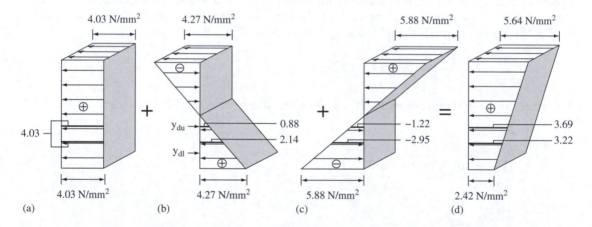

Figure 9.22 Service stress distributions at mid-span for quasi-permanent load combination: (a) axial component of prestress; (b) bending component of prestress; (c) applied loads; (d) total

Due to losses, the stress due to the axial and bending components of the prestress is reduced to 80 per cent of its value at transfer. The total stress distribution at mid-span is illustrated in Fig. 9.22. It can be seen that the service stress limits are satisfied for this example, even with the loss of prestress force. The advantage of prestressing using eccentric tendons is that not only does it prevent the compressive stresses from becoming excessively large but it also reduces the tensile stresses due to the applied loads.

Example 9.3. Simple beam with eccentric tendon

Problem: For the beam of Example 9.2, check if the stresses at the sections over the supports are within the limits specified in Example 9.1.

Solution: Consider the stresses at a section over one of the supports. There is no stress at this section due to the applied loads at either transfer or service, that is, the bending moments M_0 and M_{qp} are

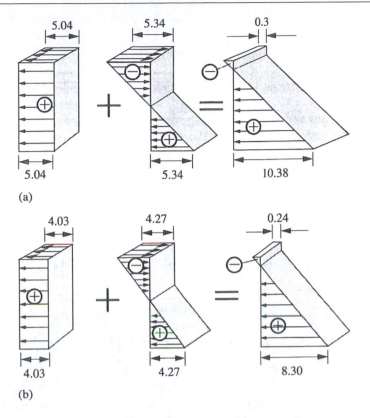

Figure 9.23 Stress distribution at supports: (a) transfer stresses; (b) service stresses

zero. However, the eccentric prestress force does cause bending at this section. The total stress distributions at transfer and service are illustrated in Fig. 9.23. It can be seen from the figure that a tensile stress is developed in the extreme top fibre at transfer and service. Tensile stresses are allowed to develop in prestressed concrete members once the decompression and crack width limits are satisfied. However, to prevent tensile stresses from becoming excessively large, tendons are generally 'profiled' or 'debonded' near supports. Both of these concepts are explained in the following section.

Profiled tendons and debonding

To take advantage of the eccentric prestress force in post-tensioned members, while at the same time preventing excessive tensile stresses from developing in the top fibre (Example 9.3), tendons are generally draped or profiled. For example, an appropriate tendon profile for the member of Fig. 9.14 is illustrated in Fig. 9.24. The maximum eccentricity of the tendon occurs at mid-span, where the tensile stress due to the applied loads is greatest. Over the supports the tendon is located at the centroid and so the prestress force is purely concentric and does not generate tensile stresses. For pre-tensioned members, where profiling is not practical, a technique known as **debonding** is often used in practice. In debonding, a plastic duct or tape is placed around some strands near the supports, as illustrated in Fig. 9.25, to prevent them from bonding with the concrete. These strands become ineffective over the length of the duct – no prestress is applied by them. This reduction in the prestress force can preventing or reduce tensile stresses due to eccentric prestress.

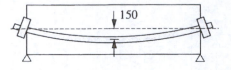

Figure 9.24 Draped tendon

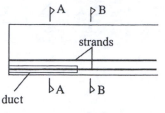

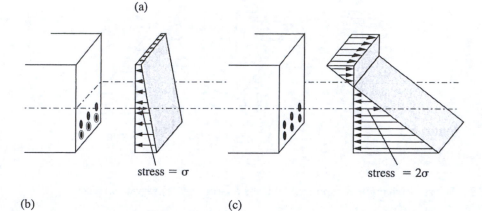

Figure 9.25 Debonded strands: (a) elevation; (b) section A–A, stress due to prestress when one group of strands is bonded; (c) section B–B, stress distribution when both groups of strands are bonded

9.4   Prestressing force and eccentricity

When designing a prestressed concrete member it is necessary to determine how much prestress force to apply, and where to locate it, at each section along the length of the member. It was shown in Section 9.3 that prestress can be applied most efficiently by placing the tendons at an eccentricity and, in the case of post-tensioned members, by varying the eccentricity of the tendons along the length of the member. For a statically determinate member with an eccentric prestress force, the stresses due to prestress in the extreme fibres are given by:

$$\sigma_t = \frac{P}{A} + \frac{Pe}{Z_t} \tag{9.8}$$

$$\sigma_b = \frac{P}{A} + \frac{Pe}{Z_b} \tag{9.9}$$

and the stresses due to prestress at the upper and lower decompression depths are given by:

$$\sigma_{du} = \frac{P}{A} + \frac{Pe}{Z_{du}} \tag{9.10}$$

$$\sigma_{dl} = \frac{P}{A} + \frac{Pe}{Z_{dl}} \tag{9.11}$$

where P is the prestress force at transfer. It can be seen from these expressions that the stress levels depend on the magnitudes of both P and e. A large stress in the extreme top fibre can be achieved using a large prestressing force and eccentricity, but as Z_b is negative, a large e reduces the stress at the bottom fibre. Thus, the process of choosing an appropriate prestress force, P, and tendon location (i.e. eccentricity, e) can prove to be quite difficult.

In this section a method is presented whereby the most appropriate value of P and e can be chosen for each critical section with relative ease. The method is based on stress limits rather than crack widths. Nevertheless, it provides a convenient method of selecting initial values for P and e. These can be subsequently changed to satisfy crack width limits if necessary.

The stresses between the upper and lower decompression depths must remain above the minimum stress limits at transfer and service, denoted p_{0min} and p_{Smin}, respectively (both 0 N/mm^2 for the decompression limit – the entire zone must remain in compression). In addition, the compressive stresses in the extreme top and bottom fibres must remain below the maximum stress limits at transfer and service, denoted p_{0max} and p_{Smax}, respectively. Thus, at transfer, to prevent decompression near the tendon:

$$p_{0min} \leq \sigma_{du} + \frac{M_0}{Z_{du}} \tag{9.12}$$

$$p_{0min} \leq \sigma_{dl} + \frac{M_0}{Z_{dl}} \tag{9.13}$$

And to prevent excessive compressive stresses anywhere:

$$\sigma_t + \frac{M_0}{Z_t} \leq p_{0max} \tag{9.14}$$

$$\sigma_b + \frac{M_0}{Z_b} \leq p_{0max} \tag{9.15}$$

where σ_{du} and σ_{dl} are the stresses at the upper and lower decompression depths **due to the prestress** at transfer, σ_t and σ_b are the stresses at the extreme top and bottom fibres **due to the prestress** at transfer and M_0 is the transfer moment.

At service, the transfer prestress, σ, drops to $\rho\sigma$ due to the losses which occur between transfer and service. The factor ρ is equal to the ratio of the prestress force at service to the prestress force at transfer (typically $\rho = 0.75$ to 0.90). Thus, for the service condition, to prevent decompression near the tendon:

$$p_{Smin} \leq \rho\sigma_{du} + \frac{M_S}{Z_{du}} \tag{9.16}$$

$$p_{Smin} \leq \rho\sigma_{dl} + \frac{M_S}{Z_{dl}} \tag{9.17}$$

And to prevent excessive compressive stresses anywhere:

$$\rho\sigma_t + \frac{M_S}{Z_t} \leq p_{Smax} \tag{9.18}$$

$$\rho\sigma_b + \frac{M_S}{Z_b} \le p_{Smax} \tag{9.19}$$

where M_S is the applied moment at service, calculated using the appropriate SLS load combination (e.g. when checking decompression, the quasi-permanent load combination should be used for members with bonded tendons in exposure classes XC2, XC3 and XC4 – Table 9.1).

By rearranging inequalities (9.12) to (9.19), the limits on the stresses due to prestress at the upper and lower decompression depths are:

$$\left(\begin{array}{c} p_{0min} - \dfrac{M_0}{Z_{du}} \\[2ex] \dfrac{1}{\rho}\left(p_{Smin} - \dfrac{M_S}{Z_{du}} \right) \end{array} \right) \le \sigma_{du} \tag{9.20}$$

$$\left(\begin{array}{c} p_{0min} - \dfrac{M_0}{Z_{dl}} \\[2ex] \dfrac{1}{\rho}\left(p_{Smin} - \dfrac{M_S}{Z_{dl}} \right) \end{array} \right) \le \sigma_{dl} \tag{9.21}$$

Similarly, the stresses due to prestress at the extreme top and bottom fibres must satisfy:

$$\sigma_t \le \left(\begin{array}{c} p_{0max} - \dfrac{M_0}{Z_t} \\[2ex] \dfrac{1}{\rho}\left(p_{Smax} - \dfrac{M_S}{Z_t} \right) \end{array} \right) \tag{9.22}$$

$$\sigma_b \le \left(\begin{array}{c} p_{0max} - \dfrac{M_0}{Z_b} \\[2ex] \dfrac{1}{\rho}\left(p_{Smax} - \dfrac{M_S}{Z_b} \right) \end{array} \right) \tag{9.23}$$

The stresses due to prestress (given by Equations (9.8)–(9.11)) are limited by inequalities (9.20)–(9.23). Inequality (9.20) can be expressed as:

$$\sigma_{du\_min} \le \sigma_{du} \tag{9.24}$$

where:

$$\sigma_{du\_min} = \text{greater of } \left(p_{0min} - \frac{M_0}{Z_{du}} \right) \text{ and } \frac{1}{\rho}\left(p_{Smin} - \frac{M_S}{Z_{du}} \right)$$

Similarly, for the lower decompression depth:

$$\sigma_{dl\_min} \le \sigma_{dl} \tag{9.25}$$

where:

$$\sigma_{dl\_min} = \text{greater of } \left(p_{0min} - \frac{M_0}{Z_{dl}} \right) \text{ and } \frac{1}{\rho}\left(p_{Smin} - \frac{M_S}{Z_{dl}} \right)$$

To ensure the compressive stress limits remain below the allowable limits at transfer and service, for the extreme top fibre, inequality (9.22) gives:

$$\sigma_t \leq \sigma_{t\_max} \tag{9.26}$$

where:

$$\sigma_{t\_max} = \text{lesser of } \left(p_{0max} - \frac{M_0}{Z_t} \right) \text{ and } \frac{1}{\rho} \left(p_{Smax} - \frac{M_S}{Z_t} \right)$$

Similarly, for the extreme bottom fibre:

$$\sigma_b \leq \sigma_{b\_max} \tag{9.27}$$

where:

$$\sigma_{b\_max} = \text{lesser of } \left(p_{0max} - \frac{M_0}{Z_b} \right) \text{ and } \frac{1}{\rho} \left(p_{Smax} - \frac{M_S}{Z_b} \right)$$

Substitution from Equations (9.8)–(9.11) for σ_{du}, σ_{dl}, σ_t and σ_b in the above inequalities yields the following four expressions:

$$\sigma_{du\_min} \leq \frac{P}{A} + \frac{Pe}{Z_{du}} \tag{9.28}$$

$$\sigma_{dl\_min} \leq \frac{P}{A} + \frac{Pe}{Z_{dl}} \tag{9.29}$$

$$\frac{P}{A} + \frac{Pe}{Z_t} \leq \sigma_{t\_max} \tag{9.30}$$

$$\frac{P}{A} + \frac{Pe}{Z_b} \leq \sigma_{t\_max} \tag{9.31}$$

where (for buildings):

$$Z_{du} = \frac{I}{y_{du}} = \frac{I}{e + (\phi_{duct}/2) + 25}$$

$$Z_{dl} = \frac{I}{y_{dl}} = \frac{I}{e - (\phi_{duct}/2) - 25}$$

Each of the inequalities (9.28) to (9.31) represents a non-linear relationship between P and e which can be illustrated on a plot of P versus e. By dividing across by P, inequalities (9.30) and (9.31) can be transformed into linear versions and plotted on graphs of e versus $1/P$. This approach is attributed to Magnel and the plots are known as Magnel diagrams. As Z_{du} and Z_{dl} are functions of e, such an approach is not possible. An example showing all four inequalities is given in Fig. 9.26. On one side of each curve the inequality is satisfied and on the other side, it is not. The feasible zone, where all four inequalities are satisfied, is bound by these curves. To determine which side of a curve represents the inequality, the origin, $(P,e) = (0,0)$, is substituted into it. If the inequality is true, all points on the side of origin are valid. Otherwise, all points on the other side of the curve are valid.

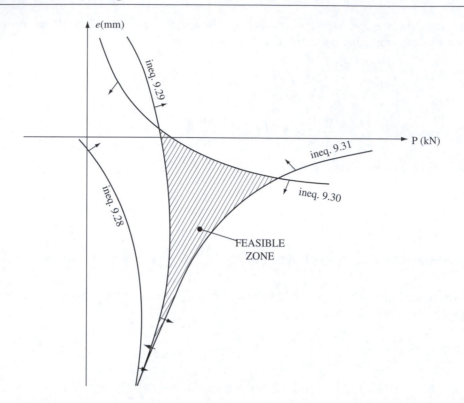

Figure 9.26 Illustration of non-linear Magnel diagram

The shaded area in the Magnel diagram, Fig. 9.26, is the zone in which all the stress limits are satisfied and is called the feasible zone. Any combination of P and e that falls within the feasible zone constitutes a valid solution. For sections where a sagging moment is induced due to an applied load, inequality (9.28) will not limit the feasible zone as the tendon will generally be below the centroid and the limit for the lower decompression zone will be critical (i.e. inequality (9.29)). This is illustrated in Fig. 9.26.

Usually, to save on prestressing steel, a solution is chosen which corresponds to a low value of P but sufficiently away from the edge of the feasible zone to allow some latitude in selecting tendon eccentricity.

Note: This procedure assumes that the section remains uncracked in service (i.e. gross second moment of area is used in calculations).

It is often helpful to draw in two further boundary lines on the Magnel diagram to show the maximum eccentricities that are physically possible. These lines, which may be more stringent than the inequalities of the feasible zone, ensure that any chosen value of e is physically within the section at all points and does not violate the requirements for cover. The lines are horizontal on a plot of P versus e, as illustrated in Fig. 9.27. To ensure adequate bond between the concrete and the tendons or strands in pre-tensioned members EC2 recommends that for pre-tensioned members, the minimum cover should not be less than 1.5 times the diameter of the strand or plain wire, while for post-tensioned members the minimum cover should not be less than the diameter of the duct. To allow for construction tolerance, up to 10 mm must be added to these minimum cover values, depending on quality assurance measures in place. This gives the **nominal cover** to reinforcement that is used for design and is specified on working drawings. For durability, the UK NA to EC2 does not specify

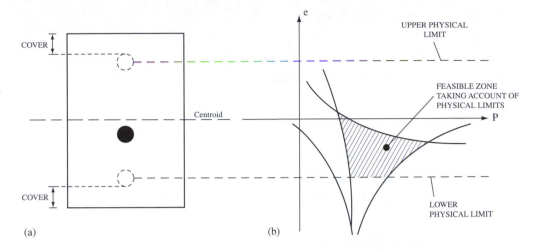

Figure 9.27 Magnel diagram with physical limits: (a) cross-section; (b) Magnel diagram

higher values of cover for prestressed concrete members, and the cover should not be less than the values specified in Table 6.3. Again, up to 10 mm should be added to the minimum cover specified in the table to give the nominal cover that is used for design.

Example 9.4 Prestress force and eccentricity

Problem: The single-span, simply supported post-tensioned beam of Fig. 9.28 carries a characteristic permanent gravity load of 5 kN/m (not including self-weight) and a characteristic variable gravity load of 20 kN/m as illustrated in the figure. This exterior beam forms part of an office building in exposure class XS1. Assume $\rho = 0.75$, a compressive cylinder strength of 35 N/mm^2 at transfer and 50 N/mm^2 at service and a duct diameter of 75 mm.

Plot the Magnel diagram for the critical section at mid-span for the member and choose an appropriate prestress force at transfer based on the decompression limits and the compressive stress limits outlined in Section 9.3. Assume that the section is uncracked in service.

Solution

In order to determine the applied bending moments, the self-weight of the beam must be calculated. For this section:

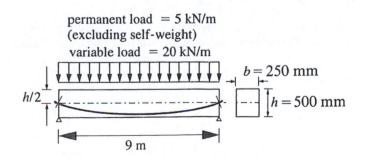

Figure 9.28 Post-tensioned beam of Example 9.4

self-weight, $q_{sw} = 25 \times (0.25 \times 0.50)$

$$= 3.13 \text{kN/m}$$

$A = 250 \times 500 = 125000 \text{ mm}^2$

$I = bh^3/12 = (250)(500)^3/12 = 2.6 \times 10^9 \text{mm}^4$

and:

$$Z_t = -Z_b = bh^2/6$$

$$= (250)(500)^2/6 = 10.42 \times 10^6 \text{mm}^3$$

At transfer, the compressive stress limit, p_{0max}, is $0.6f_{ck,0}$ ($0.6 \times 35 = 21$ N/mm^2). It is assumed that only the self-weight portion of the permanent gravity load will be present at transfer. Thus, M_0, at mid-span, is:

$$M_0 = q_{sw}L^2/8 = 3.13(9)^2/8 = 32 \text{kNm}$$

At service, different SLS load combinations must be considered. Decompression must be checked using the frequent load combination for exposure class XS1. Therefore, σ_{du_min} and σ_{dl_min} should be calculated using the frequent load combination (i.e. $M_S = M_f$). Considering the compressive stress limit at service, p_{Smax}, the compressive stress in the beam must be no greater than $0.6f_{ck}$ ($0.6 \times 50 = 30$ N/mm^2) under the characteristic combination of loads (for exposure class XS1) and no greater than $0.45f_{ck}$ ($0.45 \times 50 = 22.5$ N/mm^2) under the quasi-permanent combination of loads (otherwise non-linearity of creep must be considered). Therefore, when calculating σ_{t_max} and σ_{t_min}, the more onerous of the characteristic and quasi-permanent load combinations should be used (i.e. $M_S = M_c$ and $p_{Smax} = 30$ N/mm^2 or $M_S = M_{qp}$ and $p_{Smax} = 22.5$ N/mm^2).

With reference to Tables 3.11 and 3.13, the bending moments at mid-span due to the frequent load combination, M_f, the characteristic load combination, M_c, and the quasi-permanent load combination, M_{qp}, are:

$$M_f = q_f L^2/8 = [3.13 + 5 + (0.5)(20)](9)^2/8 = 184 \text{ kNm}$$

$$M_c = q_c L^2/8 = [3.13 + 5 + 20](9)^2/8 = 285 \text{ kNm}$$

$$M_{qp} = q_{qp} L^2/8 = [3.13 + 5 + (0.3)(20)](9)^2/8 = 143 \text{ kNm}$$

Recall that:

$$\sigma_{t\_max} = \text{lesser of} \left(p_{0max} - \frac{M_0}{Z_t} \right) \text{ and } \frac{1}{\rho} \left(p_{Smax} - \frac{M_S}{Z_t} \right)$$

where $\rho = 0.75$, $p_{0min} = 0$ N/mm^2, $p_{Smin} = 0$ N/mm^2, $p_{0max} = 21$ N/mm^2 and $p_{Smax} = 30$ or 22.5 N/mm^2 (for characteristic or quasi-permanent load combinations, respectively). Hence:

$$\left(p_{0max} - \frac{M_0}{Z_t} \right) = \left(21 - \frac{32}{10.42} \right) = 17.93$$

$$\frac{1}{\rho} \left(p_{Smax} - \frac{M_c}{Z_t} \right) = \frac{1}{0.75} \left(30 - \frac{285}{10.42} \right) = 3.53 \text{(characteristic combination)}$$

$$\frac{1}{\rho} \left(p_{Smax} - \frac{M_{qp}}{Z_t} \right) = \frac{1}{0.75} \left(22.5 - \frac{143}{10.42} \right) = 11.7 \text{(quasi-permanent combination)}$$

Therefore:

$$\sigma_{t\_max} = \text{lesser of } (17.93, \ 3.53, \ 11.7) = 3.53 \text{ N/mm}^2$$

Similarly, σ_{b_max} is given by:

$$\sigma_{b\_max} = \text{lesser of } \left(p_{0max} - \frac{M_0}{Z_b} \right) \text{ and } \frac{1}{\rho} \left(p_{Smax} - \frac{M_S}{Z_b} \right) = 24.07 \text{ N/mm}^2$$

Inequalities (9.30) and (9.31) are plotted in Fig. 9.29 and can be seen to define upper limits on P. Lower limits are governed by the decompression constraints. These limits on stresses near the tendons are given by:

$$\sigma_{du\_min} = \text{greater of } \left(p_{0min} - \frac{M_0}{Z_{du}} \right) \text{ and } \frac{1}{\rho} \left(p_{Smin} - \frac{M_S}{Z_{du}} \right)$$

$$\sigma_{dl\_min} = \text{greater of } \left(p_{0min} - \frac{M_0}{Z_{dl}} \right) \text{ and } \frac{1}{\rho} \left(p_{Smin} - \frac{M_S}{Z_{dl}} \right)$$

As these terms include Z_{du} and Z_{dl}, they are a function of eccentricity, e. Substituting these terms into inequalities (9.28) and (9.29) gives the two additional curves in Fig. 9.29. To satisfy cover requirements, physical limits for eccentricity are added to the diagram. The nominal cover for this section is governed by bond requirements, and is equal to the diameter of the duct plus 10 mm. Thus, the nominal cover is 85 mm, resulting in a physical limit on eccentricity of $(250 - 85 - 75/2 =)$ 128 mm on each side of the centroid.

From Fig. 9.29, it can be seen that the minimum value for P in the feasible zone is approximately equal to 1035 kN and this corresponds to an eccentricity of –128 mm (which is the maximum eccentricity within the physical limits of the member). This is marginally less than the minimum prestress at the top of the feasible zone ($P \approx 1040$ kN, $e \approx -48$ mm). To allow for some latitude in eccentricity, a higher value for P is chosen, say $P = 1200$ kN, which allows for a large variance in eccentricity between –53 mm and –125 mm. For this example, an eccentricity of –100 mm is

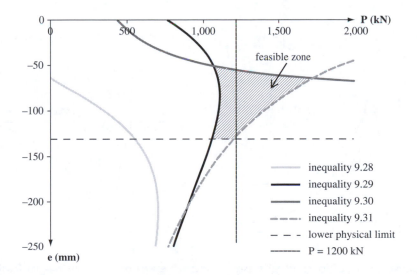

Figure 9.29 Magnel diagram for Example 9.4

chosen. These combinations of P and e can be substituted into Equations (9.8) and (9.11) to determine the stresses due to prestress. These can in turn be substituted into inequalities (9.25) and (9.26) to check that the stresses are indeed within the specified limits.

General procedure for post-tensioned members

Up to this point, we have concentrated on the prestress force and eccentricity requirements of a simply supported beam at the single critical section where the moments are greatest. However, in more typical post-tensioned beams, the critical bending moments will vary from one section to the next within each span. In addition, many post-tensioned beams have a variable depth over their length, such as the member illustrated in Fig. 9.30. In such cases, the section moduli also vary at different sections along the length of the member. It can be seen from inequalities (9.20)–(9.23) that both the magnitude of the applied moments and the section moduli determine the shape and location of the feasible zone in the Magnel diagram. Thus, for the member of Fig. 9.30(a) there is a different Magnel diagram for each section considered along its length, as illustrated in Fig. 9.30(b) and (c). However, when one tendon is to be used throughout the length of the beam, the force, P, will be the same for each section except for small differences due to losses. Thus, it will often be the case that these sections will have conflicting requirements for P and e. For such members, an appropriate

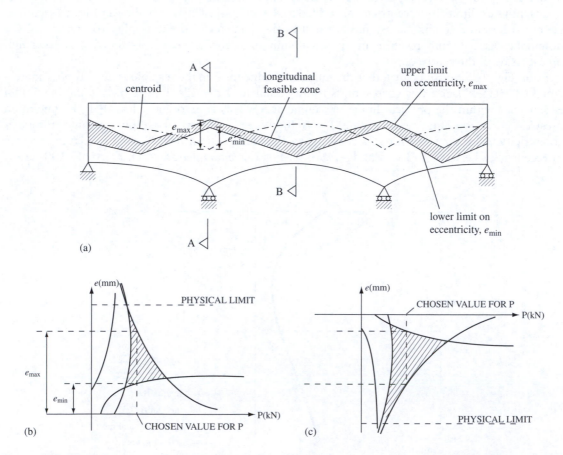

Figure 9.30 Construction of longitudinal feasible zone: (a) elevation and longitudinal feasible zone; (b) Magnel diagram at A–A; (c) Magnel diagram at B–B

prestress force and tendon profile which satisfies the prestress requirements at all sections can be found by observing the following three-step process:

1. Choose a prestress force, P, which is within the feasible zone of each Magnel diagram. If this is not possible, the section size of the member must be increased so that it is possible to find an appropriate value for P.
2. For the chosen prestress force, calculate the two limits e_{min} and e_{max} on the tendon eccentricity for each Magnel diagram, as illustrated in Fig. 9.30(b). Plot the locations of these eccentricity limits on a longitudinal section through the beam (Fig. 9.30(a)). Lines are then drawn to join the limits for adjacent sections as illustrated in the figure. These tendon limits constitute a long-itudinal feasible zone within which the tendon must lie in order to satisfy the stress limits at each section. (Identifying discrete sections and joining them with lines is of course an approximation.)
3. A tendon profile is chosen such that its centroid lies within the longitudinal feasible zone. The total number of sections along the length of a member which are considered is a matter for engineering judgement. In general, only a few sections (at points of maximum and minimum bending moment and one or two points in-between) need to be considered in order to find an acceptable tendon profile. The following example illustrates each step of the basic post-tensioned design process.

Example 9.5 Tendon profile design

Problem: For the post-tensioned beam of Fig. 9.31, the prestress requirements are to be checked at the four numbered sections. Using the Magnel diagrams given in Fig. 9.32, determine an appropriate prestress force and tendon profile, assuming a duct diameter of 75 mm. Differences in prestress force between sections generally occur due to friction losses. However, in short beams such as this one, such differences tend to be small and, for the purposes of this example, can be ignored. Provide for a minimum cover of 80 mm.

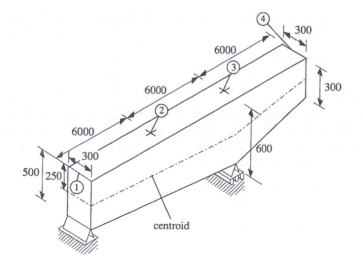

Figure 9.31 Beam of Example 9.5

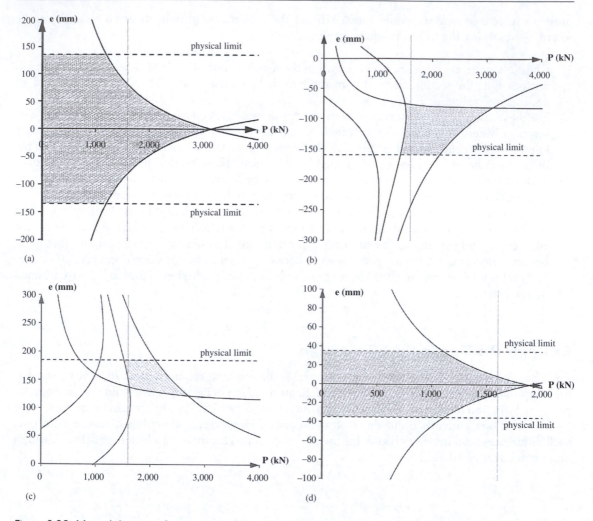

Figure 9.32 Magnel diagrams for sections of Example 9.5: (a) Section no. 1; (b) Section no. 2; (c) Section no. 3; (d) Section no. 4

Solution:

Step 1– choose value for *P*

By inspection of the four Magnel diagrams of Fig. 9.32, the most suitable value for *P* (i.e. the minimum which is within the feasible zones for all four diagrams) is found to be approximately 1,600 kN, governed by Section no. 3. It is interesting to note from Fig. 9.32(d) that this force is very close to the maximum possible at Section no. 4.

Step 2 – determine longitudinal feasible zone

For a prestress force of 1,600 kN, the eccentricity limits at each section are, from Fig. 9.32:

Section no. 1: -80 mm $\le e \le 80$ mm
Section no. 2: -76 mm $\le e \le -160$ mm
Section no. 3: 150 mm $\le e \le 185$ mm
Section no. 4: -9 mm $\le e \le 9$ mm

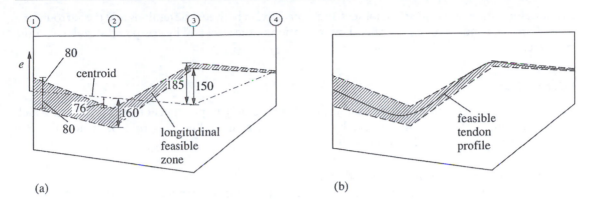

Figure 9.33 Longitudinal feasible zone for Example 9.5: (a) longitudinal section showing feasible zone; (b) feasible zone with tendon profile

The corresponding longitudinal feasible zone is illustrated in Fig. 9.33.

Step 3 – choose suitable tendon profile
A tendon profile, made up of a continuous series of lines and parabolas, is chosen which fits within the longitudinal feasible zone of Fig. 9.33(a). Note that there is little latitude in the location of the tendon in this design and hence little allowance for inaccuracy in the placing of the tendon duct on site. Thus, it may be prudent to increase the depth (by 50 mm, say) at Sections 3 and 4. A feasible solution for the current section sizes is illustrated in Fig. 9.33(b).

9.5   Losses in prestress force

As stated in Section 9.3, the design of a prestressed member involves checking the stresses in the concrete at transfer and service due to the combination of applied loads and prestressing. Due to losses of force which occur in prestressing strands and tendons, the **actual** prestress force, P, that is transferred to the concrete is not generally equal to the applied jacking force, P_{jack}, nor is it constant along the length of the member. Therefore, in order to determine the actual stress due to prestress at transfer and service, the losses in prestress must first be calculated at each design section.

The losses that occur in prestressed members can be divided into two groups in accordance with the time when they occur. Losses which occur during the prestressing process are collectively known as **immediate** losses, ΔP_i. Thus, the prestress force at section x at transfer, $P_0(x)$, is the jacking force, P_{jack}, minus the immediate losses for that section, that is:

$$\text{Immediate losses of force at section } x, \ \Delta P_i(x) = P_{jack} - P_0(x) \tag{9.32}$$

Losses which occur after the prestress is transferred to the concrete are collectively known as **time-dependent** losses, ΔP_{c+s+r} (c + s + r = creep + shrinkage + relaxation). If the prestress force at section x is reduced to a final value of $\rho P_0(x)$ after all losses have occurred, then the total time-dependent loss is equal to:

$$\Delta P_{c+s+r} = P_0(x) - \rho P_0(x) \tag{9.33}$$

In pre-tensioned members, the main source of immediate losses is from elastic shortening or deformation of the concrete. In the case of post-tensioned members, immediate losses result from elastic deformation, draw-in at anchorages and from friction between the tendons and the surrounding ducts. Time-dependent losses are caused by relaxation of the steel and by creep and shrinkage of the concrete.

Friction losses

The loss of prestress force due to friction generally arises only in post-tensioned members in which the prestressing tendons are surrounded by ducts. When such tendons are profiled, the loss of force comes from two sources of friction (Fig. 9.34):

(a)   friction due to curvature of the tendon; and
(b)   friction due to unintentional variation of the duct from its prescribed profile or 'wobble'.

The magnitude of the curvature loss is dependent on the extent of the curvature (i.e., the greater the curvature, the greater the loss). Specifically, the loss between the jack and section x along the length of the beam is an exponential function of the change in angle, θ (Fig. 9.35), between the two points. Thus, if the tendon slopes first upwards and then downwards, the corresponding changes in angle are additive.

The magnitude of the 'wobble' loss at a section depends not on the curvature of the tendon but on the distance of the section from the jacking end. The total loss in prestress force at section x due to friction is:

$$\text{total friction loss, } \Delta P_\mu(x) = P_{\text{jack}}\left(1 - e^{-\mu(\theta + kx)}\right) \tag{9.34}$$

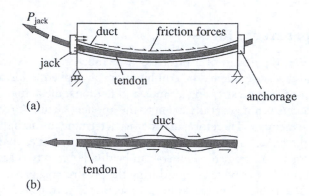

(a)

(b)

Figure 9.34 Friction losses: (a) curvature friction; (b) 'wobble' friction (exaggerated)

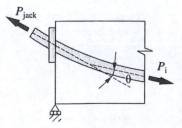

Figure 9.35 Friction curvature loss

where μ is a friction coefficient between the tendon and the duct, θ is the aggregate change in slope in radians between the jack and section x (friction curvature loss), k is an unintentional angular displacement or 'wobble' coefficient which depends on the quality of workmanship, the distance between tendon supports, the degree of vibration used in placing the concrete and the type of duct ('wobble' friction loss). The term x is the distance (in metres) from the jack to section x. EC2 recommends, in the absence of more exact data from European Technical Approval documents that μ for strands be taken as 0.19. The value for k is generally in the range 0.005 to 0.01 per metre. EC2 suggests that design values for k be taken from European Technical Approval documents.

Example 9.6 Friction losses

Problem: For the post-tensioned bridge deck of Fig. 9.36, determine the total friction losses at sections 2 and 3. Assume that the deck is jacked from both ends to a tension of 40,000 kN. Take $\mu = 0.19$ and $k = 0.01/m$. All tendons have the same profile for which details are given in Fig. 9.37 for one-half of the deck.

Solution: The slope at Section no. 1 is found by differentiation of the equation for line A:

$$y = -0.03x$$

$$\Rightarrow \frac{dy}{dx} = -0.03$$

The slope in radians is $\tan^{-1}(-0.03) = -0.03$ (the angle is very small). The slope at Section no. 2 is found from the equation for parabola B:

$$y = 1.406 \times 10^{-6}x^2 - 0.03375x + 2.5$$

$$\Rightarrow \frac{dy}{dx} = 2.812 \times 10^{-6}x - 0.03375$$

$$\Rightarrow \frac{dy}{dx}(x = 12000) = 2.812 \times 10^{-6}(12000) - 0.03375 = 0$$

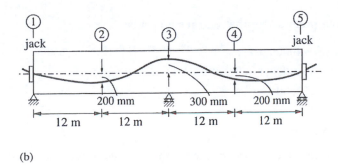

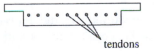

(a)

200 mm   300 mm   200 mm

12 m   12 m   12 m   12 m

(b)

Figure 9.36 Post-tensioned bridge deck: (a) typical cross-section; (b) elevation showing profile of tendons

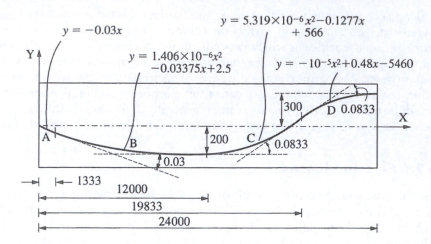

Figure 9.37 Profile details (x and y in mm)

Hence the change in slope between sections 1 and 2 is 0.03. The total friction loss at Section 2 is thus:

$$\text{Total friction loss} = P_{\text{jack}}(1 - e^{-\mu(\theta + kx)})$$
$$= 40000\left(1 - e^{-0.19(0.03 + 0.01\,(12))}\right)$$
$$= 40000(1 - 0.972)$$
$$= 1120\,\text{kN}$$

Where parabolas C and D meet, the slope is found from the equation for parabola C:

$$y = 5.319 \times 10^{-6}x^2 - 0.1277x + 566$$
$$\Rightarrow \frac{dy}{dx} = 10.638 \times 10^{-6}x - 0.1277$$
$$\Rightarrow \frac{dy}{dx}(x = 19833) = 10.638 \times 10^{-6}(19833) - 0.1277 = 0.0833$$

By differentiation of the equation for parabola D, the slope at Section no. 3 can be shown to be zero. Thus the aggregate change in angle between Sections 1 and 3 is:

$$\text{agg. change in angle} = 0.03 + 0.0833 + 0.0833$$
$$= 0.1966$$

Hence, the total friction loss over the central support is:

$$\text{total friction loss} = 40000\left(1 - e^{-0.19(0.1966 + 0.01\,(24))}\right)$$
$$= 40000(1 - 0.920)$$
$$= 3200\,\text{kN}$$

Elastic deformation losses

As the prestress is transferred to the concrete, the concrete undergoes elastic deformation (shortening) which reduces the length of the member. This can cause a slackening of the strand which results in a loss of prestress force.

Pre-tensioned members

In pre-tensioned members all the strands are stressed prior to the casting of the concrete and are released after the concrete has set. Upon release of the strands, a force P_{jack} is applied to the concrete. For a statically determinate member with eccentric strands, the stress distribution in the concrete due to prestress is illustrated in Fig. 9.38. At the level of the strands $y = e$, and the stress in the concrete adjacent to the strands is:

$$f_c = \frac{P_{\text{jack}}}{A_g} + \frac{(P_{\text{jack}}e)e}{I_g} + \frac{M_0 e}{I_g}$$

$$= P_{\text{jack}}\left(\frac{1}{A_g} + \frac{e^2}{I_g}\right) + \frac{M_0 e}{I_g}$$

Prestressing causes a strain in the concrete adjacent to the strands of:

$$\varepsilon_c = \frac{P_{\text{jack}}}{E_{\text{cm}}}\left(\frac{1}{A_g} + \frac{e^2}{I_g}\right) + \frac{M_0 e}{E_{\text{cm}} I_g}$$

where E_{cm} is the secant modulus of elasticity of concrete. This reduces the strain in the strands by the same amount, resulting in a loss of stress in the strands of:

$$\text{Loss of stress} = E_p \varepsilon_c$$

$$= P_{\text{jack}} \frac{E_p}{E_{\text{cm}}}\left(\frac{1}{A_g} + \frac{e^2}{I_g}\right) + E_p\left(\frac{M_0 e}{E_{\text{cm}} I_g}\right)$$

where E_p is the modulus of elasticity of the strands. From the above expression, the loss of prestress force due to elastic deformation in pre-tensioned members is:

$$\text{elastic deformation loss of force} = P_{\text{jack}} A_p \frac{E_p}{E_{\text{cm}}}\left(\frac{1}{A_g} + \frac{e^2}{I_g}\right) + A_p E_p\left(\frac{M_0 e}{E_{\text{cm}} I_g}\right) \tag{9.35}$$

where A_p is the total cross-sectional area of the strands.

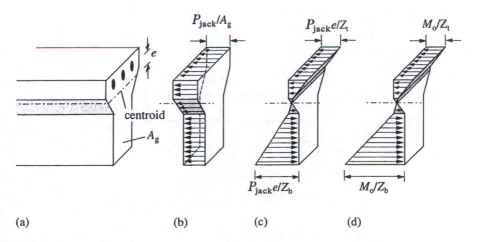

(a)   (b)   (c)   (d)

Figure 9.38 Elastic deformation loss.
Note: All the stress distributions are shown positive. In practice e and M_0 would generally be of opposite sign

Elastic deformation is an immediate loss. This is because the loss occurs during the prestressing process. The development of stress with time in the split second after a pre-tensioned strand is cut is illustrated in Fig. 9.39(a) for one strand and in Fig. 9.39(b) for a system with four strands stressed sequentially. It can be seen that the full jacking stress (before loss) is never actually applied to the concrete.

Furthermore, for an exact calculation of elastic deformation loss, the net force after this loss should be used rather than f_c. However, Equation (9.35) is conservative and is considered to be a good approximation.

Post-tensioned members

For post-tensioned members with a single tendon in which all the strands of the tendon are stressed simultaneously (using a multi-strand jack), elastic deformation (shortening) of the concrete occurs during the jacking process. However, since the deformations occur during the stressing process, additional force can be applied at the jack to compensate for loss. Thus, on the anchoring of the tendon, in such cases, there is no elastic deformation loss of prestress.

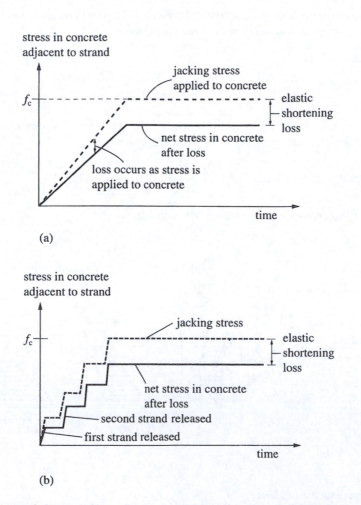

Figure 9.39 Development of elastic deformation loss with time: (a) one strand; (b) four strands

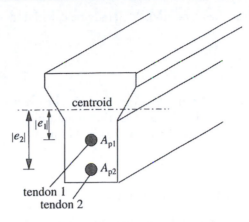

Figure 9.40 Elastic deformation of post-tensioned beam

For post-tensioned members with more than one tendon which are stressed sequentially, there is a loss in prestress due to elastic deformation. Consider the member of Fig. 9.40 which has two tendons. Assume tendon 1 is stressed first to a force P_1. Additional force is applied to tendon 1 during stressing to compensate for the elastic deformation loss due to this prestress before the strands are locked off. However, when tendon 2 is then stressed to a force P_2, there is a loss of force in tendon 1 because tendon 2 causes the concrete adjacent to tendon 1 to be shortened. The stress in the concrete adjacent to tendon 1 due to tendon 2 is:

$$f_{c1} = \frac{P_2}{A_g} + \frac{(P_2 e_2) e_1}{I_g}$$

Therefore, the loss of force in tendon 1 is equal to:

$$\text{Elastic deformation loss in tendon 1} = P_2 A_{pl} \frac{E_p}{E_{cm}} \left(\frac{1}{A_g} + \frac{e_1 e_2}{I_g} \right) \tag{9.36}$$

For the general case of a post-tensioned member with many tendons, the losses due to elastic deformation in the tendons range from zero in the last tendon stressed to a maximum in the first stressed. For a member with n identical tendons at the same eccentricity, stressing the first generates no loss. Stressing the second generates a loss, ΔP_1, of:

$$\Delta P_1 = \frac{P}{n} \frac{A_p}{n} \frac{E_p}{E_{cm}} \left(\frac{1}{A_g} + \frac{e^2}{I_g} \right)$$

in the first tendon, where P is the total prestress force for all tendons and A_p is the total area of prestressing tendon. Stressing the third brings the loss in the first tendon to $2\Delta P_1$ and generates a loss of ΔP_1 in the second. When the final tendon is stressed it brings the loss in the first tendon to $(n-1)\Delta P_1$. Hence the total loss in an n-tendon system is $\Delta P_1(1 + 2 + 3 + \ldots + (n-1))$, which can be shown to equal $\Delta P_1[n(n-1)/2]$.

On this basis, EC2 introduces a factor j for the calculation of elastic deformation loss in a post-tensioned member, where j is given by:

$$j = (n-1)/2n \tag{9.37}$$

and the total loss due to elastic deformation, ΔP_{el}, may be calculated as:

$$\Delta P_{el} = jPA_p \frac{E_p}{E_{cm}} \left(\frac{1}{A_g} + \frac{e^2}{I_g} \right) \tag{9.38}$$

As an approximation, the value of j may be taken as 0.5. The prestress force, P, used in Equation (9.38) is the force net of friction losses; that is, the magnitude of the jacking forces should be reduced by amounts equal to the total friction losses before calculation of elastic deformation losses.

Example 9.7 Elastic deformation loss

Problem: The bridge of Fig. 9.36 contains ten tendons of identical profile each comprising 20 strands of area 150 mm². Given that multi-strand jacks are used to stress the tendons in sequence, determine the total loss due to elastic deformation at the central support (section 3) and at mid-span (section 2). The jacking force and friction losses are as calculated in Example 9.6. Take $E_{cm} = 35{,}000$ N/mm², $E_p = 195{,}000$ N/mm² and ignore secondary effects (Section 9.6). For the whole bridge, the geometric properties are $I_g = 0.6 \times 10^{12}$ mm⁴ and $A_g = 5.2 \times 10^6$ mm².

Solution: From Fig. 9.36, the eccentricity of the tendons over the central support is 300 mm. From Example 9.6, the total friction loss at this section is equal to 3,200 kN. Thus, the net total force for all tendons at the central support of the beam is $40{,}000 - 3{,}200 = 36{,}800$ kN. From Equation (9.37):

$$j = (10 - 1)/2(10) = 0.45$$

Thus, the elastic deformation loss at the central support of the beam is calculated as:

$$
\begin{aligned}
\Delta P_{el} &= jPA_p \frac{E_p}{E_{cm}} \left(\frac{1}{A_g} + \frac{e^2}{I_g} \right) \\
&= (0.45)(36800)(10 \times 20 \times 150) \left(\frac{195000}{35000} \right) \left(\frac{1}{5.2 \times 10^6} + \frac{(300)^2}{0.6 \times 10^{12}} \right) \\
&= 948 \text{ kN}
\end{aligned}
$$

Similarly, the elastic deformation loss at the centre of the first span is found to be 757 kN. From the results of Examples 9.6 and 9.7, the prestress force net of friction and elastic deformation losses can be calculated at both mid-span and central support. At mid-span (Section 2 along the beam), the friction loss is 1,120 kN and the elastic deformation loss is 757 kN. Thus:

$$P_{(\text{section } 2)} = 40000 - (1120 + 757) = 38123 \text{ kN}$$

Over the central support (Section 3 along the beam), the friction loss is 3,200 kN and the elastic deformation loss is 948 kN. Thus:

$$P_{(\text{section } 3)} = 40000 - (3200 + 948) = 35852 \text{ kN}$$

Draw-in losses

After jacking the post-tensioned members to the required force, P_{jack}, the tendons are released from the jacks and the force is applied directly through the anchorages to the concrete. When wedge anchors are used, this process results in a loss of prestress due to slippage or 'draw-in' of the strands at the anchors, as illustrated in Fig. 9.41. In some instances, the movement at the anchors, Δ_s, can be up to 10 mm. Fortunately, the friction between the tendons and the ducts ensures that this loss of prestress does not extend very far from the region of the anchor in most cases, particularly in longer members. Consider the post-tensioned member of Fig. 9.42(a). If there were no friction between the tendon and the duct (i.e. if the duct were very slippery), the loss of prestress force due to draw-in of the wedges would be constant over the entire length of the member, as illustrated in Fig. 9.42(b), and would be given by:

$$\Delta P = \frac{\Delta_s}{L} E_p A_p \tag{9.39}$$

where L is the length of the member.

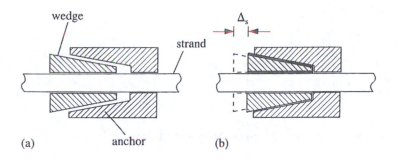

Figure 9.41 Draw-in of wedges at anchors: (a) before anchoring; (b) after anchoring

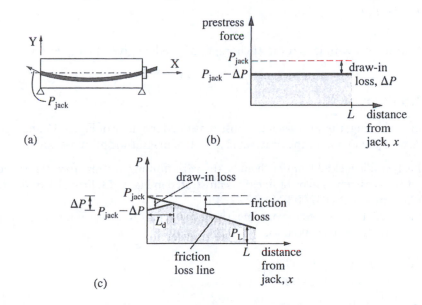

Figure 9.42 Draw-in loss: (a) beam elevation; (b) variation in applied prestress force – no friction; (c) variation in applied prestress force with friction

Consider now the real case of a similar member with friction. The variation in loss due to friction can be estimated as a linear function such as, for example, that illustrated in Fig. 9.42(c). The draw-in of the tendon results in a loss of strain of ε_{jack} at the anchorage. However, because of friction, this loss decreases along the length of the member to zero at a distance L_d from the jack. Thus, the total draw-in deformation of the tendon is:

$$\Delta_s = \frac{1}{2}\varepsilon_{jack}L_d$$

And the loss of force at the anchor is:

$$\Delta P = \varepsilon_{jack}E_pA_p$$
$$\Rightarrow \Delta P = \left(\frac{2\Delta_s}{L_d}\right)E_pA_p$$

As illustrated in the Fig. 9.42(c), the slope of the draw-in line is equal to that of the friction loss line, but of opposite sign, that is:

$$\frac{\Delta P/2}{L_d} = \frac{P_{jack} - P_L}{L}$$
$$\Rightarrow \Delta P = \frac{2(P_{jack} - P_L)}{L}L_d \tag{9.40}$$

Equating these two expressions for ΔP gives the extent of draw-in losses:

$$L_d = \sqrt{\frac{\Delta_s E_p A_p}{(P_{jack} - P_L)/L}} \tag{9.41}$$

The magnitude of the draw-in loss at the anchorage, ΔP, is then given by Equation (9.40).

Example 9.8 Draw-in loss

Problem: Upon anchoring, the strands of the post-tensioned member of Fig. 9.36 undergo a draw-in of 6 mm. Determine the extent of the draw-in loss and its magnitude at mid-span.

Solution: As this beam is jacked from both ends, the variation in prestress force after friction losses but prior to draw-in is assumed to be bi-linear, as illustrated in Fig. 9.43. From Example 9.6, the total friction loss at Section no. 3 is 3200 kN.

The friction loss will be assumed to vary linearly from the jack to this point, that is, the slope of the distribution of prestress force is taken as:

$$\text{slope} = \left(\frac{P_{jack} - P_L}{L}\right) = \left(\frac{3200000}{24000}\right)$$
$$= 133.3 \text{ N/mm}$$

Note: This is a rather crude assumption, and can sometimes result in significant inaccuracy.

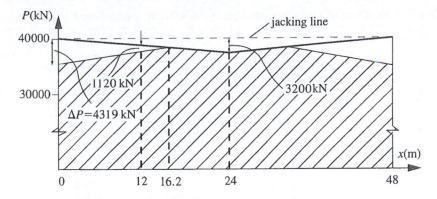

Figure 9.43 Distribution of prestress force in beam of Example 9.8

From Equation (9.41), the extent of the draw-in loss is:

$$L_d = \sqrt{\frac{\Delta_s E_p A_p}{(P_{jack} - P_L)/L}}$$
$$= \sqrt{\frac{6 \times 195000 \times (20 \times 10 \times 150)}{133.3}}$$
$$= 16227 \text{ mm}$$
$$= 16.2 \text{ m}$$

From Equation (9.40) the draw-in loss at the anchors is:

$$\Delta P = 2\frac{(P_{jack} - P_L)}{L} L_d$$
$$= 2(133.3)(16.2)$$
$$= 4319 \text{ kN}$$

From interpolation in Fig. 9.43, the draw-in loss at section no. 2 ($L = 12$ m) is given by:

draw-in loss $= 4,319 \ (16.2 - 12)/16.2 = 1,120$ kN

With reference to Fig. 9.43, the net force after friction, elastic deformation and draw-in is: $40,000 - (1,120 + 757 + 1,120) = 37,003$ kN.

Time-dependent losses

Losses in prestress that occur gradually over time are caused by deformation of the concrete due to creep and shrinkage and due to relaxation of the prestressing steel.

Shrinkage is a time-dependent strain that occurs as the concrete sets and for a period after setting. The shrinkage strain approaches a final value, ε_{cs} at infinite time. The loss of prestress force is generally constant over the entire length of the member and can be estimated as:

$$P_{cs} = \varepsilon_{cs} E_p A_p \qquad (9.42)$$

The factors which affect the magnitude of the shrinkage strain include the relative humidity during curing, the type of cement used, the concrete class and the shape and dimensions of the member.

Table 9.2 Final shrinkage strains for prestressed concrete for Class N cement

Class	Relative humidity (%)	Notional size (mm)			
		100	*200*	*300*	*≥500*
C20/25	40	0.605×10^{-3}	0.518×10^{-3}	0.460×10^{-3}	0.431×10^{-3}
	60	0.515×10^{-3}	0.442×10^{-3}	0.393×10^{-3}	0.368×10^{-3}
	80	0.325×10^{-3}	0.280×10^{-3}	0.250×10^{-3}	0.235×10^{-3}
C40/50	40	0.535×10^{-3}	0.466×10^{-3}	0.420×10^{-3}	0.397×10^{-3}
	60	0.455×10^{-3}	0.398×10^{-3}	0.360×10^{-3}	0.341×10^{-3}
	80	0.315×10^{-3}	0.279×10^{-3}	0.255×10^{-3}	0.243×10^{-3}
C60/75	40	0.485×10^{-3}	0.431×10^{-3}	0.395×10^{-3}	0.377×10^{-3}
	60	0.425×10^{-3}	0.380×10^{-3}	0.350×10^{-3}	0.335×10^{-3}
	80	0.315×10^{-3}	0.286×10^{-3}	0.268×10^{-3}	0.258×10^{-3}

Shrinkage is higher in members with a relatively large surface area, i.e. members with a large surface area per unit volume. For cross-sections, this becomes surface perimeter per unit area. EC2 defines a 'notional size' as twice the ratio of the cross-sectional area, A_g, to the perimeter length of the section, u. Values of total final shrinkage strains, based on recommendations in EC2 for the design of prestressed members, are given in Table 9.2. Shrinkage reduces with increasing notional size. It is also less when cured in humid conditions.

When maintained at a constant tensile strain, steel gradually loses its stress with time due to **relaxation**. In stringed musical instruments, for example, relaxation could contribute to strings going out of tune. This phenomenon is caused by the realignment of the steel fibres under stress. It is similar to creep except for the condition of constant strain (creep occurs with constant stress). The extent of the loss of stress in prestressing strands due to relaxation is determined by the stress to which the steel is tensioned, the ambient temperature and the class of steel. The loss of prestress which is caused by steel relaxation can range from 3 per cent to about 12 per cent of the initial force.

EC2 suggests that the loss due to relaxation be based on the 1,000 hour values (ρ_{1000}) for the three different classes of relaxation given in the code. Low relaxation wires or strands, that are commonly used, are covered by class 2. In the absence of relaxation losses from manufacturers' test certificates, the loss in prestress due to relaxation, $\Delta\sigma_{pr}$, for class 2 wires or strands can be calculated using:

$$\frac{\Delta\sigma_{pr}}{\sigma_{pi}} = 6.6\rho_{1000}e^{9.1\mu}\left(\frac{t}{1000}\right)^{0.75\,(1-\mu)} \times 10^{-6} \tag{9.43}$$

where σ_{pi} is the stress in the tendons after immediate losses, t is the time after tensioning (hours), $\mu = \sigma_{pi}/f_{pk}$, f_{pk} is the characteristic tensile strength of the prestressing steel and $\rho_{1,000}$ is the relaxation loss 1,000 hours after tensioning for a mean temperature of 20° C (can be assumed to be 2.5 per cent for class 2). According to EC2, the final or long term losses due to relaxation can be estimated for a time (after tensioning) of 500,000 hours.

The phenomenon of **creep** is essentially the same as that of relaxation. The distinction is that relaxation refers to the loss of stress under constant strain while creep is the increase of strain which occurs at constant stress. For the purposes of prestressed concrete, relaxation occurs in the steel while creep occurs in the concrete. Creep of concrete is unpredictable and can be quite substantial in prestressed members where the stress is kept constant for the design life of the structure. As for shrinkage, creep in the concrete allows the prestressing strands to slacken which results in the loss of force.

Like shrinkage strain, creep strain increases with time and approaches a final value ε_∞, at infinite time. Creep strain in concrete is usually measured in terms of the final creep coefficient, φ, which is the ratio of creep strain to linear elastic strain, that is:

$$\varphi = \frac{\varepsilon_\infty}{\sigma_c/E_c} \tag{9.44}$$

where ε_∞ is the final creep strain, σ_c is the elastic stress in the concrete and E_c is the tangent modulus of elasticity of concrete (can be taken as $1.05E_{cm}$). The magnitude of the creep coefficient depends on the size of the member, the relative humidity and on the age (or strength) of the concrete when it is first loaded. Values of the final creep coefficient, recommended by EC2 for the calculation of the total creep strain in prestressed members (and non-prestressed members) are given in Table 8.1. The final creep strain, ε_∞, at the level of the tendons is then given by:

$$\varepsilon_\infty = \frac{\varphi\sigma_c}{E_c} \tag{9.45}$$

where σ_c is the stress in the concrete at the level of the tendons due to prestress and quasi-permanent loads. From Equation (9.45), the loss in prestress force due to creep is:

$$\text{creep loss} = A_p E_p \varepsilon_\infty = A_p \frac{E_p}{E_c}\varphi\sigma_c \tag{9.46}$$

where:

$$\sigma_c = \frac{P}{A_g} + \frac{(Pe + M_{qp})e}{I_g} \tag{9.47}$$

where P is the prestress force in the tendons after immediate losses and M_{qp} is the moment due to quasi-permanent loads.

Rather than calculate each time-dependent loss separately, EC2 recommends that the losses be calculated all together using a rather expansive formula:

$$\Delta\sigma_{p,c+s+r} = \frac{\varepsilon_{cs}E_p + 0.8\Delta\sigma_{pr} + \alpha_e\varphi\sigma_c}{1 + \alpha_e \dfrac{A_p}{A_c}\left(1 + \dfrac{A_c e^2}{I_c}\right)(1 + 0.8\varphi)} \tag{9.48}$$

where: $\Delta\sigma_{p,c+p+r}$ = loss of stress in steel due to creep, shrinkage and relaxation
ε_{cs} = final shrinkage strain taken from Table 9.2
E_p = modulus of elasticity of the prestressing steel
$\Delta\sigma_{pr}$ = loss of stress in the tendons at the design section due to relaxation of the prestressing steel
α_e = modular ratio, E_p/E_{cm}
φ = final creep coefficient taken from Table 8.1
σ_c = stress in the concrete at the level of the tendons due to quasi-permanent loads plus prestress
A_p = area of prestressing steel
A_c, I_c = area and second moment of area of concrete section
e = eccentricity of tendons from centroid of section

The following example illustrates the application of the EC2 formula.

Note: The loss of prestress due to relaxation depends on the effects of creep and shrinkage on the concrete. To take this interaction into account, as an approximation, a factor of 0.8 is included in the EC2 formula (Equation (9.48)).

Example 9.9 Time-dependent losses

Problem: For the bridge of Fig. 9.36, determine the total loss of prestress force at Section no. 2 due to all effects. In addition to the data given in Examples 9.6–9.8, the following additional values may be assumed: final creep coefficient $\varphi = 1.7$, final shrinkage strain $\varepsilon_{cs} = 0.243 \times 10^{-3}$, moment at Section no. 2 due to quasi-permanent loads $M_{qp} = 6000$ kNm, characteristic tensile strength of tendons $f_{pk} = 1770$ N/mm².

Solution: The stress in the tendons after immediate losses, σ_{pi}, is given by:

$$\sigma_{pi} = \frac{P}{A_p} = \frac{37003 \times 10^3}{(10 \times 20 \times 150)}$$
$$= 1233 \text{ N/mm}^2$$

Hence the ratio of stress to characteristic strength, μ, is:

$$\mu = \sigma_{pi}/f_{pk} = 1233/1770 = 0.697$$

The loss in prestress due to relaxation, $\Delta\sigma_{pr}$, for class 2 strands is calculated as:

$$\Delta\sigma_{pr} = \sigma_{pi} 6.6 \rho_{1000} e^{9.1\mu} \left(\frac{t}{1000}\right)^{0.75\,(1-\mu)} \times 10^{-6}$$
$$= 1233(6.6)(2.5)e^{9.1(0.697)} \left(\frac{500000}{1000}\right)^{0.75(1-0.697)} \times 10^{-6}$$
$$= 1233(0.0385)$$
$$= 47.5 \text{ N/mm}^2$$

The stress, σ_c, in the concrete at the level of the tendons due to quasi-permanent loads plus prestress is calculated using Equation (9.47).

$$\sigma_c = \frac{P}{A_g} + \frac{(Pe + M_{qp})e}{I_g}$$
$$= \frac{37003 \times 10^3}{5.2 \times 10^6} + \frac{(37003 \times 10^3(-200) + 6000 \times 10^6)(-200)}{0.6 \times 10^{12}}$$
$$= 7.58 \text{ N/mm}^2$$

The total time-dependent loss of stress in the tendons is:

$$\Delta\sigma_{p,c+s+r} = \frac{\varepsilon_{cs}E_p + 0.8\Delta\sigma_{pr} + \alpha_e\varphi\sigma_c}{1 + \alpha_e \dfrac{A_p}{A_c}\left(1 + \dfrac{A_c e^2}{I_c}\right)(1 + 0.8\varphi)}$$
$$= \frac{0.243 \times 10^{-3}(195000) + (0.8)47.5 + (195000/35000)(1.7)(7.58)}{1 + \dfrac{195000}{35000}\dfrac{(200 \times 150)}{5.2 \times 10^6}\left(1 + \dfrac{5.2 \times 10^6(-200)^2}{0.6 \times 10^{12}}\right)(1 + 0.8 \times 1.7)}$$
$$= 142.6 \text{ N/mm}^2$$

Thus, the total loss of force at Section no. 2 due to time-dependent effects is:

$$\text{total time loss} = 142.6 \times (10 \times 20 \times 150)$$
$$= 4278 \text{ kN}$$

From the results of Examples 9.8 and 9.9, the applied prestress at service, ρP, can now be calculated at that section:

$$\rho P = P - 4278$$
$$= 37003 - 4278$$
$$= 32725 \text{ kN}$$

Hence, the ratio of the prestress force at service to the prestress force at transfer, ρ, is:

$$\rho = 32725/37003 = 0.88$$

9.6   Secondary effects of prestress

Up to now, the moment due to prestress, M_p, has been assumed to equal the product of the prestress force and the eccentricity, Pe. However, this assumption is only true for statically determinate prestressed structures. For statically indeterminate structures, M_p consists of two components, **primary moments** (Pe) and **secondary moments**. Secondary moments arise from reactions developed at supports during prestressing. For example, in the beam of Fig. 9.44(a), stressing the tendon causes the beam to sag and results in a downward deflection at B. It also generates the distribution of moment, known as primary moment, illustrated in Fig. 9.44(b). If this tendency to deform is prevented, such as by the central support in the two-span beam of Fig. 9.44(c–d), a reaction is caused at B which generates the distribution of bending moment illustrated in Fig. 9.44(e). The moments in this distribution are known as secondary or parasitic moments. The total distribution of moment for this beam is the sum of the primary and secondary distributions and is illustrated in Fig. 9.44(f). Since statically determinate members do not provide restraint to imposed deformations, secondary moments only apply to indeterminate structures. Although these moments are termed 'secondary', they cannot be ignored as they can often have a substantial effect on the distribution of stress in a continuous member.

The magnitude of secondary moments in a given member depends on a number of factors, including tendon curvature and geometry changes in non-prismatic members. For a qualitative understanding of secondary effects, it is useful to consider the concept of concordant tendons as described, for example, by Nawy (2003). Here, we consider one practical method for the determination of the prestress moment, M_p, in an indeterminate member known as the **equivalent load method**. In this method, the forces applied to the concrete by the prestress are represented as externally applied loads. The structure is then analysed to obtain the total (primary plus secondary) moments due to these loads. Once these moments are known, the secondary moment at a section due to prestress can readily be calculated if required by subtracting the primary moment, Pe, from the total moment. An added advantage of the equivalent load method is that, once the loads have been evaluated, they can also be used to calculate deflections due to prestress (whether or not the structure is determinate). This is important in slender members which are sometimes pre-cambered in anticipation of large deflections due to prestress and permanent gravity loads.

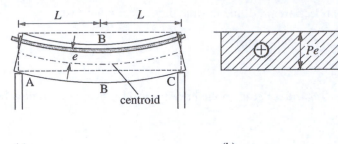

(a)

(b)

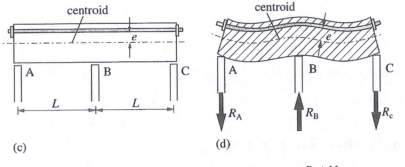

(c)

(d)

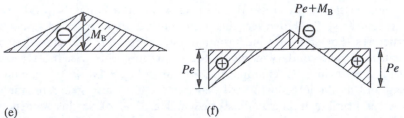

(e)

(f)

Figure 9.44 Effect of indeterminacy on prestress: (a) deformed shape of simply supported beam; (b) distribution of moment for (a), and distribution of primary moment for (c); (c) two-span beam (indeterminate); (d) deformed shape of two-span beam; (e) distribution of secondary moment; (f) total distribution of prestress moment, M_p, equal to sum of (b) and (e)

Calculation of equivalent loads

There are essentially four 'sources' of equivalent loading due to prestress in continuous members: tendon curvature, friction loss, end forces and moments, and equivalent loading due to geometry changes in beams of variable section.

Tendon curvature

In horizontal members which do not have a linear tendon profile, the prestressing tendon pushes against the concrete on the inside of the curve and, as a result, subjects the member to vertical forces. For example, tensioning of the tendon in the member illustrated in Fig. 9.45(a) results in the forces ω_1, ω_2 and ω_3 acting on the beam. Fig. 9.45(b) represents the free-body diagram for a small portion of the member of Fig. 9.45(a) over which the inclination of the tendon changes from θ_A to θ_B.

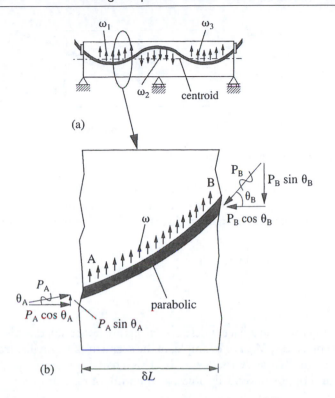

Figure 9.45 Equivalent loading due to cable curvature (a) two-span beam; (b) segment of beam

Referring to Fig. 9.45(b) and assuming that $P_A = P_B$, vertical equilibrium is satisfied with a total equivalent force, F, equal to:

$$F = P_A(\sin\theta_B - \sin\theta_A) \tag{9.49}$$

However, P_A and P_B are not equal as a result of friction and so the mean value of P_A and P_B is used in Equation (9.49) to yield:

$$F = \frac{P_A + P_B}{2}(\sin\theta_B - \sin\theta_A) \tag{9.50}$$

It should be noted that the expression for F, given by Equation (9.50), does not exactly satisfy equilibrium of vertical forces as:

$$P_B \sin\theta_B - P_A \sin\theta_A \neq \frac{P_A + P_B}{2}(\sin\theta_B - \sin\theta_A) \tag{9.51}$$

(refer to Fig. 9.45(b)). However, F is calculated using Equation (9.50) in order to satisfy equilibrium of moments as will be seen from Example 9.10. If the tendon profile takes the shape of a parabolic curve, the equivalent loading will be uniformly distributed over the length of the parabola. Thus, the equivalent uniformly distributed loading, ω, over the length δL is given by:

$$\omega = \frac{P_A + P_B}{2\delta L}(\sin\theta_B - \sin\theta_A) \tag{9.52}$$

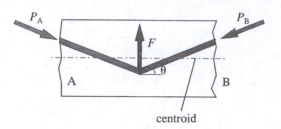

Figure 9.46 Sudden change in tendon shape

If a tendon profile changes shape sharply, such as in the symmetric member of Fig. 9.46, the equivalent loading can be represented by a concentrated load rather than by a uniformly distributed load as illustrated in that figure. The magnitude of the force, F, at the point where the slope changes is given by:

$$F = (P_A + P_B)\sin\theta \qquad (9.53)$$

Friction losses

When the prestressing forces in adjacent design sections are not equal, there is a consequent change in the moment due to prestress, M_p. For example, in the segment of beam illustrated in Fig. 9.47(a), the moment changes from $P_A e$ at A to $P_B e$ at B. This change in prestress can be modelled using equivalent, equal and opposite vertical forces, at A and B, of magnitude:

$$\frac{(P_A - P_B)e}{\delta L} \qquad (9.54)$$

The free-body diagram, which includes these forces, is illustrated in Fig. 9.47(b). Taking moments about A in this figure gives:

$$P_A e = P_B e + \frac{(P_A - P_B)e}{\delta L}\delta L$$

which is indeed true, showing that equilibrium is satisfied. When prestress forces are at different eccentricities, the mean eccentricity is used and the magnitude of the equivalent forces becomes:

$$\frac{(P_A - P_B)(e_A + e_B)}{2\delta L} \qquad (9.55)$$

as illustrated in Fig. 9.47(c). This is an approximation and, for accuracy in the results, the length δL between points where the equivalent forces are calculated should be reasonably small.

End forces and moments

In members with profiled tendons, the external force applied to the concrete by the prestress at the anchors is often inclined at an angle and/or located away from the centroid of the member. Consider the member illustrated in Fig. 9.48(a). The force, P, at the anchors can be resolved into three components: a horizontal force at the centroid ($P\cos\theta$), a vertical force ($P\sin\theta$) and a moment ($P\cos\theta$)e. As the angles involved are generally small, the horizontal force can be approximated by P, and the moment by Pe, as illustrated in Fig. 9.48(b). As for equivalent loads due to tendon curvature, an average value for prestress force may be used when it is varying due to friction.

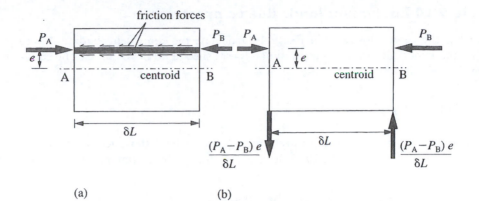

(a)   (b)

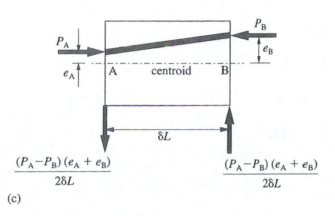

(c)

Figure 9.47 Equivalent loading due to friction losses: (a) segment with constant eccentricity; (b) free-body diagram (constant eccentricity); (c) segment with variable eccentricity

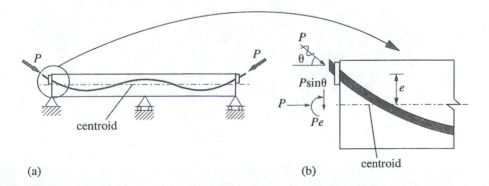

(a)   (b)

Figure 9.48 End forces and moments: (a) beam elevation; (b) segment of beam at anchor

Example 9.10 Equivalent loads due to prestress

Problem: For the beam illustrated in Fig. 9.49, calculate the equivalent loading due to prestress. Verify that moment equilibrium is (approximately) satisfied at points B and C. The equation for the parabola (in metres) is:

$$y = -0.8\left[\frac{x}{L} - \left(\frac{x}{L}\right)^2\right]$$

where the length, L, is 12 m. The prestress forces at transfer (after friction, draw-in and elastic deformation losses) are 2,400, 2,350, 2,300, 2,250 and 2,200 kN at points A, B, C, D and E, respectively.

Solution: The slope of the tendon is found by differentiation of the equation for the parabola:

$$\frac{dy}{dx} = -0.8\left(\frac{1}{L} - \frac{2x}{L^2}\right)$$

Hence the slopes at A, B and C are:

$$\text{at A}: \frac{dy}{dx}(x = 0) = -0.0667$$

$$\text{at B}: \frac{dy}{dx}\left(x = \frac{L}{4}\right) = -0.0333$$

$$\text{at C}: \frac{dy}{dx}\left(x = \frac{L}{2}\right) = 0$$

At A:

$$\sin\theta_A = \sin(\tan^{-1}(-0.0667))$$
$$= -0.0665$$

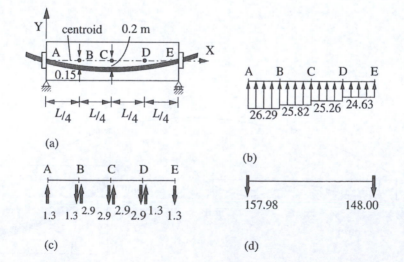

(a)

(b)

(c)

(d)

Figure 9.49 Prestressed beam of Example 9.10: (a) elevation; (b) equivalent loading due to tendon curvature (kN/m); (c) equivalent loading due to friction losses (kN); (d) equivalent loading due to end forces (kN)

The corresponding value at B is:

$$\sin \theta_B = -0.0333$$

Hence, the equivalent loads due to curvature between A and B are of magnitude:

$$\left(\frac{P_A + P_B}{2}\right)\frac{\sin \theta_B - \sin \theta_A}{\delta L} = \left(\frac{2400 + 2350}{2}\right)\frac{(0.0665 - 0.0333)}{3}$$
$$= 26.29 \text{ kN/m}$$

Similarly, the intensity between B and C, C and D and between D and E have been calculated and are as illustrated in Fig. 9.49(b).

The eccentricity at A is clearly zero. The eccentricity at B is:

$$y(x = L/4) = -0.8\left[\frac{1}{4} - \left(\frac{1}{4}\right)^2\right]$$
$$= -0.15 \text{ m}$$

Hence the equivalent forces at A and B due to friction losses between these points have magnitude:

$$\frac{(P_A - P_B)(e_A + e_B)}{2\delta L} = \frac{(2400 - 2350)(0 - 0.150)}{2 \times 3}$$
$$= -1.3 \text{ kN}$$

Similarly, the forces are calculated at each end of each segment of beam and are as illustrated in Fig. 9.49(c).

Finally, the equivalent loading due to end forces is calculated. In this case, there are end forces at A and E. Taking the average prestress force for the end segments, we get a force at A of:

$$\frac{(P_A + P_B)}{2}\sin \theta_A = \frac{(2400 + 2350)}{2}(-0.0665)$$
$$= -157.98 \text{ kN}$$

This and the force at E, found similarly, are illustrated in Fig. 9.49(d). All equivalent loading due to prestress on the beam is included in the free-body diagrams of Fig. 9.50. Taking moments about B in the diagram of Fig. 9.50(a) gives:

$$M_B = -(157.98 - 1.3) \times 3 + (26.29 \times 3) \times 1.5$$
$$= -352 \text{ kNm}$$

As the beam is determinate, this equals the product of prestress force and eccentricity at B:

$$M_B = 2350 \times (-0.150)$$
$$= -353 \text{kNm}$$

Similarly, taking moments about C in the free-body diagram of Fig. 9.50(b) gives:

$$M_C = -(157.98 - 1.3) \times 6 + 1.7 \times 3 + (26.29 \times 3) \times 4.5 + (25.82 \times 3) \times 1.5$$
$$= -464 \text{kNm}$$

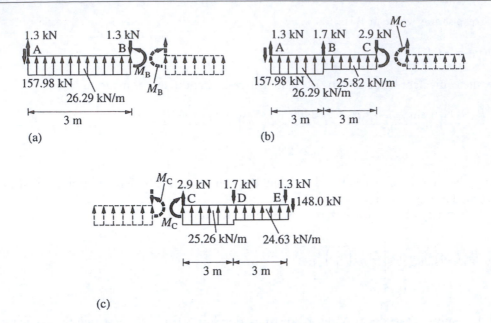

Figure 9.50 Free-body diagrams showing total equivalent loading due to prestress

Alternatively, taking moments about C in the free-body diagram of Fig. 9.50(c) gives $M_C = -455$ kNm. The product of prestress force and eccentricity at C is $(2300 \times -0.200) = -460$ kNm. Thus, the moment, as calculated using the equivalent loading, approximately equals the actual moment due to prestress, M_p.

Geometry change

The depth of prestressed members is frequently increased over internal supports in order to increase the second moment of area at points of peak moment. Fig. 9.51(a) illustrates a typical member with variable depth. In such non-prismatic members, the depth of the centroid below a horizontal datum line, D, varies along its length. This deviation has the effect of generating secondary bending moments in the member. The loading equivalent to a change in the depth, δD, of the centroid between two design sections is a pair of equal and opposite vertical forces of magnitude:

$$\frac{(P_A + P_B)\delta D}{2\delta L} \tag{9.56}$$

as illustrated in Fig. 9.51(b).

For indeterminate beams or slabs, analysis of the member under these loads, using a hand method such as moment distribution or a computer-based method, gives the total moments due to prestress at transfer. The moments due to prestress at service are then derived by reducing the transfer moments by an amount equal to the time-dependent losses. Simple factoring in this way is less accurate than calculating the service moments due to prestress using the equivalent loads. However, it is considered sufficiently accurate for design purposes.

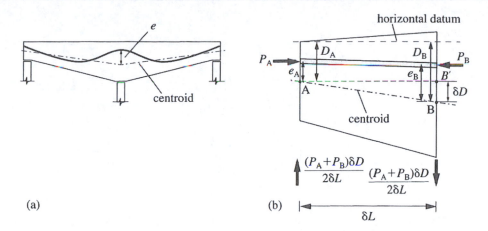

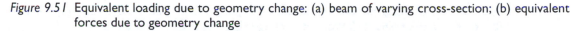

Figure 9.51 Equivalent loading due to geometry change: (a) beam of varying cross-section; (b) equivalent forces due to geometry change

9.7 Ultimate moment capacity of prestressed concrete

Once a prestressed member has been designed to satisfy stress limits at transfer and service, it must then be checked for other serviceability limit states (i.e. crack widths and deflection) and the ultimate limit states. The ultimate limit state checks consist of ensuring that the member has adequate strength in bending, shear and torsion under ultimate loads. This section deals only with the design of prestressed members for bending since the behaviour and design of prestressed members in shear and torsion is dealt with in Chapter 11.

As for ordinary reinforced concrete members, the ultimate moment capacity of a prestressed section is calculated by equilibrium of forces acting at the section. Consider the section of Fig. 9.52 (a), which has a single tendon at an eccentricity e from the centroid of the section. For members where the neutral axis is within the section, failure at the section under ultimate loads occurs when the strain in the extreme fibre in compression reaches its ultimate value, ε_{cu}. (*Note*: Refer to Section 10.4 for recommended strain distributions where the entire section is in compression.) In the absence of applied loads, the strain in the concrete is that caused by the prestress force (after all losses), ρP. The distribution of strain due to prestress (axial and bending components) for the section of Fig. 9.52(a) is illustrated in Fig. 9.52(b). The compressive strain in the concrete due to prestress, ε_{ce}, at the level of the prestressing tendon is given by:

$$\varepsilon_{ce} = \frac{1}{E_c}\left(\frac{\gamma_p \rho P}{A_g} + \frac{\gamma_p \rho M_p e}{I_g}\right) \tag{9.57}$$

where M_p is the moment due to prestress and γ_p is the partial factor for prestress ($\gamma_{P,fav} = 0.9$ for favourable actions, $\gamma_{P,unfav} = 1.1$ for unfavourable actions and $\gamma_{P,unfav} = 1.2$ for local effects). In a determinate structure, $M_p = Pe$. In an indeterminate structure, secondary effects will, in general, change M_p from Pe. However, plastic hinges may have formed in which case the structure could once again be treated as determinate. In the absence of a more detailed study, a conservative section design can be found by considering both possible values for M_p.

As the tendon in Fig. 9.52 is at a positive eccentricity, the moment due to prestress is positive (i.e. sagging) and the corresponding applied moment would normally be negative (i.e. hogging) as prestress is generally placed so as to oppose applied moment (causing a favourable effect). However, the less usual situation of positive eccentricity and positive moment is first considered in Fig. 9.52(c).

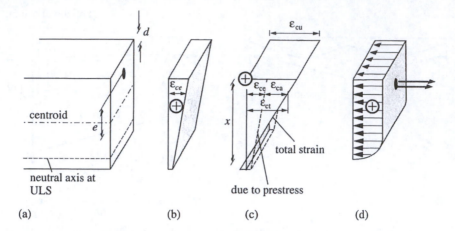

Figure 9.52 Strain and stress distributions for positive applied moment and eccentricity: (a) section;
(b) strain due to prestress; (c) total strain (prestress + applied); (d) total stress

The corresponding ultimate stress distribution is illustrated in Fig. 9.52(d). The strain in the tendon due to prestress alone at ULS is (contraction positive):

$$\text{tendon strain due to prestress} = -\frac{\gamma_p \rho P}{A_p E_p} \tag{9.58}$$

where A_p is the area of the prestressing tendon and E_p is its modulus of elasticity. When only prestress is present, the strain in the concrete adjacent to the tendon is ε_{ce}. When the ultimate moment is applied, the distribution of strain changes from this base level to that illustrated in Fig. 9.52(c). Thus, the compression in the concrete adjacent to the tendon increases by the strain due to applied moment, ε_{ca}, and the tendon slackens accordingly to an ultimate strain of:

$$\varepsilon_{pu} = -\frac{\gamma_p \rho P}{A_p E_p} + \varepsilon_{ca} \tag{9.59}$$

As the total strain in the concrete, ε_{ct}, can be determined by similar triangles and the strain due to prestress is known, it is convenient to express the tendon strain ε_{pu} as:

$$\varepsilon_{pu} = -\frac{\gamma_p \rho P}{A_p E_p} + \varepsilon_{ct} - \varepsilon_{ce} \tag{9.60}$$

The more usual case, where moments due to prestress and applied loads are of opposite sign, is illustrated in Fig. 9.53. As before, the strain in the tendon due to prestress alone is given by Equation (9.58) and the corresponding base-level concrete strain is given in Fig. 9.53(b). The applied moment has the effect of reversing the compression in the concrete adjacent to the tendon and causing a tension there, that is, ε_{ca} is negative (tensile) in Equation (9.59). The effect of the applied moment is, in fact, to increase the tension in the tendon. For example, if the strain in the tendon due to prestress is (factored and after losses) -4850×10^{-6}, $\varepsilon_{ce} = 350 \times 10^{-6}$ and $\varepsilon_{ct} = -200 \times 10^{-6}$, then $\varepsilon_{ca} = -200 \times 10^{-6} - 350 \times 10^{-6} = -550 \times 10^{-6}$ and the ultimate strain in the tendon will be:

$$\begin{aligned}
\varepsilon_{pu} &= -4850 \times 10^{-6} + (-550 \times 10^{-6}) \\
&= -5400 \times 10^{-6}
\end{aligned}$$

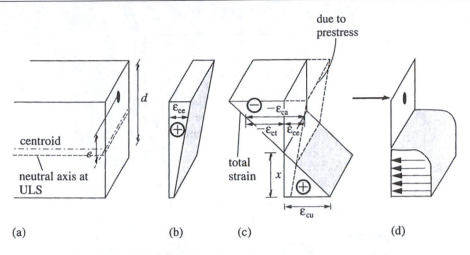

Figure 9.53 Strain and stress distributions for negative applied moment and positive eccentricity: (a) section: (b) strain due to prestress; (c) total strain (prestress + applied); (d) total stress

It is sometimes more convenient to express Equation (9.60) in terms of absolute values of strains. Thus, when prestress and applied moment are of opposite sign, Equation (9.60) can be written as:

$$\varepsilon_{pu} = -\frac{\gamma_p \rho P}{A_p E_p} + |\varepsilon_{ct}| - |\varepsilon_{ce}| \qquad (9.61)$$

By considering similar triangles in Fig. 9.53(c), the total ultimate strain, ε_{ct}, in the concrete at the level of the tendon is found to be (contraction positive):

$$\varepsilon_{ct} = -\frac{\varepsilon_{cu}(d - x)}{x} \qquad (9.62)$$

EC2 recommends that ε_{cu} be taken as 0.0035 (for $f_{ck} \leq 50$ N/mm²). Hence, Equation (9.62) becomes:

$$\varepsilon_{ct} = -\frac{0.0035(d - x)}{x} \qquad (9.63)$$

Once the magnitude of the strain in the steel at ULS has been established using Equation (9.61), the ultimate moment capacity of a section, M_{ult}, is determined by taking moments about, say, the neutral axis. (Note that the neutral axis at ULS is not located at the centroid as it is at SLS). EC2 allows any of the stress–strain relationships for concrete of Fig. 8.21 to be used for the calculation of the compressive force at the section. The following example illustrates the steps involved in the calculation of M_{ult}.

Example 9.11 Ultimate moment capacity

Problem: For the section illustrated in Fig. 9.54, determine the ultimate capacity to resist sag moment. Assume the EC2 equivalent rectangular stress block for concrete. The moduli of elasticity are $E_p = 195,000$ N/mm² and $E_c = 35,000$ N/mm², $f_{pk} = 1,770$ N/mm², $f_{p0.1k} = 1,520$ N/mm² and $f_{ck} = 40$ N/mm². The area of tendon is $A_p = 2,400$ mm² and it is stressed to a prestress force of 3,023 kN at transfer. The time-dependent losses are 21 per cent implying a loss ratio, ρ, of 0.79.

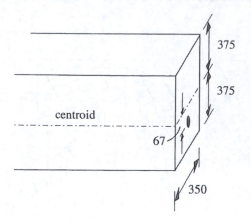

Figure 9.54 Section of Example 9.11

Solution: The stress and strain distributions are illustrated in Fig. 9.55. The gross second moment of area is given by:

$$I_g = \frac{350 \times 750^3}{12} = 12.3 \times 10^9 \text{ mm}^4$$

Since the prestress causes a favourable effect in this case, $\gamma_{P,fav} = 0.9$ is used. Assuming no secondary effects, Equation (9.57) gives the compressive strain in the concrete due to prestress, ε_{ce}, at the level of the prestressing tendon as:

$$\varepsilon_{ce} = \frac{1}{E_c}\left(\frac{\gamma_p \rho P}{A_g} + \frac{\gamma_p \rho M_p e}{I_g}\right)$$

$$= \frac{1}{35000}\left(\frac{0.9 \times 0.79 \times 3023 \times 10^3}{350 \times 750} + \frac{0.9 \times 0.79 \times 3023 \times 10^3 \times (-67)^2}{12.3 \times 10^9}\right)$$

$$= 256 \times 10^{-6}$$

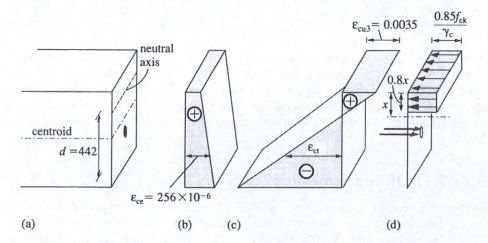

Figure 9.55 Strain and stress distributions for Example 9.11: (a) section; (b) strain due to prestress; (c) total strain; (d) total stress

From Equation (9.63), the total strain in the concrete at the level of the tendon at ULS is equal to:

$$\varepsilon_{ct} = -\frac{0.0035(d - x)}{x} = -\frac{0.0035(442 - x)}{x}$$

Hence, from Equation (9.60), the total strain in the tendon at the ultimate limit state is:

$$\varepsilon_{pu} = -\frac{\gamma_p \rho P}{A_p E_p} + \varepsilon_{ct} - \varepsilon_{ce}$$

$$= -\frac{0.9 \times 0.79 \times 3023 \times 10^3}{2400 \times 195000} - \frac{0.0035(442 - x)}{x} - 256 \times 10^{-6}$$

$$= -4849 \times 10^{-6} - \frac{0.0035(442 - x)}{x}$$

EC2 specifies that design may be carried out based on the actual stress/strain behaviour of a strand, or using either the horizontal or inclined branch in Fig. 9.56. Initially, assume the steel has not yielded (i.e. $\varepsilon_{pu} < f_{p0.1k}/\gamma_s E_p$). Hence, the total force in the tendon, F_p, is:

$$F_p = A_p(E_p \varepsilon_{pu})$$

$$= 2400(195000)\left(-4849 \times 10^{-6} - \frac{0.0035(442 - x)}{x}\right)$$

The compressive force, F_c, acting on the concrete in compression is (Fig. 9.55(d)):

$$F_c = 0.8x(350)\frac{0.85 f_{ck}}{\gamma_c} = 0.8x(350)\frac{0.85(40)}{1.5}$$

$$= 6347x$$

By equilibrium, $F_p + F_c = 0$. Thus:

$$2400(195000)\left(-4849 \times 10^{-6} - \frac{0.0035(442 - x)}{x}\right) + 6347x = 0$$

Multiplying by x and rearranging gives:

$$6.347x^2 - 631x - 723996 = 0$$

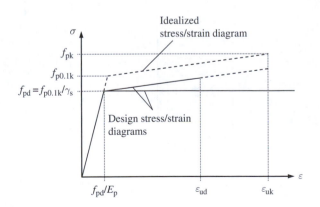

Figure 9.56 Typical stress/strain relationship for prestressing steel (from BS EN 1992-1-1)

The only positive root of this quadratic equation is:

$$x = 391 \text{ mm}$$

To check that the steel has not yielded, substitute $x = 391$ mm into the expression for ε_{pu} to give:

$$\varepsilon_{pu} = \left(-4849 \times 10^{-6} - \frac{0.0035(442 - x)}{x} \right)$$
$$= -0.0053$$

The yield strain (see Fig. 9.56) is $-f_{p0.1k}/\gamma_s E_p = -(1520)/(1.15 \times 195000) = -0.00678$.

Thus, the assumption that the steel had not yielded is correct. Hence, the ultimate moment capacity is:

$$M_{ult} = F_p z = F_c z = 6347 x z$$

From Fig. 9.55:

$$z = (d - 0.4x) = (442 - 0.4(391)) = 286 \text{ mm}$$

Thus:

$$M_{ult} = (6347)(391)(286) = 710 \times 10^6 \text{ Nmm}$$
$$= 710 \text{ kNm}$$

Problems

Section 9.3

9.1 For the simply supported post-tensioned beam of Fig. 9.57, check if the applied level of prestress satisfies concrete compressive stress limits and the decompression limit for buildings at mid-span. The 15 m long beam is post-tensioned by applying a prestress force of 1,500 kN.

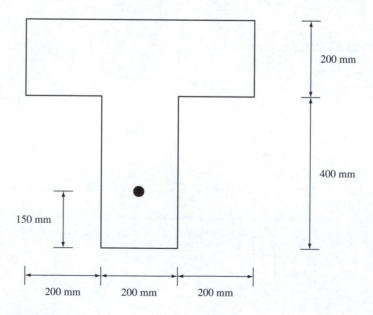

Figure 9.57 Beam of Problem 9.1

The tendon is bonded and has a diameter of 75 mm. The beam forms part of a structural scheme for a lecture theatre and has a permanent gravity load of 5 kN/m (not including self-weight) and a variable gravity load of 6 kN/m. Assume exposure class XC3, $f_{ck,0} = 35$ N/mm^2, $f_{ck} = 50$ N/mm^2 and $\rho = 0.8$.

Section 9.4

9.2 Show from first principles that the prestressed section of Fig. 9.58 is adequate, given exposure class XC3, $M_0 = 112$ kNm, $M_{qp} = 500$ kNm, $f_{ck,0} = 40$ N/mm^2, $f_{ck} = 55$ N/mm^2, $\rho = 0.8$ and $\phi = 70$ mm.

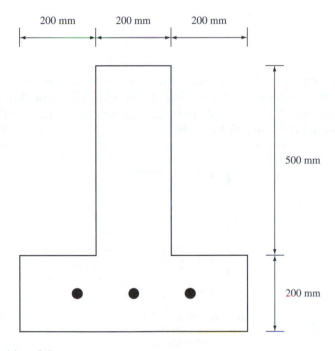

Figure 9.58 Beam of Problem 9.2

9.3 Construct the Magnel diagram for the beam of Fig. 9.57.

Section 9.5

9.4 Calculate the immediate and time-dependent losses for the beam illustrated in Fig. 9.59. The beam is post-tensioned with a prestress force of 2,200 kN using multi-strand jacking from both ends. Take $A_p = 1,800$ mm^2, $\mu = 0.19$ and $k = 0.01$/m, $\Delta_s = 6$ mm, $\varphi = 1.7$, $\varepsilon_{cs} = 0.243 \times 10^{-3}$, $M_{qp} = 450$ kNm and $f_{pk} = 1,770$ N/mm^2.

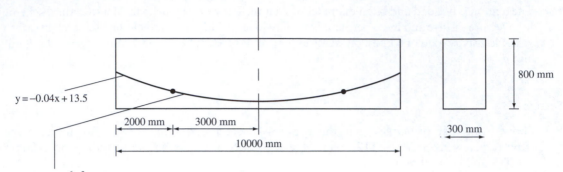

Figure 9.59 Beam of Problem 9.4

Section 9.7

9.5 Determine the ultimate moment capacity of the pre-tensioned beam illustrated in Fig. 9.60 given that the strands are stressed to 70 per cent of their ultimate strength at transfer and the total time-dependent loss is 18 per cent. Assume no secondary effects and use the stress/strain diagram for strand illustrated in Fig. 9.56.

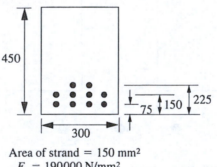

Area of strand = 150 mm²
E_p = 190000 N/mm²
f_{pk} = 1770 N/mm²
f_{ck} = 45 N/mm²

Figure 9.60 Beam of Problem 9.5

Chapter 10

Combined axial force and bending of reinforced concrete members

10.1 Introduction

Axially loaded members are classified as those that carry their load primarily in tension or compression. Since tension members are not commonly used in concrete structures (remember concrete is weak in tension), this chapter deals for the most part with the design of compression members. The majority of compression/tension members carry a portion of their load in bending. This may be due to the load not being applied at the centroid of the member (i.e. load is applied eccentrically), as illustrated in Fig. 10.1(a). Alternatively, bending moments in a compression member may result from unbalanced moments in the members connected to its ends, as illustrated in Fig. 10.1(c). The result of such bending moments in axially loaded members is to reduce the range of axial force that the member can safely carry. For this reason, it is essential that the effects of bending in axially loaded members are considered. In this chapter, one-dimensional compression members are often referred to as columns and two-dimensional members as walls. Strictly speaking, a column is a particular type of compression member which is vertical. However, the term 'column' is used loosely to describe all one-dimensional compression members, irrespective of their orientation.

10.2 Classification of compression members (columns)

Braced/unbraced columns

As was stated above, the axial load-carrying capacity of a member depends on the magnitude of bending moment in it. An unbraced structure is one in which frame action is used to resist horizontal (wind) loads. In such a structure, the horizontal loads are transmitted to the foundations through bending action in the beams and columns. The moments in the columns due to this bending can substantially reduce their axial (vertical) load-carrying capacity. Unbraced structures are generally quite flexible and allow horizontal displacement as illustrated in Fig. 10.2. When frame action is the principal means of resisting horizontal forces, the displacements are generally large enough to significantly influence the column moments. In such cases, the structure is termed a **sway frame**. Columns in a sway structure undergo lateral displacement between their ends, as illustrated in Fig. 10.2, which reduces their resistance to buckling. For this reason, such columns, termed **unbraced** columns, are treated differently from a design viewpoint than other columns. In a column, the bending moment is increased by an additional amount $N\Delta$, where N is the axial force and Δ is the relative horizontal displacement of the ends of the column. Thus, to maximize the axial load capacity of columns, non-sway structures should be used whenever possible. Fully braced (non-sway) structures are difficult to achieve in practice but EC2 allows a member to be considered as braced if it is assumed to not contribute to the overall horizontal stability of the structure in analysis and design. A column within such a non-sway structure is considered to be braced and the second-order moment on such columns, $N\Delta$, is relatively small.

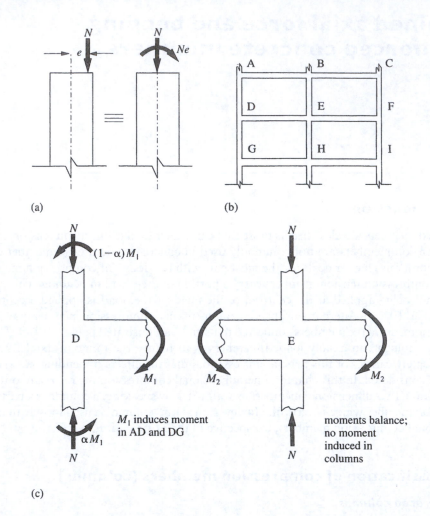

Figure 10.1 Columns subjected to axial load and moment: (a) eccentric load; (b) frame; (c) balanced and out-of-balance moments

Slender/short columns

When an unbalanced moment or a moment due to eccentric loading (Fig. 10.1) is applied to a column, the member responds by bending, as illustrated in Fig. 10.3. If, as shown, the deflection at the centre of the member is δ, then at the centre there is a force N and a total moment of $M + N\delta$. The second-order bending component, $N\delta$, is due to the extra eccentricity of the axial force which results from the deflection. If the column is short and squat, δ is small and this second-order moment is negligible. If, on the other hand, the column is long and slender, δ is large and $N\delta$ must be calculated and added to the applied moment, M. A column in which the moment $N\delta$ is negligible is commonly known as a **short column**, while a column in which $N\delta$ is not negligible is known as a **slender column**. It should be noted that in the case of tension members, there is no second-order bending component. EC2 allows second-order effects to be ignored if they are less than 10 per cent of the corresponding first-order moment, or if the slenderness of the member is below a certain value.

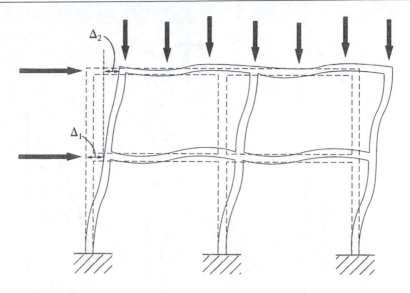

Figure 10.2 Sway frame

Slenderness

The significance of the $N\delta$ term (i.e. whether a column is short or slender) is normally determined from a **slenderness ratio**, which is a function of the parameters that determine the lateral deflection of the column. In EC2, the slenderness ratio is defined by:

$$\lambda = l_0/i \tag{10.1}$$

where l_0 is the effective length (explained below) of the member and i is the radius of gyration of the uncracked concrete section. The radius of gyration is equal to:

$$i = \sqrt{I/A} \tag{10.2}$$

where I is the second moment of area of the section (usually taken as the gross value, I_g) and A is its cross-sectional area (usually taken as the gross area, A_g). By definition, second-order moments in a column can be ignored if:

$$\lambda < \lambda_{lim} \tag{10.3}$$

where:

$$\lambda_{lim} = 20\text{ABC}\sqrt{A_c f_{cd}/N_{Ed}} \tag{10.4}$$

The constant, A, is given by:

$$A = 1/(1 + 0.2\varphi_{ef})$$

where φ_{ef} is the effective creep ratio (see Section 10.6). A can be taken as 0.7 if φ_{ef} is not known. The second constant in Equation (10.4) is,

$$B = \sqrt{1 + 2\omega}$$

where $\omega = A_s f_{yd}/(A_c f_{cd})$

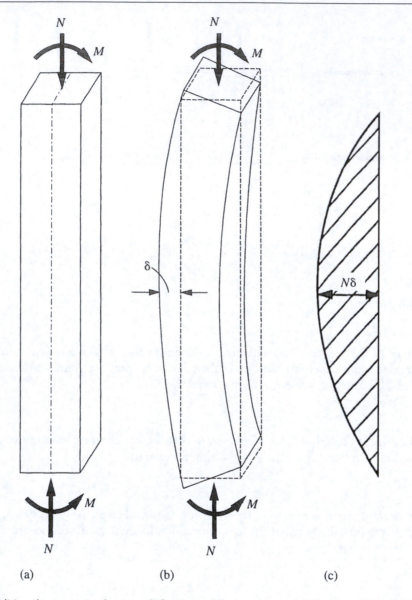

Figure 10.3 Additional moment due to deflection: (a) geometry and loading; (b) deflected shape; (c) additional moment, $N\delta$

A_s = cross-sectional area of steel reinforcement
f_{yd} = design value of yield strength of reinforcement ($= f_{yk}/\gamma_s$)
A_c = cross-sectional area of concrete
f_{cd} = design value of compressive strength of concrete ($= \alpha_{cc}f_{ck}/\gamma_c$)

B can be taken as 1.1 if ω is not known. The constant, C, is given by:

$$C = 1.7 - r_m$$

where $r_m = M_{01}/M_{02}$, and

M_{01} and M_{02} are first-order moments at the two ends of the member ($|M_{02}| > |M_{01}|$). This constant, C can be taken as 0.7 if r_m is not known.

The other term in Equation (10.4), N_{Ed}, is the design value of the applied axial force. In summary, columns that satisfy inequality (10.3) are classified as **short columns** while those that do not satisfy it are classed as **slender columns**.

Effective length of columns

The effective length of a column can be looked on as the length which is effective against buckling. The greater the effective length, the more likely the column is to buckle. The effective length is dependent on the support conditions at the ends of the column, whether or not the column is braced and, of course, the actual length (height) of the member. Theoretically, the effective length of a column is equal to the distance between the points of contraflexure in the member.

A braced column pinned at both ends (pinned-pinned) in a non-sway frame buckles in a straightforward manner and its effective length, l_0, is equal to its actual length, l, as illustrated in Fig. 10.4(a). However, a fixed-fixed column (no rotation at either end) in a non-sway frame is much less likely to buckle and its effective length is substantially less. The points of contraflexure in a braced fixed-fixed column occur at distances of $l/4$ from the fixed ends, as illustrated in Fig. 10.4(b). Thus, the effective length of a fixed-fixed braced column is half its actual length. For a fixed-fixed column in a sway frame (i.e. unbraced), such as that illustrated in Fig. 10.5(a), there is only one point of contraflexure which occurs at mid-span. However, another point of contraflexure exists on an imaginary line produced through the member as illustrated in the figure. Thus, the effective length of the unbraced column of Fig. 10.5(a) is equal to its actual length. For an unbraced column, fixed against rotation at one end and pinned at the other (Fig. 10.5(b)), the effective length is equal to twice the actual length. Notice that the effective length in an unbraced frame is always greater than that in a braced frame. In real structures,

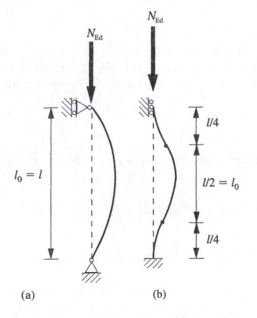

(a)   (b)

Figure 10.4 Effective lengths of columns in non-sway structures: (a) pinned-pinned column; (b) fixed–fixed column

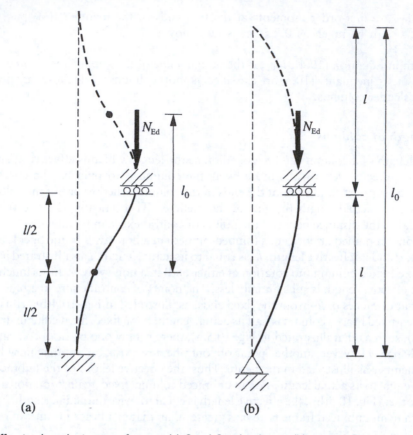

Figure 10.5 Effective lengths in sway frames: (a) fixed-fixed column; (b) pinned-fixed column

the ends of columns are neither pinned nor fixed but are actually something in-between. The actual rotational restraint provided by adjoining members at the end of a column can be expressed by a function of their stiffnesses. The relative stiffness of the restraining members is given by:

$$k = \frac{K_{\text{col}}}{\sum K_{\text{beam}}} \tag{10.5}$$

where $K_{\text{col}} = EI/l$ (assuming adjacent columns do not contribute to stiffness) and referring to Appendix A:

$$K_{\text{beam}} = \frac{4EI}{l} \qquad \text{(for a beam with opposite end fixed)} \tag{10.6}$$

$$K_{\text{beam}} = \frac{3EI}{l} \qquad \text{(for a beam with opposite end simply supported)} \tag{10.7}$$

According to EC2, the stiffness of restraining members (such as beams) should take account of cracking. Therefore, as discussed in Chapter 5, when analysing frames, the ratio of I to L for a beam is taken as one-half of its actual value (Fig. 5.15). The length of the beam can be taken as the effective length (typically measured from centre to centre of columns – see EC2 for more details).

Once the relative stiffness, k, for each end of the column is known, the effective length can be computed as:

$$l_0 = 0.5l\sqrt{\left(1 + \frac{k_1}{0.45 + k_1}\right)\left(1 + \frac{k_2}{0.45 + k_2}\right)} \qquad \text{(for braced members)} \qquad (10.8)$$

$$l_0 = l \times \max\left\{\sqrt{1 + 10\frac{k_1 k_2}{k_1 + k_2}}; \left(1 + \frac{k_1}{1 + k_1}\right) \times \left(1 + \frac{k_2}{1 + k_2}\right)\right\} \text{(for unbraced members)} \quad (10.9)$$

where k_1 and k_2 are the relative stiffnesses for the two ends.

Example 10.1 Classification of column

Problem: For the frame of Fig. 10.6, determine whether the column is short or slender. It may be assumed that the same grade of concrete is used for all members (i.e. E_c constant throughout) and that $f_{ck} = 30$ N/mm^2. The applied design axial force is 525 kN.

Solution: For members AB and BC, the second moment of area is:

$$I_b = bh^3/12 = 0.3\,(0.6)^3/12 = 5.4 \times 10^{-3}\,\text{m}^4$$

For member BD:

$$I_{col} = bh^3/12 = (0.4)^4/12 = 2.13 \times 10^{-3}\,\text{m}^4$$

The effective length of the beams, l_{eff}, is assumed to be equal to:

$$l_{eff} = 7.8 - 0.2 - 0.2 = 7.4\,\text{m}$$

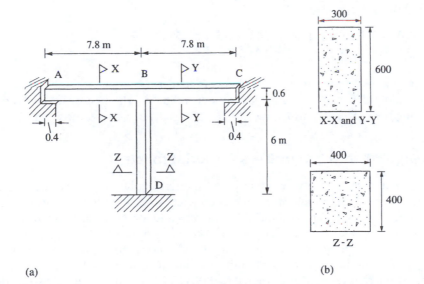

(a)   (b)

Figure 10.6 Column of Example 10.1: (a) geometry; (b) cross-sectional dimensions

The length of the column, l, measured between the centres of connections is given by:

$$l = 6.0 + 0.3 = 6.3 \, \text{m}$$

Using Equations (10.5) and (10.7), the rotational stiffness at joint B is:

$$k_B = \frac{K_{col}}{\sum K_{beam}} = \frac{\dfrac{E_c I_{col}}{l}}{2\left(\dfrac{\frac{1}{2} \times 3E_c I_b}{l_{eff}}\right)}$$

Since E_c is equal for all members, the relative stiffness of the column at B, k_B, becomes:

$$k_B = \frac{\dfrac{I_{col}}{l}}{\dfrac{2 \times 1.5 I_b}{l_{eff}}} = \frac{I_{col}}{l} \times \frac{l_{eff}}{2 \times 1.5 I_b} = \frac{2.13 \times 10^{-3}}{6.3} \times \frac{7.4}{2 \times 1.5(5.4 \times 10^{-3})} = 0.1544$$

The support at the base of the column is fully fixed. Therefore, the rotational stiffness of joint D is $k_D = 0$. However, EC2 recommends that k not be taken less than 0.1. The effective length of the column is given by Equation (10.8), that is:

$$l_0 = 0.5(6.3)\sqrt{\left(1 + \frac{0.1544}{0.45 + 0.1544}\right)\left(1 + \frac{0.1}{0.45 + 0.1}\right)} = 3.84 \, \text{m}$$

The slenderness ratio, λ, is given by Equation (10.1) as:

$$\lambda = \frac{l_0}{i} = \frac{l_0}{\sqrt{I/A}} = \frac{3.84}{\sqrt{2.13 \times 10^{-3}/0.4^2}} = 33.3$$

From Inequality (10.3), a column is short if λ is less than λ_{lim}. Since no information is provided for the calculation of A, B or C (from Equation (10.4)), standard values are assumed:

$$\lambda_{lim} = 20ABC\sqrt{A_c f_{cd}/N_{Ed}}$$

$$= 20(0.7)(1.1)(0.7)\sqrt{(400)^2(30 \times 0.85/1.5)/525 \times 10^3}$$

$$= 24.5$$

As λ is greater than λ_{lim}, the column is classified as 'slender'.

   Note: The value of C can vary significantly depending on the first-order end moments of the compression member. The assumed value of C = 0.7 results in a conservative value of λ_{lim}.

10.3   Design of short members for axial force

The reinforcement detail in a typical reinforced concrete column is illustrated in Fig. 10.7. In compression, both the longitudinal steel and the concrete contribute to the resistance of the applied axial force. The links serve to confine the concrete and prevent buckling of the longitudinal reinforcement in compression. For the design of compression members subjected to approximately concentric loading (i.e. purely compression force), EC2 limits the mean compressive strain in the concrete to ε_{c3} for a bi-linear stress distribution. As illustrated in Fig 8.21(b), ε_{c3} is the strain where the maximum strength is reached and is equal to 0.00175 for a bi-linear stress distribution. Based on the stress/strain curve of Fig. 8.22(a) or (b), reinforcement with strength $f_{yk} = 500 \, \text{N/mm}^2$ will also yield close to this strain (assuming $E_s = 200{,}000\ \text{N/mm}^2$, $f_{yd}/E_s = 0.00217$). The ultimate capacity to resist compressive force, N_{ult}, is equal to:

$$N_{ult} = \alpha_{cc}\frac{f_{ck}}{\gamma_c}A_c + \frac{f_{yk}}{\gamma_s}A_s \tag{10.10}$$

where α_{cc} = coefficient defined in Section 8.5, taken as 0.85 for flexure and axial loading
$\qquad f_{ck}$ = characteristic compressive cylinder strength of the concrete
$\qquad A_c$ = cross-sectional area of concrete
$\qquad A_s$ = cross-sectional area of longitudinal reinforcement
$\qquad f_{yk}$ = yield strength of reinforcement
$\qquad \gamma_c$ = partial factor of safety for concrete
$\qquad \gamma_s$ = partial factor of safety for steel

Once N_{ult} is reached, failure occurs by crushing of the concrete or local buckling of the longitudinal reinforcement between the links.

Note: Due to geometric imperfections and tolerances in construction, axially loaded members must be designed to resist some moments, so members are not usually designed to resist pure axial force.

In tension, it is common practice to ignore the small tensile strength of concrete and to assume that resistance is only provided by the reinforcement, that is $N_{ult} = (f_{yk}/\gamma_s)\,A_s$. Although tension does not often arise in reinforced concrete members, it does occur in some circumstances. For example, pure tension does occur in deep beams as described in Section 10.7.

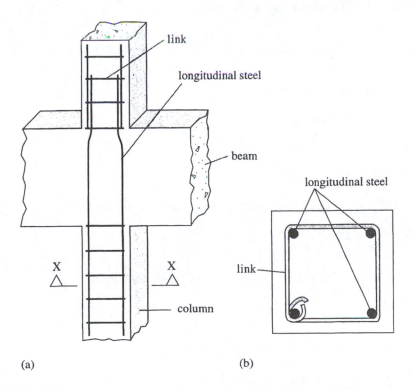

(a)   (b)

Figure 10.7 Typical connection reinforcement: (a) part elevation; (b) section X–X

Detailing requirements

Most codes of practice place restrictions on the maximum and minimum areas of reinforcement provided in columns. As for beams, a maximum percentage, $A_{s,max}$, of longitudinal steel is specified to ensure that the steel yields before the concrete crushes at the ultimate limit state, thus ensuring ductile behaviour. It is also important to limit the area of reinforcement so that there will be room for concrete to pass between the bars, particularly at laps. Limits on the minimum percentage of longitudinal reinforcement $A_{s,min}$ ensure that the steel does not yield under service loads and that there is sufficient reinforcement to prevent sudden failure when the concrete reaches its tensile strength (under ULS loads). Upper limits are also commonly specified on the spacing of links to prevent buckling of the longitudinal reinforcement. For an applied design axial compressive force N_{Ed}, the limits for reinforcement in columns recommended by EC2 are as follows:

(a) $A_{s,min}$ = greater of $0.1 N_{Ed}/(f_{yk}/\gamma_s)$ and $0.002 A_c$
(b) $A_{s,max} = 0.04\,A_c$ outside lap locations and $0.08\,A_c$ at lap locations
(c) $\phi_{min} = 12$ mm (minimum allowable diameter of longitudinal reinforcing bars)
(d) the maximum spacing of links is the lesser of 20ϕ, 400 mm and the least cross-sectional dimension of the column, where ϕ is the diameter of the smallest longitudinal bars within the column.

These detailing requirements apply to all columns, short or slender, braced or unbraced.

Example 10.2 Short circular column

Problem: Calculate the ultimate capacity to resist compressive and tensile force for a 300 mm diameter, short circular column with eight 16 mm diameter bars, where $f_{yk} = 500$ N/mm^2 and $f_{ck} = 35$ N/mm^2.

Solution: The total area of reinforcement is:

$$A_s = 8\pi (16)^2/4 = 1608\ \text{mm}^2$$

The cross-sectional area of concrete is:

$$A_c = \pi (300)^2/4 - 1608 = 69078\ \text{mm}^2$$

From Equation (10.10) the ultimate capacity to resist compressive force is:

$$N_{ult} = \frac{0.85 f_{ck}}{\gamma_c} A_c + \frac{f_{yk}}{\gamma_s} A_s$$

Taking $\gamma_c = 1.5$ and $\gamma_s = 1.15$ gives a capacity to resist compression of:

$$N_{ult} = \frac{0.85(35)}{1.5}(69078) + \frac{500}{1.15}(1608) = 2069\ \text{kN}$$

In tension, the ultimate axial load is simply given by:

$$N_{ult} = -\frac{f_{yk}}{\gamma_s} A_s = -\frac{500}{1.15}(1608) = -699\ \text{kN}$$

Example 10.3 Short rectangular column

Problem: Design a short rectangular column to carry a design ultimate compressive force of 2,500 kN given $f_{ck} = 40$ N/mm^2 and $f_{yk} = 500$ N/mm^2.

Solution: From Equation (10.10):

$$N_{ult} = \frac{0.85f_{ck}}{\gamma_c} A_c + \frac{f_{yk}}{\gamma_s} A_s$$

Assume an area of longitudinal reinforcement, A_s equal to $0.02A_g$ (see Chapter 6 for preliminary design recommendations). For a design load of 2,500 kN, Equation (10.10) then becomes:

$$2500 \times 10^3 \leq \frac{0.85f_{ck}}{\gamma_c} (1 - 0.02)A_g + \frac{f_{yk}}{\gamma_s} 0.02A_g$$

$$\Rightarrow 2500 \times 10^3 \leq \frac{0.85(40)}{1.5} (1 - 0.02)A_g + \frac{500}{1.15} 0.02A_g$$

$$\Rightarrow A_g \geq 80883 \text{ mm}^2$$

$$\Rightarrow \sqrt{A_g} \geq 284 \text{ mm}$$

Therefore, take a square section of side length 300 mm. For this section, $A_g = 90{,}000$ mm^2. The exact area of reinforcement required is found using Equation (10.10), that is:

$$2500 \times 10^3 \leq \frac{0.85(40)}{1.5} (90\,000 - A_s) + \frac{500}{1.15} A_s$$

$$\Rightarrow 2500 \times 10^3 \leq 2.04 \times 10^6 + 412A_s$$

$$\Rightarrow A_s \geq 1117 \text{ mm}^2$$

Thus, adopt four 20 mm diameter bars of total area $A_s = 1{,}257$ mm^2.

10.4   Design of short members for axial force and uniaxial bending

When the load applied to a column is eccentric about one axis only, as illustrated in Fig. 10.8, a bending moment results about that one axis and the column is said to be subjected to uniaxial bending. Uniaxial bending also occurs when unbalanced moments are transferred from beams or slabs to the column and are confined to a single axis of the column. As stated in Section 10.1, the presence of this form of bending in axially loaded members reduces the axial load capacity of the member. At the serviceability limit state, the applied axial compression in columns tends to close up any cracks caused by bending. Thus, the presence of compression with bending is actually beneficial at SLS and crack widths are unlikely to be excessive. It is therefore the combined effect of axial compression and bending at the ultimate limit state that tends to govern the design. In the case of tension members, cracking at SLS may cause problems and does need to be checked.

Imperfections

To cover dimensional inaccuracies and any uncertainty as to the location of the applied axial load, an eccentricity, e_i, should always be assumed. The recommended value of e_i in EC2 for isolated braced columns is:

$$e_i = \frac{l_0}{400} \tag{10.11}$$

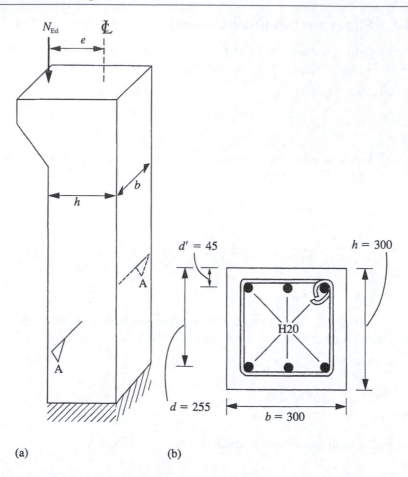

Figure 10.8 Column with eccentric load: (a) geometry and loading; (b) section A–A

In the case of eccentrically applied loads, e_i should be added to the known eccentricity of the load. However, a minimum eccentricity, e_0, is also specified in EC2 and is given by $h/30$, but should be no less than 20 mm. Thus, the design moment acting on a short member due to an applied design force N_{Ed} at an eccentricity e becomes:

$$M_{Ed} = N_{Ed}(e + e_i) \geq N_{Ed}e_0 \qquad (10.12)$$

Ultimate strength

The ultimate strength of short members subjected to combined axial force and bending is evaluated using the same techniques as those described in Chapter 8 for members in pure bending. As an example, consider the member of Fig. 10.9 which is subjected to a design compressive force N_{Ed} at an eccentricity e about the member centre line. To highlight the similarity with beam design, the column with its eccentric load is plotted on its side in Fig. 10.9(a) and the equivalent loading of N_{Ed} and $M_{Ed} = N_{Ed}(e + e_i)$ is illustrated in Fig. 10.9(b). The strain and stress distributions due to the combined effects of M_{Ed} and N_{Ed} are illustrated in Fig. 10.10(a–b). To simplify design, EC2 also allows the use of the simplified stress diagrams of Fig. 8.21 in place of the actual stress distribution.

As the section is on the point of failure, the maximum compressive strain has reached its ultimate value and the distribution of stress is also that for a section on the point of failure. The precise

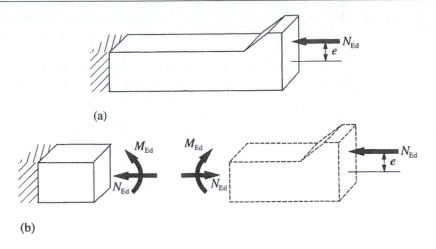

(a)

(b)

Figure 10.9 Equivalent loading on eccentrically loaded column: (a) column with original loading; (b) equivalent loading

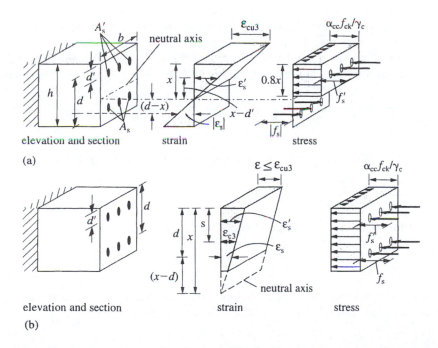

Figure 10.10 Alternative strain and stress distributions: (a) case (1) assuming rectangular stress distribution; (b) case (2) assuming rectangular stress distribution and the strain in the bottom fibre >0.0007 (refer to Fig. 8.21(c))

location of the neutral axis depends on the relative magnitude of the design compressive force N_{Ed} and the moment M_{Ed}. In broad terms, two design cases can be distinguished:

1. The neutral axis lies between the rows of reinforcement as normally occurs in beams (Fig. 10.10(a)). In this case, the reinforcement is in tension on one side of the neutral axis and in compression on the other side. The ultimate limit state of failure is reached when the strain in the concrete

reaches a specified ultimate value, ε_{cu2} (parabolic-rectangular stress distribution) or ε_{cu3} (bi-linear or rectangular stress distribution). The values for ε_{cu2} and ε_{cu3} are taken as 0.0035 by EC2.

2. The neutral axis lies outside the section and the entire section is in compression (Fig. 10.10(b)). In this case, all of the reinforcement (both layers) is in compression. Where there is no tension in the section, the ultimate limit state of failure is reached when the strain at a certain depth, s, reaches, ε_{c2} or ε_{c3}, depending on the assumed stress distribution (ε_{c2} for parabolic-rectangular and ε_{c3} for bi-linear). The values for ε_{c2} and ε_{c3} are specified in EC2 as 0.002 and 0.00175, respectively. The depth, s, is found by interpolation:

$$s = \left(1 - \frac{\varepsilon_{c2}}{\varepsilon_{cu2}}\right)x = \left(1 - \frac{0.002}{0.0035}\right)x = \frac{3}{7}x \qquad \text{(parabolic-rectangular stress distribution)}$$

$$s = \left(1 - \frac{\varepsilon_{c3}}{\varepsilon_{cu3}}\right)x = \left(1 - \frac{0.00175}{0.0035}\right)x = 0.5x \quad \text{(bi-linear or rectangular stress distribution)}$$

Each location of the neutral axis corresponds to a unique combination of ultimate moment resistance M_{ult} and ultimate axial resistance N_{ult}. The stresses in the reinforcement and the concrete can be determined for any given location of the neutral axis and, from this, the respective values of moment and axial force that would cause this distribution of stress can be calculated.

Consider the case where the neutral axis lies between the rows of reinforcement, at a depth x from the extreme fibre in compression as illustrated in Fig. 10.10(a). The compressive force in the concrete, F_c, is given by:

$$F_c = 0.8xb\frac{\alpha_{cc}f_{ck}}{\gamma_c} \tag{10.13}$$

By similar triangles of the strain diagram, the strain in the 'top' layer of reinforcement is:

$$\varepsilon'_s = \frac{\varepsilon_{cu3}(x - d')}{x} \tag{10.14}$$

Similarly, the strain in the 'bottom' layer of reinforcement is:

$$\varepsilon_s = -\frac{\varepsilon_{cu3}(d - x)}{x} \tag{10.15}$$

When the neutral axis lies outside the section, assuming a rectangular stress distribution, by similar triangles of the strain diagram, the strain in the 'top' layer of reinforcement is:

$$\varepsilon'_s = \frac{\varepsilon_{c3}(x - d')}{x - s} \tag{10.16}$$

Similarly, the strain in the 'bottom' layer of reinforcement is:

$$\varepsilon_s = \frac{\varepsilon_{c3}(x - d)}{x - s} \tag{10.17}$$

The corresponding steel stresses, f'_s and f_s are found from the stress/strain diagram of Fig. 8.22(a) or (b). The compressive force in the top reinforcement is then:

$$f'_s A'_s \tag{10.18}$$

and in the bottom reinforcement is:

$$f_s A_s \tag{10.19}$$

When the neutral axis lies between the rows of reinforcement, f_s is negative giving a negative compressive force, that is, a tensile force. By equilibrium of all axial forces, the design axial resistance, with this distribution of stress, must be:

$$N_{ult} = F_c + f'_s A'_s + f_s A_s \tag{10.20}$$

where f'_s and f_s are both considered positive when compressive. Similarly, by taking moments about a point on the section, the design moment of resistance in the presence of this axial force can be found. Moments can be taken about any point and it is conventional, for pure bending, to take moments about the neutral axis. However, in the case of combined axial force and bending, this would involve a component due to N_{Ed} which is conventionally assumed to be applied at the centre line (an axial force applied at any point is equivalent to an axial force at the centre line plus a moment). Accordingly, it is more convenient to take moments about the **centre line**. Hence, where the neutral axis lies between the rows of reinforcement:

$$M_{ult} = F_c\left(\frac{h}{2} - 0.4x\right) + f'_s A'_s\left(\frac{h}{2} - d'\right) - f_s A_s\left(d - \frac{h}{2}\right) \tag{10.21}$$

The third term is subtracted from the others because a compressive force in the 'bottom' reinforcement acts against the other components of moment. When the neutral axis lies between the layers of reinforcement, f_s is negative and this term becomes positive. For members where the neutral axis lies outside the section, if the strain in the bottom fibre is greater than 0.0007, the compressive force in the concrete, F_c, acts at the mid-depth of the section (i.e. at $h/2$), and the first term of Equation (10.21) is zero. If the strain is less than 0.0007, refer to Fig. 8.21(c) to determine the shape of the stress diagram.

In order to design a section to carry a design moment and axial resistance, the procedure outlined above must be repeated using different values of x until the value for N_{ult} equals the applied design force, N_{Ed}. Then M_{ult} is calculated and compared with the applied design moment, M_{Ed}. This procedure can be laborious and as a result it is common practice to use a design aid known as an **interaction diagram**. The following example illustrates the construction of one such interaction diagram.

Example 10.4 Construction of interaction diagram

Problem: Construct the interaction diagram for the column of Fig. 10.8, assuming that all six bars are 20 mm diameter (total area equals 1,885 mm^2), the characteristic tensile strength of the reinforcing bars is $f_{yk} = 500$ N/mm^2 and that the characteristic strength of the concrete is $f_{ck} = 35$ N/mm^2. Use the EC2 equivalent rectangular stress block for concrete in compression.

Solution: An interaction diagram is a chart or graph illustrating the capacity of a column to resist a range of combinations of force and moment. It is found by assuming a number of strain distributions (i.e. assume a number of values for x) and calculating in each case the combination of force, N_{ult} and moment, M_{ult} that would cause that strain distribution. These results form isolated points on a plot of N_{ult} versus M_{ult} and, once sufficient points have been determined, they can be joined to form a design curve representing the general solution. For the first point on the interaction diagram, assume that the column is in pure compression under the action of a force at the centre line (i.e. $M_{ult} = 0$ and x is infinite). As stated in Section 10.3, the strain at failure is then uniform across the section and equal

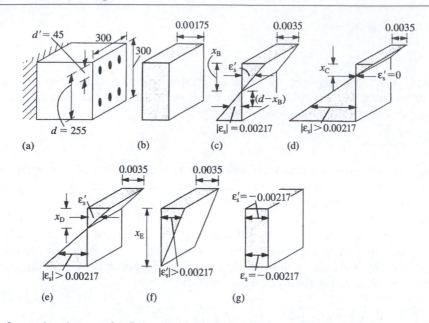

Figure 10.11 Strain distributions for Example 10.4: (a) elevation and section; (b) pure compression (A); (c) balanced failure (B); (d) x = 45 (C); (e) x = 100 (D); (f) x = 300 (E); (g) pure tension (F)

to 0.00175 (ε_{c3}, assuming equivalent rectangular stress distribution), as illustrated in Fig. 10.11(b). From Equation (10.10), the ultimate axial force capacity is:

$$
\begin{aligned}
N_{ult} &= \frac{0.85f_{ck}}{\gamma_c} A_c + \frac{f_{yk}}{\gamma_s} A_s \\
&= \frac{0.85(35)}{1.5} \left[(300)^2 - 1885 \right] + \frac{500}{1.15} 1885 \\
&= 2567 \text{ kN}
\end{aligned}
$$

This result forms the first point, point A, on the interaction diagram for the column as illustrated in Fig. 10.12.

Next, consider the case when the strain in the tension reinforcement is on the point of yielding at the same time as concrete is failing in compression. This is referred to as balanced design and the corresponding force capacity is denoted N_{bal} (see Section 8.6). Assuming $E_s = 200,000 \text{ N/mm}^2$, the strain in the bottom reinforcement is:

$$
\varepsilon_s = -\frac{f_{yk}}{\gamma_s E_s} = -0.00217
$$

The complete strain distribution for this case is illustrated in Fig. 10.11(c). By similar triangles:

$$
\begin{aligned}
\frac{0.00217}{d - x_B} &= \frac{0.0035}{x_B} \\
\Rightarrow x_B &= \frac{0.0035(255)}{(0.00217 + 0.0035)} = 157 \text{ mm}
\end{aligned}
$$

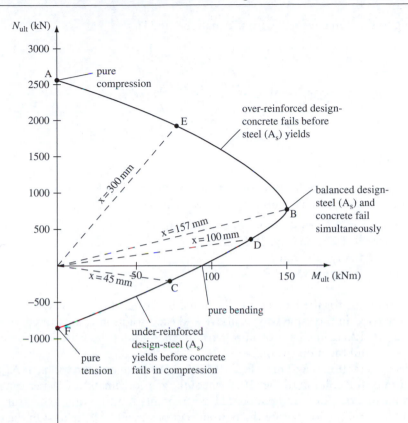

Figure 10.12 Interaction diagram of Example 10.4

The strain in the top reinforcement, A'_s, is found by similar triangles:

$$\frac{\varepsilon'_s}{x_B - d'} = \frac{0.0035}{x_B}$$

$$\Rightarrow \varepsilon'_s = \frac{0.0035(157 - 45)}{157} = 0.0025$$

As this exceeds the yield strain of 0.00217, the top reinforcement has yielded in compression.
The compressive force in the concrete, F_c, is given by Equation (10.13):

$$F_c = 0.8xb\frac{\alpha_{cc}f_{ck}}{\gamma_c}$$

$$= 0.8(157)(300)\left(\frac{0.85 \times 35}{1.5}\right)$$

$$= 747320 \, \text{N}$$

The ultimate compressive resistance is given by Equation (10.20), that is:

$$N_{ult} = F_c + f'_s A'_s + f_s A_s$$

$$= 747320 + (f_{yk}/\gamma_s)\left(\frac{1885}{2}\right) - (f_{yk}/\gamma_s)\left(\frac{1885}{2}\right)$$

$$= 747 \, \text{kN}$$

Corresponding to this force we have, from Equation (10.21):

$$M_{ult} = F_c\left(\frac{h}{2} - 0.4x\right) + f'_s A'_s\left(\frac{h}{2} - d'\right) - f_s A_s\left(d - \frac{h}{2}\right)$$

$$= 747320[150 - (0.4)(157)] + \left(\frac{500}{1.15}\right)\left(\frac{1885}{2}\right)(150 - 45)$$

$$- \left(\frac{-500}{1.15}\right)\left(\frac{1885}{2}\right)(255 - 150)$$

$$= 151 \text{ kNm}$$

This result is represented by point B on the interaction diagram of Fig. 10.12.

A further four design cases have been considered as follows:

(i)   point C: $x = 45$ mm (zero strain in top reinforcement)
(ii)   point D: $45 < x < 157$ mm (say, 100 mm)
(iii) point E: $157 < x$ (say, 300 mm)
(iv)   point F: member in pure tension.

The strain distributions for these cases are illustrated in Fig. 10.11(d–g), respectively. The solution in each case is derived in the same way as above and each solution is illustrated on the interaction diagram of Fig. 10.12. In the case of pure tension, the ultimate tensile force is simply given by $-(f_{yk}/\gamma_s)(A_s + A'_s)$ and the moment is zero.

The complete interaction diagram of Fig. 10.12 represents all combinations of N_{ult} and M_{ult} that can cause failure of the column of Fig. 10.8 where the reinforcement is 2.1 per cent of gross area. All combinations of axial force and moment between points A and B on the diagram will cause the concrete to fail in compression before the bottom reinforcement, A_s, yields. On the other hand, all combinations between points B and F will result in the tensile yielding of A_s before the concrete fails in compression.

For a complete design chart, interaction diagrams are plotted for other percentages of reinforcement in Fig. 10.13 up to the maximum of 8 per cent allowed in EC2. If the depth, breadth, location of reinforcement (d or d') or concrete strength of the member is altered, the interaction diagram must be redrawn, that is, other design charts must be used. Interaction diagrams are published for a wide range of cross-sectional geometries and concrete strengths and are used extensively in design offices for the design of columns.

Example 10.5 Application of design chart

Problem: Using an appropriate design chart, calculate the area of reinforcement required to resist a design compressive force that varies in the range 440–600 kN combined with a uniaxial design moment in the range 110–145 kNm. The column is short and has the dimensions and material properties given for Example 10.4.

Solution: The ratio N_{ult}/bh ranges between:

$$\frac{440 \times 10^3}{(300)^2} = 4.89 \text{ and } \frac{600 \times 10^3}{(300)^2} = 6.67$$

Similarly, the ratio M_{ult}/bh^2 ranges between:

$$\frac{110 \times 10^6}{(300)^3} = 4.07 \text{ and } \frac{145 \times 10^6}{(300)^3} = 5.37$$

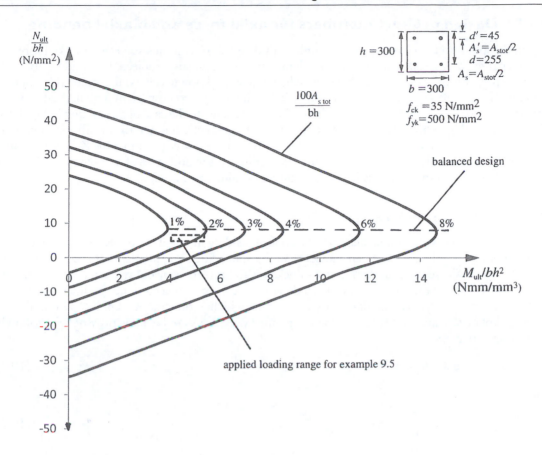

Figure 10.13 Interaction diagram for different percentages of reinforcement

This range of combinations of N_{ult}/bh and M_{ult}/bh^2 is indicated in Fig. 10.13. It can be seen that for this example (and indeed throughout the interaction diagram) increasing the moment increases the required area of reinforcement. However, this is not true of applied force. As can be seen in the figure, for this problem the force/moment combination for which the required area of reinforcement is maximum is the largest moment with the **smallest** force in the specified range. The total area of reinforcement required to provide a capacity of $N_{ult}/bh = 4.89$ and $M_{ult}/bh^2 = 5.37$ is about 2.4 per cent or $(0.024(300)^2 =) 2,160$ mm^2, of which half should be placed at each face.

The minimum area of reinforcement required by EC2 is:

$$A_{s,min} = \text{greater of } 0.1N_{Ed}/\left(f_{yk}/\gamma_s\right) \text{ and } 0.002A_c$$

$$= \text{greater of } \frac{0.1 \times 600 \times 10^3}{500/1.15} \text{ and } 0.002\left(300^2 - 2160\right)$$

$$= 176\,\text{mm}^2$$

Clearly this does not govern design.

10.5   Design of short members for axial force and biaxial bending

Up to this point in the chapter, we have only considered short columns subjected to axial force and bending about a single axis. In fact, many columns are simultaneously subjected to bending about two (usually perpendicular) axes. Such bending, when it arises, is known as biaxial bending. Like uniaxial bending, biaxial bending is caused by eccentric loading of the column or by unbalanced moments transferred to the column from the members joined to its ends. In the case of the column in Fig. 10.14, the axial force N_{Ed} is applied at eccentricities e_y and e_z to the centre line and the member is subjected to a moment $M_{Edy} = N_{Ed}e_z$ about the Y-axis and to a moment $M_{Edz} = N_{Ed}e_y$ about the Z-axis. In a similar manner, biaxial bending can result from unbalanced moments in both axes of the column. This form of biaxial bending commonly arises in corner columns, as illustrated in Fig. 10.15.

Substantially uniaxial bending

For many members with biaxial bending, the applied bending moment (or the eccentricity of the axial load) about one of the axes is much smaller than the applied bending moment (or eccentricity) about the other axis. Where such cases arise, it is sufficiently accurate to carry out two separate designs, each involving bending about just one axis. Each design is carried out in accordance with Section 10.4.

For members with rectangular cross-sections in which moments are applied about the two principal planes, Y and Z, EC2 allows separate (uniaxial) checks to be made provided the following conditions are met:

$$\lambda_y/\lambda_z \le 2 \quad \text{and} \quad \lambda_z/\lambda_y \le 2 \tag{10.22}$$

and:

$$\frac{e_y/h_y}{e_z/h_z} \le 0.2 \quad \text{or} \quad \frac{e_z/h_z}{e_y/h_y} \le 0.2 \tag{10.23}$$

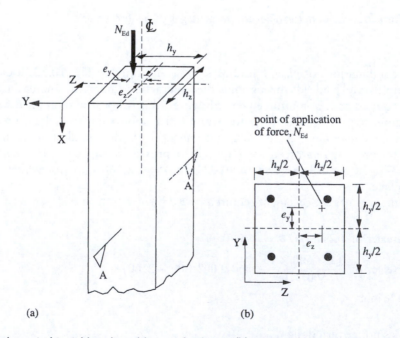

(a)   (b)

Figure 10.14 Column in biaxial bending: (a) top of column; (b) section A–A

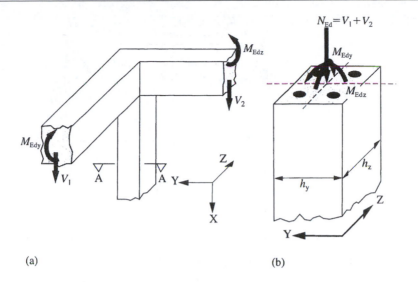

(a)   (b)

Figure 10.15 Corner column: (a) geometry and loading; (b) section A–A

Condition (10.22) ensures that the relative slenderness about the Y and Z planes are similar and condition (10.23) ensures that the member is substantially uniaxial. Condition (10.23) is satisfied if the point of application of the axial load is located within the shaded area of Fig. 10.16. M_{Edy} and M_{Edz} are the design moments about the Y and Z-axes, resulting from the transfer of unbalanced moments. The equivalent eccentricities are: $e_y = M_{Edz}/N_{Ed}$ and $e_z = M_{Edy}/N_{Ed}$. According to EC2, when carrying out separate (uniaxial) checks, only one of e_y or e_z needs to include eccentricity due to imperfections (e_i). It should be taken into account in the direction where it has the most onerous effect.

For members where both applied moments (or eccentricities) are substantial, however, the biaxial (combined) effect of the moments must be considered. The method of design of such members is described in the remainder of this section.

Ultimate capacity

The ultimate capacity of short members subjected to combined axial load and biaxial bending is evaluated using the same principles as those used for members with uniaxial bending. For sections

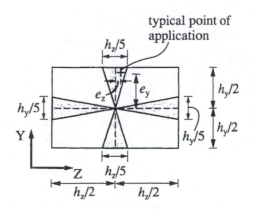

Figure 10.16 Zone in which bending is substantially uniaxial

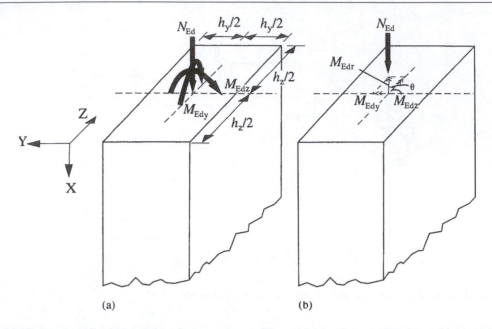

Figure 10.17 Resolution of biaxial bending moments: (a) applied force and moments; (b) resolution of moment vectors

where the neutral axis lies within the section, failure occurs when the strain in the concrete reaches its ultimate value of $\varepsilon_{cu3} = 0.0035$. For cases where the neutral axis lies outside the section, failure occurs when the strain at a depth of $0.5h$ reaches a value of $\varepsilon_{c3} = 0.00175$ (assuming rectangular stress distribution). Consider the case of the eccentrically loaded column of Fig. 10.17(a). By resolution of the moment vectors, the design bending moments M_{Edy} and M_{Edz} at any section can be replaced by a single resultant moment, M_{Edr}, as illustrated in Fig. 10.17(b). The magnitude and direction of the resultant moment are given respectively by:

$$M_{Edr} = \sqrt{\left(M_{Edy}\right)^2 + \left(M_{Edz}\right)^2} \tag{10.24}$$

$$\theta = \tan^{-1}\left(M_{Edy}/M_{Edz}\right) \tag{10.25}$$

When moments are due to eccentric loading, Equation (10.24) can be written as:

$$M_{Edr} = \sqrt{\left(N_{Ed}e_z\right)^2 + \left(N_{Ed}e_y\right)^2} = N_{Ed}\sqrt{\left(e_z\right)^2 + \left(e_y\right)^2}$$

and Equation (10.25) becomes:

$$\theta = \tan^{-1}\left(N_{Ed}e_z/N_{Ed}e_y\right) = \tan^{-1}\left(e_z/e_y\right)$$

The section is then designed to resist the **uniaxial** moment, M_{Edr} and the axial force, N_{Ed}. However, the calculation of the design axial resistance and moment capacities is complicated by the fact that the neutral axis is inclined at an angle θ to the Z-axis of the section.

A typical distribution of strain and stress for the section of Fig. 10.17(b) is illustrated in Fig. 10.18. As for uniaxial bending, strain increases linearly with distance from the neutral axis. Thus, in an elevation **perpendicular to the plane of the neutral axis**, the distribution of strain is triangular. This is illustrated in Fig. 10.19. The strains in the reinforcing bars, now all at different distances from the

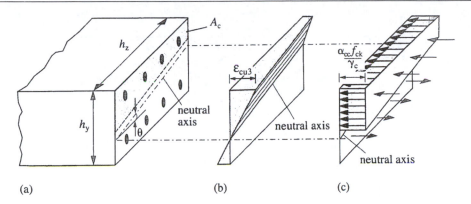

Figure 10.18 Distributions of strain and stress due to N_{Ed} and M_{Edz} : (a) elevation and section; (b) strain distribution; (c) stress distribution

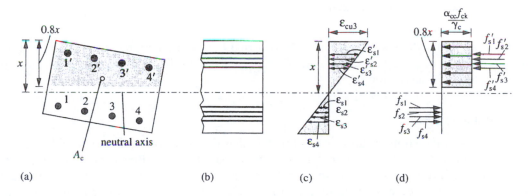

Figure 10.19 Elevation perpendicular to plane of neutral axis: (a) section; (b) elevation; (c) strain; (d) stress

neutral axis, are found from the strain diagram of Fig. 10.19(c). The corresponding stresses are then determined from the stress/strain relationship and, from these, the forces can be calculated. The compressive force in the concrete, F_c, is equal to $(\alpha_{cc}f_{ck}/\gamma_c)A_c$, where A_c is the area of concrete in compression (the shaded area of Fig. 10.19(a)).

It should be remembered when calculating F_c that the shape of the compression zone depends on the inclination of the neutral axis as is illustrated in Fig. 10.20. When the shape of the zone is as illustrated in Fig. 10.19(a), we have, by equilibrium:

$$N_{ult} = F_c + \sum_{i=1}^{4} f'_{si}A'_{si} + \sum_{i=1}^{4} f_{si}A_{si} \tag{10.26}$$

Taking moments of the internal forces about the centre then gives the ultimate moment of resistance for bending about this point.

As for columns with uniaxial bending, the design of a section with biaxial bending and axial force involves iteration of the procedure outlined above for various values of x until the value of N_{ult} equals the applied design axial force, N_{Ed}. Then the ultimate moment of resistance is compared with the applied moment. For this reason, this method of designing short members with combined axial

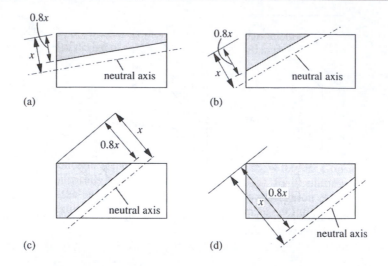

Figure 10.20 Shapes of compression zone

force and biaxial bending is impractical unless carried out on a computer. In practice, computer programs are used or the design of such members is carried out using three-dimensional interaction diagrams known as interaction surfaces.

Interaction surfaces

In Section 10.4, interaction diagrams were introduced as the most convenient method for checking the force/moment capacity of short columns with uniaxial bending. Recall that such interaction diagrams are derived by assuming a range of strain distributions (values for x), deriving the design force and moment for each distribution and using the results to draw the design curve. The more strain distributions chosen, the more accurate the design curve.

In a manner similar to the construction of interaction diagrams, the general solution for the moment/force capacity of short columns with biaxial bending can be represented by a three-dimensional design surface, known as an interaction surface. Like an interaction diagram, an interaction surface is found by assuming a number of strain distributions for the design section. However, for a member with biaxial bending, this means assuming a number of inclinations of the neutral axis in addition to a number of values for x.

Consider a member subjected to a design axial force N_{Ed}, and moments M_{Edy} and M_{Edz}. The resultant moment, M_{Edr}, is given by Equation (10.24) and its inclination to the Z-axis, θ, is given by Equation (10.25). By definition, the inclination of the neutral axis is also equal to θ. Say, first of all, that $M_{Edz} = 0$, that is $M_{Edr} = M_{Edy}$ and $\theta = 90°$. The interaction diagram corresponding to this condition is found as before (by assuming different values for x) and is illustrated in Fig. 10.21(a). Next, assume $M_{Edy} = 0$, that is $M_{Edr} = M_{Edz}$ and $\theta = 0°$. The interaction diagram in this case is illustrated in Fig. 10.21(b). Next, consider a case where both M_{Edy} and M_{Edz} are non-zero, say $M_{Edy} = M_{Edz}$. In this case, $M_{Edr} = \sqrt{2}M_{Edy}$ and $\theta = 45°$. The interaction diagram in this case is illustrated in Fig. 10.21(c). By taking other values for θ, deriving the interaction diagram for each by assuming a range of values for x and plotting the diagrams together, the complete interaction surface for the section is found to be as illustrated in Fig. 10.21(d). Each point on this surface represents a unique set of N_{Ed}, M_{Edy} and M_{Edz} for the particular section which can (just) be resisted (i.e. where $N_{Ed} = N_{ult}$, $M_{Edy} = M_{ult,y}$ and $M_{Edz} = M_{ult,z}$). All points within the interaction surface represent 'safe' designs.

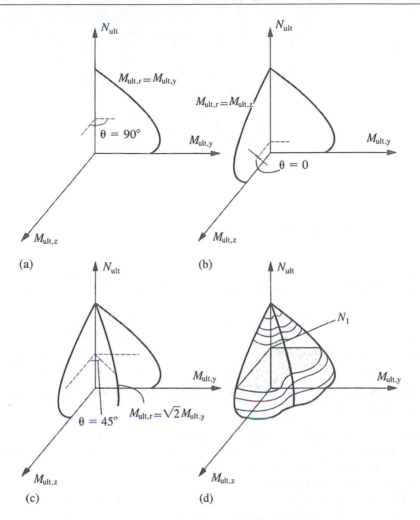

Figure 10.21 Construction of interaction surface: (a) uniaxial bending, $M_{\text{ult,r}} = M_{\text{ult,y}}$; (b) uniaxial bending, $M_{\text{ult,r}} = M_{\text{ult,z}}$; (c) $M_{\text{ult,r}} = \sqrt{2} M_{\text{ult,y}}$; (d) interaction surface

A horizontal section through the interaction surface of Fig. 10.21(d) is considered where the design axial force, $N_{\text{Ed}} = N_1$. The interaction curve obtained, shown in Fig. 10.22, represents all combinations of M_{Edy} and M_{Edz} which will cause failure in the presence of the axial load, N_1. The precise shape of this interaction curve varies for different values of N_1. For small N_1, the curve is approximately linear while for large N_1, the curve is almost circular. Therefore, an expression defining the precise shape of the interaction surface for the general case is not easily found. An approximate expression for the safe zone inside the curve is presented in EC2:

$$\left(\frac{M_{\text{Edz}}}{M_{\text{ult, z}}}\right)^{a} + \left(\frac{M_{\text{Edy}}}{M_{\text{ult, y}}}\right)^{a} \leq 1.0 \tag{10.27}$$

Note: N_{ult}, $M_{\text{ult,z}}$ and $M_{\text{ult,y}}$ are denoted N_{Rd}, M_{Rdz} and M_{Rdy} in EC2.

In inequality (10.27), the exponent 'a' is equal to 2 for circular and elliptical cross-sections. For rectangular cross-sections the value of 'a' is a function of the ratio, $N_{\text{Ed}}/N_{\text{ult}}$, and can be taken from Table 10.1.

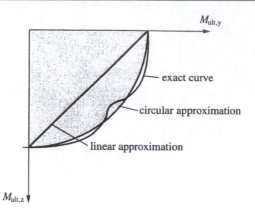

Figure 10.22 Section through interaction surface $N_{Ed} = N_1$

Table 10.1 Values for exponent 'a' in Equation (10.27) for rectangular cross-sections (from BS EN 1992-1-1)

N_{Ed}/N_{ult}	0.1	0.7	1.0
a*	1.0	1.5	2.0

* Use linear interpolation for intermediate values

10.6 Design of slender members for axial force and uniaxial bending

Recall from Section 10.2 that a slender member is defined by EC2 as one for which the second-order moments due to deflection of the member must be accounted for in design. Columns which do not satisfy inequality (10.3) are classed as slender. The principal effect of the second-order moment in a slender column is to reduce significantly its load-carrying capacity.

Consider a typical column subject to a compressive force, N_{Ed}, and a moment, M, as illustrated in Fig. 10.23(a). In this case, both end moments are equal (i.e. $M = M_{01} = M_{02}$). This is equivalent to the column illustrated in Fig. 10.23(b) where the force is applied at an eccentricity. In a short column, the distribution of internal moment due to such an eccentric force is uniform over the member as illustrated in Fig. 10.23(c). Allowing for inaccuracies in the placement of the load, the magnitude of the so-called 'first-order' moment is given by Equation (10.12), that is:

$$M_{01} = M_{02} = N_{Ed}(e + e_i) \geq N_{Ed}e_0$$

Now, if the member is slender it will deflect as illustrated in Fig. 10.23(d) with a maximum deflection of e_2, occurring at mid-span in this case. This lateral deflection increases the eccentricity of the axial force and consequently increases the bending moments in the member. According to EC2, the total design moment, M_{Ed}, in a slender column is equal to:

$$M_{Ed} = M_{0Ed} + M_2 \tag{10.28}$$

where M_{0Ed} is the first-order moment including moment due to imperfections ($M_{0Ed} = M_{01} = M_{02}$ in Fig. 10.23) and M_2 is the second-order moment, ($M_2 = N_{Ed}e_2$). Note that M_{02} and M_2 are not the same thing – M_{02} is the larger of the two first order end moments of the member whereas M_2 is the second order moment.

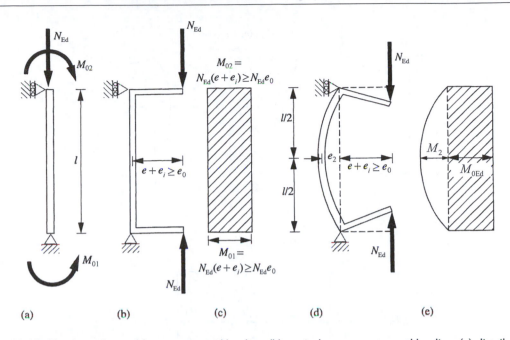

Figure 10.23 Slender column: (a) geometry and loading; (b) equivalent geometry and loading; (c) distribution of first-order moment; (d) deflection due to applied moment; (e) distribution of total moment

The maximum value of M_{Ed} depends on the distribution of M_{0Ed} and M_2 along the length of the member. For the column in Fig. 10.23(a), the distribution of first-order moments is uniform (Fig. 10.23(c)) since the end moments are equal and the maximum moment due to second-order effects occurs at mid-span. The total bending moment diagram for the member of Fig. 10.23(a) is then as illustrated in Fig. 10.23(e).

In most practical design situations the applied end moments are unequal and, in some cases, of opposite eccentricity. These two general cases are illustrated in Fig. 10.24 and Fig. 10.25. From these figures, it can be seen that a member with unequal end moments in which the end eccentricities are of the same sign undergoes single curvature (as illustrated in Fig. 10.24(c)) while a member having eccentricities of opposite sign undergoes double curvature (as illustrated in Fig. 10.25(c)). For a member in single curvature, the distributions of first- and second-order moment combine to give the total moment distribution of Fig. 10.24(e). For a member in double curvature, the total moment distribution is illustrated in Fig. 10.25(e).

Where end moments are unequal and/or of opposite eccentricity, the maximum in-span moment does not necessarily occur at mid-span, but can be calculated using the distribution of M_{0Ed} and M_2 along the length of the member (Equation (10.28)). However, EC2 states that for members where no loads are applied between their ends, M_{01} and M_{02} may be replaced by an **equivalent** first-order end moment, M_{0e}:

$$M_{0e} = 0.6M_{02} + 0.4M_{01} \geq 0.4M_{02} \tag{10.29}$$

In this case, the first-order design moment, M_{0Ed}, can be taken as the equivalent first-order end moment, M_{0e}, rather than the greater of M_{01} and M_{02} (i.e. $M_{0Ed} = M_{0e}$). Thus:

$$M_{Ed} = M_{0e} + M_2$$

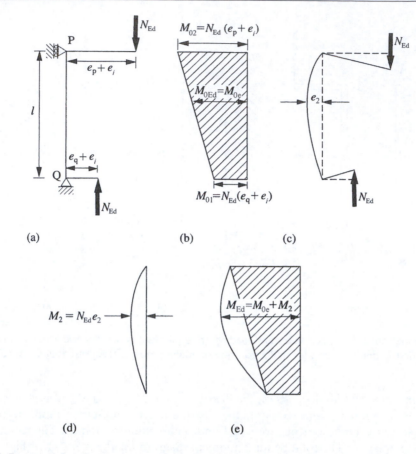

Figure 10.24 Eccentric loading with eccentricities of like sign: (a) geometry and loading; (b) first-order moment; (c) deflected shape; (d) second-order moment; (e) total moment

Concise Eurocode 2, advises that two additional checks should be carried out when a column is classified as slender. It is recommended that the design bending moment, M_{Ed}, in a braced structure should be calculated as:

$$M_{Ed} = \text{greater of } (M_{0e} + M_2; M_{02}; M_{01} + 0.5M_2) \tag{10.30}$$

The effect of slenderness on the capacity of a member to resist load can be visualized using interaction diagrams. Fig. 10.26 shows the interaction diagram for a typical short member. The line OP in the figure represents the path followed when the force is increased while the eccentricity of that force is kept constant. When there is no lateral deflection, e_2, the behaviour of a column under increasing axial force can be represented by the linear force/moment path OP. The corresponding path for a short column, in which the deflection (e_2) is small, is represented by OA. The slope of the line OA remains nearly constant up to the point of failure because the moment, $N_{Ed}e_2$, due to deflection is negligible. This form of failure is commonly known as **material failure** since the member remains stable until its constituent material reaches its ultimate capacity.

For a moderately slender braced member, the lateral deflection is not negligible and the total moment is given by Equation (10.30), where $M_2 = N_{Ed}e_2$. The behaviour of such a member can be represented by the force/moment path OB of Fig. 10.26. For small N_{Ed}, the lateral deflection is small

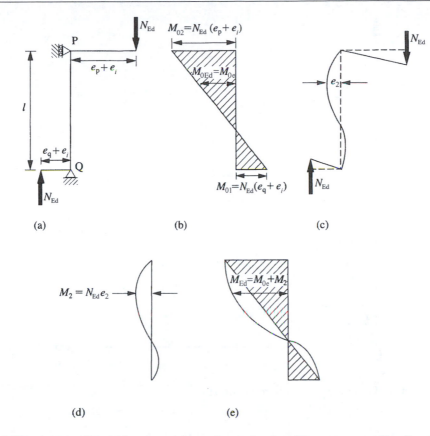

Figure 10.25 Eccentric loading with eccentricities of opposite sign: (a) geometry and loading; (b) first-order moment; (c) deflected shape; (d) second-order moment; (e) total moment

and the contribution of e_2 is negligible. As N_{Ed} is increased, however, the lateral deflection increases more rapidly and causes the force/moment path to become significantly non-linear. Material failure of the member occurs when the interaction diagram is reached at B. At failure, the horizontal distance between the force/moment paths, OB and OP, is equal to $N_{Ed}e_2$. Also, it can be seen from Fig. 10.26 that the design axial resistance at point B is less than at point A, indicating that slenderness has reduced the load-carrying capacity of the column.

The behaviour of a very slender member is represented by the force/moment path OC. In this case, the eccentricity due to second-order effects, e_2, increases so rapidly that it reaches a peak load before a material failure can occur. When the load reaches the peak value, the member becomes unstable and buckles, since any increase in deflection results in a reduction in the force capacity. This type of failure, known as **stability failure**, is rare in reinforced concrete members.

Quantifying e_2

In order to calculate the maximum lateral deflection, e_2, in a slender member caused by the applied loading, we must first consider the curvature of the deflected member. Recall that for a section in bending, the elastic moment curvature relationship is given by:

$$\frac{M}{I} = \frac{E}{R} = E\kappa \tag{10.31}$$

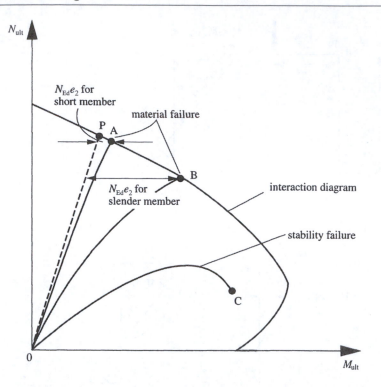

Figure 10.26 Effect of slenderness on force/moment path

where $\kappa = 1/R$ is the curvature. It has been shown by Cranston (1972) that, for a reinforced concrete column under the action of ultimate loads, the maximum curvature, κ_{max}, which occurs at the critical section, depends only on the effective length of the member and the depth of the section.

The precise curvature distribution in slender columns is not known as the cross-section may be cracked to a greater or lesser extent. However, it is reasonable to assume that the shape is somewhere between a triangular and a rectangular distribution as illustrated in Fig. 10.27. The triangular distribution assumes only one critical section (at the apex of the triangle as illustrated in Fig. 10.27(c)) while the rectangular distribution assumes that the member is critical along its entire length. For these two curvature distributions, the maximum deflection can readily be calculated using corollaries to the moment area theorems (Section 4.5) known as the curvature-area theorems. The deflections are found by simply replacing $M/(EI)$ in the moment-area theorems by κ. Hence Equation (4.28) gives the deviation at B, t_{AB}, of the tangent to point A, as illustrated in Fig. 4.51:

$$t_{AB} = \bar{x}_B \int_A^B \kappa dx \qquad (10.32)$$

where $\bar{x}_B$ is the distance from point B to the centroid of the portion of the κ diagram between A and B. As point A, by definition, is the point of maximum deflection, the tangent at A is vertical. Thus:

$$t_{AB} = e_2$$

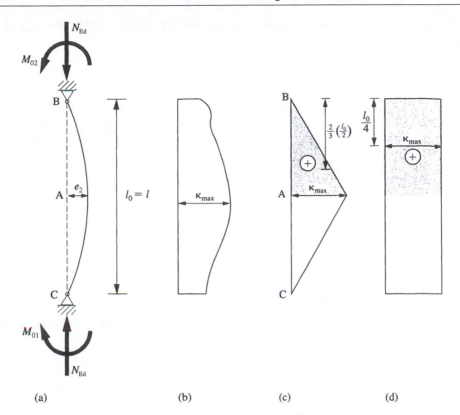

(a)   (b)   (c)   (d)

Figure 10.27 Curvature distributions for slender moment: (a) geometry and loading; (b) actual curvature distribution; (c) triangular curvature distribution; (d) rectangular curvature distribution

For the triangular curvature diagram of Fig. 10.27(c), Equation (10.32) then yields:

e_2 = moment of κ diagram between A and B about point B

$$= \left(\frac{1}{2}\kappa_{max}\frac{l_0}{2}\right)\left(\frac{2}{3}\frac{l_0}{2}\right)$$

$$\Rightarrow e_2 = \frac{\kappa_{max}l_0^2}{12} \tag{10.33}$$

Alternatively, for the rectangular curvature diagram of Fig. 10.27(d), Equation (10.32) yields:

$$e_2 = \left(\kappa_{max}\frac{l_0}{2}\right)\left(\frac{l_0}{4}\right)$$

$$\Rightarrow e_2 = \frac{\kappa_{max}l_0^2}{8} \tag{10.34}$$

From Equations (10.33) and (10.34), a logical estimate of the actual maximum deflection of the member is therefore:

$$e_2 = \frac{\kappa_{max}l_0^2}{10} \tag{10.35}$$

and this is the expression given in EC2.

For members with constant symmetrical cross-sections, EC2 recommends that κ_{max} be approximated by:

$$\kappa_{max} = K_r K_\varphi \left(\frac{f_{yd}}{E_s 0.45d} \right) \tag{10.36}$$

where K_r = reduction factor which takes account of the decrease of the curvature due to the presence of axial force, K_φ = factor to account for creep, f_{yd} = design yield strength of longitudinal reinforcement ($= f_{yk}/\gamma_s$), E_s = modulus of elasticity of reinforcing steel and d = effective depth.

The factor K_r is defined by:

$$K_r = \frac{n_u - n}{n_u - n_{bal}} \leq 1.0 \tag{10.37}$$

where n = the relative axial force ($= N_{Ed}/(A_c f_{cd})$):

$$n_u = 1 + \omega,$$
$$\omega = A_s f_{yd}/(A_c f_{cd})$$

n_{bal} = value of n at maximum moment resistance (may be taken as 0.4).

Note that it is always conservative to assume that $K_r = 1$. The creep factor K_φ is defined by:

$$K_\varphi = 1 + \beta \varphi_{ef} \tag{10.38}$$

where:

$$\beta = 0.35 + \left(\frac{f_{ck}}{200} \right) - \left(\frac{\lambda}{150} \right)$$

The effective creep coefficient, φ_{ef}, is given by:

$$\varphi_{ef} = \varphi_{(\infty, t0)} M_{0Eqp}/M_{0Ed}$$

where $\varphi_{(\infty,t0)}$ = final creep coefficient (Table 8.1), M_{0Eqp} = the first-order bending moment for the SLS quasi-permanent load combination and M_{0Ed} = the first-order bending moment for the ULS design load combination. The application of Equations (10.35) to (10.38) is illustrated by the following example.

Example 10.6 Slender braced column with equal end moments

Problem: Determine the quantity of reinforcement required in a slender braced column of dimensions $l = 6$ m, $b = h = 300$ mm, to resist a compressive force of 1,650 kN and first-order design moment of $M_{0Ed} = 80$ kNm. The first-order bending moment for the SLS quasi-permanent load combination, M_{0Eqp}, is 50 kNm. The cover is 25 mm, $l_0 = 0.7l$, $\varphi_{(\infty,t0)} = 2.0$, $f_{ck} = 35$ N/mm^2 and $f_{yk} = 500$ N/mm^2.

Solution: The basic procedure for the design of slender braced columns with equal end moments consists of the following six steps.

Step 1 – Find K_φ
Determine the interaction diagram to be used. From this diagram, read off N_{bal}, which is used to determine n_{bal}. Assuming, as in Example 10.5, that the reinforcement consists of 20 mm diameter longitudinal bars (concentrated at the ends of the cross-section) and 10 mm diameter links, we have:

$$d = 300 - 25 - 10 - \frac{1}{2}(20) = 255 \text{ mm}$$

Therefore, the design chart of Fig. 10.13 can be used. The value for N_{bal} is the axial force corresponding to the maximum moment. From the design chart of Fig. 10.13:

$$N_{bal}/bh = 8.5$$
$$\Rightarrow N_{bal} = 8.5\,(300)^2 = 765 \text{ kN}$$

Thus:

$$n_{bal} = N_{bal}/(A_c f_{cd}) = 765 \times 10^3/(300^2 \times 0.85 \times 35/1.5)$$
$$n_{bal} = 0.429$$

The slenderness ratio, λ, is given by Equation (10.1) as:

$$\lambda = \frac{l_0}{\sqrt{I/A}} = \frac{l_0}{\sqrt{(bh^3/12)/(bh)}} = \frac{l_0}{\sqrt{(h^2/12)}}$$
$$= \frac{0.7 \times 6000}{\sqrt{300^2/12}} = 48.5$$

The factor to account for creep is calculated using:

$$K_\varphi = 1 + \beta\varphi_{ef}$$

where:

$$\beta = 0.35 + \left(\frac{f_{ck}}{200}\right) - \left(\frac{\lambda}{150}\right) = 0.35 + \left(\frac{35}{200}\right) - \left(\frac{48.5}{150}\right) = 0.202$$

The effective creep coefficient, φ_{ef}, is given by:

$$\varphi_{ef} = \varphi_{(\infty,t0)} M_{0Eqp}/M_{0Ed} = 2.0 \times 50/80 = 1.25$$
$$\Rightarrow K_\varphi = 1 + (0.202)(1.25) = 1.25$$

Step 2 – Find M_{Ed}

Assume initially that $K_r = 1$ when calculating e_2 and derive the total design moment.
   Taking $K_r = 1$, Equation (10.36) yields:

$$\kappa_{max} = K_r K_\varphi \left(\frac{f_{yd}}{E_s 0.45d}\right)$$
$$= 1.0 \times 1.25 \times \left(\frac{500/1.15}{200000 \times 0.45 \times 255}\right)$$
$$= 23.7 \times 10^{-6} \text{ mm}^{-1}$$

The second-order eccentricity, e_2, is given by Equation (10.35), that is:

$$e_2 = \frac{\kappa_{max} l_0^2}{10}$$
$$= 23.7 \times 10^{-6} \left(\frac{(0.7 \times 6000)^2}{10}\right)$$
$$= 42 \text{ mm}$$

The second-order moment corresponding to this additional eccentricity is:

$$M_2 = (1650 \times 10^3)(42)$$
$$= 69.3 \times 10^6 \text{ Nmm}$$
$$= 69.3 \text{ kNm}$$

Thus, the total design moment from Equation (10.28) is equal to:

$$M_{Ed} = M_{0Ed} + M_2 = 80 + 69.3 = 149.3 \text{ kNm}$$

Step 3 – Find $A_{s,tot}$
Determine the required area of reinforcement from the interaction diagram:

$$\frac{N_{Ed}}{bh} = \frac{1650 \times 10^3}{(300)^2} = 18.3$$

$$\frac{M_{Ed}}{bh^2} = \frac{149.3 \times 10^6}{(300)^3} = 5.53$$

From Fig. 10.13, $100 A_{s,tot}/(bh) \approx 3.4$ (i.e. 3.4 per cent). Therefore:

$$A_{s,tot} = 3.4(300)^2/100 = 3060 \text{ mm}^2$$

Step 4 – Calculate K_r
Based on this value for the area of reinforcement, recalculate K_r using Equation (10.37):

$$K_r = \frac{n_u - n}{n_u - n_{bal}} \le 1.0$$

where $n = N_{Ed}/(A_c f_{cd}) = (1650 \times 10^3)/\left((300^2 - 3060) \times 0.85 \times 35/1.5\right) = 0.957$, and:

$$n_u = 1 + \frac{A_s f_{yd}}{A_c f_{cd}} = 1 + \frac{3060 \times 500/1.15}{(300^2 - 3060) \times 0.85 \times 35/1.5} = 1.77$$

Thus:

$$K_r = \frac{1.77 - 0.957}{1.77 - 0.429} = 0.61$$

Step 5 – Iterate
If the new K_r is almost equal to the old value, the design is complete. If not, return to step 2, using the new K_r value to recalculate e_2. Since the new K_r does not equal the old value (of unity), we must return to step 2 and repeat the design process. Starting again with $K_r = 0.62$, the curvature becomes $\kappa_{max} = 14.5 \times 10^{-6}$, the deflection becomes 26 mm and the second-order moment becomes 42 kNm. Accordingly, the required area of reinforcement reduces to 2.7 per cent or $A_{s,tot} = 2,430 \text{ mm}^2$. Recalculation then yields $K_r = 0.57$. The process is repeated in this way until there is convergence of K_r. As six H25 bars provide an area in excess of 2,520 mm^2 and it seems unlikely that the required

area will reduce to a level for which 4H25 or 6H20 will be adequate, no further iteration is performed in this example.

Note: As H25 bars are required, the original calculation of *d* is now incorrect by 3 mm. However, as the area provided is well in excess of that required, this seems unlikely to affect the end result.

Step 6 – Check limits
Ensure that the completed design satisfies the detailing requirements of Section 10.3. The maximum area of reinforcement allowed is 4 per cent of the gross area outside lap locations. The area provided by 6H25 is:

$$\frac{6 \times \pi \times 25^2/4}{300^2} 100 = 3.3 \text{ per cent}$$

This is less than the maximum allowed. The minimum allowable reinforcement is:

$$
\begin{aligned}
A_{s,\min} &= \text{greater of } 0.1 N_{Ed}/\left(f_{yk}/\gamma_s\right) \text{ and } 0.002 A_c \\
&= \text{greater of } \frac{0.1 \times 1650 \times 10^3}{500/1.15} \text{ and } 0.002\left(300^2 - 2945\right) \\
&= 380 \text{ mm}^2
\end{aligned}
$$

It can be seen that the solution is well within the required limits.

Example 10.7 Slender braced column with unequal end moments

Problem: Determine the quantity of reinforcement required in a slender braced column of dimensions $l = 5.5$ m, $b = h = 300$ mm to resist a compressive force of 1400 kN and first-order end moments of $M_{02} = 145$ kNm and $M_{01} = -120$ kNm. The first-order bending moment for the SLS quasi-permanent load combination, M_{0Eqp}, is 35 kNm. The cover is 25 mm, $l_0 = 0.66l$, $\varphi_{(\infty,t0)} = 2.0$, $f_{ck} = 35$ N/mm^2 and $f_{yk} = 500$ N/mm^2.

Solution: To design this member, and indeed all members with unequal end moments, the procedure of Example 10.6 is followed. For step 2, however, the first-order design moment is calculated using Equation (10.29).

Step 1 – Find K_φ
Assume, as for Example 10.6, that the reinforcement for this member consists of 20 mm diameter longitudinal bars and 10 mm diameter links with 25 mm cover to the links. The effective depth is then, as before, $d = 255$ mm and the design chart of Fig. 10.13 applies. As before, $n_{bal} = 0.429$. The slenderness ratio, λ, is given by Equation (10.1) as:

$$\lambda = \frac{l_0}{\sqrt{I/A}} = \frac{0.66 \times 5500}{\sqrt{300^2/12}} = 42$$

The factor to account for creep is:

$$K_\varphi = 1 + \beta \varphi_{ef}$$

where:

$$\beta = 0.35 + \left(\frac{f_{ck}}{200}\right) - \left(\frac{\lambda}{150}\right) = 0.35 + \left(\frac{35}{200}\right) - \left(\frac{42}{150}\right) = 0.245$$

In this example, the end moments are not equal so the first-order design moment, M_{0Ed}, is taken as an equivalent first-order end moment, M_{0e}:

$$M_{0Ed} = M_{0e} = 0.6M_{02} + 0.4M_{01} \geq 0.4M_{02}$$
$$= \text{greater of } [0.6(145) + 0.4(-120)] \text{ and } 0.4(145)$$
$$= \text{greater of } 39 \text{ and } 58$$
$$= 58 \text{ kNm}$$

The effective creep coefficient, φ_{ef}, is given by:

$$\varphi_{ef} = \varphi_{(\infty,t0)}M_{0Epq}/M_{0Ed} = 2.0 \times 35/58 = 1.21$$
$$\Rightarrow K_{\varphi} = 1 + (0.245)(1.21) = 1.30$$

Step 2 – Find M_{Ed}
Assuming initially that $K_r = 1$, Equation (10.36) gives:

$$\kappa_{max} = 1.0 \times 1.3 \times \left(\frac{500/1.15}{200000 \times 0.45 \times 255} \right)$$
$$= 24.6 \times 10^{-6} \text{ mm}^{-1}$$

Therefore, from Equation (10.35):

$$e_2 = \frac{(\kappa_{max})l_0^2}{10}$$
$$= 24.6 \times 10^{-6} \left(\frac{(0.66 \times 5500)^2}{10} \right)$$
$$= 32 \text{ mm}$$

The second-order moment corresponding to this additional eccentricity is:

$$M_2 = (1400 \times 10^3)(32)$$
$$= 44.8 \text{ kNm}$$

From Equation (10.30), the total design moment is given by:

$$M_{Ed} = \text{greater of } (M_{0e} + M_2; M_{02}; M_{01} + 0.5M_2)$$
$$= \text{greater of } (58 + 44.8; 145; 120 + 0.5(44.8))$$
$$= \text{greater of } (102.8; 145; 142.4)$$
$$= 145 \text{ kNm}$$

Step 3 – Find $A_{s,tot}$:

$$\frac{N_{Ed}}{bh} = \frac{1400 \times 10^3}{(300)^2} = 15.6$$
$$\frac{M_{Ed}}{bh^2} = \frac{145 \times 10^6}{(300)^3} = 5.37$$

From the design chart of Fig. 10.13, $100A_{s,tot}/(bh) \approx 2.8$ (i.e. 2.8 per cent). Therefore:

$$A_{s,tot} = 0.028(300)^2 = 2520 \text{ mm}^2$$

Step 4 – Calculate K_r
Calculate K_r using Equation (10.37):

$$K_r = \frac{n_u - n}{n_u - n_{bal}} \leq 1.0$$

where $n = N_{Ed}/(A_c f_{cd}) = (1400 \times 10^3)/\left((300^2 - 2520) \times 0.85 \times 35/1.5\right) = 0.807$, and:

$$n_u = 1 + \frac{A_s f_{yd}}{A_c f_{cd}} = 1 + \frac{2520 \times 500/1.15}{(300^2 - 2520) \times 0.85 \times 35/1.5} = 1.631$$

Thus:

$$K_r = \frac{1.631 - 0.807}{1.631 - 0.429} = 0.69$$

Step 5 – Iterate
The new K_r does not equal the old K_r and so we must return to step 2 and repeat the design process. Starting again with $K_r = 0.69$ does not alter the design moment, that is, M_{02} is still maximum. Therefore, the value of $A_{s,tot} = 2,520$ mm^2 is taken as the final design requirement. Six H25 bars provide an area of 2,945 mm^2.

Step 6 – Check limits
Ensure that the completed design satisfies the detailing requirements of Section 10.3. The maximum area of reinforcement allowed is 4 per cent of the gross area outside lap locations. The area provided by 6H25 is:

$$\frac{6 \times \pi \times 25^2/4}{300^2} 100 = 3.3 \text{ per cent}$$

which is less than the maximum allowed. The minimum allowable reinforcement is:

$$
\begin{aligned}
A_{s,min} &= \text{greater of } 0.1 N_{Ed}/(f_{yk}/\gamma_s) \text{ and } 0.002 A_c \\
&= \text{greater of } \frac{0.1 \times 1400 \times 10^3}{500/1.15} \text{ and } 0.002(300^2 - 2945) \\
&= 322 \text{ mm}^2
\end{aligned}
$$

It can be seen that the solution is well within the required limits.

Design of slender unbraced members

In a frame structure, the stability of the frame against lateral loads depends on the moment capacity of the individual columns. As stated in Section 10.2, columns in such sway frames are termed 'unbraced' because, unlike braced columns, the ends are not fixed against lateral displacement. In a sway frame all the columns at a particular level must displace by the same amount. For this reason it is necessary to consider the entire structure when designing the columns.

A simple sway frame is illustrated in Fig. 10.28. The first-order moments due to lateral load and the resulting deflections of the members are illustrated in Fig. 10.28(b–c), respectively. The additional second-order moments due to the deflection are illustrated in Fig. 10.28(d). The maximum second-order moment in each column occurs at the end of greater stiffness, say, in this case, at the bottom of each column. The second-order moment at the other end of each column can, in the general case, be reduced by an amount in proportion to the ratio of the joint stiffnesses at either end

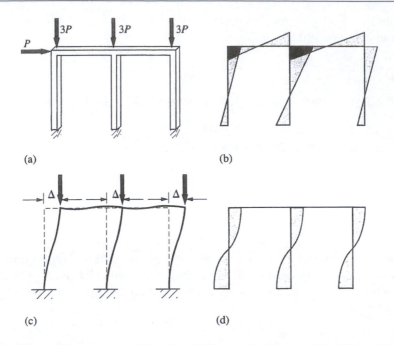

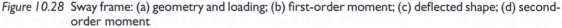

Figure 10.28 Sway frame: (a) geometry and loading; (b) first-order moment; (c) deflected shape; (d) second-order moment

as defined by Equation (10.5). The magnitude of second-order moment in a column at a particular floor level is given by:

$$M_2 = N_{Ed}\Delta \qquad (10.39)$$

where N_{Ed} is the design axial force and Δ is the lateral deflection at that level (see Fig. 10.28(c)).

It can be seen from Fig. 10.28 that the total bending moment will be maximum at the ends of the columns rather than within the span, as is often the case with braced members. Thus, to design an unbraced slender member, the design procedure of Example 10.6 is followed, taking the design moment as the greater of the two total end moments.

10.7 Design of reinforced concrete deep beams

Recall that a deep beam was defined as one in which the span is less than three times the depth. In concrete buildings, deep beams are most commonly found in shear wall systems (deep cantilevers) and are often located on the perimeter of framed structures where they provide stiffness against horizontal loads (see Section 2.3). The applied loads in deep beams are carried by membrane action, which effectively means the load is carried through tension and compression zones. Fig. 10.29(a) illustrates the stress trajectories (load paths) by which load is carried to the supports of a simply supported deep beam subject to a central point load. It can be seen from Fig. 10.29(a) that the compressive stress trajectories are mainly confined to direct paths between the applied force and the supports. Also, the tensile stress trajectories are roughly horizontal and are concentrated near the bottom face of the member. The intensity and trajectory of the internal stresses in a deep beam can be determined using sophisticated linear elastic methods of analysis (such as the finite element method). Even when such programs are available, they do not usually take account of the redistribution of

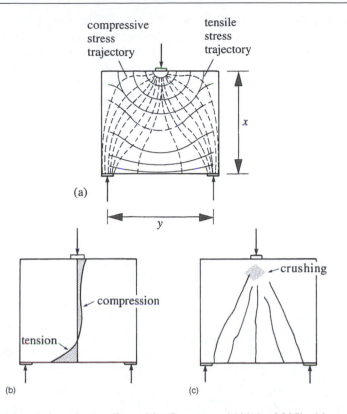

Figure 10.29 Simply supported deep beam (from MacGregor and Wight 2005): (a) stress trajectories; (b) elastic stress distribution; (c) crack pattern

stress which occurs when concrete cracks in tension. For the member of Fig. 10.29(a), the distribution of internal stress at a section directly below the concentrated load is illustrated in Fig. 10.29(b). In contrast to a simply supported beam in flexure (i.e. where the span/depth ratio exceeds three), it can be shown that the tensile stress at the bottom of a deep beam is approximately constant along most of the length of the member.

The results of a linear elastic analysis are only valid while the deep beam remains uncracked. In practice, however, tensile cracks develop in most deep beams at between one-third and one-half of the ultimate load. The crack pattern for the simply supported member of Fig. 10.29(a) is illustrated in Fig. 10.29(c). It can be seen that the tensile cracks form along lines which are perpendicular to the tensile stress trajectories (and hence parallel to the compressive stress trajectories). Thus, to prevent failure of the beam, steel reinforcement is required along the bottom of the member. If sufficient tension reinforcement is provided, failure of the beam can only occur when the compressive strength of the concrete is exceeded on the diagonals between the applied load and the supports. However, it is usually the tension reinforcement requirement which governs the design of deep beams.

Deep beam models for design

Linear elastic methods of analysis are not commonly used in the design of deep beams due to their inherent complexity and because they are only valid while the member remains uncracked and the reinforcement unyielded. In practice, most deep beams are idealized as statically determinate strut and tie (truss) models. EC2 does not place any restriction on the type of truss models that are used. In any model, the force in each strut/tie is established from considerations of equilibrium with the applied loads at ULS. Reinforcement is then provided to carry the tension in the ties. In addition, the compressive stress in the struts is compared with the compressive strength of the concrete in those zones. According to EC2, the design strength for a concrete strut, $\sigma_{Rd,max}$, without transverse tension, is equal to $f_{cd} = \alpha_{cc} f_{ck}/\gamma_c$. For concrete struts with transverse tension:

$$\sigma_{Rd,\,max} = 0.6v'f_{cd}$$

where:

$$v' = 1 - f_{ck}/250$$

The design strength of transverse reinforcement ties is equal to $f_{yd} = f_{yk}/\gamma_s$. Refer to EC2 for the design of nodes where concentrated forces may develop.

Papers such as that of Kotsovos (1988) deal with the design of reinforced concrete deep beams using strut and tie models and books such as MacGregor and Wight (2005) deal with strut and tie analysis in detail. Examples of simplified strut and tie models for deep beams and shear walls are

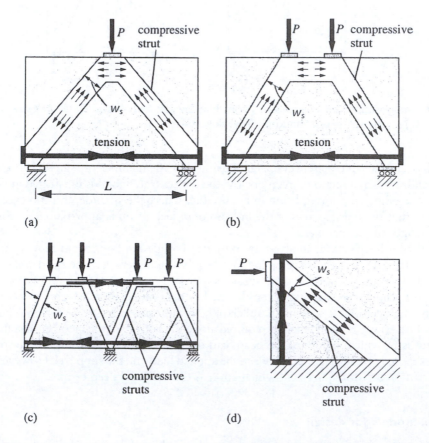

Figure 10.30 Deep beam strut/tie models: (a) simply supported beam with concentrated load; (b) simply supported beam with two concentrated loads; (c) model for continuous beam; (d) model for shear wall

presented in Fig. 10.30, where the width of the strut is denoted w_s. The simply supported deep beam of Fig. 10.30(a), consists of two inclined compressive struts (of concrete) and a single horizontal tie (of reinforcement). Fig. 10.30(b) illustrates the proposed model for a simply supported deep beam subject to two concentrated loads. This type of arrangement may also be used to design a simply supported member subject to a uniform load, where the uniform load is replaced with two equivalent point loads. Similar models can be derived for continuous deep beams (Fig. 10.30(c)) and for shear walls which are, in effect, deep cantilevers (Fig. 10.30 (d)). The following example illustrates the procedure by which the models of Fig. 10.30 are used in design.

Example 10.8 Simply supported deep beam

Problem: The simply supported deep beam of Fig. 10.31(a) is 200 mm thick has a characteristic compressive cylinder strength, f_{ck}, of 30 N/mm². If the beam is subject to two 338 kN loads 5 m from the centre of each support, calculate the required width of the compressive struts, w_s, and the required area of reinforcement. Design of the nodes is not required.

Solution: Fig. 10.31(b) illustrates one half on the strut and tie model. Let T equal the tensile force in the tie, let C_1 equal the compressive force in the horizontal portion of strut between the two concentrated loads and C_2 equal the compressive force in the diagonal strut. Based on a truss analogy, the forces in the struts and ties can be determined using Fig. 10.31(c). As in MacGregor

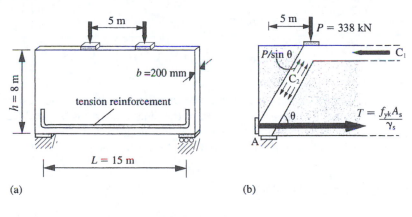

(a)   (b)

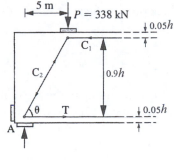

(c)

Figure 10.31 Deep beam of Example 10.8: (a) geometry; (b) strut/tie model; (c) simplified model

and Wight (2005), it is initially assumed that the centroid of the horizontal strut is $0.05h$ from the top surface of the member and the centroid of the tie is $0.05h$ from the bottom surface (see Fig. 10.31(c)). This may need to be altered during the design of the nodes, in which case a further check on the capacity of the struts and ties would be required. With reference to Fig. 10.31(c) the angle, θ, between the tie and the diagonal strut is calculated as:

$$\theta = \tan^{-1}(0.9 \times 8000/5000) = 55°$$

Analysing the strut and tie model using a truss analogy, the compressive force in the diagonal strut can be calculated as:

$$C_2 = \frac{P}{\sin\theta} = \frac{338\,\text{kN}}{\sin 55°} = 413\,\text{kN}$$

The forces in the horizontal strut and tie are equal and are calculated from moment equilibrium as 237 kN (i.e. $C_1 = T = 237$ kN).

The compressive strength of the struts is given by:

$$\sigma_{\text{Rd,max}} = \left(\frac{\alpha_{cc}f_{ck}}{\gamma_c}\right) = \left(\frac{0.85 \times 30}{1.5}\right) = 17\,\text{N/mm}^2 \tag{10.40}$$

Hence, the required area of the compressive struts are:

$$A_1 = \frac{C_1}{\sigma_{\text{Rd,max}}} = \frac{237 \times 10^3}{17} = 13,941\,\text{mm}^2$$

$$A_2 = \frac{C_2}{\sigma_{\text{Rd,max}}} = \frac{413 \times 10^3}{17} = 24,294\,\text{mm}^2$$

The deep beam is 200 mm thick. Therefore, the required width of the horizontal strut is $w_{s1} = 13,941/200 = 70$ mm and the required width of the diagonal strut is $w_{s2} = 24,294/200 = 121$ mm. Note: The width of the strut may also be affected by the node design.

The force in the tie is 237 kN and the required area of reinforcement is given by:

$$A_s = \frac{T}{f_{yk}/\gamma_s} = \frac{237 \times 10^3}{500/1.15} = 545\,\text{mm}^2$$

This can be provided using six bars (three on each face) of 12 mm diameter at a spacing of 150 mm, that is, in a zone 300 mm deep.

Detailing of deep beams

Tensile reinforcing bars in deep beams, corresponding to the ties in the design model, should be fully anchored at their ends by using hooks or some form of anchoring device. In addition, the United Kingdom National Annex (UK NA) to EC2 recommends that deep beams are provided with an orthogonal mesh of reinforcement near each face having a minimum area of 0.2 per cent in each face. If the struts are particularly slender, the area of reinforcement required in these regions may, of course, be even greater. Spacing of the bars of the mesh should not exceed the lesser of twice the thickness of the deep beam or 300 mm.

Problems

Section 10.2

10.1 The roof of a building is supported on vertical cantilevers with clear heights of 8 m and rectangular cross-sections of dimensions 500 mm × 1000 mm. Determine the slenderness ratios given that the structure is braced.

10.2 Repeat Problem 10.1 assuming an unbraced structure.

10.3 The column of Problem 10.1 is subjected to a design axial force of 7000 kN. Determine if the column is short or slender as defined in EC2.

Section 10.4

10.4 A rectangular column of breadth 350 mm, depth 700 mm and effective depth 650 mm is constructed from concrete with $f_{ck} = 40$ N/mm^2. Eight H25 bars placed in the mid-sides and corners of the column give a total area of reinforcement of 3,927 mm^2. Determine if the column has the capacity to resist a combination of $N_{Ed} = 2,000$ kN and $M_{Ed} = 750$ kNm.

10.5 Check the capacity of the section of Problem 10.4 for an applied ultimate axial force of 2,500 kN at an eccentricity (including e_i) of 300 mm.

10.6 Find four key points on the interaction diagram for the section illustrated in Fig. 10.32 given that $f_{ck} = 35$ N/mm^2. The external loads are applied at the centres of the adjoining beams as illustrated.

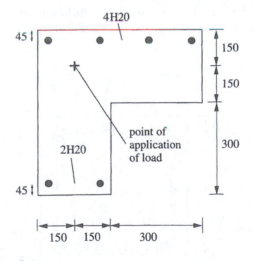

Figure 10.32 Section of Problem 10.6

10.7 Construct the interaction diagram for the octagonal section illustrated in Fig. 10.33, given that $f_{yk} = 500$ N/mm^2 and $f_{ck} = 35$ N/mm^2.

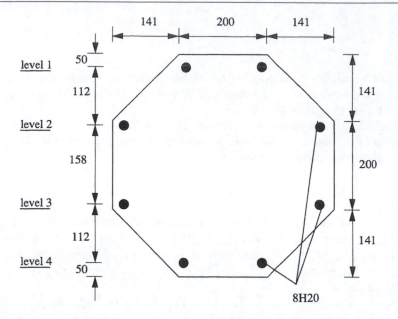

Figure 10.33 Section of Problem 10.7

Section 10.5

10.8 The square column illustrated in Fig. 10.34 is subjected to a design axial force of 400 kN combined with design moments of $M_{Edy} = 500$ kNm and $M_{Edz} = 375$ kNm. Check the capacity of the section if the concrete has a cylinder strength of 40 N/mm^2.

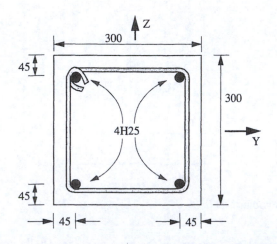

Figure 10.34 Column of Problem 10.8

10.9 Construct an interaction surface for the column whose section is illustrated in Fig. 10.35 given that $f_{ck} = 30$ N/mm^2.

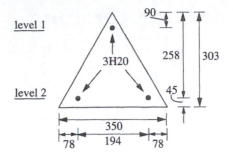

Figure 10.35 Section of Problem 10.9

Section 10.6

10.10   Determine the quantity of reinforcement required in a slender braced column of dimensions, $l = 5.5$ m, $b = h = 300$ mm, to resist a compressive force of 1,400 kN and first-order end moments of 145 kNm and 100 kNm. The effective length is $l_0 = 0.66l$ and the characteristic strengths are, $f_{ck} = 35$ N/mm^2 and $f_{yk} = 500$ N/mm^2.

10.11   For the deep cantilever illustrated in Fig. 10.36 calculate the required width of the compressive struts and the required area of reinforcement to resist the factored ultimate loads given. The wall may be assumed to be braced against lateral buckling.

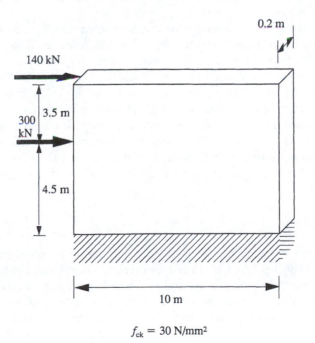

$f_{ck} = 30$ N/mm^2

Figure 10.36 Cantilever of Problem 10.11

Chapter 11

Design for shear and torsion

11.1 Introduction

It can readily be shown that shear force in a beam, V, is related to moment, M, by:

$$V = \frac{dM}{dx}$$

where x is the distance along the beam. Hence, the magnitude of shear equals the slope of the bending moment diagram. It follows that, where there is moment in a beam, there must also be shear, although not necessarily in the same part (see, e.g. Fig. 11.1). In fact, moment tends to be greatest near mid-span or at internal supports. Shear force tends to be large at all supports, internal and external, and small or zero at mid-span. Torsion is moment about a beam's own axis. It is treated in this chapter because it causes shear stresses in a beam and is resisted by mechanisms similar to those which resist shear.

Both shear and torsion are resisted by the concrete itself and by shear/torsion reinforcement where this is present. Reinforcement generally takes the form of shear links, as illustrated in Fig. 11.2(a). Shear tends to cause roughly diagonal cracking and links are provided to carry shear force across such cracks. While links can be inclined (between 45° and 90° to longitudinal reinforcement), it is far more usual for them to be vertical. Torsion tends to cause spiral cracking. Thus, closed links, as illustrated in Fig. 11.2(b), are used to control torsional cracking on all surfaces of the beam including the top surface. Torsion links must be anchored sufficiently by means of hooked ends (Fig. 11.2(b)) or laps. Longitudinal reinforcement is also required for the resistance of torsion. Ordinary longitudinal reinforcement can be used to resist shear if it is 'bent-up' as illustrated in Fig. 11.2(c). However, this is no longer widespread practice in Europe and North America.

Shear

As shear failure in unreinforced members is sudden, a minimum amount of shear reinforcement must always be provided in beams (even when not theoretically required). Fortunately, shear does not generally govern the design of reinforced and prestressed concrete members, that is, the section dimensions are not normally dictated by the requirements for shear. In contrast to beams, slabs do not normally have shear reinforcement and EC2 does not require reinforcement when it is not theoretically required. This is because slabs generally have a higher degree of redundancy than beams; that is, alternative load paths are available should part of the slab be defective. This may also reflect the fact that it is difficult to provide shear reinforcement in floor slabs due to their small depth. In some circumstances, particularly for flat slab construction, shear can govern the required slab depth and/or require the provision of shear reinforcement.

bending moment diagram   bending moment diagram   bending moment diagram

shear force diagram   shear force diagram   shear force diagram

(a)   (b)   (c)

Figure 11.1 Bending moment and corresponding shear force diagrams: (a) simply supported, uniformly loaded; (b) simply supported with point load; (c) three-span uniformly loaded

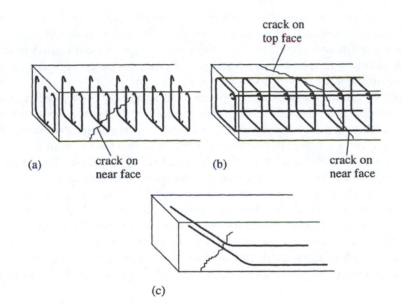

Figure 11.2 Shear and torsion reinforcement: (a) links for resistance of shear; (b) closed links and longitudinal bars for resistance of torsion; (c) bent-up longitudinal bars for resistance of shear

While shear can cause failure of the member at the ultimate limit state (ULS), it does not normally cause any adverse effects at the serviceability limit state. Therefore, provided that rules are adhered to regarding the spacing of shear reinforcement to control cracking, it is generally sufficient to perform a ULS shear safety check only.

Torsion

Torsion occurs frequently in structures but generally its importance is secondary to that of moment and shear. Two types of torsion are commonly identified. **Equilibrium torsion** is that which is required to maintain equilibrium of the member. In such situations, the external load has no other

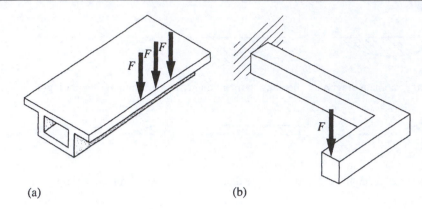

(a)   (b)

Figure 11.3 Examples of equilibrium torsion: (a) eccentrically loaded member; (b) member with change in direction of longitudinal axis

option but to be carried by torsion; that is, there exists no alternative load path or mechanism by which the load can be transferred through the member. Examples of equilibrium torsion are an eccentrically loaded beam and a beam in which there is a change in the direction of its longitudinal axis (as illustrated in Fig. 11.3). Equilibrium torsion is of primary interest in design because failure of the member is inevitable if it has insufficient torsional strength.

The other type of torsion, known as **compatibility torsion**, arises in indeterminate structures having rigidly connected members. It results from the compatibility of deformations of members meeting at a joint. Due to the monolithic nature of their construction, most members in concrete structures undergo a certain degree of compatibility torsion. Perhaps the most common example is an external drop beam supporting a floor slab, as illustrated in Fig. 11.4(a). Bending due to applied loads causes rotation at the edges of the slab which in turn causes rotation of the edge beam (Fig. 11.4(b)). If the edge beam were free to rotate, no torsion would result. However, if it resists rotation because it is attached to columns at its ends, torsion does result. Although such torsion may result in the formation of large cracks at the joint, more serious consequences are unlikely if the member possesses adequate ductility to redistribute the torsional moments. Further, torsional stiffness tends to be less than bending stiffness. This causes an edge beam in such a situation to offer less resistance to rotation with the result that the bulk of the load is transferred from the edge beam to its supporting columns in

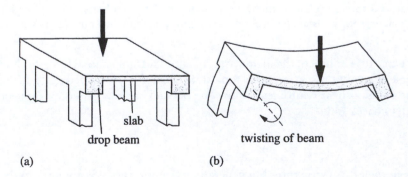

slab

drop beam   twisting of beam

(a)   (b)

Figure 11.4 Example of compatibility torsion: (a) floor slab and drop beams; (b) deflected shape of part of structure

the form of bending. For these reasons, compatibility torsion is generally of secondary interest in design and can often be ignored.

Specifically for EC2, it is usually unnecessary to design for compatibility torsion at the ultimate limit state. However, a minimum amount of suitably spaced shear reinforcement must be provided (according to EC2) to control cracking at the serviceability limit state. Links and longitudinal bars also act as torsional reinforcement, as will be shown in subsequent sections.

11.2   Types of cracking

Types of cracking can be identified by reference to the arrangement of loading illustrated in Fig. 11.5. At the neutral axis of this beam, the axial stress due to bending is zero and a small segment of concrete is subjected to shear stress only (Fig. 11.5(b)). By vertical equilibrium, there exist two equal and opposite vertical stresses, v, acting on either side of the segment illustrated in the figure. However, these two stresses tend to cause clockwise rotation of the element and so two further balancing stresses, as illustrated in Fig. 11.5(c), are required for overall equilibrium of the element. The forces corresponding to these four shear stresses can be resolved into components at 45° to the horizontal as illustrated in Fig. 11.5(d). Combining components, it can be seen that there is a total compressive force of $2v\delta x/\sqrt{2}$ which equals $\sqrt{2}\,v\delta x$ in one direction and a total tensile force of $\sqrt{2}\,v\delta x$ in the other. As these act on a length $\sqrt{2}\,\delta x$ (diagonal of the square) the corresponding stresses are $\sqrt{2}\,v\delta x/\sqrt{2}\,\delta x$ which reduces to v. These stresses, known as principal stresses, are illustrated in Fig. 11.5(e) where it can be seen that the tensile stresses tend to cause cracking at 45°

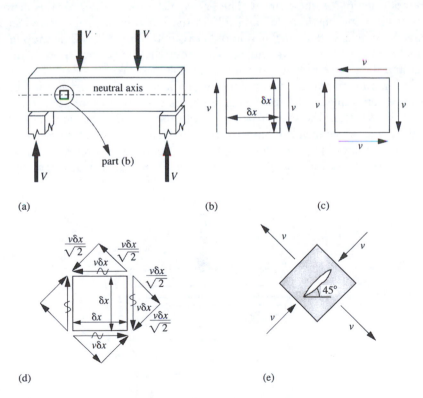

Figure 11.5 Stresses of segment of concrete: (a) simply supported beam; (b) small segment of concrete; (c) shear stresses; (d) resolution of forces; (e) principal stresses

to the horizontal. This cracking occurs when the principal tensile stress exceeds the tensile strength of the concrete.

The segment of concrete considered above was at the neutral axis of the section where axial stress is, by definition, of zero magnitude. For segments at other points, axial stress exists as a result of moment and, in the case of prestressed members, the prestress force. Sections where shear stress predominates throughout the depth are not common but do occur in some cases where the bending moment is negligible or where the web width of the member is small. In such situations, cracking, known as shear-web cracking, occurs, as illustrated in Fig. 11.6.

For a more comprehensive understanding of the behaviour of members with shear, we must consider the combined effects of stresses due to shear, moment and prestress force. For the uncracked rectangular member of Fig. 11.7(a), the distribution of axial stress due to bending is illustrated in Fig. 11.7(b). For such a rectangular section, the distribution of shear stress is parabolic, varying from zero at the top and bottom to a maximum at the centre as illustrated in Fig. 11.7(c). Consider a small segment, A, at the bottom of the member (Fig. 11.7(a)). From Fig. 11.7(b–c), the shear stress, v, in this segment is zero and the bending stress, σ, is at its maximum value. The equivalent principal stresses, illustrated in Fig. 11.7(d), act on a vertical plane. Thus, a vertical flexural crack will form at this point when the tensile stress due to moment exceeds the tensile strength of the concrete.

A small segment at the neutral axis of the member (such as B in Fig. 11.7(a)) was considered above. At this point, the shear stresses are maximum and the bending stress is zero. The corresponding principal stresses, illustrated in Fig. 11.7(e), act on a plane 45° to the horizontal. Therefore, a diagonal crack inclined at 45° tends to develop at this point under the principal tensile stresses. For a segment, C, between the neutral axis and the bottom of the member, a combination of bending and shear stresses is acting on the element. The principal tensile stresses for this element act on a plane inclined at an angle between 45° and 90° to the horizontal, as illustrated in Fig. 11.7(f). Therefore, diagonal cracks with an inclination of between 45° and 90° will develop in this element. Clearly, the inclination of the cracks will decrease towards the neutral axis as the shear stress becomes larger and the axial stress due to bending approaches zero.

By considering small segments at other points within the member, the orientation of the principal stress planes can be determined. These planes can be represented by stress paths through the member, as illustrated in Fig. 11.7(g). The stress paths represent the direction of the principal compressive stresses through an uncracked member. Since the principal tensile stresses act normal to the principal compressive stresses, these lines also represent the lines along which tension cracks will tend to develop in the member. Thus, for a member with shear and bending, the cracks tend to be curved and

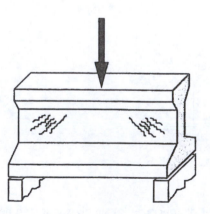

Figure 11.6 Shear-web cracking

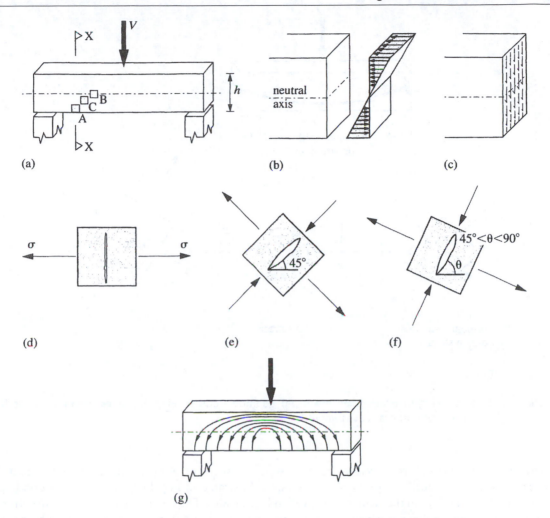

Figure 11.7 Orientation of principal stresses: (a) geometry and loading; (b) axial stress distribution at
section X–X; (c) shear stress distribution at section X–X; (d) segment A of part (a);
(e) segment B of part (a); (f) segment C of part (a); (g) stress paths

vary in slope from 90° at the extreme fibre (point of pure bending) to 45° at the neutral axis (point of
pure shear). This form of cracking is commonly known as **shear-flexure cracking**. In most reinforced
and prestressed concrete members, flexural cracks form before the principal stresses at the neutral
axis are large enough for shear-web cracking to occur. Thus, shear-flexure cracking is generally the
more common type of failure.

   In prestressed concrete members, compressive prestress force is applied which reduces or elim-
inates tensile stress at the serviceability limit state. However, at the ultimate limit state, the applied
loads are significantly larger and, hence, tensile stress does occur. Thus, prestressing has the effect of
increasing the applied load at which flexural cracks form. In addition, the prestress force affects the
orientation of inclined cracks. Consider the introduction of a prestressed tendon to the member of
Fig. 11.7(a), as illustrated in Fig. 11.8(a). Segment B is located at the centroid where direct stress due
to bending (from loading and eccentricity of prestress) is zero. Now, if the prestress force in the
tendon at service is ρP, then the axial component of prestress (compressive stress) at B is $\sigma = \rho P/A_g$,

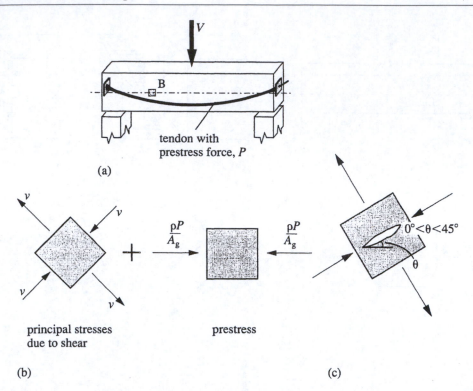

Figure 11.8 ULS cracking in prestressed beam: (a) geometry and loading; (b) stresses on segment B; (c) orientation of crack at segment B

where A_g is the (gross) cross-sectional area. Thus, the segment is now subjected to horizontal compressive stress in addition to the shear stress, as illustrated in Fig. 11.8(b). The corresponding principal stresses acting on the element are given in Fig. 11.8(c). The compressive stress has the effect of reducing the tensile shear stress and, as a consequence, cracks tend to form at an angle of between $0°$ and $45°$ making them flatter than the cracks which form at this point in an ordinary reinforced beam. It will be seen in subsequent sections that these flatter inclined cracks improve the shear strength of members because a greater number of links (shear reinforcement) cross the cracks.

Although the stress paths described above do roughly model the pattern by which shear-flexure cracks develop in practice, they are by no means exact. For instance, the model does not account for the redistribution of the stresses which occurs when cracks are formed and does not account for the relative magnitudes of the shear, bending and compressive stresses which determine the precise inclination of the cracks. Thus, the shear strength of a concrete member cannot be predicted solely by calculating its principal tensile stresses in its uncracked state. In practice, the prediction of shear strength relies on the application of empirical formulae based on observed experimental results.

Average shear stress

In order to determine the capacity of a member to resist shear force, it is necessary to be able to relate the shear stress within the cracked member to the corresponding applied design shear force, V_{Ed}. For an uncracked linear elastic material, the distribution of shear stress over a cross-section is given by the formula:

$$v_y = \frac{V_{Ed}Q}{Ib} \tag{11.1}$$

where v_y = shear stress at a distance y from the centroid of the section

V_{Ed} = design shear force acting on the section

Q = first moment of area of the portion of the section, with cross-sectional area A_y, lying beyond the point where the shear stress is measured (see Fig. 11.9); $Q = A_y\bar{y}$, where $\bar{y}$ is the distance from the centroid of the section to the centroid of the area, A_y

I = second moment of area of the section

b = breadth of the member at the point where the shear stress is calculated

Examples of the respective stress distributions due to a shear force, V_{Ed}, for a rectangular section and an I-section, as derived from Equation (11.1), are illustrated in Fig. 11.10. In each case, the maximum shear stress, v_{max}, (which occurs at the centroid) and the average shear stress across the section, v_{Ed}, are indicated in the figure. Since the proportion of shear carried by the flanges of the I-section is small,

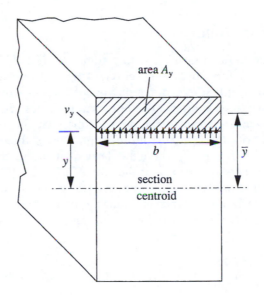

Figure 11.9 Interpretation of Equation (11.1)

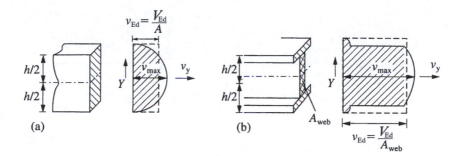

Figure 11.10 Distributions of shear stress: (a) rectangular section; (b) I-section

the average shear stress for the section is approximated by dividing V_{Ed} by the cross-sectional area of the web. The same approximation can be made for other forms of flanged section.

As stated above, Equation (11.1) is only valid for uncracked linear elastic sections. Concrete does not fall within this category as it generally cracks under the action of bending and shear. Furthermore, at the ultimate limit state, the distribution of stress through the section is highly non-linear. Consequently, the distribution of shear stress in a cracked concrete section is somewhat different from that in an uncracked section. A good approximation of the **average** design shear stress in a concrete section is given by:

$$v_{Ed} = \frac{V_{Ed}}{b_w d} \tag{11.2}$$

where d = effective depth of the section and b_w = breadth of the section or, in the case of flanged sections, the minimum web breadth.

Shear strength is a term commonly used to refer to the average shear stress at which failure of a member occurs. Thus, shear strength is the capacity of a member to resist shear force, divided by the area, $b_w d$.

11.3　Types of shear failure

Inclined cracks must develop in a member before complete shear failure can occur. As described above, inclined cracks can form by shear-web cracking or, more commonly, by shear-flexure cracking. It is possible to identify several different types of shear failure in reinforced and prestressed members, each of which exhibits particular failure characteristics. The type of shear failure which occurs in a particular member depends on various factors including its geometry, the load config-uration and the quantity of longitudinal reinforcement. One of the most significant factors is the shear span/effective depth ratio (a_v/d), where the shear span, a_v, is defined as the distance from the edge of the support to the edge of the load, as illustrated in Fig. 11.11(a).

Much of the research into the shear behaviour of concrete beams and slabs, from which empirical formulae have been derived, has been carried out using the load arrangement illustrated in Fig. 11.11(b). It has been found that shear span/effective depth ratios can be divided into the following four general categories:

Category I:　　$0 < a_v/d \leq 1$
Category II:　　$1 < a_v/d \leq 2.5$
Category III:　　$2.5 < a_v/d \leq 6$
Category IV:　　$6 < a_v/d$

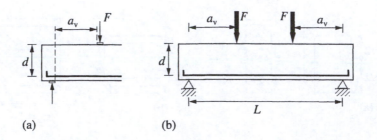

Figure 11.11 Shear span in experimental beam: (a) illustration of shear span; (b) geometry and loading

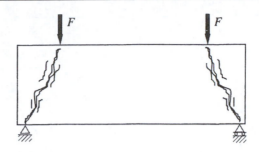

Figure 11.12 Deep beam failure (category I)

Members with very short shear spans or which have a large effective depth, that is, deep beams, fall into category I. The type of failure associated with such members is commonly known as **deep beam failure**, illustrated in Fig. 11.12. Diagonal shear-web cracks form almost in a direct line between the applied load and the support due to the splitting action of the compressive force between the two points. Since the shear force cannot be transmitted across the cracks, the member exhibits a truss-like behaviour with the longitudinal reinforcement (or prestressing tendon) acting as a tie and the compression zones acting as struts (Fig. 11.13). Most commonly, the mode of failure in such a member is anchorage failure at the end of the reinforcement due to the large tensile force. If the reinforcement is bent at the ends, as illustrated in Fig. 11.14, failure can be delayed until the concrete in front of the hook fails in compression.

Members in category II (where a_v/d is larger) behave in a similar manner to category I members initially; that is, shear-web cracks develop in the region between the loaded sections and the supports. However, unlike deep beam failure, the crack often then propagates along the tension reinforcement, destroying the bond between the reinforcement and the surrounding concrete, as illustrated in Fig. 11.15. This form of failure is known as **shear bond failure**. Alternatively, category II members can fail due to dowel failure of the reinforcement where it is crossed by the inclined crack, as illustrated in Fig. 11.16 or by crushing failure of the concrete at the points of application of the loads (Fig. 11.17). The latter type of failure is known as **shear compression failure**, which often occurs explosively and at a load substantially less than that for deep beam failure.

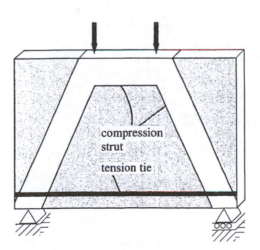

Figure 11.13 Truss model for deep beams

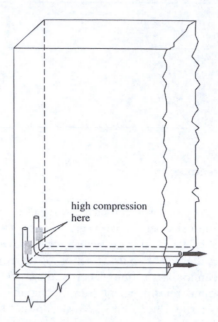

high compression
here

Figure 11.14 Failure of bond in category I members

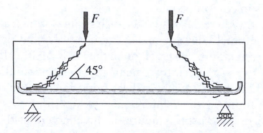

F F

45°

Figure 11.15 Shear bond failure (category II)

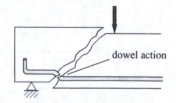

dowel action

Figure 11.16 Dowel failure (category II)

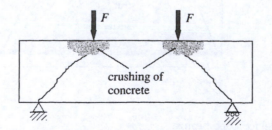

F F

crushing of
concrete

Figure 11.17 Shear compression failure (category II)

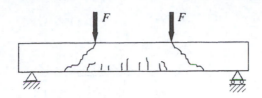

Figure 11.18 Diagonal tension failure (category III)

Members with an a_v/d ratio in category III are most likely to develop flexural cracks before the compressive force is great enough to develop shear-web cracks. The flexural cracks nearest the support (where the shear force is greatest) develop into inclined shear-flexure cracks and propagate towards the applied load as illustrated in Fig. 11.18, splitting the member. This type of failure is most usually known as **diagonal tension failure**. The load at which diagonal tension failure occurs is approximately half that for shear compression/shear bond failure. Since the majority of beams fall within this category it is one of the most common types of failure.

Category IV members are so slender that they tend to fail in pure flexure, as illustrated in Fig. 11.19, that is, the longitudinal reinforcement yields and the concrete above it crushes before shear cracking occurs.

Up to this point, we have only considered the shear failure of members which exhibit beam-type behaviour. This is applicable to beams and to one-way spanning slabs. It is also applicable to two-way spanning slabs supported on walls or beams such as in beam and slab construction. For such two-way spanning slabs, the shear in each of the two coordinate directions is considered separately (see Fig. 11.20).

Another type of shear failure, known as **punching shear failure**, can arise under concentrated loads or at isolated column supports in flat slab construction. The mechanism associated with punching shear failure is illustrated in Fig. 11.21. It is characterized by inclined cracks propagating around edges of the concentrated load or support to form a pyramidal or conically shaped wedge of concrete. Thus, the connection between the slab and the load or support is effectively destroyed. This type of failure can occur suddenly with little or no warning. In general, flat slabs have a punching shear capacity less than their capacity for regular shear but it is necessary to check for both forms of failure.

Subscript notation:
Subscript $_{Ed}$ is added to general notation to indicate a design load effect. For example, V_{Ed} is the design value of applied shear force. Subscript $_{Rd}$ denotes design resistance (e.g. V_{Rd} is the design shear resistance). Additional subscripts are added for more specific notation. For example, $V_{Rd,c}$ is the design shear resistance of a concrete member without shear reinforcement. Similarly, in this chapter, the subscript $_w$ has been added to notation to indicate shear (e.g. f_{ywd} is the design yield strength of shear reinforcement and A_{sw} is the area of shear reinforcement).

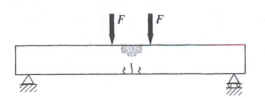

Figure 11.19 Flexural failure (category IV)

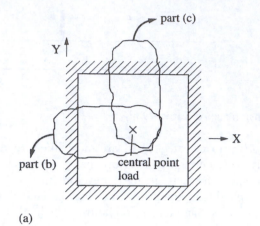

(a)

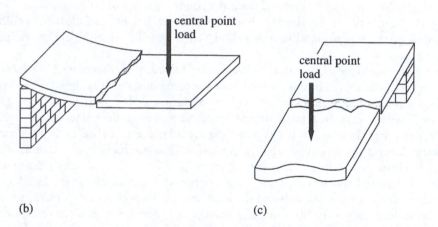

(b)   (c)

Figure 11.20 Two-way spanning slab: (a) plan; (b) failure on face parallel to Y-axis; (c) failure on face parallel to X-axis

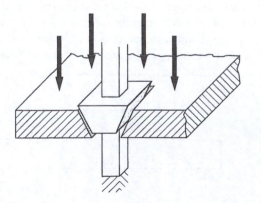

Figure 11.21 Punching shear failure

11.4   Shear strength of members without shear reinforcement

For the design of reinforced and prestressed concrete members with shear, a knowledge of the natural shear resistance of the concrete is essential. To date, most estimates of the shear strength of members without shear reinforcement are based on empirical equations derived from extensive experimental work. These account for the different influencing parameters which affect the shear capacity. Before such equations are presented, however, it is first necessary to illustrate the mechanisms of shear transfer in a cracked section and to review briefly the parameters which influence the capacity to resist shear.

Mechanisms of shear transfer

Consider a typical member in bending and shear, which is reinforced with longitudinal steel against bending. Under its applied loads, the member will crack in one of the characteristic manners, such as that illustrated in Fig. 11.22.

The shear force is transmitted through the cracked member by a combination of three mechanisms. The first is **dowel action** of the reinforcement (ordinary reinforcement or tendons/strands in the case of a prestressed concrete member). This results from the resistance of the reinforcement to local bending as illustrated in Fig. 11.23(a) and the resistance of the concrete to localized crushing near the reinforcement. The second mechanism by which cracked members resist shear is known as **aggregate interlock** and results from the forces transmitted across the crack by interlocking pieces of aggregate (Fig. 11.23(b)). The third mechanism by which cracked members resist shear is through the resistance of the concrete in the uncracked portion of the beam where the axial stress is compressive.

The free-body diagram for the portion of the member between the left-hand support and the first inclined crack is illustrated in Fig. 11.24. The total capacity to resist shear force, $V_{Rd,c}$, comes from the combined effect of the three mechanisms illustrated in that figure. It has been found that dowel action is generally the first to reach its capacity, followed by failure of the aggregate interlock mechanism followed by shear failure of the concrete in the compression zone. However, the precise proportion of the total shear force carried by each mechanism is difficult to establish and, consequently, the shear strength of the concrete is most often represented by a single expression which accounts for all three mechanisms together.

Factors affecting shear capacity

The shear capacity of a member is strongly dependent on the shear span/effective depth ratio, with the capacity decreasing with increasing a_v/d as was seen in the previous section. The strength (grade)

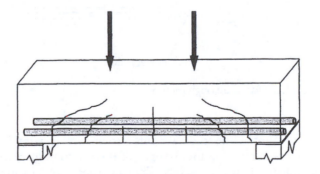

Figure 11.22 Member with longitudinal reinforcement

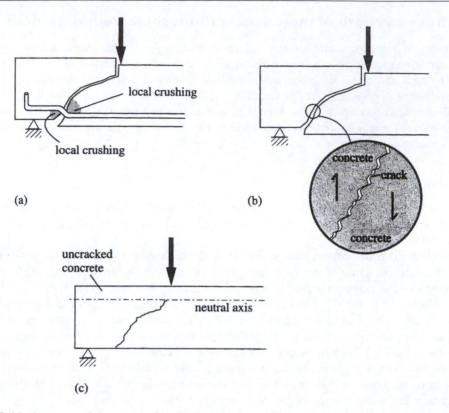

Figure 11.23 Mechanisms of shear transfer: (a) dowel action; (b) aggregate interlock; (c) shear stresses in uncracked concrete

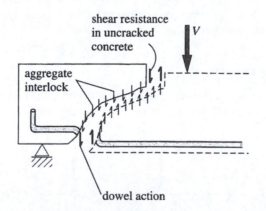

Figure 11.24 Transfer of shear force through member

of the concrete and the amount of longitudinal reinforcement are also dominant factors affecting the shear capacity of members. Increasing the strength of the concrete increases the capacity through each of the three mechanisms of shear resistance. The total shear capacity increases approximately in proportion to $\sqrt{f_{ck}}$. The area of longitudinal reinforcement determines the dowel shear resistance at a cracked section. In addition, an increase in the area of reinforcement effectively increases the aggregate interlock resistance by reducing the width of the inclined crack. The introduction of compressive axial forces, including prestressing, usually enhances the shear capacity of a member while tensile axial forces have the effect of reducing the capacity.

The size of the member is also a major factor affecting its shear capacity. Larger members are, in fact, less effective at resisting shear than smaller members; that is, the shear strength (average shear stress required to cause failure) is significantly less in larger members. The influences of all the parameters described above on the shear strength of a concrete member are incorporated into one empirically derived formula in the next section.

Recommended design formula for shear strength

The expression for the design shear strength of reinforced concrete members without shear reinforcement, with the nationally determined parameters recommended by the UK NA to EC2, is:

$$V_{Rd,c} = [(0.18/\gamma_c)k(100\rho_1 f_{ck})^{1/3} + 0.15\sigma_{cp}]b_w d \tag{11.3}$$

with a minimum value of:

$$\left(0.035k^{3/2}f_{ck}^{1/2} + 0.15\sigma_{cp}\right)b_w d \tag{11.4}$$

where k is a 'depth factor' that is given by:

$$k = 1 + \sqrt{\frac{200}{d}} \leq 2.0 \quad (d \text{ in mm})$$

and ρ_1 = lesser of the longitudinal tension reinforcement ratio, $A_{sl}/b_w d$, and 0.02. A_{sl} is the area of tensile reinforcement which extends (d + required anchorage length) past the section considered. $\sigma_{cp} = N_{Ed}/A_c < 0.2f_{cd}$, where N_{Ed} is the factored design axial force either from an external force or from a prestress force (compression positive), A_c is the area of the concrete cross-section and $f_{cd} = \alpha_{cc}f_{ck}/\gamma_c$ (α_{cc} can be taken as 1.0 for shear, or conservatively as 0.85 for all phenomena).

Equation (11.3) may also be used for regions of single span prestressed concrete members which have cracked in bending. An alternative formula is given in EC2 for regions of single span prestressed members uncracked in bending.

Shear enhancement

In a member where inclined shear-flexure cracks form, the cracks closest to the supports extend outwards from those supports at an inclination of between 30° and 45°, extending a maximum distance of approximately $2d$, as illustrated in Fig. 11.25. In such circumstances, concentrated loads that are located within this distance from the supports will be transferred more by direct compression to the supports than by shear mechanisms. Thus, the resistance to shear force in these regions is effectively increased.

EC2 allows for this effective 'enhancement' of shear resistance near supports. Where concentrated loads are applied within a distance of $0.5d \leq a_v \leq 2d$ from the edge of the support, the contribution of this load to the shear force, V_{Ed}, may be reduced by multiplying by a factor, $\beta = a_v/2d$. However, this is only allowed when the following conditions are satisfied:

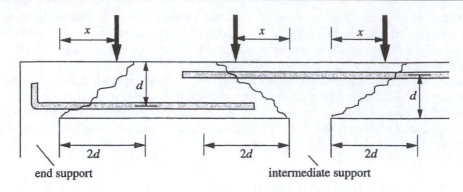

Figure 11.25 Applied load near supports

(a) This reduction to V_{Ed} can be applied when checking $V_{Rd,c}$ in Equation (11.3). However, this applied design shear force V_{Ed} (not reduced) must not be greater than $V_{Rd,max}$, where:

$$V_{Rd,max} = 0.5 b_w d \upsilon f_{cd} \qquad (11.5)$$

and:

$$\upsilon = 0.6 \left[1 - \frac{f_{ck}}{250}\right] \qquad (11.6)$$

(b) The loading and support reactions are such that they cause only diagonal compression in the member. Thus, the support must be provided at the bottom of the beam when the loading is applied from the top.

(c) Longitudinal tension reinforcement must be anchored at the support.

Note: EC2 recommends that a value of $a_v = 0.5d$ be used when $a_v \leq 0.5d$.

Alternatively, beams with loads applied near to supports may be designed using strut and tie models (see Section 10.7). Uniformly loaded members do not involve a concentrated load near the support. However, enhancement does apply for that portion of the load which is applied near the support. In addition, the design shear force, V_{Ed}, does not need to be checked between the face of the support and a distance d from that point. However, the shear force at the support must be less than $V_{Rd,max}$, given by Equation (11.5).

Example 11.1 Shear enhancement

Problem: Check the capacity of the continuous member, illustrated in Fig. 11.26, to resist the given distribution of shear force between A and B. The characteristic cylinder strength of the concrete is $f_{ck} = 35$ N/mm². *Note*: The contribution of the load at C to the shear force between A and C is 92 kN.

Solution: The shear force at point A is 203 kN and this extends to point C. The load at C is applied 750 mm from the edge of the support (i.e. $a_v = 750$ mm), which is between $0.5d (= 250$ mm)

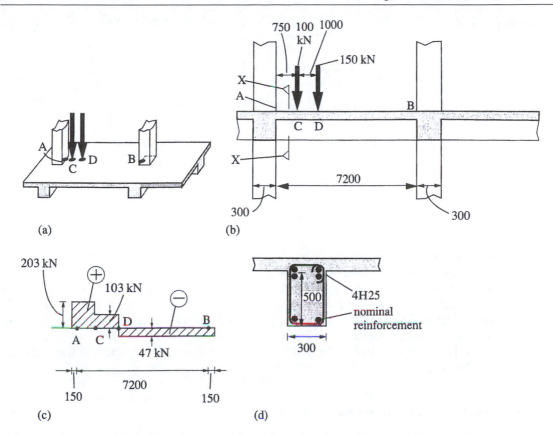

Figure 11.26 Member of Example 11.1: (a) pictorial view; (b) elevation (dimensions in mm); (c) shear force diagram; (d) section X–X

and $2d$ ($= 1,000$ mm). Therefore, the contribution of this load to the shear force may be reduced by a factor of:

$$\beta = a_v/2d$$
$$= 750/(2 \times 500) = 0.75$$

The contribution of this load to the shear force between A and C is 92 kN (shear force at A due to point load of 100 kN at C). Thus, the contribution of other loads to the shear force between A and C is 111 kN. This results in a reduced total design shear force between A and C of:

$$V_{Ed}(\text{reduced}) = \beta(92) + 111 = (0.75)(92) + 111 = 180 \text{ kN}$$

The area of four 25 mm diameter bars is $4(\pi\, 25^2/4) = 1,963$ mm^2. Hence, the reinforcement ratio for the (longitudinal) top reinforcement is:

$$\rho_l = \frac{A_{sl}}{b_w d} = \frac{1963}{300 \times 500} = 0.013$$

(This reinforcement is assumed to extend the required distance on both sides of the point under consideration.)

The depth factor, k, is given by:

$$k = 1 + \sqrt{\frac{200}{d}} = 1 + \sqrt{\frac{200}{500}} = 1.63 \leq 2.0$$

Thus, the shear force capacity at this point is, from Equation (11.3):

$$V_{Rd,c} = \left[(0.18/1.5)(1.63)(100 \times 0.013 \times 35)^{1/3} + 0 \right] 300 \times 500 = 105 \, \text{kN}$$

with a minimum value of:

$$\left(0.035(1.63)^{3/2}(35)^{1/2} + 0 \right) 300 \times 500 = 65 \, \text{kN}$$

Thus, shear reinforcement will be required since the design shear force of V_{Ed}(reduced) $= 180 \, \text{kN}$, is greater than the shear capacity of the concrete section, $V_{Rd,c} = 105 \, \text{kN}$.

At the point of application of the second load, D, the maximum shear is 103 kN. This point is beyond 2.0 d from the face of the support, so no reduction in shear force applies. As the capacity ($V_{Rd,c}$) exceeds the applied shear of 103 kN at this point, minimum shear reinforcement is sufficient here. If the area of longitudinal reinforcement remains the same between D and B, the shear capacity is also 105 kN and the applied shear is 47 kN. Thus, minimum shear reinforcement is also sufficient for this part of the beam.

Members with inclined prestressing tendons

In prestressed members with inclined tendons, the vertical component of the prestress force is $\rho P(\sin \theta_p)$, where ρ is the loss ratio and θ_p is the inclination of the tendon (Fig. 11.27(e)). As mentioned in Chapter 9, the profile of a tendon often follows the shape of the bending moment diagram (see Fig. 11.27). When this is the case, the vertical component of prestress frequently acts in the opposite direction to the applied shear force, V_{Ed}. When this occurs, the magnitude of the applied shear force is reduced by the prestress and the net magnitude is given by:

$$V_{net} = V_{Ed} - \gamma_p \rho P(\sin \theta_p) \tag{11.7}$$

where γ_p is the partial factor for prestress (the UK NA to EC2 specifies $\gamma_{P,fav} = 0.9$ for favourable actions, $\gamma_{P,unfav} = 1.1$ for unfavourable actions). Thus, the introduction of a profiled prestressing tendon can effectively increase the shear strength of a member. It should, however, be noted that there can be situations where prestress acts with the applied shear and increases rather than reduces the design shear force.

When post-tensioned tendons are placed in ungrouted ducts, the nominal web width for the calculation of shear capacity should be reduced by $1.2\Sigma\phi$, where ϕ is the diameter of one duct and the sum of all duct diameters, $\Sigma\phi$, is determined at the most unfavourable depth. When ducts are grouted (which is normally the case), EC2 specifies that for $\phi > b_w/8$, the nominal web width is equal to the actual web width minus $0.5\Sigma\phi$. This allows for the fact that the grout does offer some resistance to shear. Where $\phi \leq b_w/8$ the nominal web width is equal to the actual web width. The reduction in the web width due to the presence of post-tensioning ducts can significantly affect the requirements for shear, particularly in flanged members where the web width may not be very wide.

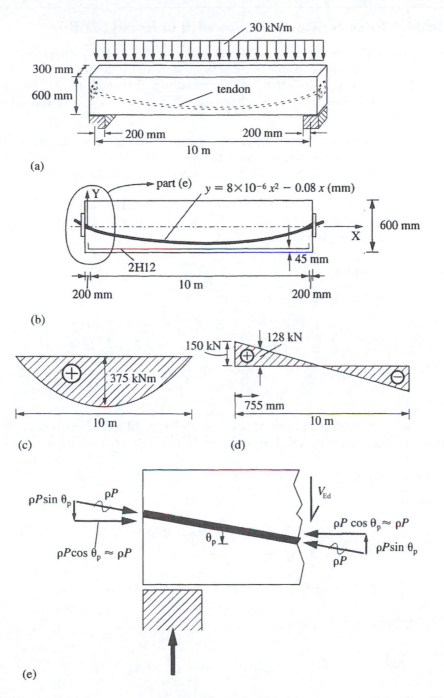

Figure 11.27 Post-tensioned prestressed beam of Example 11.2: (a) geometry and loading; (b) elevation showing tendon and ordinary reinforcement; (c) bending moment diagram; (d) shear force diagram; (e) free-body diagram near left support (elevation)

Example 11.2 Post-tensioned beam with uniform loading

Problem: Determine the shear capacity of the beam of Fig. 11.27 where the area of pre-stressing strand is 1,800 mm^2 and the prestress force after all losses is $pP = 1,400$ kN. The factored ULS loading is 30 kN/m and the resulting distributions of moment and shear are given in Fig. 11.27(c–d). The duct diameter is 75 mm and the characteristic cylinder strength is 40 N/mm^2.

Solution: For the calculation of effective depth, the ISE manual recommends that inclined prestressing strands should be ignored. Hence, the effective depth is, from Fig. 11.27(b), $d = 600 - 45 = 555$ mm. Allowing for enhancement of shear capacity near the support in uniformly loaded members, it is necessary only to check the shear at a distance of d (= 555 mm) from the face of the support or, in this case, 755 mm from the centre of the support. At that point, the applied shear force (by linear interpolation in Fig. 11.27(d)) is 128 kN. The slope of the tendon profile at this point is found by differentiating the equation given in Fig. 11.27(b):

$$y = 8 \times 10^{-6}x^2 - 0.08x$$
$$\Rightarrow \frac{dy}{dx} = 16 \times 10^{-6}x - 0.08$$

At $x = 755$, the slope is:

$$\frac{dy}{dx}(x = 755) = -0.0679$$

Hence the tendon is inclined at an angle of $\tan^{-1}(-0.0679) = 3.88°$ and the vertical component of prestress is $1,400\sin(3.88°) = 95$ kN. The net applied shear force is therefore:

$$V_{net} = 128 - 0.9 \times 95$$
$$= 43\text{kN}$$

Assuming the prestressed single span section is cracked in bending at ULS, the shear capacity is given by:

$$V_{Rd,c} = [(0.18/\gamma_c)k(100\rho_1 f_{ck})^{1/3} + 0.15\sigma_{cp}]b_w d$$

where the depth factor, k, is given by:

$$k = 1 + \sqrt{\frac{200}{d}} = 1 + \sqrt{\frac{200}{555}} = 1.6 \le 2.0$$

Allowing for the presence of a grouted duct of 75 mm diameter where $\phi > b_w/8$, the nominal web width is:

$$b_w = 300 - 0.5 \times 75$$
$$= 263 \text{ mm}$$

As the ordinary reinforcement consists of two 12 mm diameter bars and $A_p = 1,800$ mm^2, the longitudinal reinforcement ratio is:

$$\rho_l = \text{lesser of} \left(\frac{A_{sl}}{b_w d} \text{ and } 0.02 \right)$$

$$= \text{lesser of} \left(\frac{2(\pi 12^2/4) + 1800}{263 \times 555} \right) \text{and } 0.02$$

$$= \text{lesser of} \left(\frac{226 + 1800}{263 \times 555} \right) \text{and } 0.02$$

$$= 0.01388$$

This reinforcement is assumed to extend the required distance on both sides of the point under consideration.

The prestress in the member is:

$$\sigma_{cp} = \frac{\gamma_p \rho P}{A_c}$$

$$= \frac{0.9 \times 1400 \times 10^3}{300 \times 600 - 226 - 1800}$$

$$= 7.08 \, \text{N/mm}^2$$

Hence, the shear capacity is, from Equation (11.3):

$$V_{Rd,c} = \left[(0.18/1.5)(1.6)(100 \times 0.01388 \times 40)^{1/3} + 0.15(7.08) \right] 263 \times 555 = 262 \, \text{kN}$$

As $V_{Rd,c}$ greatly exceeds the net applied force of 43 kN, there is ample capacity to resist shear and minimum shear reinforcement is sufficient.

11.5   Shear strength of members with shear reinforcement

When the applied design shear force, V_{Ed}, exceeds the shear capacity of the concrete, $V_{Rd,c}$, shear reinforcement must be provided. Even when the applied shear in beams is less than the shear capacity, a minimum quantity of shear reinforcement must be provided.

Shear links, such as those illustrated in Fig. 11.2(a), are the most common form of shear reinforcement. Links are generally placed vertically in the web of the member as illustrated and should be fixed around the longitudinal reinforcement. For variable depth members, vertical links are no longer perpendicular to the longitudinal axis of the beam and special care must be taken in their design. Another form of shear reinforcement is bent-up bars (Fig. 11.2(c)), which are bars of bottom longitudinal reinforcement bent-up at their ends to cross the cracks developed by shear stresses. For beams, however, EC2 recommends that bent-up bars should not be used for shear reinforcement except in combination with links and, furthermore, at least 50 per cent of the shear reinforcement should be in the form of links. More importantly perhaps, bent-up bars, while effective in terms of material, have a high associated labour cost and tend to add to the overall cost of construction.

The minimum quantity of shear reinforcement for beams, recommended by EC2, is given by:

$$\rho_{w,min} = \frac{0.08\sqrt{f_{ck}}}{f_{yk}}$$ (11.8)

This minimum value may be disregarded in minor structural elements such as lintels of less than 2 m span. For slabs, it is not necessary to provide any shear reinforcement provided $V_{Rd,c} > V_{Ed}$. However, where links are to be provided, a minimum area of reinforcement (calculated using Equation (11.8)) should be provided. Shear reinforcement is not effective in slabs with a depth of less than 200 mm, so where shear reinforcement is to be provided, the depth should be at least 200 mm. The minimum quantity of reinforcement is specified in terms of the shear reinforcement ratio, ρ_w. For vertical links in prismatic members, this is the ratio of the cross-sectional area of link to the area of concrete in plan (Fig. 11.28), that is:

$$\rho_w = \frac{A_{sw}}{sb_w}$$ (11.9)

where A_{sw} = area of shear reinforcement within length, s; for links with two legs each, the combined area from both legs is used in the calculation of A_{sw}
s = spacing of shear reinforcement
b_w = web breadth or, in the case of rectangular sections, actual breadth

Mechanism of shear transfer

Shear reinforcement, like flexural reinforcement, does not prevent cracks from forming in a member. Its purpose is to ensure that the member will not undergo shear failure before the full bending capacity of the member is reached. When inclined cracks form in a member with shear reinforcement, the bars that cross the cracks contribute to the shear resistance of the member, as illustrated in Fig. 11.29. The shear capacity is provided by the reinforcement and the concrete (contribution from dowel action, aggregate interlock and the shear stresses in the uncracked concrete) as illustrated in Fig. 11.29.

If the applied shear force is sufficiently large, the shear reinforcement will reach its yield strength. Beyond this point, the reinforcement behaves plastically and the cracks open more rapidly. As the

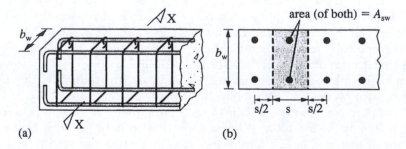

Figure 11.28 Definition of ρ_w: (a) pictorial view; (b) section X–X (plan view)

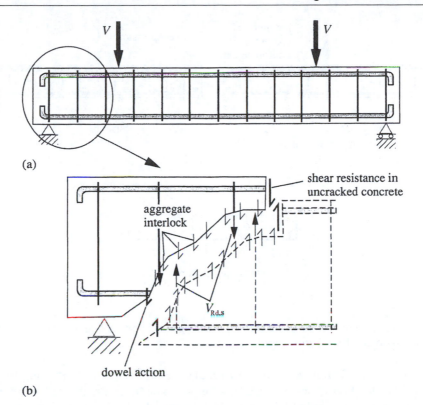

(a)

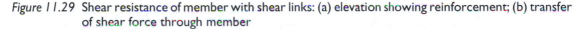

(b)

Figure 11.29 Shear resistance of member with shear links: (a) elevation showing reinforcement; (b) transfer of shear force through member

cracks widen, the proportion of shear resisted by aggregate interlock is reduced, forcing an increase in dowel action and shear stress in the uncracked portion of the section. Failure finally occurs by dowel splitting or by crushing of the concrete in the compression zone.

Truss model for members with shear reinforcement

To quantify the behaviour and strength of members with shear reinforcement, an equivalent truss model is frequently used to represent the behaviour of members in shear. The simplest and earliest of the available models is described here as it clearly demonstrates the basic principles. When subjected to combined moment and shear, a member with shear reinforcement, such as that illustrated in Fig. 11.30(a), develops inclined cracks in the same way as a member without shear reinforcement. For the cracked member illustrated, compressive stress develops in the concrete at the top of the section and tensile stress develops in the longitudinal reinforcement at the bottom of the section as illustrated in Fig. 11.30(b). In addition, compressive forces are developed along inclined paths between the cracks and tensile forces are developed in each of the links crossing the cracks. All of these internal forces are illustrated in Fig. 11.30(b), and together they are similar in pattern to the forces generated in a truss such as that illustrated in Fig. 11.30(c).

For the general truss analogy, the tension members corresponding to the shear reinforcement are inclined at an angle, α, to the horizontal, i.e., to the main reinforcement. For vertical links, this angle, α, is 90° as illustrated in Fig. 11.30(c). In a similar manner, the compression

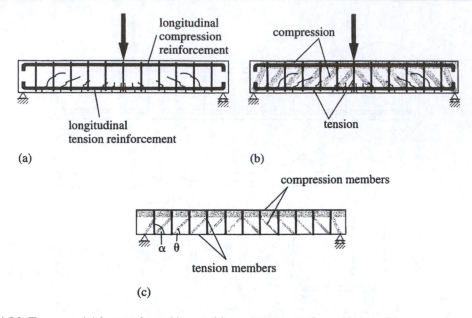

Figure 11.30 Truss model for reinforced beam: (a) cracking in reinforced beam; (b) zones of compression between cracks; (c) truss model

members are inclined at an angle, θ, to the horizontal where $0° < \theta < 90°$. In the simplest of the truss models, the inclination of the concrete 'struts' is assumed to be constant, generally with $\theta = 45°$. EC2 allows θ to vary between 22° and 45°. In more complex models, the inclination of the concrete struts is assumed to vary along the length of the member. This latter model is considered to be more accurate and its use can result in significant savings in shear reinforcement. An example of such a model is illustrated in Fig. 11.31.

To determine the capacity of the reinforcement to resist shear force, $V_{Rd,s}$, consider the analogous truss of Fig. 11.32(a), having constant concrete strut inclinations. The forces acting on a portion of the truss to the left of section X–X (a section taken parallel to the concrete strut) are illustrated in Fig. 11.32(b). For the purpose of clarity, the truss illustrated in Fig. 11.32(a–b) is simplified to show only one link joining adjacent compression struts. In reality, the spacing of links is typically less than the spacing of the compression struts and there are a number of links crossing a section such as X–X. Hence, the total tensile force that can safely be carried by the shear reinforcement crossing this section is:

$$V_{Rd,s} = nA_{sw}f_{ywd} \tag{11.10}$$

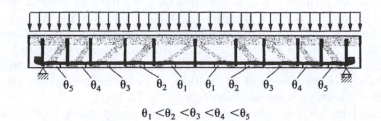

$$\theta_1 < \theta_2 < \theta_3 < \theta_4 < \theta_5$$

Figure 11.31 Truss with struts at various angles

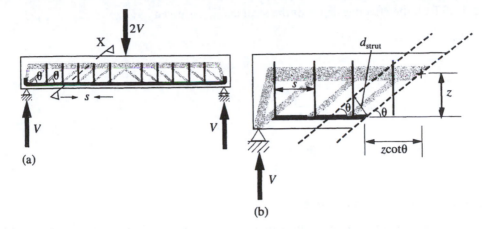

Figure 11.32 Details of truss model: (a) elevation; (b) elevation of beam left of X–X

where n is the total number of links or bent-up bars crossing the section, A_{sw} is the cross-sectional area of each link, f_{ywd} is the design yield strength of the links ($= f_{ywk}/\gamma_s$), f_{ywk} is the characteristic yield strength of the links and γ_s the material factor of safety for steel ($\gamma_s = 1.15$). If the spacing of the links is given by s, the total number of bars crossing section X–X and the total number of compression struts is:

$$n = z \cot\theta / s \tag{11.11}$$

where z is the lever arm between the tensile force in the longitudinal bottom reinforcement and the horizontal compression force in the concrete (according to EC2, z can be approximated as $0.9d$). Thus, Equation (11.10) becomes:

$$V_{Rd,s} = \frac{A_{sw}f_{ywd}(z \cot\theta)}{s} \tag{11.12}$$

The compressive force acting on the concrete struts is assumed to generate a uniform stress distribution in each strut. This will cause compression failure in the strut when the applied stress equals the compressive strength of the concrete. The maximum force that can be taken by each strut is:

$$C_{strut} = \alpha_{cw}\nu_1 f_{cd} b_w d_{strut} \tag{11.13}$$

where α_{cw} is a factor which takes account of the state of stress in the compression chord ($= 1.0$ for non-prestressed structures), f_{cd} is the compressive strength of the concrete struts ($f_{cd} = \alpha_{cc}f_{ck}/\gamma_c$), $\alpha_{cc} = 1.0$ for shear and d_{strut} is the depth, as illustrated in Fig. 11.32(b). The **strength reduction factor**, ν_1, reflects the reduced strength of concrete cracked in shear. The shear force required to cause failure of the compressive struts, $V_{Rd,max}$, is the vertical component of C_{strut} multiplied by the number of struts and is given by:

$$V_{Rd,max} = C_{strut}(\sin\theta)n = \alpha_{cw}\nu_1 f_{cd} b_w d_{strut}(\sin\theta)n \tag{11.14}$$

From Fig. 11.32(b), the effective depth of the struts, d_{strut}, is given by:

$$d_{strut} = s(\sin\theta) \tag{11.15}$$

$$\Rightarrow V_{Rd,max} = \alpha_{cw}\nu_1 f_{cd} b_w s(\sin^2\theta)n$$

$$= \alpha_{cw}\nu_1 f_{cd} b_w s(\sin^2\theta)\frac{z(\cot\theta)}{s}$$

$$V_{Rd,max} = \alpha_{cw}\nu_1 f_{cd} b_w z(\cos\theta)(\sin\theta) \tag{11.16}$$

Design formulae recommended by EC2

Where shear reinforcement is required, other than the minimum, EC2 recommends the **variable strut inclination method**, which allows the designer to select the inclination of the notional concrete struts. EC2 allows any values for θ within the range $22° < \theta < 45°$, provided certain conditions on curtailment are satisfied. This design method for members with vertical shear links is outlined here.

According to the method outlined in EC2, the section is designed assuming that all shear is taken by the shear reinforcement. However, the maximum shear in the section is limited by the capacity of the concrete struts. Thus, the shear capacity of a section with shear reinforcement is the lesser of $V_{Rd,s}$ (Equation (11.12)) and $V_{Rd,max}$ (Equation (11.16)). If the diagonal strut has sufficient capacity, the required area of shear reinforcement can then be determined.

The first step is to check if the capacity of the strut is sufficient for a strut angle between 22° and 45° (*Note*: A shallower angle requires less shear reinforcement but results in a lower value of concrete strut capacity). If the capacity of the concrete strut is not sufficient within this range, the section should be resized or the concrete strength should be increased. If the section is adequate, the required strut angle, θ, is determined by equating the design shear force and the maximum shear resistance:

$$V_{Ed} = V_{Rd,max} = \alpha_{cw}\nu_1 f_{cd} b_w z(\cos\theta)(\sin\theta)$$

If the stress in the shear reinforcement is greater than $0.8 f_{yk}$, $\nu_1 = \nu$, which is given by Equation (11.6). For lower levels of stress in the shear reinforcement, alternative values are provided in EC2. The required area of shear reinforcement can then be determined using θ, ensuring that the capacity of the section due to shear reinforcement is greater than the applied design shear force, V_{Ed}. For vertical links, Equation (11.12) then gives:

$$\frac{A_{sw}}{s} \geq \frac{V_{Ed}}{f_{ywd}z(\cot\theta)} \tag{11.17}$$

It should be noted, when using these equations, that the spacing of the reinforcement, s, is restricted by detailing rules given in EC2. An expression is also provided in EC2 to determine the additional tensile force in the longitudinal reinforcement due to shear in the member:

$$\Delta F_{td} = 0.5 V_{Ed}(\cot\theta - \cot\alpha)$$

The total tensile force in the longitudinal reinforcement due to moment and shear (i.e. $M_{Ed}/z + \Delta F_{td}$) should not be taken as greater than $M_{Ed,max}/z$, where $M_{Ed,max}$ is the maximum moment along the beam.

Example 11.3 Reinforced concrete beam

Problem: Using an assumed strut angle of 22° determine the shear capacity at sections A–A and B–B for the member illustrated in Fig. 11.33. The characteristic strength of the links is 500 N/mm² and the characteristic cylinder strength for the concrete is 35 N/mm². The effective depth for both hogging and sagging is $d = 650$ mm.

Solution: According to EC2, the shear capacity of a section with shear reinforcement is the lesser of:

$$V_{Rd,s} = \frac{A_{sw}f_{ywd}z(\cot\theta)}{s} \quad \text{and} \quad V_{Rd,max} = \alpha_{cw}\nu_1 f_{cd}b_w z(\cos\theta)(\sin\theta)$$

At section A–A, the shear reinforcement consists of 10 mm diameter shear links at 150 mm centres. There are two legs in each link and the cross-sectional area of the two legs is:

$$A_{sw} = 2(\pi 10^2/4)$$
$$= 157 \text{ mm}^2$$

From Equation (11.12) the capacity of the shear reinforcement is:

$$V_{Rd,s} = \frac{A_{sw}f_{ywd}z(\cot\theta)}{s}$$
$$= \frac{(157)(500/1.15)(0.9 \times 650)(\cot 22)}{150}$$
$$= 659 \text{ kN}$$

From Equation (11.16) the capacity of the section due to the diagonal concrete strut is:

$$V_{Rd,max} = \alpha_{cw}\nu_1 f_{cd}b_w z(\cos\theta)(\sin\theta)$$

Assuming the stress in the shear reinforcement is greater than $0.8f_{yk}$:

$$\nu_1 = 0.6\left[1 - \frac{f_{ck}}{250}\right] = 0.6\left[1 - \frac{35}{250}\right] = 0.516$$

and:

$$V_{Rd,max} = (1.0)(0.516)(35/1.5)(300)(0.9 \times 650)(\cos 22)(\sin 22)$$
$$= 734 \text{ kN}$$

Clearly this does not govern the design, and the shear capacity at A–A is 659 kN.

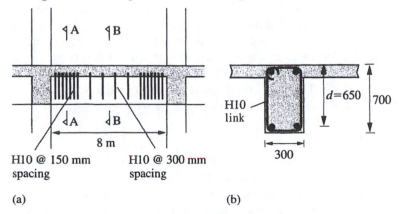

Figure 11.33 Beam of Example 11.3: (a) elevation; (b) sections A–A and B–B

At B–B, the capacity of the links is:

$$V_{Rd,s} = \frac{A_{sw}f_{ywd}z(\cot\theta)}{s}$$
$$= \frac{(157)(500/1.15)(0.9 \times 650)(\cot 22)}{300}$$
$$= 329\,kN$$

Hence the final shear capacity at B–B is 329 kN.

Shear enhancement

As discussed in Section 11.4, the resistance of members to shear force in regions near supports is effectively increased. For members with shear reinforcement, EC2 allows for this 'enhancement' of shear resistance near supports. Where concentrated loads are applied within a distance of $0.5d \leq a_v \leq 2d$ from the edge of the support, the contribution of this load to the shear force, V_{Ed}, may be reduced by multiplying by a factor, $\beta = a_v/2d$. However, this is only allowed when the following conditions are satisfied:

(a) $V_{Ed} \leq A_{sw}f_{ywd}$

where $A_{sw}f_{ywd}$ is the capacity provided by the shear reinforcement within the central 75 per cent of the zone between the edge of the support and the load (i.e. within the central $0.75\,a_v$ zone; see Fig. 11.11 for illustration of a_v).

(b)   The reduction to V_{Ed} can be applied when calculating the required area of shear reinforcement (Equation (11.17)). However, V_{Ed} (not reduced) must not be greater than $V_{Rd,max}$ (Equation (11.16)).

(c)   The loading and support reactions are such that they cause only diagonal compression in the member. Thus, the support must be provided at the bottom of the beam when the loading is applied from the top.

(d)   Longitudinal tension reinforcement must be anchored at the support.

Note: EC2 recommends that a value of $a_v = 0.5d$ be used when $a_v \leq 0.5d$.

Interface shear

In flanged members subjected to bending, a shear stress develops in the flange and between the flange and the web. This is the process by which the compressive axial stress is transmitted from the web to the flange and, particularly in the case of thin flanges, its effect can be quite significant. The phenomenon is illustrated in Fig. 11.34 which shows a segment of beam where moment is increasing from zero at A to a high (sag) value at B. The interface shear stresses illustrated are necessary for equilibrium of the segment of flange abcd. Reinforcement requirements can be calculated on the basis of a truss analogy such as that illustrated in Fig. 11.34(b). In this case, the truss is used to transfer load in the horizontal plane between the flange and the web parts of the beam. Refer to EC2 for more details.

Variable depth members

Up to now, only members of prismatic (constant) cross-section have been considered. For members of variable depth, such as illustrated in Fig. 11.35, the capacity to resist shear is generally different from the shear capacity of prismatic members. This is due to the direction of the forces which resist the applied moment. The force in the tension reinforcement, F_s, clearly acts parallel to that reinforcement. The force in the concrete acts at some angle, α_2 between the angle of the top chord and the neutral

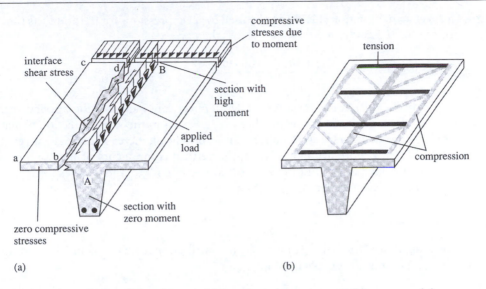

Figure 11.34 Interface shear in flanged beam: (a) interface shear stresses; (b) truss model

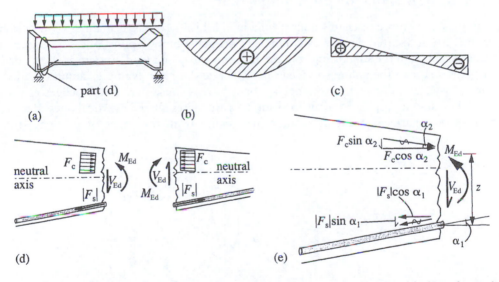

Figure 11.35 Beam of variable depth: (a) elevation; (b) bending moment diagram; (c) shear force diagram; (d) part elevation near the left end of the beam; (e) resolution of forces

axis. These forces are resolved into horizontal and vertical components in Fig. 11.35(e). It can be seen from this figure that the horizontal components, $F_s \cos \alpha_1$ and $F_c \cos \alpha_2$, resist the applied design moment. However, the vertical components act to increase the applied shear, V_{Ed}. Thus, the total shear to be resisted by the member is:

$$\text{Force to be resisted} = V_{Ed} + |F_s| \sin \alpha_1 + F_c \sin \alpha_2 \tag{11.18}$$

The moment is related to the horizontal components by:

$$M_{Ed} = (|F_s| \cos \alpha_1)z = (F_c \cos \alpha_2)z \tag{11.19}$$

where z is the lever arm. Equation (11.19) can be used to get expressions for $|F_s|$ and F_c as functions of M_{Ed}. Substituting for $|F_s|$ and F_c in Equation (11.18) then gives:

$$\text{Force to be resisted} = V_{Ed} + \frac{M_{Ed} \tan \alpha_1}{z} + \frac{M_{Ed} \tan \alpha_2}{z} \qquad (11.20)$$

For the example of Fig. 11.35, the variation in depth effectively increases the applied shear. This happens when the effective depth of the member is decreasing in the same direction as moment is numerically increasing. In other cases, the effect can be beneficial, increasing the shear resistance of the member (see EC2 for more details). In EC2, the vertical components of the compressive and tensile forces in the inclined top and bottom chords are denoted V_{ccd} and V_{td}, respectively, and the total shear resistance of the member, V_{Rd}, is given by:

$$V_{Rd} = V_{Rd,s} + V_{ccd} + V_{td}$$

The sum of the design shear force, V_{Ed}, and the shear contribution from the inclined chords must also be less than the maximum shear capacity, $V_{Rd,max}$.

11.6 Design of slabs for punching shear

The punching shear failure mechanism illustrated in Fig. 11.36 can occur in any two-way spanning members that are supported directly by columns or are acted on by heavy concentrated loads. The cross-sectional shape of the supporting column or the concentrated load, known as the **loaded area**, determines the shape of the failure surface in the concrete slab. For example, circular loaded areas cause conical wedges to form and rectangular loaded areas cause pyramidal wedges to form (see Fig. 11.21). For design purposes, the inclined surfaces are represented by vertical ones. It is assumed that these vertical surfaces occur at a constant distance from the edges of the loaded area as

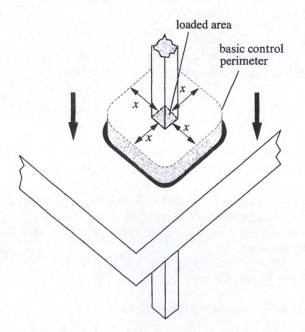

Figure 11.36 Punching shear failure at column support

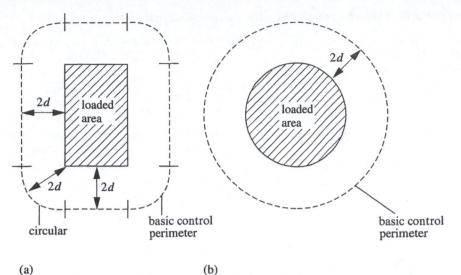

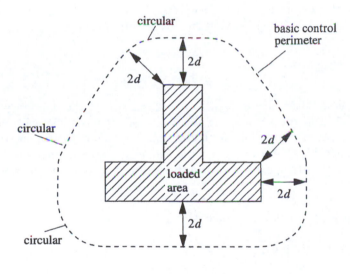

Figure 11.37 Basic control perimeters for typical loaded areas specified in EC2. Perimeters are located a constant distance, 2*d* from edge of the loaded area

illustrated in Fig. 11.37. The corresponding perimeter, illustrated in both figures, is known as the **basic control perimeter**.

EC2 recommends that the basic control perimeter for members with constant depth is normally located a distance 2*d* from the loaded area, where *d* is the average effective depth of longitudinal reinforcement spanning in two perpendicular directions. In order to minimize the length, u_1, of the basic control perimeter, EC2 recommends that the corners of the perimeter be rounded off. Some examples of basic control perimeters for loaded areas of different shapes are illustrated in Fig. 11.37. Refer to EC2 for guidance on the calculation of u_1 for edge or perimeter columns.

Checking for punching shear in slabs with constant depth

As for regular shear, the average applied shear stress in a member with punching shear is given by Equation (11.2) that is:

$$v_{Ed} = \frac{V_{Ed}}{b_w d}$$

However, in the case of members with punching shear, the breadth of the section in shear is equal to the length of the control perimeter. Further, with punching shear, there are two effective depths, one for reinforcement running parallel to the Z-axis and one for reinforcement running parallel to the Y-axis (the layers of reinforcement will be at different depths as one must rest on the other). Thus, for punching shear, the average design shear stress becomes:

$$v_{Ed} = \frac{V_{Ed}}{u_1 d_{eff}} \tag{11.21}$$

where u_1 is the length of the basic control perimeter and d_{eff} is the average of the two effective depths.

The shear stress must be checked at the column perimeter, (length of column perimeter denoted u_0), where the design shear stress, v_{Ed}, must be less than the maximum shear capacity of the section, given by:

$$v_{Rd,max} = 0.5 v f_{cd}$$

where v is given by Equation (11.6). The design shear stress, v_{Ed}, at the column perimeter is calculated using Equation (11.21) and substituting the length of the basic control perimeter, u_1, with the length of the column perimeter, u_0. The shear stress must also be checked at the basic control perimeter to ensure that the punching shear strength of the slab, $v_{Rd,c}$, is greater than the design shear stress, v_{Ed}. The punching shear strength for both prestressed and reinforced concrete members without shear reinforcement is:

$$v_{Rd,c} = \left[(0.18/\gamma_c) k \, (100 \rho_1 f_{ck})^{1/3} + 0.1 \sigma_{cp} \right] \geq \left(0.035 k^{3/2} f_{ck}^{1/2} + 0.1 \sigma_{cp} \right) \tag{11.22}$$

where k is a 'depth factor' that is given by:

$$k = 1 + \sqrt{\frac{200}{d}} \leq 2.0 \quad (d \text{ in mm}) \tag{11.23}$$

$$\rho_1 = \sqrt{\rho_{lz}\rho_{ly}} \leq 0.02 \quad (d \text{ in mm}) \tag{11.24}$$

and ρ_{lz}, ρ_{ly} are the longitudinal tension reinforcement ratios in the Z- and Y-directions respectively, to be calculated over a slab width equal to the column width plus $3d$ on each side of the column. σ_{cp} is the design axial stress (where the slab is stressed by different amounts in two directions the average stress should be used).

When the punching shear force due to applied loads is eccentric to the loaded area or the force is combined with a moment as would happen in an edge or corner column (see Fig. 11.38), the adverse effect of the applied load is significantly increased. This effect is allowed for in EC2 by the multiplication of the applied force, V_{Ed}, by a factor β (Note: Should not be confused with the β factor for shear enhancement). For braced structures where adjacent spans do not differ by more than

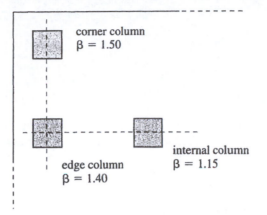

corner column
β = 1.50

internal column
β = 1.15

edge column
β = 1.40

Figure 11.38 Factor β to allow for eccentricity of loading

25 per cent, β can be taken from Fig. 11.38. (Note that even in an internal column, the variable portion of the load can be applied asymmetrically, which results in moment in the column.) For other cases, refer to the recommendations given in EC2.

Example 11.4 Punching shear in slab of constant depth

Problem: Check the slab of Fig. 11.39 for punching shear at an internal column (300 × 300 mm) given that the ultimate factored reaction in the internal column due to permanent and variable loading is 400 kN. The characteristic cylinder strength of the concrete in the slab is 30 N/mm². The top reinforcement in both directions consists of 16 mm diameter reinforcing bars at 150 mm centres.

Solution: As moment over the support will be hogging, the effective depths are measured from the bottom surface to the top layers of reinforcement. For reinforcement parallel to the Z- and Y-axes, the respective effective depths are, from Fig. 11.39:

$$200 - 30 - 16/2 = 162 \text{ mm}$$

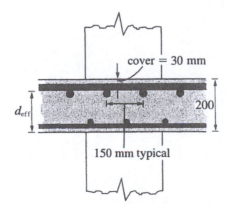

cover = 30 mm

d_{eff}

200

150 mm typical

Figure 11.39 Slab of Example 11.4

and:

$$200 - 30 - 16/2 - 16 = 146 \text{ mm}$$

Hence:

$$d_{\text{eff}} = (162 + 146)/2 = 154 \text{ mm}$$

First, it is necessary to check the punching shear stress at the column perimeter. The length of the column perimeter is 4×300 mm $= 1,200$ mm. Thus, the shear stress is given by:

$$v_{\text{Ed}} = \frac{\beta V_{\text{Ed}}}{u_0 d_{\text{eff}}} = \frac{1.15 \times 400 \times 10^3}{1200 \times 154} = 2.49 \text{ N/mm}^2$$

The maximum shear resistance is given by:

$$\begin{aligned} v_{\text{Rd,max}} &= 0.5 \times 0.6 \left[1 - \frac{f_{\text{ck}}}{250} \right] \times f_{\text{cd}} \\ &= 0.5 \times 0.6 \left[1 - \frac{30}{250} \right] \times \frac{30}{1.5} \\ &= 5.28 \text{ N/mm}^2 \end{aligned}$$

As the shear capacity at the face of the column is greater than the shear stress, the section sizes are adequate. However, the shear capacity at the basic control perimeter must also be checked.

The longitudinal tension reinforcement ratios in each direction are calculated over a slab width equal to the column width plus $3d$ on each side of the column (i.e. over a width $3(154) + 300 + 3(154) = 1,224$ mm). The area of top reinforcement over a width of 1,224 mm in each direction is:

$$\frac{\pi 16^2}{4} \times \frac{1224}{150} = 1641 \text{ mm}^2$$

Hence, the reinforcement ratios are:

$$\rho_{\text{lz}} = \rho_{\text{ly}} = \frac{1641}{154 \times 1224} = 0.0087$$

and, from Equation (11.24), the effective reinforcement ratio is:

$$\rho_l = \sqrt{\rho_{\text{lz}} \rho_{\text{ly}}} = 0.0087 \leq 0.02$$

From Equation (11.23), the depth factor is given by:

$$\begin{aligned} k &= 1 + \sqrt{\frac{200}{d}} \leq 2.0 \\ &= 1 + \sqrt{\frac{200}{154}} = 2.14 > 2.0 \\ \Rightarrow k &= 2.0 \end{aligned}$$

The capacity to resist punching shear force is, from Equation (11.22):

$$\begin{aligned} v_{\text{Rd,c}} &= (0.18/\gamma_c) k (100 \rho_l f_{\text{ck}})^{1/3} + 0.1\sigma_{\text{cp}} \\ &= (0.18/1.5)(2.0)(100 \times 0.0087 \times 30)^{1/3} \\ &= 0.71 \text{ N/mm}^2 \end{aligned}$$

with a minimum value of:

$$\left(0.035k^{3/2}f_{ck}^{1/2} + 0.1\sigma_{cp}\right) = 0.035\,(2.0)^{3/2}\,(30)^{1/2}$$

$$= 0.54\,\text{N/mm}^2$$

$$\Rightarrow v_{Rd,c} = 0.71\,\text{N/mm}^2$$

The basic control perimeter illustrated in Fig. 11.40 consists of four straight segments, each of length 300 mm, and four quarter segments of a circle of total length equal to the perimeter of a circle (with radius $= 2d$). Hence:

$$\text{length} = 4 \times 300 + 2\pi \times 308$$

$$= 3135\ \text{mm}$$

Thus, the shear stress at the basic control perimeter is given by:

$$v_{Ed} = \frac{\beta V_{Ed}}{u_1 d_{eff}} = \frac{1.15 \times 400 \times 10^3}{3135 \times 154} = 0.95\,\text{N/mm}^2$$

As $v_{Ed} > v_{Rd,c}$, the slab is clearly not adequate without shear reinforcement.

Slabs with variable depth

For the member of Example 11.4, the factored applied shear force exceeded the punching shear capacity of the concrete. Therefore, punching shear failure will occur at the loaded area unless the capacity to resist it is increased. This can be achieved in a number of ways, including:

1. Increasing the depth of the member over its entire area.
2. Introducing shear reinforcement near the loaded areas.
3. Providing column heads or 'drops', as illustrated in Fig. 11.41, to thicken the member locally near the loaded area.
4. Introducing a gradual increase in depth which gives a deeper section near the loaded area.

A common form of variable-depth slab is when the slab is thickened locally around the column. Two alternative cases are illustrated in Fig. 11.42: a sudden increase in depth (on left) or a gradual increase in depth (on right). Whichever option is chosen, a check of the shear capacity should be made at two control sections, one within the column head and one outside the column head. For example, the section within the column head should be taken on a perimeter a distance of $l_{int} = 2d_H$ from the face of the column, where d_H is the effective depth of the column head

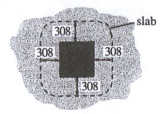

Figure 11.40 Basic control perimeter Example 11.4

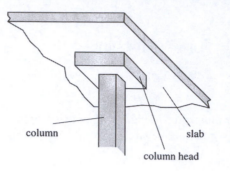

Figure 11.41 Column head

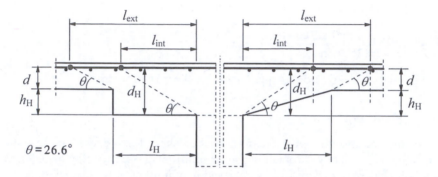

Figure 11.42 Critical perimeters for slab with column head where $l_H > 2(d + h_H)$

(see Fig. 11.42). The section outside the column head should be located on a perimeter a distance of $l_{ext} = 2d + l_H$ from the face of the column, where d is the effective depth of the slab and l_H is the distance from the face of the column to the edge of the column head. The adequacy of the two control sections is then checked as described above. If $l_H < 2h_H$, then only one control perimeter needs to be checked. Expressions are provided in EC2 to determine the location of the control perimeter.

Example 11.5 Flat slab with column heads

Problem: The column of Example 11.4 is provided with a square head of side length 1,400 mm and depth (additional to that of the slab) of 100 mm, as illustrated in Fig. 11.43. Check the adequacy of the member against punching shear at the columns.

Solution: The effective depth within the column head is increased by 100 mm over that in Example 11.4. Hence, the average effective depth value is:

$$d_H = 154 + 100 = 254\,\text{mm}$$

Hence, the first perimeter to be checked is at a distance of $l_{int} = 2(254) = 508$ mm from the face of the column. The length of the perimeter illustrated is:

$$u_{int} = 300 \times 4 + 2 \times \pi \times 508$$
$$= 4392\,\text{mm}$$

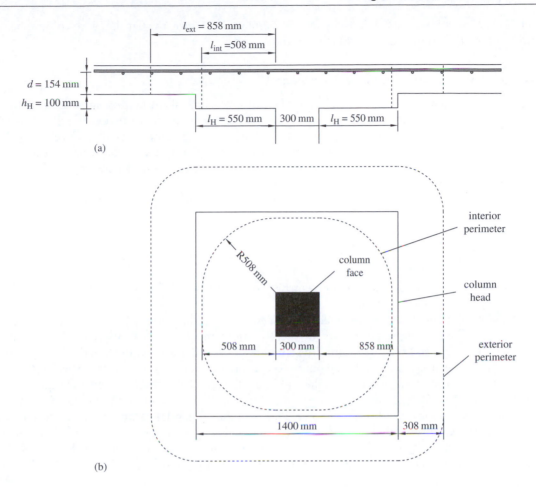

Figure 11.43 Column head of Example 11.5: (a) section; (b) plan showing interior and exterior control perimeters

The shear stress at the interior control perimeter is given by:

$$v_{Ed} = \frac{\beta V_{Ed}}{u_{int} d_H} = \frac{1.15 \times 400 \times 10^3}{4392 \times 254} = 0.41 \, \text{N/mm}^2$$

In Example 11.4, the punching capacity, $v_{Rd,c}$, is equal to 0.71 N/mm^2. Increasing the depth by 100 mm changes r_1 and k, resulting in a reduction in the capacity to 0.57 N/mm^2. However, it is still greater than the shear stress at the interior control perimeter. The second perimeter to be considered is at a distance of $l_{ext} = 2(154) + 550 = 858$ mm from the face of the column and 308 mm from the face of the column head. The length of the perimeter is:

$$u_{ext} = 1400 \times 4 + 2 \times \pi \times 308$$
$$= 7535 \, \text{mm}$$

Thus, the shear stress at the exterior control perimeter is given by:

$$v_{Ed} = \frac{\beta V_{Ed}}{u_{ext} d_H} = \frac{1.15 \times 400 \times 10^3}{7535 \times 154} = 0.40 \, \text{N/mm}^2$$

which is also less than the shear capacity of 0.71 N/mm². Therefore, the flat slab with column heads is sufficient to resist the shear stresses in the slab.

Members with punching shear reinforcement

For deeper slabs such as foundation pads and pile caps, punching shear reinforcement is commonly provided in the form of links. For floor slabs, the relatively small depth makes it difficult to fix conventional links. Proprietary systems which overcome these difficulties have been developed. One such system consists of shear studs, as illustrated in Fig. 11.44. This system incorporates vertical steel rods capped with a circular plate. The capped rods are welded to strips of flat steel plate. They have been found to be very effective and easy to fix on site.

The outer control perimeter where shear reinforcement is not required is denoted u_{out} and is calculated using:

$$u_{\text{out}} = \frac{\beta V_{\text{Ed}}}{v_{\text{Rd,c}}d} \qquad (11.25)$$

The outermost perimeter of shear reinforcement should be placed at a maximum distance of $1.5d$ inside this perimeter (see Fig. 11.45).

EC2 assumes that shear reinforcement is placed in a radial arrangement, as illustrated in Fig. 11.45(a). The spacing requirements for shear reinforcement state that inside the basic control perimeter, the tangential spacing of link legs should not exceed $1.5d$ and outside the basic control perimeter, the tangential spacing should not exceed $2d$. For a radial arrangement of shear reinforcement, the links can be arranged such that the maximum tangential spacing between link legs does not exceed $2d$ (Fig. 11.45(a)). However, where shear links are fixed around longitudinal bars in two perpendicular directions, as illustrated in Fig. 11.45(b), the tangential spacing may be greater than $2d$. In such cases, an effective outer perimeter (denoted $u_{\text{out,ef}}$) is determined, with gaps in the perimeter where the tangential spacing of shear reinforcement is greater than $2d$ (Fig. 11.45(b)). Additional detailing rules for punching shear reinforcement are provided in EC2 and in the UK NA to EC2.

As for members reinforced against regular shear, if the shear capacity of the concrete, $V_{\text{Rd,c}}$, exceeds the applied shear force, βV_{Ed}, then no reinforcement is necessary. If, on the other hand, βV_{Ed}

(a)

(b)

Figure 11.44 Shear rail reinforcement: (a) shear studs attached to base plate; (b) shear studs in place around a column. (Photographs courtesy of Hy-Ten Ltd.)

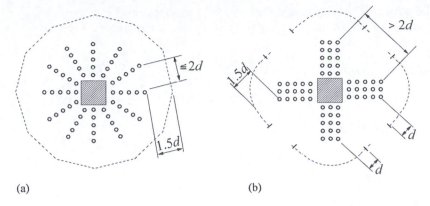

Figure 11.45 Control perimeters at internal columns: (a) outer control perimeter, u_{out}; (b) effective outer control perimeter, $u_{out,ef}$

exceeds $V_{Rd,c}$, then punching shear reinforcement must be provided. For members where punching shear reinforcement is required, EC2 specifies the minimum area of a link leg, which is calculated using:

$$A_{sw,min} = \frac{0.08\sqrt{f_{ck}}(s_r s_t)}{1.5 f_{yk}} \tag{11.26}$$

where s_r is the radial spacing of perimeters of shear reinforcement and s_t is the tangential spacing of shear reinforcement.

For a slab with vertical shear reinforcement, the shear capacity is given by:

$$v_{Rd,cs} = 0.75 v_{Rd,c} + 1.5(d/s_r) A_{sw} f_{ywd,ef}(1/(u_1 d)) \tag{11.27}$$

where $v_{Rd,c}$ is given by Equation (11.22), A_{sw} is the area of one perimeter of shear reinforcement, d is the mean effective depth of the longitudinal reinforcement, u_1 is the length of the basic control perimeter (at $2d$) and $f_{ywd,ef}$ is given by:

$$f_{ywd,ef} = 250 + 0.25 d \le f_{ywd} \tag{11.28}$$

The shear capacity provided by shear studs and other proprietary forms of reinforcement should be determined by testing according to the relevant European Technical Approval.

Rearranging Equation (11.27) and substituting $v_{Rd,cs}$ with v_{Ed}, the required area of shear reinforcement (for one perimeter) can be determined using:

$$A_{sw} = \frac{(v_{Ed} - 0.75 v_{Rd,c})s_r u_1}{1.5 f_{ywd,ef}} \tag{11.29}$$

Example 11.6 Slab with punching shear reinforcement

Problem: Determine the quantity of shear reinforcement required to prevent punching failure due to a load of 3,800 kN (ULS factored) from an interior square column 400 × 400 mm. The average effective depth of the slab in this region is 550 mm, $k = 1.6$, $f_{ck} = 35$ N/mm^2 and the average effective reinforcement ratio is $\rho_l = 0.018$.

Solution: The punching shear stress at the column perimeter must be determined. The length of the column perimeter is 4×400 mm $= 1,600$ mm. Thus, the shear stress is given by:

$$v_{Ed} = \frac{\beta V_{Ed}}{u_0 d} = \frac{1.15 \times 3800 \times 10^3}{1600 \times 550} = 4.97 \, \text{N/mm}^2$$

The maximum shear resistance of the concrete is given by:

$$v_{Rd,max} = 0.5 \times 0.6 \left[1 - \frac{f_{ck}}{250} \right] \times f_{cd}$$

$$= 0.5 \times 0.6 \left[1 - \frac{35}{250} \right] \times \frac{35}{1.5}$$

$$= 6.02 \, \text{N/mm}^2$$

Thus, the slab has sufficient capacity to resist the punching shear stress at the face of the column.

The shear capacity must also be checked at the basic control perimeter, u_1, which is $2d$ ($= 1,100$ mm) from the face of the column. The length of the basic control perimeter (Fig. 11.37) is:

$$u_1 = 4 \times 400 + 2 \times \pi \times 1100 = 8512 \, \text{mm}$$

The shear stress at the basic control perimeter is given by:

$$v_{Ed} = \frac{\beta V_{Ed}}{u_1 d} = \frac{1.15 \times 3800 \times 10^3}{8512 \times 550} = 0.93 \, \text{N/mm}^2$$

The capacity of the concrete to resist punching shear force is, from Equation (11.22):

$$v_{Rd,c} = (0.18/\gamma_c) k \, (100 \rho_1 f_{ck})^{1/3}$$

$$= (0.18/1.5)(1.6)(100 \times 0.018 \times 35)^{1/3}$$

$$= 0.76 \, \text{N/mm}^2$$

with a minimum value of:

$$\left(0.035 k^{3/2} f_{ck}^{1/2} + 0.1 \sigma_{cp} \right) = 0.035 \, (1.6)^{3/2} (35)^{1/2}$$

$$= 0.42 \, \text{N/mm}^2$$

$$\Rightarrow v_{Rd,c} = 0.76 \, \text{N/mm}^2$$

As $v_{Ed} > v_{Rd,c}$, punching shear reinforcement is required.

The perimeter where punching shear reinforcement is not required can be calculated using Equation (11.25):

$$u_{out} = \frac{\beta V_{Ed}}{v_{Rd,c} d} = \frac{1.15 \times 3800 \times 10^3}{0.76 \times 550} = 10455 \, \text{mm}$$

This outer perimeter is the same shape as the basic control perimeter, so the distance from the face of the column to the outer control perimeter is calculated as:

$$\text{distance} = \frac{10455 - (4 \times 400)}{2\pi} = 1409 \, \text{mm}$$

This distance is less than $3d$ $(= 1,650$ mm$)$ so the shear reinforcement is to be placed between $0.3d$ $(= 165$ mm$)$ and $1.5d$ $(= 825$ mm$)$ from the face of the column. The maximum tangential spacing of shear reinforcement, $s_{t,max}$, placed within the basic control perimeter is $1.5d$ and the maximum radial spacing, $s_{r,max}$, is $0.75d$ $(= 413$ mm$)$. Thus, three perimeters of reinforcement are required, say at 250 mm, 500 mm and 750 mm from the face of the column. Taking $s_{t,max} = 300$ mm and $s_{r,max} = 250$ mm, the minimum area of leg link is given by Equation (11.26):

$$A_{sw,min} \geq \frac{0.08\sqrt{35}(250 \times 300)}{1.5(500)} = 47 \text{ mm}^2$$

H8 links are sufficient with a leg area of 50 mm^2. The required area of shear reinforcement for a perimeter, $2d$ from the column face, is calculated from Equation (11.29) as:

$$A_{sw} = \frac{(v_{Ed} - 0.75v_{Rd,c})s_r u_1}{1.5 f_{ywd,ef}}$$

where $f_{ywd,ef}$ is given by Equation (11.28):

$$f_{ywd,ef} = 250 + 0.25d = 388 \leq f_{ywd}(= 435 \text{ N/mm}^2)$$

$$\Rightarrow A_{sw} = \frac{(v_{Ed} - 0.75v_{Rd,c})s_r u_1}{1.5 f_{ywd,ef}}$$

$$= \frac{(0.93 - 0.75 \times 0.76)(250 \times 8512)}{1.5 \times 388}$$

$$= 1316 \text{ mm}^2$$

The required area of reinforcement is 1,316 mm^2 for the basic control perimeter, which could be provided by 28 link legs around the perimeter at a tangential spacing of 300 mm.

11.7 Torsional stresses in uncracked members

Members subjected to a torsional moment, or torque, develop shear stresses. In general, these tend to increase in magnitude from the longitudinal axis of the member to its surface. If the shear stresses are sufficiently large, cracks will propagate through the member and, if torsion reinforcement is not provided, the member will collapse suddenly.

The elastic behaviour of uncracked concrete members with torsion, particularly non-circular members, is difficult to model precisely. In a circular member subjected to a torque T, such as that of Fig. 11.46(a), the circumferential shear stress at a given cross-section varies linearly from the long-itudinal axis of the member to a maximum value, τ_{max}, at the periphery of the section (Fig. 11.46(c)). The stress at any distance r from the longitudinal axis of a circular member is given by:

$$\tau_r = rT/I_p \tag{11.30}$$

where I_p is the polar second moment of area of the section and is equal to $\pi\phi^4/32$, where ϕ is the member diameter. The maximum shear stress, τ_{max}, is found by setting $r = \phi/2$ in Equation (11.30).

For a non-circular member, the distribution of shear stress is not so straightforward. The rectangular member of Fig. 11.47(a), for instance, has the stress distribution illustrated in Fig. 11.47(b) at any given section when subjected to the torque illustrated. Unlike in the circular member, the stress distribution in a rectangular member is non-linear. The maximum shear stress occurs at the mid-point of the longer side and the shear stress at the corners of the section is zero

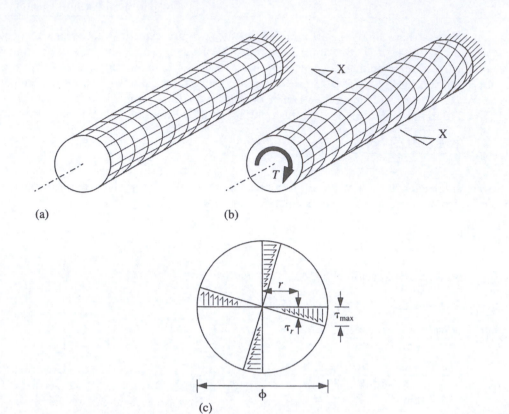

(a)   (b)

(c)

Figure 11.46 Member of circular section subjected to torsion: (a) original geometry with reference grid drawn on surface; (b) deformed shape; (c) section X–X

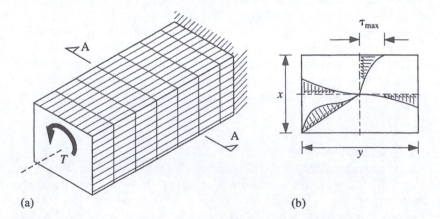

(a)   (b)

Figure 11.47 Member of rectangular section subjected to torsion: (a) geometry and loading; (b) section A–A ($y > x$)

indicating that the corners of the section are not distorted under torsion. Analytical studies have shown that the maximum shear stress in a rectangular section is given by:

$$\tau_{max} = \frac{T}{\alpha x^2 y} \tag{11.31}$$

where x and y are the lengths of the shorter and longer sides, respectively. The value of the parameter α depends on the relative values of x and y. For a square section, $\alpha = 0.208$, while for a section with $x/y = 0.1$, $\alpha = 0.312$.

The stress distribution in thin-walled hollow members is much easier to determine than for solid non-circular members. The shear stress in the walls is reasonably constant and is given by:

$$\tau_{max} = \frac{T}{2A_k t} \tag{11.32}$$

where t is the thickness of the wall of the member and A_k is the area within a perimeter bounded by the centre line of the wall (Fig. 11.48(b)). For members with varying thickness, such as the member illustrated in Fig. 11.48(a), the shear stress is maximum where the thickness of the wall is minimum.

Failure of concrete members with torsion

Consider the rectangular member of Fig. 11.49 subjected to a torque T_{Ed}. Since there are no other external forces (and ignoring self-weight) the member is considered to be in pure torsion. The torque causes the member to twist and to develop shear stresses. Consider small elements on each face of the member, as illustrated in the figure. As for members with applied shear, shear stresses act on the sides of each element in the directions shown in Fig. 11.50(a). The equivalent principal stresses, inclined at an angle of 45° to the horizontal, are illustrated in Fig. 11.50(b). In the same way as for shear, the

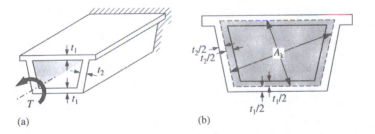

Figure 11.48 Thin walled hollow section: (a) hollow bridge of box section; (b) definition of A_k

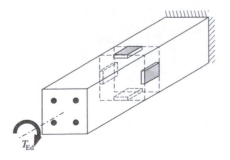

Figure 11.49 Elements in member subjected to torsion

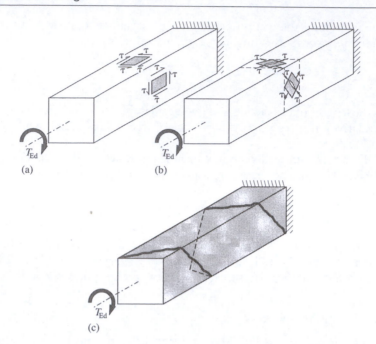

Figure 11.50 Stresses and cracking due to torsion: (a) shear stresses; (b) principal stresses; (c) spiral cracking

principal tensile stresses induce cracks at an angle of 45°. However, in the case of torsion, they form a spiral all around the member, as illustrated in Fig. 11.50(c). Since the shear stresses in members with torsion are greatest at the surface, these cracks develop inwards from the surface of the member.

The member illustrated in Fig. 11.51 is subjected to a vertical force, V_{Ed}, in addition to the applied torque. This results in a combination of bending, shear and torsion and alters the orientation of the inclined cracks in a similar way to that described in Section 11.2. As for members with shear, the introduction of prestress has the effect of delaying the onset of torsional cracking and altering the orientation of the inclined cracks.

For members with no form of reinforcement to prevent the opening of torsional cracks, failure of the member will occur almost as soon as the cracking begins. Torsional failure of a member without reinforcement is prevented only if the shear strength of the concrete exceeds the shear stress due to applied torsion. In practice, the shear strength is increased slightly through dowel action by the longitudinal reinforcing bars which cross the cracks.

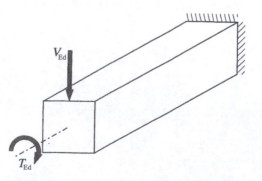

Figure 11.51 Combined shear, moment and torsion

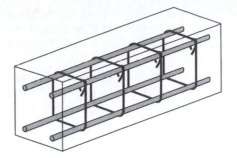

Figure 11.52 Torsion reinforcement

The torsional strength of a concrete member can be significantly increased by providing suitable torsion reinforcement across the cracks. This is usually provided in the form of 'closed' four-sided links, as illustrated in Fig. 11.52, in combination with longitudinal bars distributed around the periphery of the section. This reinforcement controls the propagation of cracks and ensures that if failure occurs due to yielding of the reinforcement, it is not sudden.

To quantify the behaviour of members with such torsional reinforcement, an equivalent space truss model, similar to the plane truss model for shear, can be used. This theory, developed by Lampert and Collins (1972), assumes that solid members can be designed as equivalent hollow members. Extensive tests indicate that this is a fair assumption since it has been found that the presence of the concrete at the centre of the member does not have a very significant effect on its torsional resistance. Thus, members are designed as equivalent thin-walled members. The equivalent thickness of the wall, $t_{ef,i}$, is commonly taken as:

$$t_{ef,i} = A_g/u \qquad\qquad (11.33)$$

where A_g is the gross cross-sectional area of the member and u is the length of the outer perimeter.

The space truss model proposed by Lampert and Collins is illustrated in Fig. 11.53 for the member of Fig. 11.52. The legs of the closed links act as tension members, the longitudinal steel bars act as

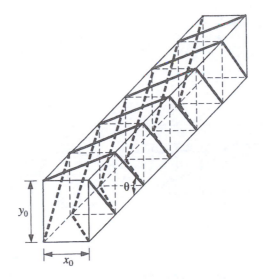

Figure 11.53 Space truss model

continuous top and bottom chords and the concrete in compression between the cracks acts as compression struts. The concrete struts are inclined at an angle θ which, like shear, varies in the range $22° < \theta < 45°$. The truss dimensions, x_0 and y_0, are measured from centre to centre of the notional thin walls, such that $A_k = x_0 \times y_0$.

Referring to Fig. 11.54 (a), the number of links crossing a side face of height y_0 is $y_0(\cot \theta)/s$ where s is the link spacing. Hence, assuming that reinforcement has yielded, the total vertical shear force transmitted across the cracks by the links on one side face is:

$$V_1 = \left(\frac{y_0(\cot \theta)}{s} \right) A_{sw} \left(\frac{f_{yk}}{\gamma_s} \right) \tag{11.34}$$

where A_{sw} is the area of one leg of the link. Similarly the shear force transferred across the top or bottom face is:

$$V_2 = \left(\frac{x_0(\cot \theta)}{s} \right) A_{sw} \left(\frac{f_{yk}}{\gamma_s} \right) \tag{11.35}$$

The shear force on the side wall of a thin-walled member is the product of the average stress due to applied load and the surface area. Hence the shear on a side wall is (see Fig. 11.54(b)):

$$V_1 = \tau t_{ef,i} y_0$$

where $t_{ef,i}$ is the effective wall thickness. Substituting for τ, from Equation (11.32) gives:

$$V_1 = \left(\frac{T_{Ed}}{2A_k t_{ef,i}} \right) t_{ef,i} y_0$$

$$\Rightarrow V_1 = \frac{T_{Ed}}{2x_0} \tag{11.36}$$

where T_{Ed} is the applied design torsion. Hence:

$$T_{Ed} = 2x_0 V_1 \tag{11.37}$$

Similarly it can be shown that:

$$T_{Ed} = 2y_0 V_2 \tag{11.38}$$

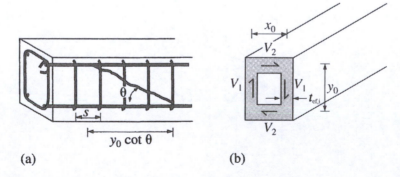

(a)   (b)

Figure 11.54 Details of truss model: (a) links transferring forces across crack; (b) equivalent thin walled member

Substituting from Equation (11.34) into Equation (11.37) or from Equation (11.35) into Equation (11.38), gives the torsion at which the links yield:

$$T_{Ed} = 2x_0 \left(\frac{y_0(\cot\theta)}{s} \right) A_{sw} \left(\frac{f_{yk}}{\gamma_s} \right)$$

$$\Rightarrow \quad T_{Ed} = \left(\frac{2A_k(\cot\theta)}{s} \right) A_{sw} \left(\frac{f_{yk}}{\gamma_s} \right) \tag{11.39}$$

Thus the link reinforcement required to resist a torsion of T_{Ed} is:

$$\frac{A_{sw}}{s} = \frac{T_{Ed}}{2A_k(\cot\theta)(f_{yk}/\gamma_s)} \tag{11.40}$$

In addition to the link reinforcement, longitudinal reinforcement is required to resist torsion. As can be seen in the truss model of Fig. 11.55(a), diagonal compression struts join the vertical members of the truss. Equilibrium at a joint of the truss where these members meet is considered in Fig. 11.55(b–c).

To satisfy equilibrium of vertical forces, the compressive force in the diagonal members, C, must equal $V_1/\sin\theta$. Horizontal force equilibrium implies an axial force of $N = C\cos\theta = V_1\cot\theta$. Substitution from Equation (11.34) gives:

$$N = \left(\frac{y_0}{s} \right) A_{sw} \left(\frac{f_{yk}}{\gamma_s} \right) (\cot^2\theta) \tag{11.41}$$

The total force in the longitudinal members from all four joints at a given cross-section is:

$$2V_1\cot\theta + 2V_2\cot\theta = 2(V_1 + V_2)\cot\theta$$

If the total area of longitudinal reinforcement is A_{sl}, then:

$$A_{sl}\frac{f_{yk}}{\gamma_s} = 2(V_1 + V_2)\cot\theta$$

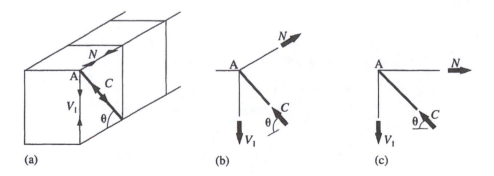

(a) (b) (c)

Figure 11.55 Equilibrium at joint in truss model: (a) part of truss model; (b) free-body diagram for joint A; (c) elevation of free-body diagram at A

Substituting from Equations (11.37) and (11.38) gives:

$$A_{sl}\frac{f_{yk}}{\gamma_s} = 2\left(\frac{T_{Ed}}{2x_0} + \frac{T_{Ed}}{2y_0}\right)\cot\theta$$

$$= T_{Ed}\frac{(x_0 + y_0)}{x_0 y_0}\cot\theta$$

$$= T_{Ed}\frac{u_k}{2A_k}\cot\theta$$

where A_k is the area within a perimeter bounded by the centre line of the wall (i.e. $A_k = x_0 y_0$) and u_k is the perimeter of area A_k (i.e. $u_k = 2(x_0 + y_0)$).

$$\Rightarrow A_{sl} = T_{Ed}\frac{u_k\cot\theta}{2A_k(f_{yk}/\gamma_s)} \tag{11.42}$$

This is the total area (all bars) of longitudinal reinforcement required to resist an applied torsion of T_{Ed}. It is additional to whatever longitudinal reinforcement is required to resist bending moment. Alternatively, Equation (11.42) can be rearranged to give the maximum torsion possible without leading to yielding of the longitudinal reinforcement:

$$T_{Ed} = \frac{2A_{sl}A_k(f_{yk}/\gamma_s)}{u_k\cot\theta} \tag{11.43}$$

Regardless of how much link and longitudinal reinforcement is provided, the torsion must not cause crushing of the concrete in the diagonal struts. As mentioned above, equilibrium at the joints of the truss requires a compressive force in the struts of $C = V_1/\sin\theta$. This force is resisted by stresses in the concrete between the diagonal cracks. The surface area of concrete to which this is applied is, from Fig. 11.56, $y_0(\cos\theta)t_{ef,i}$, where $t_{ef,i}$ is the effective thickness of the notional wall. Hence the stress in the struts is:

$$\frac{V_1/\sin\theta}{y_0(\cos\theta)t_{ef,i}} = \frac{V_1}{y_0 t_{ef,i}(\sin\theta)(\cos\theta)}$$

This must not exceed the compressive strength of the concrete, vf_{ck}/γ_c, that is:

$$\frac{V_1}{y_0 t_{ef,i}(\sin\theta)(\cos\theta)} \le \frac{vf_{ck}}{\gamma_c} \tag{11.44}$$

Figure 11.56 Breadth of compression struts

where v is given by Equation (11.6). Substitution for V_1 from Equation (11.36) gives:

$$\frac{T_{Ed}/2x_0}{y_0 t_{ef,i}(\sin\theta)(\cos\theta)} \leq \frac{vf_{ck}}{\gamma_c}$$

Rearranging, the torsion that would cause crushing of the concrete struts, $T_{Rd,max}$, is:

$$T_{Rd,max} = \frac{2A_k t_{ef,i}(\sin\theta)(\cos\theta)vf_{ck}}{\gamma_c} \tag{11.45}$$

11.8   Design of members for torsion in accordance with EC2

Where the static equilibrium of a structure relies on the torsional resistance of individual members, that is, in the case of equilibrium torsion, EC2 stipulates that a full design for torsion is necessary. The torsional resistance of members is calculated on the basis of an equivalent thin-walled closed section (i.e. the truss analogy) as described above. As for shear, the strut inclination angle, θ, can have any value in the range $22° \leq \theta \leq 45°$. The equivalent wall thickness is given by Equation (11.33) but must not be less than twice the distance between the outer edge of the section and the centre of the longitudinal reinforcement. In the case of hollow members, the equivalent wall thickness should not exceed the actual wall thickness.

For sections of complex (solid) shape, such as T-sections, the torsional resistance can be calculated by dividing the section into individual elements of simple (say, rectangular) shape. The torsional resistance of the section is equal to the sum of the capacities of the individual elements, each modelled as an equivalent thin-walled section.

Members with pure torsion

For members with pure equilibrium torsion, EC2 requires that:

(a)   the applied ultimate torque, T_{Ed}, does not exceed the torsional capacity, T_{Rd}, as dictated by the quantities of link and longitudinal reinforcement present; and
(b)   the applied ultimate torque, T_{Ed}, does not exceed the level that would cause crushing of the compressive struts, $T_{Rd,max}$.

The longitudinal reinforcement limits the capacity for torsion to that given by Equation (11.43), while the link reinforcement limits the capacity to the value given by Equation (11.39). Alternatively, Equations (11.42) and (11.40) can be used to determine the areas of longitudinal and link reinforcement required to resist a torque, T_{Ed}. The torque that would cause crushing of the compression struts, $T_{Rd,max}$, is calculated from Equation (11.45).

Example 11.7 Member with pure torsion

Problem: Determine the maximum torque which can be applied to the member of Fig. 11.57 given that $f_{ck} = 30$ N/mm^2 and the yield strength for the longitudinal reinforcement is $f_{yk} = 500$ N/mm^2.

Solution: The total area of longitudinal reinforcement available to resist torsion is:

$$A_{sl} = 4(\pi 16^2/4) = 804 \text{ mm}^2$$

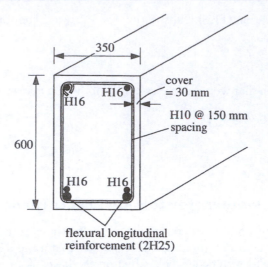

Figure 11.57 Beam of Example 11.7

From Fig. 11.57, the dimensions of the analogous thin-walled section are:

$$t_{ef,i} = \frac{A_g}{u} = \frac{350 \times 600}{2(350 + 600)} = 110\ mm$$

$$x_0 = 350 - t_{ef,i} = 240\ mm$$

$$y_0 = 600 - t_{ef,i} = 490\ mm$$

$$A_k = 240 \times 490 = 117,600\ mm^2$$

$$u_k = 2(240 + 490) = 1460\ mm$$

Hence assuming a compression strut angle of 22°, Equation (11.43) gives:

$$T_{Rd} = \frac{2A_{sl}A_k(f_{yk}/\gamma_s)}{u_k \cot \theta}$$

$$= \frac{2(804)(117,600)(500/1.15)}{(1460)\cot 22} = 22,752,139\ Nmm$$

$$= 23\ kNm$$

Similarly, Equation (11.39) gives the torsional capacity as dictated by the area of the shear link reinforcement. The area of one leg is:

$$A_{sw} = \pi 10^2/4 = 78.5\ mm^2$$

Hence:

$$T_{Rd} = \left(\frac{2A_k(\cot \theta)}{s}\right) A_{sw}\left(\frac{f_{yk}}{\gamma_s}\right)$$

$$= \left(\frac{2(117,600)(\cot 22)}{150}\right)(78.5)\left(\frac{500}{1.15}\right) = 132,458,039\ Nmm$$

$$= 132\ kNm$$

From Equation (11.6), v is given by:

$$v = 0.6\left[1 - \frac{f_{ck}}{250}\right] = 0.6\left[1 - \frac{30}{250}\right] = 0.528$$

Hence the torque that would cause crushing of the compression struts is, from Equation (11.45):

$$
\begin{aligned}
T_{Rd,max} &= \frac{2A_k t_{ef,i}(\sin\theta)(\cos\theta)vf_{ck}}{\gamma_c} \\
&= \frac{2(117600)(110)(\sin 22)(\cos 22)(0.528)(30)}{1.5} = 94{,}893{,}223 \text{ Nmm} \\
&= 95 \text{ kNm}
\end{aligned}
$$

Thus, with a compression strut inclined at an angle of 22°, the torsion capacity is governed by the area of longitudinal reinforcement. However, the strut inclination angle can have any value in the range $22° \leq \theta \leq 45°$. Hence, $\cot\theta$ can vary in the range $2.5 \geq \theta \geq 1.0$. By trial and error (or by equating the two equations for T_{Rd}), an optimum value for $\cot\theta$ can be found. Taking $\cot\theta = 1.0$ ($\theta = 45°$), the longitudinal reinforcement dictates a torsional capacity of 56 kNm and the link reinforcement dictates a capacity of 54 kNm. The corresponding value for $T_{Rd,max}$ is 137 kNm. It can therefore be concluded that this beam has the capacity to resist a torsion of 54 kNm.

Members with combined actions

For members subjected to combined moment and torsion, EC2 recommends that the requirements for each action be determined separately and that the following rules are applied:

(a) In the flexural tension zone, the longitudinal reinforcement required for torsion should be provided in addition to the amount required for moment.
(b) In the flexural compression zone, the area of longitudinal reinforcement required for torsion may be reduced in proportion to the compressive force in the concrete due to moment.

For members with combined torsion and shear, the design torque, T_{Ed}, and the design shear force, V_{Ed}, should satisfy the condition:

$$\frac{T_{Ed}}{T_{Rd,max}} + \frac{V_{Ed}}{V_{Rd,max}} \leq 1.0 \tag{11.46}$$

where $T_{Rd,max}$ and $V_{Rd,max}$, are the torque and shear force respectively that would, acting alone, cause crushing of the concrete struts (given by Equations (11.45) and (11.16), respectively). The calculations for the design of links may be made separately for torsion and shear. However, the angle, θ, for the concrete struts must be the same in both cases. The requirements for shear and torsion are, of course, additive.

Problems

Section 11.4

11.1   The ordinary reinforced section illustrated in Fig. 11.58 is subjected to a sagging moment. Determine whether link reinforcement other than the minimum is required to resist an applied factored shear force of 100 kN, given that the concrete cylinder strength is $f_{ck} = 35$ N/mm².

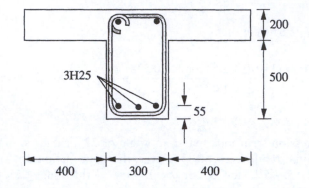

Figure 11.58 Section of Problem 11.1

11.2   For the beam illustrated in Fig. 11.59, determine if shear reinforcement is required at B, given that the factored ULS shear is 300 kN and the factored ULS moment is 500 kNm. The concrete cylinder strength is $f_{ck} = 40$ N/mm².

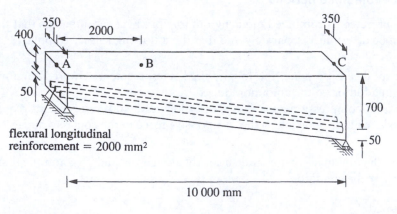

Figure 11.59 Beam of Problem 11.2

Section 11.5

11.3   Calculate the shear reinforcement required for the beam and loading of Problem 11.2.
11.4   For the section illustrated in Fig. 11.60, calculate the shear reinforcement required. The total factored ULS shear force is 600 kN. Check for crushing of concrete in the compression zone of the section.

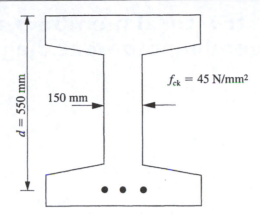

Figure 11.60 Section of Problem 11.4

Section 11.6

11.5 Determine the capacity of the 200 mm slab illustrated in Fig. 11.61 to resist punching of
the corner column given that it is reinforced with H12 bars at 150 mm centres in the
Y-direction (inner layer) and H16 at 200 mm centres in the X-direction (outer layer).
Assume $f_{ck} = 35$ N/mm^2 and that the cover is 30 mm.

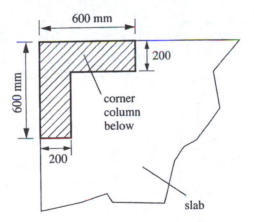

Figure 11.61 Plan view of portion of slab of Problem 11.5

Appendix A

Stiffness of structural members and associated bending moment diagrams

No.	Force/moment per unit displacement	Bending moment diagram (positive sag)

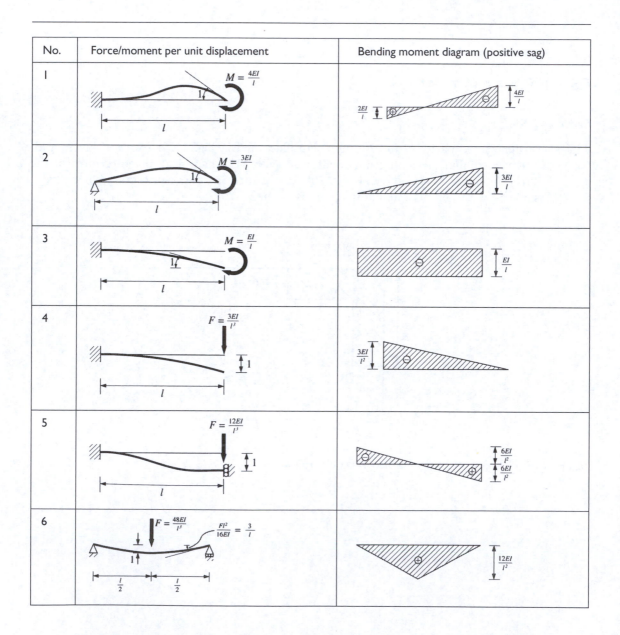

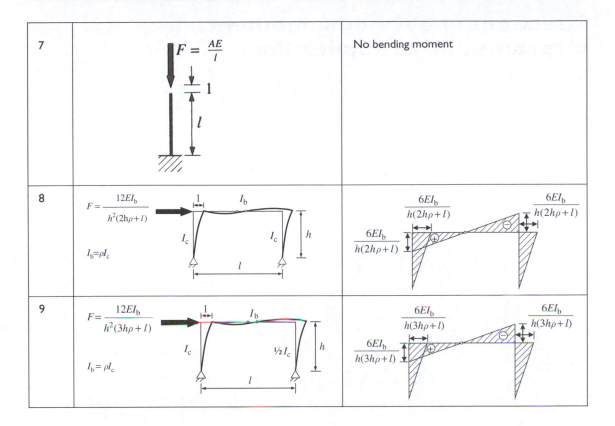

7	$F = \dfrac{AE}{l}$	No bending moment
8	$F = \dfrac{12EI_b}{h^2(2h\rho + l)}$ $\quad I_b = \rho I_c$	
9	$F = \dfrac{12EI_b}{h^2(3h\rho + l)}$ $\quad I_b = \rho I_c$	

Appendix B

Reactions and bending moment diagrams due to applied load

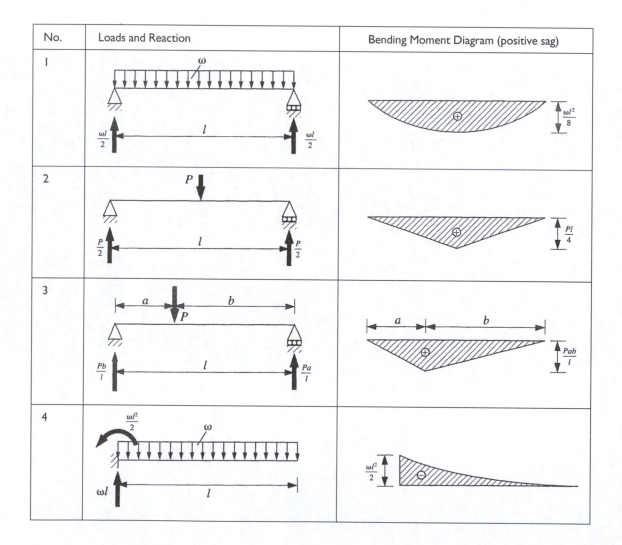

No.	Loads and Reaction	Bending Moment Diagram (positive sag)
1		$\frac{\omega l^2}{8}$
2		$\frac{Pl}{4}$
3		$\frac{Pab}{l}$
4		$\frac{\omega l^2}{2}$

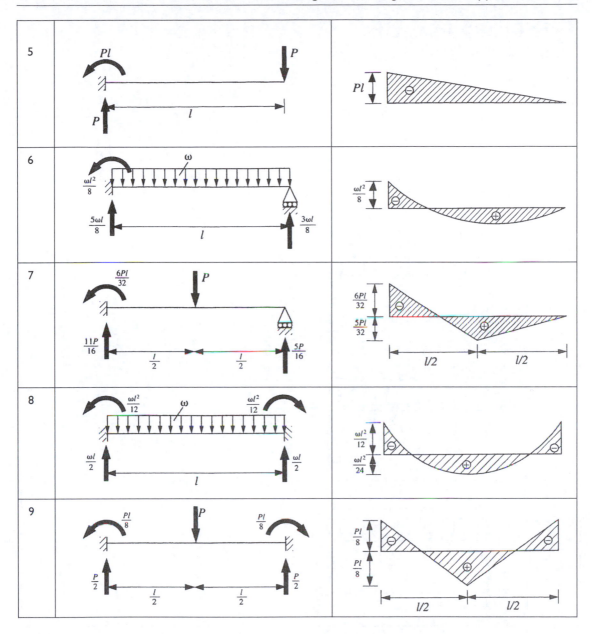

Appendix C

Tributary lengths

No.	Loading, Reactions and Tributary Lengths
1	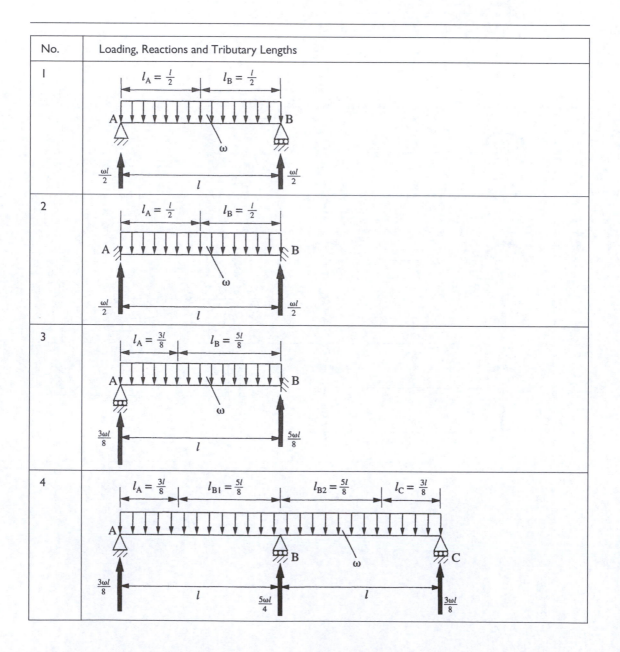
2	
3	
4	

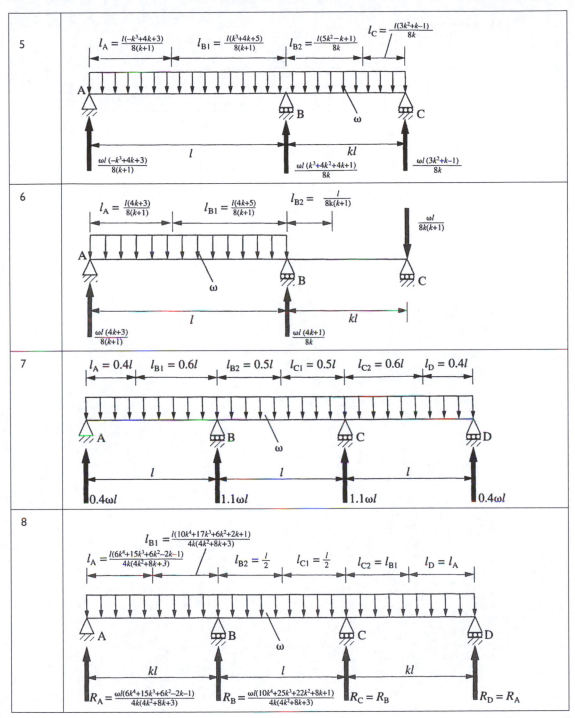

5

$$l_A = \frac{l(-k^3+4k+3)}{8(k+1)} \qquad l_{B1} = \frac{l(k^3+4k+5)}{8(k+1)} \qquad l_{B2} = \frac{l(5k^2-k+1)}{8k} \qquad l_C = \frac{l(3k^2+k-1)}{8k}$$

A   ω   C

$$\frac{\omega l(-k^3+4k+3)}{8(k+1)} \qquad \frac{\omega l(k^3+4k^2+4k+1)}{8k} \qquad \frac{\omega l(3k^2+k-1)}{8k}$$

6

$$l_A = \frac{l(4k+3)}{8(k+1)} \qquad l_{B1} = \frac{l(4k+5)}{8(k+1)} \qquad l_{B2} = \frac{l}{8k(k+1)}$$

$$\frac{\omega l}{8k(k+1)}$$

A   ω   B   C

$$\frac{\omega l(4k+3)}{8(k+1)} \qquad \frac{\omega l(4k+1)}{8k}$$

7

$$l_A = 0.4l \qquad l_{B1} = 0.6l \qquad l_{B2} = 0.5l \qquad l_{C1} = 0.5l \qquad l_{C2} = 0.6l \qquad l_D = 0.4l$$

A   ω   B   C   D

$$0.4\omega l \qquad 1.1\omega l \qquad 1.1\omega l \qquad 0.4\omega l$$

8

$$l_{B1} = \frac{l(10k^4+17k^3+6k^2+2k+1)}{4k(4k^2+8k+3)}$$

$$l_A = \frac{l(6k^4+15k^3+6k^2-2k-1)}{4k(4k^2+8k+3)} \qquad l_{B2} = \frac{l}{2} \qquad l_{C1} = \frac{l}{2} \qquad l_{C2} = l_{B1} \qquad l_D = l_A$$

A   ω   B   C   D

$$R_A = \frac{\omega l(6k^4+15k^3+6k^2-2k-1)}{4k(4k^2+8k+3)} \qquad R_B = \frac{\omega l(10k^4+25k^3+22k^2+8k+1)}{4k(4k^2+8k+3)} \qquad R_C = R_B \qquad R_D = R_A$$

(continued on the next page)

9

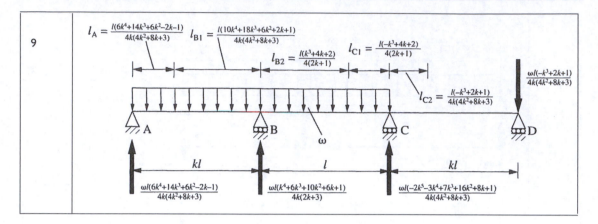

$$l_A = \frac{l(6k^4+14k^3+6k^2-2k-1)}{4k(4k^2+8k+3)}$$

$$l_{B1} = \frac{l(10k^4+18k^3+6k^2+2k+1)}{4k(4k^2+8k+3)}$$

$$l_{B2} = \frac{l(k^3+4k+2)}{4(2k+1)}$$

$$l_{C1} = \frac{l(-k^3+4k+2)}{4(2k+1)}$$

$$l_{C2} = \frac{l(-k^3+2k+1)}{4k(4k^2+8k+3)}$$

$$\frac{\omega l(-k^3+2k+1)}{4k(4k^2+8k+3)}$$

$$\frac{\omega l(6k^4+14k^3+6k^2-2k-1)}{4k(4k^2+8k+3)}$$

$$\frac{\omega l(k^4+6k^3+10k^2+6k+1)}{4k(2k+3)}$$

$$\frac{\omega l(-2k^5-3k^4+7k^3+16k^2+8k+1)}{4k(4k^2+8k+3)}$$

kl l kl

Appendix D

Formulae for analysis of continuous beams (from Reynolds and Steedman 1988)

Internal moment at support $= \alpha QUF$ (base span length) where:

$\alpha = 1$ for uniform loading or 1.5 for central point loading
$Q =$ Support moment coefficient, given below for each support
$U =$ Moment multiplier given below ($=$ unity if all spans are equal)
$F =$ Total applied load
Base span length $=$ length of 'base span', as identified below

Note: $K_i =$ ratio of span i to base span length.

For load arrangements other than those shown, superimpose the moments due to individual loadings. For moments between supports, superimpose the bending moment diagrams for simply supported spans given in Appendix B (Nos 1 and 2).

	Loading	Support Moment Coefficients			Moment Multipliers	
		Q_A	Q_B	Q_C		
Two-span	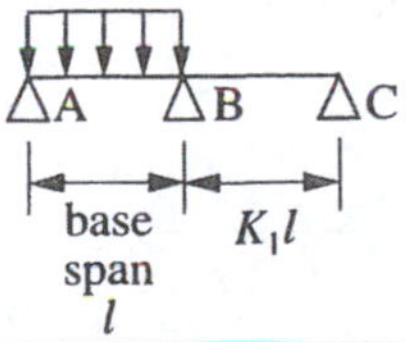 (uniform load on base span) A, B, C	–	−0.063	–	$U_B = \dfrac{2}{1 + K_1}$	
	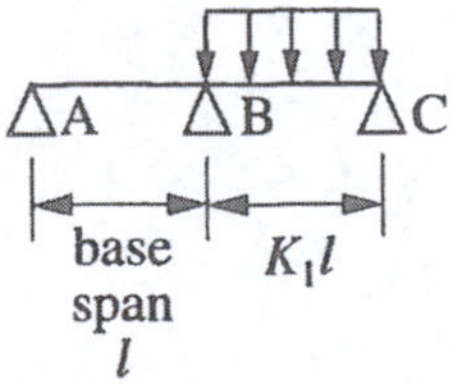 (uniform load on K_1l span) A, B, C	–	−0.063	–	$U_B = \dfrac{2K_1^2}{1 + K_1}$	
	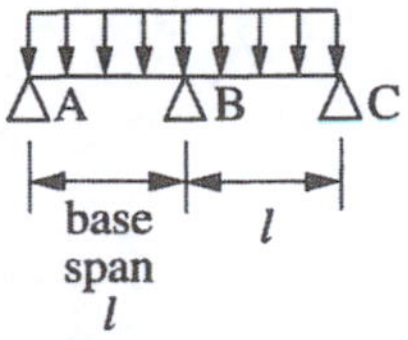 (uniform load on both spans) A, B, C	–	−0.125*	–		

(*continued on the next page*)

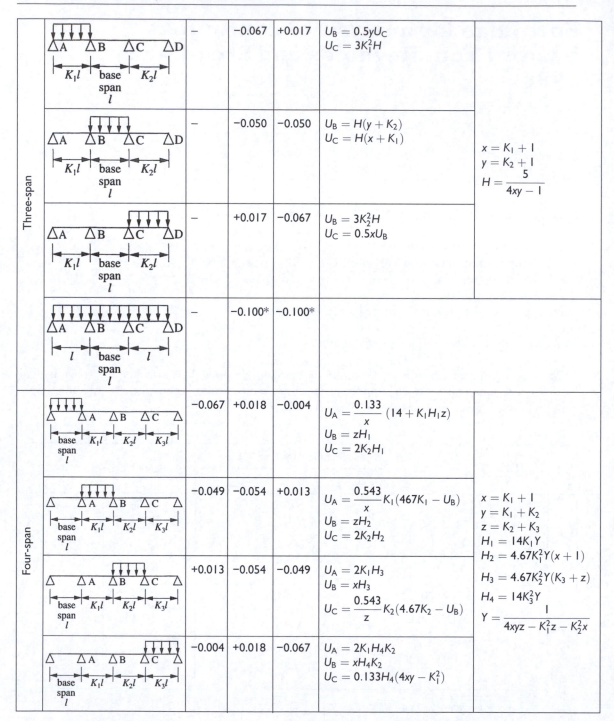

Three-span

	–	−0.067	+0.017	$U_B = 0.5yU_C$ $U_C = 3K_1^2H$
	–	−0.050	−0.050	$U_B = H(y + K_2)$ $U_C = H(x + K_1)$
	–	+0.017	−0.067	$U_B = 3K_2^2H$ $U_C = 0.5xU_B$
	–	−0.100*	−0.100*	

$x = K_1 + 1$
$y = K_2 + 1$
$H = \dfrac{5}{4xy - 1}$

Four-span

	−0.067	+0.018	−0.004	$U_A = \dfrac{0.133}{x}(14 + K_1H_1z)$ $U_B = zH_1$ $U_C = 2K_2H_1$
	−0.049	−0.054	+0.013	$U_A = \dfrac{0.543}{x}K_1(467K_1 - U_B)$ $U_B = zH_2$ $U_C = 2K_2H_2$
	+0.013	−0.054	−0.049	$U_A = 2K_1H_3$ $U_B = xH_3$ $U_C = \dfrac{0.543}{z}K_2(4.67K_2 - U_B)$
	−0.004	+0.018	−0.067	$U_A = 2K_1H_4K_2$ $U_B = xH_4K_2$ $U_C = 0.133H_4(4xy - K_1^2)$

$x = K_1 + 1$
$y = K_1 + K_2$
$z = K_2 + K_3$
$H_1 = 14K_1Y$
$H_2 = 4.67K_1^2Y(x + 1)$
$H_3 = 4.67K_2^2Y(K_3 + z)$
$H_4 = 14K_3^2Y$
$Y = \dfrac{1}{4xyz - K_1^2z - K_2^2x}$

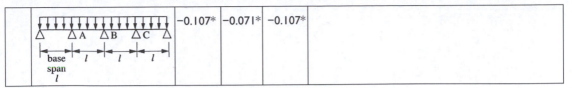

	−0.107*	−0.071*	−0.107*	

* The coefficient for all spans equally loaded is only applicable when all spans are of equal length

Example of application of continuous beam formulae

Find the bending moment diagram for the beam illustrated in Fig. D1.

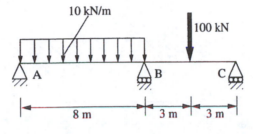

Figure D1

First analyse for the uniformly distributed loading (see Fig. D2):

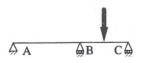

Figure D2

$$K_1 = 6/8 = 0.75$$

$$U_B = \frac{2}{1 + K_1} = 1.143$$

$$M_B = \alpha Q U F (\text{base span length})$$
$$= (1)(-0.063)(1.143)(10 \times 8)(8) = -46.1 \text{ kNm}$$

Then analyse for the point load (Fig. D3):

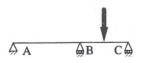

Figure D3

$K_1 = 6/8 = 0.75$

$U_B = \dfrac{2K_1^2}{1 + K_1} = 0.643$

$M_B = \alpha QUF(\text{base span length})$

$\quad = (1.5)(-0.063)(0.643)(100)(8) = -48.6 \, \text{kNm}$

Total moment at B is $-46.1 - 48.6 = -94.7$. Superimposing the diagrams from Appendix B gives Fig. D4:

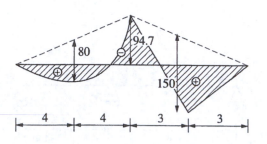

Figure D4

Appendix E

Slab design moment equations

Formulae have been developed by Wood and Armer for design moments in slabs (see, for example, the paper by Wood 1968). When reinforcement is parallel to the coordinate axes, the required moment capacities per unit breadth are:

$$m_x^* = m_x + |\tan \alpha||m_{xy}| \tag{E.1}$$

$$m_y^* = m_y + \frac{|m_{xy}|}{|\tan \alpha|} \tag{E.2}$$

where m_x, m_y and m_{xy} are the applied moments per unit breadth and α is the angle between the failure plane and the Y-axis, as illustrated in Fig. E.1. Selection of the moment capacities dictates the orientation of the failure plane. Hence, any value for α may be selected provided Equations E.1 and E.2 are satisfied. Selecting $\alpha = 45°$ results in the following simplified form of the equations:

$$m_x^* = m_x + |m_{xy}| \tag{E.3}$$

$$m_y^* = m_y + |m_{xy}| \tag{E.4}$$

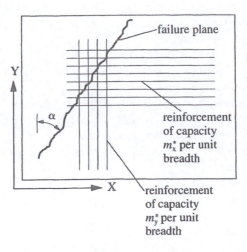

Fig. E.1 Plan view of portion of slab

Appendix F

General notation for chapters 8–11

General reinforced concrete

A_c = cross-sectional area of concrete
$A_{c,eff}$ = effective tension area of the concrete
A_g = gross cross-sectional area
A_s = area of longitudinal tension reinforcement
A'_s = area of longitudinal compression reinforcement
b = breadth of section
b_{eff} = effective breadth of the flange
b_w = web breadth
c = cover to longitudinal reinforcement
d = effective depth equal to the distance from extreme fibre in compression to the centre of the tension reinforcement
E_c = modulus of elasticity (Young's modulus) for concrete
E_{cm} = secant modulus of elasticity for concrete (Table 8.2)
E_s = modulus of elasticity for steel
$E_{c,eff}$ = effective elastic modulus for concrete, taking account of the effect of creep
f_{bd} = ultimate design bond strength (maximum shear stress that can act at the interface of the two materials)
f_c = stress in concrete
f_{ck} = characteristic cylinder compressive strength of concrete (capacity to resist compressive stress)
f_{ctk} = characteristic tensile strength of concrete
f_{ctm} = mean tensile strength of concrete
f_s = stress in tension reinforcement (compression positive)
f_{sr} = stress in tension steel assuming a cracked section, due to loading which causes initial cracking
f_{tk} = characteristic tensile strength of steel
f_{yk} = characteristic yield strength of steel (capacity to resist stress)
$f_{0.2k}$ = 0.2 per cent proof stress for cold worked steel (stress at which yielding is said to occur)
F_c = total compressive force in (gross) concrete section
F_{disp} = compressive force that would have been in concrete had it not been displaced by the compression reinforcement
F_s = force in tension reinforcement
F'_s = force in compression reinforcement
h = total section depth
I_c = cracked second moment of area
I_g = gross second moment of area (neglecting presence of reinforcement)
I_u = uncracked second moment of area (allowing for the presence of reinforcement)

l_{bd} = design anchorage length
$l_{b,min}$ = minimum anchorage length
$l_{b,rqd}$ = basic required anchorage length
M, M_{Ed} = applied design moment (factored ULS)
M_{Cr} = cracking moment equal to moment at which section first cracks
M_{ult} = design ultimate moment capacity
$s_{r,max}$ = maximum crack spacing
w_k = design crack width
x = distance from extreme fibre in compression to neutral axis
x_u = depth to the uncracked neutral axis
y = distance from neutral axis measured in direction of extreme fibre in compression
z = lever arm of internal forces
α_{cc} = coefficient which takes account of long-term effects on the compressive strength of concrete and of the unfavourable effects resulting from the way in which the load is applied
α_e = effective modular ratio (equal to E_s/E_c)
γ_c = partial factor of safety for concrete strength (equal to 1.5)
γ_s = partial factor of safety for the strength of steel (equal to 1.15)
δ = elastic deflection
δ_u = deflection assuming an uncracked section
δ_c = deflection assuming a fully cracked section
$\varepsilon(y)$ = strain at a distance y above the centroid
ε_c = strain in concrete (compression positive)
ε_{cm} = mean strain in the concrete between cracks
ε_{cu2}, ε_{cu3} = strain at which concrete crushes (equal to 0.0035)
ε_{c2}, ε_{c3} = strain at which characteristic compressive strength is reached (0.002 and 0.00175, respectively)
ε_s = strain in tension reinforcement
ε_s' = strain in compression reinforcement
ε_{sm} = mean strain in the tension steel allowing for tension stiffening
ζ = distribution factor
$\rho_{p,eff}$ = effective reinforcement ratio (equal to $A_s/A_{c,eff}$ for reinforced concrete)
φ = creep coefficient (Table 8.1)
ϕ = bar diameter

Additional notation for chapter 9

A_p = area of prestressing steel
e = eccentricity of prestress
E_p = elastic modulus of prestressing steel
$f_{ck,0}$ = characteristic compressive cylinder strength of concrete at transfer
f_{pk} = characteristic tensile strength of prestressing steel
$f_{p,0.1k}$ = characteristic 0.1 per cent proof stress of prestressing steel
M_c = moment due to characteristic load combination (SLS)
M_f = moment due to frequent load combination (SLS)
M_0 = moment due to applied loading at transfer
M_p = moment due to prestress
M_{qp} = moment due to quasi permanent load combination (SLS)

M_S = moment due to applied loading at service (SLS)

p_{0min}, p_{0max} = minimum and maximum permissible total stresses in concrete at transfer

p_{Smin}, p_{Smax} = minimum and maximum permissible total stresses in concrete at service (SLS)

P = prestress force

P_{jack} = applied jacking force

P_0 = prestress force at transfer

x = distance along length of beam

y_b = distance from the centroid to the extreme bottom fibre (negative below centroid)

y_{dl} = distance from the centroid to the lower decompression depth (negative below centroid)

y_{du} = distance from the centroid to the upper decompression depth (negative below centroid)

y_t = distance from the centroid to the extreme top fibre

Z_b, Z_t = elastic section moduli for bottom and top fibres (Z_b negative)

Z_{dl}, Z_{du} = elastic section moduli for lower and upper decompression depths

γ_p = partial factor of safety for prestress

ΔP_{c+s+r} = time-dependent losses

ΔP_i = immediate losses

θ = angle of inclination of prestressing tendon

ρ = ratio of prestress force at service to prestress force at transfer

σ_b, σ_t = stress due to prestress at transfer at bottom and top fibres

σ_{dl}, σ_{du} = stress due to prestress at lower and upper decompression depths

σ_{dl_min} = lower allowable limit on σ_{dl}

σ_{du_min} = lower allowable limit on σ_{du}

σ_{b_max} = upper allowable limit on σ_b

σ_{t_max} = upper allowable limit on σ_t

Additional notation for chapter 10

$A_{s,min}$, $A_{s,max}$ = minimum and maximum total areas of reinforcement

e = eccentricity of applied force

e_i = additional eccentricity due to imperfections

e_0 = minimum eccentricity

e_2 = lateral deflection due to second-order effects

f_{cd} = design value of compressive strength of concrete (equal to $\alpha_{cc}f_{ck}/\gamma_c$)

f_{yd} = design value of yield strength of reinforcement (equal to f_{yk}/γ_s)

i = radius of gyration (equal to $\sqrt{I/A}$)

l_0 = effective length

M_{0e} = equivalent first-order end moment

M_{0Ed} = first-order moment (including moment due to imperfections)

M_{0Eqp} = first-order bending moment for the SLS quasi-permanent load combination

M_{01}, M_{02} = first-order end moments ($|M_{02}|>|M_{01}|$)

M_2 = second-order moment (due to deflection e_2)

N, N_{Ed} = design value of the applied axial force (factored ULS)

N_{ult} = ultimate capacity to resist compressive force

κ = curvature (equal to M/EI)

λ = slenderness ratio (Equation 10.1)

$\sigma_{Rd,max}$ = design strength for a concrete strut

Additional notation for chapter 11

a_v = shear span equal to the distance from the edge of the support to the edge of the load

A_k = area within a perimeter bounded by the centre line of the wall

A_{sl} = area of tensile reinforcement which extends (d + required anchorage length) past the section considered

A_{sw} = area of shear reinforcement

$A_{sw,min}$ = minimum area of a link leg

d_{eff} = average effective depth (of two layers of reinforcement in coordinate directions)

d_H = effective depth of the column head

f_{ywd} = design yield strength of shear reinforcement (equal to f_{ywk}/γ_s)

f_{ywk} = characteristic yield strength of the links

k = depth factor

s = spacing of shear reinforcement

l_{ext} = distance from the face of the column to the exterior control perimeter

l_{int} = distance from the face of the column to the interior control perimeter

l_H = distance from the face of the column to the edge of the column head

s_r = radial spacing of perimeters of shear reinforcement

s_t = tangential spacing of shear reinforcement

t = thickness of the wall of a member

$t_{ef,i}$ = equivalent thickness of the wall of a member

T, T_{Ed} = applied design torque (factored ULS)

$T_{Rd,max}$ = design value of torque that would cause crushing of the concrete struts

u_{ext} = length of exterior control perimeter

u_{int} = length of interior control perimeter

u_{out} = outer control perimeter where shear reinforcement is not required

u_0 = length of column perimeter

u_1 = length of basic control perimeter

v_{Ed} = average design shear stress (equal to $V_{Ed}/b_w d$)

V_{Ed} = applied design shear force

V_{Rd} = design shear resistance

$V_{Rd,c}$ = design shear resistance of a concrete member without shear reinforcement

$V_{Rd,max}$ = design value of shear force to cause crushing of the compression struts

$V_{Rd,s}$ = design shear resistance due to shear reinforcement

β = shear enhancement reduction factor or factor to allow for eccentricity of loading

ΔF_{td} = additional tensile force in the longitudinal reinforcement due to shear in the member

θ = angle of inclination of compression strut

ρ_w = shear reinforcement ratio (equal to A_{sw}/sb_w)

$\rho_{w,min}$ = minimum quantity of shear reinforcement for beams

τ_{max} = maximum torsional shear stress

References

Bhatt, P. and Nelson, H.M. 1990. *Marshall and Nelson's Structures*. 3rd ed. Harlow: Longman Scientific and Technical.

Brooker, O. 2006. *Concrete Buildings Scheme Design Manual*. Surrey: The Concrete Centre.

Brooker, O. and Hennessy, R. 2008. *Residential Cellular Concrete Buildings*. Surrey: The Concrete Centre.

BS EN 1990 (2002)[*] *Eurocode – Basis of Structural Design*. London: British Standards Institution.

BS EN 1991-1-1 (2002)[*] *Eurocode 1: Actions on Structures, Part 1–1: General Actions – Densities, Self-weight, Imposed Loads for Buildings*. London: British Standards Institution.

BS EN 1991-1-3 (2003)[*] *Eurocode 1: Actions on Structures, Part 1–3: General Actions – Snow Loads*. London: British Standards Institution.

BS EN 1991-1-4 (2005)[*] *Eurocode 1: Actions on Structures, Part 1–4: General Actions – Wind Actions*. London: British Standards Institution.

BS EN 1992-1-1 (2004)[*] *Eurocode 2: Design of Concrete Structures – Part 1–1: General Rules and Rules for Buildings*. London: British Standards Institution.

BS EN 1992-1-2 (2004)[*] *Eurocode 2: Design of Concrete Structures – Part 1–2: General Rules – Structural Fire Design*. London: British Standards Institution.

The Concrete Centre 2004. *High Performance Buildings – Using Tunnel Form Concrete Construction*. Surrey: The Concrete Centre.

Cranston, W.B. 1972. *Analysis and Design of Reinforced Concrete Columns* (Research Report No. 20). Slough: Cement and Concrete Association.

Goodchild, C.H., Webster, R.M. and Elliott, K.S. 2009. *Economic Concrete Frame Elements to Eurocode 2*. Surrey: The Concrete Centre.

Gordon, J.E. 1991. *Structures: or Why Things Don't Fall Down*. Middlesex: Penguin (adapted Fig. 14 from Chapter 9, p. 188).

Hambly, E.C. 1991. *Bridge Deck Behaviour*. 2nd ed. London: E&FN Spon (Figure 3.8, p. 62).

Heyman, J. 1971. *Plastic Design of Frames, Vol. 2: Applications*. Cambridge: Cambridge University Press.

Hillerborg, A. 1975. *Strip Method of Design*. Wexham Springs, Slough: Cement and Concrete Association.

Hillerborg, A. 1996. *Strip Method Design Handbook*. London: E&FN Spon.

ISE 1988. *Stability of Buildings*. London: Institution of Structural Engineers.

ISE 2002. *Manual for the Design of Steelwork Building Structure*. London: Institution of Structural Engineers.

ISE 2006. *Manual for the Design of Reinforced Concrete Building Structures to Eurocode 2*. London: Institution of Structural Engineers.

Kotsovos, M.D. 1988. Design of reinforced concrete deep beams. *The Structural Engineer*, 66 (2), 28–32.

Lampert, P. and Collins, M.P. 1972. Torsion bending and confusion – an attempt to establish the facts. *ACI Journal Proceedings*, 69 (8), 500–504.

MacGregor, J.G. and Wight, J.K. 2005. *Reinforced Concrete – Mechanics and Design*. 4th ed. New Jersey: Prentice Hall Inc.

Mosley, B., Bungey, J. and Hulse, R. 2007. *Reinforced Concrete Design to Eurocode 2*. 6th ed. New York: Palgrave Macmillan.

NA to BS EN 1991-1-3 (2003)[*] *UK National Annex to Eurocode 1: Actions on Structures, Part 1–3: General Actions – Snow Loads*. London: British Standards Institution.

NA to BS EN 1991-1-4 (2005)[*] *UK National Annex to Eurocode 1: Actions on Structures, Part 1–4: General Actions – Wind Actions*, London: British Standards Institution.

NA to BS EN 1992-1-1 (2004)[*] *UK National Annex to Eurocode 2: Design of Concrete Structures – Part 1–1: General Rules and Rules for Buildings.* London: British Standards Institution.

Narayanan, R.S. and Beeby, A. 2005. *Designers' Guide to EN1992-1-1 and EN1992-1-2, Eurocode 2: Design of Concrete Structures. General Rules and Rules for Buildings and Structural Fire Design.* London: Thomas Telford.

Narayanan, R.S. and Goodchild, C.H. 2006. *Concise Eurocode 2.* Surrey: The Concrete Centre.

Nawy, E.G. 2003. *Prestressed Concrete: A Fundamental Approach.* 2nd ed. New Jersey: Prentice Hall Inc.

O'Brien, E.J. and Keogh, D.L. 1999. *Bridge Deck Analysis.* London: E&FN Spon.

Reddy, J.N. 2006. *Theory and Analysis of Elastic Plates and Shells.* 2nd ed. Boca Raton, FL: CRC Press.

Reynolds, C.E. and Steedman, J.C. 1988. *Reinforced Concrete Designer's Handbook.* 10th ed. London: E&FN Spon (Table 43, p. 169).

Taranath, B.S. 2009. *Reinforced Concrete Design of Tall Buildings.* Boca Raton, FL: CRC Press.

Wood, R.H. 1968. The reinforcement of slabs in accordance with a predetermined field of moments. *Concrete,* 69–76.

Websites

[*]British Standards can be obtained in PDF or hard copy formats from the BSI online shop at: www.bsigroup.com/Shop or by contacting BSI Customer Services for hardcopies only: Tel: +44 (0)20 8996 9001; Email: cservices@bsigroup.com.

Concast website is available at: http://www.concast.ie

Index